Jolien D. E. Creighton and
Warren G. Anderson

Gravitational-Wave Physics and Astronomy

Related Titles

Stahler, S. W., Palla, F.

The Formation of Stars

2004
ISBN: 978-3-527-40559-6

Roos, M.

Introduction to Cosmology

2003
ISBN: 978-0-470-84910-1

Liddle, A.

An Introduction to Modern Cosmology

2003
ISBN: 978-0-470-84835-7

Jolien D. E. Creighton and Warren G. Anderson

Gravitational-Wave Physics and Astronomy

An Introduction to Theory, Experiment and Data Analysis

WILEY-VCH
VCH

WILEY-VCH Verlag GmbH & Co. KGaA

The Authors

Dr. Jolien D. E. Creighton
University of Wisconsin–Milwaukee
Department of Physics
P.O. Box 413
Milwaukee, WI 53201
USA
jolien@uwm.edu

Dr. Warren G. Anderson
University of Wisconsin–Milwaukee
Department of Physics
P.O. Box 413
Milwaukee, WI 53201
USA
warren@gravity.phys.uwm.edu

Cover
Post-Newtonian apples created by Teviet Creighton. Hubble ultra-deep field image (NASA, ESA, S. Beckwith STScl and the HUDF Team).

Library of Congress Card No.: applied for

British Library Cataloguing-in-Publication Data:
A catalogue record for this book is available from the British Library.

Bibliographic information published by the Deutsche Nationalbibliothek
The Deutsche Nationalbibliothek lists this publication in the Deutsche Nationalbibliografie; detailed bibliographic data are available on the Internet at http://dnb.d-nb.de.

Typesetting le-tex publishing services GmbH, Leipzig
Cover Design Adam-Design, Weinheim
Printing and Binding Fabulous Printers Pte Ltd

Printed in Singapore
Printed on acid-free paper

ISBN Print 978-3-527-40886-3

ISBN ePDF 978-3-527-63605-1
ISBN oBook 978-3-527-63603-7
ISBN ePub 978-3-527-63604-4

JDEC: To my grandmother.

WGA: To my parents, who never asked me to stop asking why, although they did stop answering after a while, and to family, Lynda, Ethan and Jacob, who give me the space I need to continue asking.

Contents

Preface

During the writing of this book we often had to escape the office for week-long mini-sabbaticals. We would like to thank the Max-Planck-Institut für Gravitationsphysik (Albert-Einstein-Institute) in Hannover, Germany, for hosting us for one of these sabbaticals, the Warren G. Anderson Office of Gravitational Wave Research in Calgary, Alberta for hosting a second one, the University of Minnesota for hosting a third and the University of Cardiff for our final retreat.

We thank, in no particular order (other than alphabetic), Bruce Allen, Patrick Brady, Teviet Creighton, Stephen Fairhurst, John Friedman, Judy Giannakopoulou, Brennan Hughey, Lucía Santamaría Lara, Vuk Mandic, Chris Messenger, Evan Ochsner, Larry Price, Jocelyn Read, Richard O'Shaughnessy, Bangalore Sathyaprakash, Peter Saulson, Xavier Siemens, Amber Stuver, Patrick Sutton, Ruslan Vaulin, Alan Weinstein, Madeline White and Alan Wiseman for a great deal of assistance.

This work was supported by the National Science Foundation grants PHY-0701817, PHY-0600953 and PHY-0970074.

Calgary, June 2011 *J.D.E.C.*

List of Examples

Introduction

This work is intended both as a textbook for an introductory course on gravitational-wave astronomy and as a basic reference on most aspects in this field of research.

As part of the syllabus of a course on gravitational waves, this book could be used to follow a course on General Relativity (in which case, the first chapter could be greatly abbreviated), or as an introductory graduate course (in which case the first chapter is required reading for what follows). Not all material would be covered in a single semester.

Within the text we include examples that elucidate a particular point described in the main text or give additional detail beyond that covered in the body. At the end of each chapter we provide a short reference section that contains suggested further reading. We have not attempted to provide a complete list of work in the field, as one might have in a review article; rather we provide references to seminal papers, to works of particular pedagogic value, and to review articles that will provide the necessary background for researchers. Each chapter also has a selection of problems.

Please see http://www.lsc-group.phys.uwm.edu/~jolien for an errata for this book. If you find errors that are not currently noted in the errata, please notify jolien@uwm.edu.

Conventions

We use bold sans-serif letters such as $\mathsf{\mathbf{T}}$ and $\mathsf{\mathbf{u}}$ to represent generic tensors and spacetime vectors, and italic bold letters such as $\boldsymbol{v}$ to represent purely spatial vectors. When writing the components of such objects, we use Greek letters for the indices for tensors on spacetime, $T_{\alpha\beta}$ and u^α, while we use Latin letters for the indices for spatial vectors or matrices, for example v^i and M_{ij}. Spacetime indices normally run over four values, so $\alpha \in \{0, 1, 2, 3\}$, while spatial indices normally run over three values, $i \in \{1, 2, 3\}$, unless otherwise specified. We employ the Einstein summation convention where there is an implied sum over repeated indices (known as *dummy indices*), so that $T_{\alpha\mu} u^\mu = \sum_{\mu=0}^{3} T_{\alpha\mu} u^\mu$ and $M_{ij} v^j = \sum_{j=1}^{3} M_{ij} v^j$. In these examples, the indices α and i are not contracted

Gravitational-Wave Physics and Astronomy, First Edition. Jolien D. E. Creighton, Warren G. Anderson.

and are called *free indices* (that is, these are actually four equations in the first case and three equations in the second case since α can have the values 0, 1, 2, or 3, while i can have the values 1, 2, or 3).

We distinguish between the covariant derivative, ∇_α, and the three-space gradient operator $\boldsymbol{\nabla}$, which is the operator $\partial/\partial x^i$ in Cartesian coordinates. The Laplacian is $\boldsymbol{\nabla}^2 = \partial^2/\partial x^2 + \partial^2/\partial y^2 + \partial^2/\partial z^2$ in Cartesian coordinates, and the flat-space d'Alembertian operator is $\Box = -c^{-2}\partial^2/\partial t^2 + \boldsymbol{\nabla}^2$ in Cartesian coordinates.

Our spacetime sign convention is $-, +, +, +$ so that flat spacetime in Cartesian coordinates has the line element $ds^2 = -c^2\,dt^2 + dx^2 + dy^2 + dz^2$. The sign conventions of common tensors follow that of Misner *et al.* (1973) and Wald (1984).

The *Fourier transform* of some time series $x(t)$ is used to find the frequency series $\tilde{x}(f)$ according to

$$\tilde{x}(f) = \int_{-\infty}^{\infty} x(t) e^{-2\pi i f t} dt \,, \tag{0.1}$$

while

$$x(t) = \int_{-\infty}^{\infty} \tilde{x}(f) e^{2\pi i f t} df \tag{0.2}$$

is the *inverse Fourier transform.*

References

Misner, C.W., Thorne, K.S. and Wheeler, J.A. (1973) *Gravitation*, Freeman, San Francisco.

Wald, R.M. (1984) *General Relativity*, University of Chicago Press.

1
Prologue

1.1
Tides in Newton's Gravity

A brief review of Newtonian gravity is useful not only as a limit of weak-field relativistic gravity, but also as a reminder of the principles upon which general relativity was formulated. Newtonian gravity is conveniently formulated in a fixed rectilinear coordinate system in terms of an absolute time coordinate. In such coordinates as these, Newton's laws of motion and gravitation describe the motion of a body of mass m falling freely about another body of mass M by the force

$$\boldsymbol{F} = m\frac{d^2\boldsymbol{x}}{dt^2} = -\frac{GMm}{\|\boldsymbol{x}-\boldsymbol{x}'\|^3}(\boldsymbol{x}-\boldsymbol{x}') , \tag{1.1}$$

where $\boldsymbol{x}$ is the position of the body with mass m, $\boldsymbol{x}'$ is the position of the body with mass M, t is the absolute time coordinate, and $G \simeq 6.673 \times 10^{-11}\,\mathrm{m^3\,kg^{-1}\,s^{-2}}$ is Newton's gravitational constant. Famously, the quantity m cancels and

$$\frac{d^2\boldsymbol{x}}{dt^2} = -\frac{GM}{\|\boldsymbol{x}-\boldsymbol{x}'\|^3}(\boldsymbol{x}-\boldsymbol{x}') . \tag{1.2}$$

If there is a continuous distribution of matter then we can sum up all contributions to the acceleration from all pieces of the distribution to obtain

$$\frac{d^2\boldsymbol{x}}{dt^2} = -G\int_{\text{body}} \frac{\boldsymbol{x}-\boldsymbol{x}'}{\|\boldsymbol{x}-\boldsymbol{x}'\|^3}\rho(\boldsymbol{x}')d^3\boldsymbol{x}' = \boldsymbol{\nabla}\left[G\int_{\text{body}}\frac{\rho(\boldsymbol{x}')}{\|\boldsymbol{x}-\boldsymbol{x}'\|}d^3\boldsymbol{x}'\right], \tag{1.3}$$

where ρ is the mass distribution (density) and $\boldsymbol{\nabla}$ is the gradient operator in $\boldsymbol{x}$. Therefore, the *acceleration* of the body (with respect to the Newtonian system of rectilinear coordinates) is

$$\boldsymbol{a} = \frac{d^2\boldsymbol{x}}{dt^2} = -\boldsymbol{\nabla}\Phi(\boldsymbol{x}) , \tag{1.4}$$

where

$$\Phi(\boldsymbol{x}) := -G\int_{\text{body}}\frac{\rho(\boldsymbol{x}')}{\|\boldsymbol{x}-\boldsymbol{x}'\|}d^3\boldsymbol{x}' \tag{1.5}$$

Gravitational-Wave Physics and Astronomy, First Edition. Jolien D. E. Creighton, Warren G. Anderson.

is the *Newtonian potential*. The Newtonian potential satisfies the *Poisson equation*

$$\nabla^2 \Phi(\boldsymbol{x}) = -G \int \rho(\boldsymbol{x}') \nabla^2 \frac{1}{\|\boldsymbol{x} - \boldsymbol{x}'\|} d^3 x' = 4\pi G \rho(\boldsymbol{x}) , \qquad (1.6)$$

where we have used

$$\nabla^2 \frac{1}{\|\boldsymbol{x} - \boldsymbol{x}'\|} = -4\pi \delta(\boldsymbol{x} - \boldsymbol{x}') . \qquad (1.7)$$

Because the mass of the falling body does not enter into the equations of motion, any two bodies will fall the same way. If you can only see nearby free-falling bodies, you cannot tell whether you're falling or not. You feel the same if you are freely falling toward some massive object as you would if you were in no gravitational field whatsoever. The gravitational acceleration describes the motion of the falling body with respect to the absolute Newtonian coordinates – but is there any way for a freely falling observer to know if they are accelerating or not?

Einstein codified the observation that freely falling objects fall together as a principle known as the *equivalence principle*: a freely falling observer could always set up a local (freely falling) frame in which all the laws of physics are the same as they would be if that observer were not in a gravitational field. The coordinate acceleration $\boldsymbol{a}$ does not have any physical importance (as it does in Newtonian gravity) because one can always choose a frame of reference – freely falling with the observer – in which the observer is at rest.

Example 1.1 Coordinate acceleration in non-inertial frames of reference

An inertial frame of reference in Newtonian mechanics is any frame of reference that can be related to the absolute Newtonian frame of reference by a uniform velocity and a constant translation of position. That is, if $\boldsymbol{x}$ is the location of a particle in one inertial frame of reference, then another inertial frame of reference will have $\boldsymbol{x}' = \boldsymbol{x} - \boldsymbol{x}_0 - \boldsymbol{v}t$ for some constant vectors $\boldsymbol{x}_0$ and $\boldsymbol{v}$. Inertial frames preserve the form of Newton's second law since $\boldsymbol{a}' = d^2\boldsymbol{x}'/dt^2 = d^2\boldsymbol{x}/dt^2 = \boldsymbol{a}$.

In non-Cartesian coordinates, however, the form of the coordinate acceleration is different. For example, for a two-dimensional system we could express the location of a particle in polar coordinates $r = (x^2 + y^2)^{1/2}$ and $\phi = \arctan(y/x)$. In these coordinates, the coordinate velocity of a particle is given by $dr/dt = \boldsymbol{v} \cdot \boldsymbol{e}_r$ and $d\phi/dt = r^{-1}\boldsymbol{v} \cdot \boldsymbol{e}_\phi$ where $\boldsymbol{e}_r$ and $\boldsymbol{e}_\phi$ are unit vectors in the r- and ϕ-directions, and the equations of motion for the particle are $F_r = m[d^2r/dt^2 + r(d\phi/dt)^2]$ and $F_\phi = m[d^2\phi/dt^2 + 2r^{-1}(dr/dt)(d\phi/dt)]$. Even when there is no force on the particle, $\boldsymbol{F} = 0$, there is still a coordinate acceleration in that d^2r/dt^2 and $d^2\phi/dt^2$ do not vanish except for purely radial motion. This merely arises because of the choice of non-Cartesian coordinates – the geometrical form of Newton's second law, $\boldsymbol{F} = m\boldsymbol{a}$ still holds.

A non-inertial frame is a frame that is accelerating relative to an inertial frame. A common example is a uniformly rotating reference frame with angular velocity vector $\boldsymbol{\omega}$. In such a reference frame, Newton's second law has the form $\boldsymbol{F} = m\boldsymbol{a} +$

$m\boldsymbol{\omega} \times (\boldsymbol{\omega} \times \boldsymbol{r}) + 2m\boldsymbol{\omega} \times \boldsymbol{v}$ where the two additional terms, the *centrifugal force*, $m\boldsymbol{\omega} \times (\boldsymbol{\omega} \times \boldsymbol{r})$ and the *Coriolis force*, $2m\boldsymbol{\omega} \times \boldsymbol{v}$, arise because the frame of reference is non-inertial. These are known as fictitious forces.

A *freely falling frame of reference* in Newtonian theory is a non-inertial frame of reference because it is accelerating relative to the absolute set of Newtonian coordinates. The following coordinate transformation relates a freely falling frame of reference (primed coordinates) at point $\boldsymbol{x}_0$ with the absolute Newtonian coordinates (unprimed): $\boldsymbol{x}' = \boldsymbol{x} - \boldsymbol{x}_0 - \frac{1}{2}\boldsymbol{g}t^2$, where $\boldsymbol{g} = -\boldsymbol{\nabla}\Phi(\boldsymbol{x}_0)$ is a constant. It is straightforward to see that $\boldsymbol{a}' = d^2\boldsymbol{x}'/dt^2 = -\boldsymbol{\nabla}[\Phi(\boldsymbol{x}) - \Phi(\boldsymbol{x}_0)]$ which vanishes at point $\boldsymbol{x}_0$.

In fact, there *is* a way to tell if you are falling. If there is another object that is some small distance away from you then its acceleration will be slightly different. Suppose $\boldsymbol{\zeta}$ is the vector pointing from you to the other object. The acceleration of that object is

$$\boldsymbol{a}(\boldsymbol{x} + \boldsymbol{\zeta}) = \boldsymbol{a}(\boldsymbol{x}) + (\boldsymbol{\zeta} \cdot \boldsymbol{\nabla})\boldsymbol{a}(\boldsymbol{x}) + O(\zeta^2) \tag{1.8}$$

and so the relative acceleration or *tidal acceleration* is

$$\Delta a_i = -\zeta^j \frac{\partial^2 \Phi}{\partial x^i \partial x^j} = -\mathcal{E}_{ij}\zeta^j\,, \tag{1.9}$$

where

$$\mathcal{E}_{ij} := \frac{\partial^2 \Phi}{\partial x^i \partial x^j} \tag{1.10}$$

is known as the *tidal tensor field*. The tidal acceleration is not really local since it depends on the separation $\boldsymbol{\zeta}$ between falling bodies. The tidal field, however, is a local quantity, and it encodes the presence of the gravitational field. We will see later that in General Relativity, the tidal field is a measure of the spacetime curvature.

In the above expressions, the indices i and j run over the three spatial coordinates $\{x^1, x^2, x^3\}$ or equivalently $\{x, y, z\}$ and ζ^i is the ith component of the vector $\boldsymbol{\zeta}$. (The three components of the vector are ζ^1, ζ^2 and ζ^3 so we would write $\zeta^i = [\zeta^1, \zeta^2, \zeta^3]$.) The tidal field is a rank-2 tensor having nine components: $\mathcal{E}_{11}, \mathcal{E}_{12}, \mathcal{E}_{13}, \mathcal{E}_{21}, \mathcal{E}_{22}, \mathcal{E}_{23}, \mathcal{E}_{31}, \mathcal{E}_{32}$ and $\mathcal{E}_{33}$. It is symmetric: $\mathcal{E}_{12} = \mathcal{E}_{21}$, $\mathcal{E}_{13} = \mathcal{E}_{31}$ and $\mathcal{E}_{23} = \mathcal{E}_{32}$, or, more concisely, $\mathcal{E}_{ij} = \mathcal{E}_{ji}$. Einstein's summation convention is being used here: there is an implicit summation over repeated indices. That is, the expression

$$\mathcal{E}_{ij}\zeta^j$$

is short-hand for

$$\sum_{j=1}^{3} \mathcal{E}_{ij}\zeta^j = \mathcal{E}_{i1}\zeta^1 + \mathcal{E}_{i2}\zeta^2 + \mathcal{E}_{i3}\zeta^3\,.$$

For example, if two objects are separated in the x^3- or z-direction, so that ζ^1 and ζ^2 both vanish, then the three components of the tidal acceleration are

$$\Delta a_1 = -\mathcal{E}_{13}\zeta^3\,, \quad \Delta a_2 = -\mathcal{E}_{23}\zeta^3\,, \quad \text{and} \quad \Delta a_3 = -\mathcal{E}_{33}\zeta^3\,.$$

Example 1.2 Tidal acceleration

Consider a body falling toward the Earth. The Newtonian potential is

$$\Phi = -\frac{G M_\oplus}{(x^2 + y^2 + z^2)^{1/2}} \,. \tag{1.11}$$

The tidal field component $\mathcal{E}_{11}$ is

$$\mathcal{E}_{11} = \frac{\partial^2 \Phi}{\partial x^2} = -G M_\oplus \left[3\frac{x^2}{(x^2 + y^2 + z^2)^{5/2}} - \frac{1}{(x^2 + y^2 + z^2)^{3/2}} \right], \tag{1.12}$$

the tidal field component $\mathcal{E}_{12}$ is

$$\mathcal{E}_{12} = \frac{\partial^2 \Phi}{\partial x \partial y} = -G M_\oplus \left[3\frac{x y}{(x^2 + y^2 + z^2)^{5/2}} \right], \tag{1.13}$$

and so forth. The components can be written concisely as

$$\mathcal{E}_{ij} = -\frac{G M_\oplus}{r^5} \left[3x_i x_j - \delta_{ij} r^2 \right], \tag{1.14}$$

where $r = (x^2 + y^2 + z^2)^{1/2}$ and δ_{ij} is the *Kronecker delta,*

$$\delta_{ij} := \begin{cases} 1 & i = j \\ 0 & i \neq j \,, \end{cases} \tag{1.15}$$

and so $x_i = \delta_{ij} x^j$.

Suppose that a reference body is on the z-axis at a distance $r = z$ from the centre of the Earth. Then the tidal tensor is

$$\mathcal{E}_{ij} = \frac{G M_\oplus}{r^3} \begin{bmatrix} 1 & 0 & 0 \\ 0 & 1 & 0 \\ 0 & 0 & -2 \end{bmatrix} . \tag{1.16}$$

Consider a nearby second body that is also on the z-axis, a distance Δz farther from the centre of the Earth. The relative tidal acceleration of this body is

$$\Delta a_i = -\mathcal{E}_{ij} \xi^j = -\mathcal{E}_{i3} \Delta z \,. \tag{1.17}$$

The only non-vanishing component is the z-component:

$$\Delta a_3 = 2\frac{G M_\oplus}{r^3} \Delta z \,. \tag{1.18}$$

A third body is next to the reference body, lying a small distance Δx away on the x-axis. The relative tidal acceleration of this body is

$$\Delta a_i = -\mathcal{E}_{ij} \xi^j = -\mathcal{E}_{i1} \Delta x \tag{1.19}$$

and the only non-vanishing component is the x-component:

$$\Delta a_1 = -\frac{G M_\oplus}{r^3} \Delta x \,. \tag{1.20}$$

Notice that a collection of freely falling objects will be pulled apart along the direction in which they are falling while being squeezed together in the orthogonal directions.

Unlike the coordinate acceleration, the tidal acceleration has intrinsic physical meaning. We witness ocean tides caused by the Moon and the Sun. These tides dissipate energy on the Earth. That is, tidal forces can do work. To compute the work, consider an extended body (say, the Earth) moving within a tidal field produced by another body (say, the Moon). An element of the extended body, located at a position $\boldsymbol{x}$ and having mass $\rho(\boldsymbol{x})d^3\boldsymbol{x}$, experiences a tidal force

$$F_i = -\mathcal{E}_{ij} x^j \rho(\boldsymbol{x}) d^3\boldsymbol{x} \,. \tag{1.21}$$

If the element is moving through the tidal field with velocity $\boldsymbol{v}$ then there is an amount $F_i v^i$ of work per unit time done on that element. Summing over all elements that comprise the body yields the total amount of tidal work:

$$\begin{aligned}\frac{dW}{dt} &= -\int\limits_{\text{body}} \mathcal{E}_{ij} v^i x^j \rho(\boldsymbol{x}) d^3\boldsymbol{x} \\ &= -\frac{1}{2}\mathcal{E}_{ij}\frac{d}{dt}\int\limits_{\text{body}} x^i x^j \rho(\boldsymbol{x}) d^3\boldsymbol{x} \\ &= -\frac{1}{2}\mathcal{E}_{ij}\frac{dI^{ij}}{dt}\,,\end{aligned} \tag{1.22}$$

where

$$I^{ij} := \int\limits_{\text{body}} x^i x^j \rho(\boldsymbol{x}) d^3\boldsymbol{x} \tag{1.23}$$

is the *quadrupole tensor.* Note that this tensor is closely related to the *moment of inertia tensor*

$$\mathcal{I}_{ij} := \left(\delta_{ij}\delta_{kl} - \delta_{ik}\delta_{jl}\right) I^{kl} = \int\limits_{\text{body}} \left(r^2\delta_{ij} - x_i x_j\right) \rho(\boldsymbol{x}) d^3\boldsymbol{x} \tag{1.24}$$

and also to the (traceless) *reduced quadrupole tensor*

$$\text{I\kern-0.4em-}_{ij} := \left(\delta_{ik}\delta_{jl} - \frac{1}{3}\delta_{ij}\delta_{kl}\right) I^{kl} = \int\limits_{\text{body}} \left(x_i x_j - \frac{1}{3}r^2\delta_{ij}\right) \rho(\boldsymbol{x}) d^3\boldsymbol{x} \,. \tag{1.25}$$

Here $r^2 = \|\boldsymbol{x}\|^2 = \delta_{ij} x^i x^j$.

Tidal work can also be performed by a dynamical system with a time-changing tidal field $\mathcal{E}_{ij}(t)$. The work performed by such a system on another body with a quadrupole tensor I^{ij} is found by integrating Eq. (1.22) by parts:

$$W = -\frac{1}{2}\,\mathcal{E}_{ij} I^{ij}\Big|_0^T + \frac{1}{2}\int\limits_0^T \frac{d\mathcal{E}_{ij}}{dt} I^{ij} dt \,. \tag{1.26}$$

The first term is bounded, while the second term secularly increases with time and represents a transfer of energy from the dynamical system that is producing the time-changing tidal field to the other body. For example, the source of the time-changing tidal field might be a rotating dumbbell or a binary system of two stars in orbit about each other. Over a long time (large T) the secularly growing term will dominate, and we can write the work done *by* the dynamical source *on* the body with moment of inertia tensor I^{ij} as

$$\frac{dW}{dt} \approx \frac{1}{2}\frac{d\mathcal{E}_{ij}}{dt} I^{ij} \,. \tag{1.27}$$

1.2 Relativity

The *special theory of relativity* postulates that there is no preferred inertial frame: local measurements of physical quantities are the same no matter which inertial frame the measurement is made in. This is the *principle of relativity*. In particular, measurements of the speed of light in any inertial frame will always yield the same value, $c := 299\,792\,458\ \mathrm{m\,s^{-1}}$. The consequence of this is that the Newtonian separation of space and time must be abandoned. Consider a spaceship travelling at a constant speed v in the x-direction relative to the Earth (see Figure 1.1). Within the spaceship, an experimental determination of the speed of light is made in which a photon is emitted from a source in the y-direction, reflected by a mirror a distance $\frac{1}{2}\Delta y$ away from the source, and received back at the source. The time-of-flight $\Delta\tau$ is measured and the speed of light $c = \Delta y/\Delta\tau$ is computed. For an observer on the Earth, however, the distance travelled by the photon is $[(\Delta x)^2 + (\Delta y)^2]^{1/2}$, where $\Delta x = v\Delta t$ and Δt is the amount of time the observer on the Earth determines it takes the photon to travel from the emitter to the receiver. Since the observer on Earth must measure the same speed of light, $c = [(\Delta x)^2 + (\Delta y)^2]^{1/2}/\Delta t$, we see that

$$c^2 = \frac{(\Delta x)^2 + (\Delta y)^2}{(\Delta t)^2} = \frac{(\Delta x)^2 + (c\Delta\tau)^2}{(\Delta t)^2} \,, \tag{1.28}$$

where we have used $\Delta y = c\Delta\tau$, and so

$$c^2(\Delta\tau)^2 = c^2(\Delta t)^2 - (\Delta x)^2 \,. \tag{1.29}$$

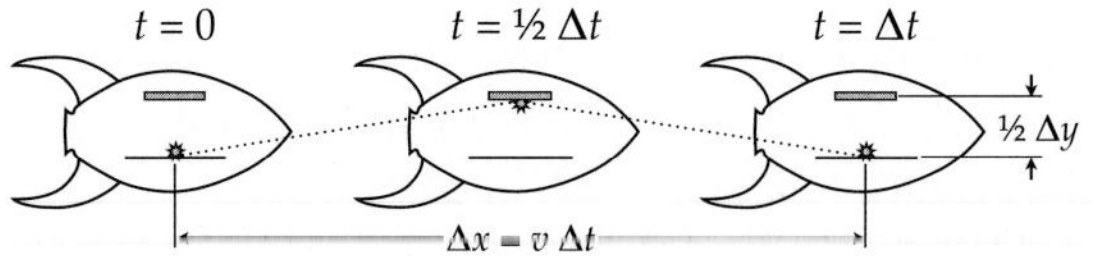

Figure 1.1 A measurement of the speed of light, performed in a rocket moving at speed v relative to the Earth, as seen by an observer on the Earth. A flash of light is produced at $t = 0$. The light travels a vertical distance $\frac{1}{2}\Delta y$, reflects off of the mirror and returns to the source after a time Δt (as measured by the observer on the Earth). The rocket has moved a horizontal distance $\Delta x = v\Delta t$ in this time.

The usual time dilation formula $\Delta t = \gamma \Delta \tau$, where $\gamma = (1 - v^2/c^2)^{-1/2}$ is the Lorentz factor, follows by setting $\Delta x = v \Delta t$. This relationship between how time is measured within the moving frame of the spaceship to how time is measured on Earth is not particular to the experiment with the photon: time really does move differently in the different inertial frames of reference.

Equation (1.29) relates the amount of time $\Delta \tau$ between two events, as recorded in an inertial frame in which the two events occur at the same spatial position (which is known as the *proper time* between the two events), to the amount of time Δt between the same two events as seen in an inertial frame in which the two events are separated by a spatial distance Δx. Since the notion of an absolute time is lost in special relativity, we understand time to simply be a new coordinate which, along with the three spatial coordinates, depends on the frame of reference. Together, the time and space coordinates are used to identify points (or events) on a four-dimensional *spacetime*. For rectilinear coordinates in an inertial frame, we define an invariant interval $(\Delta s)^2$ between two points in spacetime, (t, x, y, z) and $(t + \Delta t, x + \Delta x, y + \Delta y, z + \Delta z)$, by

$$(\Delta s)^2 := -c^2(\Delta t)^2 + (\Delta x)^2 + (\Delta y)^2 + (\Delta z)^2 \,, \tag{1.30}$$

which has the same form as the Pythagorean theorem except for the factor of $-c^2$ in front of the square of the time interval. This equation is just a generalization of Eq. (1.29) with $(\Delta s)^2 := -c^2(\Delta \tau)^2$.

Special relativity is incompatible with Newtonian gravity because Newton's law of gravitation defines a force between two distant bodies in terms of their separation at a given instant in time. However, in special relativity, there is no unique notion of simultaneity. In addition, different frames of reference will make different measurements of the Newtonian gravitational force, a result that is at odds with the principle of relativity.

The *general theory of relativity* provides a description of gravity in terms of a curved spacetime. This is discussed in Chapter 2. In general relativity, the inertial frames of reference are freely falling frames, and the principle of relativity is then taken to hold in such frames of reference. Tidal acceleration is the physical manifestation of gravitation, but measurement of a tidal field requires a somewhat extended apparatus.

Of course, Newtonian gravity must be recovered in some limit of general relativity: this limit is when $GM/(c^2R) \ll 1$ and $v/c \ll 1$ where M is the characteristic mass of the system, R is the characteristic size of the system, and v is the characteristic speed of bodies in the system. And since in Newtonian gravity a changing tidal field is capable of producing work on distant bodies, this must be true in general relativity as well. This means that in order to ensure that energy is conserved, energy must be radiated from the gravitating system that is producing the changing tidal field to the rest of the universe, because there is no way that the bodies on which the work is done can create an instantaneous reactive force on the gravitating system – this would be incompatible with relativity. The radiation is called gravitational radiation.

2
A Brief Review of General Relativity

The intent of this chapter is to provide a brief review of General Relativity and to introduce the concepts and notation that are required for the discussion of gravitational waves in subsequent chapters. The review will not be comprehensive as there are many excellent introductory texts on General Relativity: Hartle (2003) is a clear, physics-first introduction to the subject, and Schutz (2009) is another excellent text for a first course in General Relativity. The classic Misner *et al.* (1973) is a complete reference book. Advanced texts include Wald (1984) and Weinberg (1972) which have very different approaches but are both essential reading.

The *principle of relativity* – a foundation of Einstein's theory of Special Relativity – suggests that there is no preferred frame of reference or state of motion. Physical theory needs to be formulated in a manner in which physical quantities are invariant under a class of transformations known as *Poincaré transformations*. That is, physics is invariant under translations, rotations and boosts. Special relativity can be elegantly formulated on a four-dimensional *spacetime* in which the three normal spatial dimensions and a time dimension are combined.

To describe relativistic gravity, Einstein extended the principle of relativity to a new principle, the *principle of general covariance*, which demands that there is no preferred coordinate system at all. For example, a freely falling observer can always construct a freely falling frame of reference and any physical experiment carried out in that frame of reference must give the same results as a similar experiment carried out by an observer who is not in any gravitational field whatsoever. Einstein described gravity in terms of a curved spacetime in which particles naturally follow the straightest possible lines – not necessarily straight lines in some predetermined coordinate system – and the physical effects of gravity can then be understood in terms of the curvature of spacetime. For example, the tidal field is related to the curvature tensor. The curvature is produced, to some extent, by the masses in spacetime.

Gravitational-Wave Physics and Astronomy, First Edition. Jolien D. E. Creighton, Warren G. Anderson.

2.1
Differential Geometry

General relativity is formulated on a four-dimensional *manifold* – a four dimensional surface on which our physical theory is described. The manifold of general relativity is called spacetime because three of the dimensions correspond to the observed three dimensions of space and the fourth dimension of the manifold corresponds to what we perceive as time. The structure of the manifold can be quite complicated in principle, but for our purposes it is not necessary to consider general situations.

2.1.1
Coordinates and Distances

Like the surface of the Earth, the manifold of spacetime can be covered with patches or *charts* on which coordinates can be constructed. The set of overlapping charts that covers all of spacetime is called an *atlas*. Unlike Newtonian theory, there is no intrinsically physical set of coordinates or charts. Physical theory in general relativity is formulated in a covariant way so that the physical quantities are invariant under changes of coordinates.

There is a particularly useful class of coordinate choices are called *normal coordinates*. Normal coordinates are the closest things to inertial coordinates in flat-space, and so the reference frame described by normal coordinates is called a *locally inertial frame*. Normal coordinates can typically be constructed over a region with a size comparable to the curvature scale of spacetime. We will use the fact that, because of the equivalence principle, we can always find normal coordinates in the vicinity of any spacetime point and, in these coordinates, much of our flat-space intuition will hold.

The distance between two points that are sufficiently close together is a geometric invariant and so it is the same regardless of what set of coordinates are adopted. The two points need to be close together so that there is a unique notion of what path is taken from one point to the other over which we construct the distance. Therefore we write Pythagoras' formula in its differential form: consider two points, $\mathcal{P}$ and $\mathcal{Q}$ that are infinitesimally close together. These points are labelled by the coordinates $x^\alpha_\mathcal{P}$ and $x^\alpha_\mathcal{Q}$ respectively, and the infinitesimal coordinate difference between the two points is $dx^\alpha = x^\alpha_\mathcal{Q} - x^\alpha_\mathcal{P}$. The squared distance between the two points, ds^2, is computed by

$$ds^2 = g_{\mu\nu}(x^\alpha)dx^\mu dx^\nu \,, \tag{2.1}$$

where $g_{\mu\nu}(x^\alpha)$ is the *metric* tensor of spacetime, which is a function of spacetime coordinates x^α. Note that the index α runs over four values in a four-dimensional spacetime, and by convention we take values to be $\{0, 1, 2, 3\}$ so that x^1, x^2 and x^3 are the three spatial coordinates and x^0 is the single time coordinate. The metric determines the distance between any two neighbouring points in spacetime and therefore determines all of the geometry of the spacetime.

In a flat spacetime or *Minkowski spacetime*, we use the symbol $\eta_{\alpha\beta}$ for the metric. In the standard rectilinear coordinates the distance between any two points in Minkowski spacetime is

$$ds^2 = \eta_{\mu\nu} dx^\mu dx^\nu = -c^2 dt^2 + dx^2 + dy^2 + dz^2 \quad \text{(rectilinear coordinates)} . \tag{2.2}$$

A transformation to a new set of coordinates is specified by the four functions $x'^\alpha(x^\mu)$ relating the new primed coordinates with the original unprimed coordinates. Under this transformation,

$$dx^\mu = \frac{\partial x^\mu}{\partial x'^\alpha} dx'^\alpha , \tag{2.3}$$

where $x^\mu(x'^\alpha)$ is the inverse transformation. Since the squared distance element ds^2 is invariant under such transformations,

$$ds^2 = g_{\mu\nu} dx^\mu dx^\nu = g_{\mu\nu} \frac{\partial x^\mu}{\partial x'^\alpha} \frac{\partial x^\nu}{\partial x'^\beta} dx'^\alpha dx'^\beta = g'_{\alpha\beta} dx'^\alpha dx'^\beta , \tag{2.4}$$

where

$$g'_{\alpha\beta} = g_{\mu\nu} \frac{\partial x^\mu}{\partial x'^\alpha} \frac{\partial x^\nu}{\partial x'^\beta} . \tag{2.5}$$

In fact, any physical quantity does not depend on the choice of the coordinate system; the freedom of coordinate redefinition $x'^\alpha(x^\mu)$ therefore represents the *gauge freedom* of our geometric description of gravity, and coordinate transformations are also *gauge transformations.*

For an infinitesimal coordinate transformation (or infinitesimal gauge transformations) of the form

$$x^\alpha \to x'^\alpha = x^\alpha + \xi^\alpha(x^\mu) , \tag{2.6}$$

where ξ is a displacement vector we see that

$$dx^\alpha \to dx'^\alpha = dx^\alpha + \frac{\partial \xi^\alpha}{\partial x^\mu} dx^\mu \tag{2.7}$$

and therefore

$$g_{\alpha\beta} \to g'_{\alpha\beta} = g_{\alpha\beta} - g_{\alpha\mu} \frac{\partial \xi^\mu}{\partial x^\beta} - g_{\mu\beta} \frac{\partial \xi^\mu}{\partial x^\alpha} + O(\xi^2) . \tag{2.8}$$

Example 2.1 Transformation to polar coordinates

Given the two-dimensional flat-space metric in rectilinear coordinates,

$$ds^2 = g_{\mu\nu} dx^\mu dx^\nu = dx^2 + dy^2 , \tag{2.9}$$

one can transform into polar coordinates $r = (x^2 + y^2)^{1/2}$ and $\phi = \arctan(y/x)$. The inverse transformation is $x = r\cos\phi$ and $y = r\sin\phi$ so $dx =$

$\cos\phi\, dr - r\sin\phi\, d\phi$ and $dy = \sin\phi\, dr + r\cos\phi\, d\phi$ and hence

$$\begin{aligned} ds^2 &= dx^2 + dy^2 = (\cos\phi\, dr - r\sin\phi\, d\phi)^2 + (\sin\phi\, dr + r\cos\phi\, d\phi)^2 \\ &= dr^2 + r^2 d\phi^2 = g'_{\mu\nu} dx'^{\mu} dx'^{\nu} \,. \end{aligned} \tag{2.10}$$

Therefore we have $g'_{rr}(r,\phi) = 1$, $g'_{\phi\phi}(r,\phi) = r^2$, and $g'_{r\phi}(r,\phi) = g'_{\phi r}(r,\phi) = 0$.

Example 2.2 Volume element

Under the coordinate transformation of Eq. (2.3), the metric transforms according to Eq. (2.5). The metric is a 4×4 matrix whose determinant is related to the *volume element* of spacetime. To see this, we take the determinant of Eq. (2.5):

$$\det \mathbf{g}' = \det \left| \frac{\partial(\mathbf{x})}{\partial(\mathbf{x}')} \right|^2 \det \mathbf{g} \,, \tag{2.11}$$

where $\mathbf{J} = \partial(\mathbf{x})/\partial(\mathbf{x}')$ is the *Jacobian matrix* $J^{\alpha}_{\beta} = \partial x^{\alpha}/\partial x'^{\beta}$. Recall that the *Jacobian determinant* arises in a change of variables in integral calculus: under the coordinate transformation $\mathbf{x} \to \mathbf{x}'$, the measure changes as

$$d^4x' = \det \left| \frac{\partial(\mathbf{x}')}{\partial(\mathbf{x})} \right| d^4x \,. \tag{2.12}$$

Since we can always locally perform a coordinate transformation to a locally inertial Cartesian frame in which the metric is $g'_{\alpha\beta} = \eta_{\alpha\beta} = \mathrm{diag}[-c^2, 1, 1, 1]$, which has determinant $\det \boldsymbol{\eta} = -c^2$, and for which the volume element is $dV = c\,d^4x' = c\,dt'dx'dy'dz'$, we see that

$$\begin{aligned} dV &= c\,d^4x' = c \det \left| \frac{\partial(\mathbf{x}')}{\partial(\mathbf{x})} \right| d^4x = c\sqrt{\frac{\det \mathbf{g}}{\det \boldsymbol{\eta}}}\, d^4x \\ &= (-\det \mathbf{g})^{1/2} d^4x \,. \end{aligned} \tag{2.13}$$

Therefore $|\det \mathbf{g}|^{1/2} d^4x$ is the volume element at a location in spacetime.

As an example in two dimensions, consider the metric of two-dimensional flat-space in polar coordinates that was found in Example 2.1: $g_{\alpha\beta} = \mathrm{diag}[1, r^2]$. The volume element is therefore $(\det \mathbf{g})^{1/2} d^2x = r\,dr\,d\phi$.

2.1.2 Vectors

Geometric constructs such as vectors and tensors need to be generalized from their normal flat-space definition (e.g. a vector as going from one point in space to another) to a generalized definition that can be ported to curved manifolds.

The neighbourhood of any point of spacetime can be thought of as a flat four-dimensional vector space. Any curvature of the manifold thus resides in the way these *tangent spaces* are put together. Because of the curvature of spacetime, vectors cannot generally be constructed as arrows connecting distant points on the manifold. But we know from the equivalence principle that in a sufficiently small neighbourhood of a point in spacetime – small enough that the tidal effects of gravity are negligible, which really means the size of the neighbourhood is much smaller than the curvature scale of spacetime – our usual intuition about vectors must be valid. Therefore we are motivated to define a vector in terms of differential quantities that are restricted to some neighbourhood of a point on spacetime.

In differential geometry we describe a vector as a *directional derivative*: Imagine a curve in spacetime that is described parametrically by the four functions $x^\alpha(t)$ where t is the curve parameter. Let $F(x^\alpha)$ be some function on spacetime. Given this function, a function $f(t) = F(x^\alpha(t))$ can be defined on the curve – it is a function of the curve parameter. The derivative of $f(t)$ with respect to t is

$$\frac{df}{dt} = \frac{dx^\mu}{dt}\frac{\partial F}{\partial x^\mu} = u^\mu \frac{\partial F}{\partial x^\mu}\,, \tag{2.14}$$

where

$$u^\alpha := \frac{dx^\alpha}{dt} \tag{2.15}$$

are the components of the *tangent vector*, $\mathbf{u} = d/dt$, to the curve. That is, the derivative along a curve passing through a point in spacetime (a directional derivative) is associated with a set of components u^α which is our old notion of a vector.

Conversely, in flat space, we can interpret a vector with components v^α as the directional derivative d/ds along the curve

$$x^\alpha(s) = x^\alpha(0) + v^\alpha s\,. \tag{2.16}$$

This establishes that, in flat space, the two notions of what a vector is are identical.

Example 2.3 How are directional derivatives like vectors?

Consider the point $\mathcal{P} = (3, 4)$ in Cartesian coordinates, that is $x^\mu = [x, y] = [3, 4]$. This point may be reached from the origin by the curve (a straight line)

$$\begin{aligned} x &= 3s \\ y &= 4s \end{aligned} \tag{2.17}$$

with $s = 1$. The vector $\mathbf{v} = d/ds$ has components $v^\mu = [dx/ds, dy/ds] = [3, 4]$, which is just the usual notion of the vector from the origin to the point x^μ.

The directional derivative of the generic function $f(t)$ in Eq. (2.14) clearly does not depend on the coordinate system in which it is evaluated. This shows us how to transform the components u^α of the vector $\mathbf{u}$: let $x'^\beta(x^\alpha)$ be a new set of coordinates defined in terms of our original set. Then

$$\mathbf{u} = u^\mu \frac{\partial}{\partial x^\mu} = u^\mu \frac{\partial x'^\nu}{\partial x^\mu} \frac{\partial}{\partial x'^\nu} = u'^\nu \frac{\partial}{\partial x'^\nu} \tag{2.18}$$

so

$$u'^\alpha = \frac{\partial x'^\alpha}{\partial x^\mu} u^\mu \,. \tag{2.19}$$

The *inner product* between any two vectors, say **u** and **v**, is defined in terms of the metric tensor as

$$\mathbf{u} \cdot \mathbf{v} := g_{\mu\nu} u^\mu v^\mu \,. \tag{2.20}$$

In particular, this inner product also defines the squared length of a vector, $\|\mathbf{u}\|^2 := g_{\mu\nu} u^\mu u^\nu$. This definition of an inner product agrees with our flat-space intuition: if dx^α is the vector from point $x^\alpha_{\mathcal{P}}$ to $x^\alpha_{\mathcal{Q}}$, $dx^\alpha = x^\alpha_{\mathcal{P}} - x^\alpha_{\mathcal{Q}}$, then the length interval between these two points, ds^2, is simply $(\mathbf{dx}) \cdot (\mathbf{dx}) = g_{\mu\nu} dx^\mu dx^\nu$ (cf. Eq. (2.1)). As a notational device, we use the metric tensor to "raise" or "lower" indices, that is, we define $u_\alpha = g_{\alpha\mu} u^\mu$ and $u^\alpha = g^{\alpha\mu} u_\mu$ where $g^{\alpha\beta}$ is the inverse of the matrix $g_{\alpha\beta}$ so that $g^{\alpha\mu} g_{\mu\beta} = \delta^\alpha_\beta$, which is the Kronecker delta. Using this notation we can write the inner product as $\mathbf{u} \cdot \mathbf{v} = u^\mu v_\mu = u_\mu v^\mu$.

2.1.3
Connections

Having established that vectors can be understood as directional derivatives in some neighbourhood of a point in spacetime, we now wish to examine how vectors at different points in spacetime can be related. It is here that we will see the curvature of spacetime manifest itself.

In flat space, we know how to compare two vectors at different points: simply slide one vector until the ends coincide. When sliding a vector it is important not to change its orientation, length and so on, but this is easily done in flat space by making sure that the vector is moved by *parallel transport*, that is at every stage in the process the displaced vector remains parallel to what it was before the displacement.

A similar process of parallel transport can be done in curved spacetime since each infinitesimal displacement can be done just as it was in flat space (thanks again to the equivalence principle!). But for series of displacements that go between two points that are not infinitesimally close, it is now important to first specify the path along which we will parallel transport our vector – we will see that if we parallel transport a vector from one point to another along different paths then the resulting vector will be different in general.

Consider some curve $\mathcal{P}(t)$ parameterized by the parameter t. At any point on the curve there is the tangent vector $\mathbf{u} = d/dt$ whose components are $u^\alpha = \partial x^\alpha / \partial t$ in a coordinate system in which $x^\alpha(t)$ are the points along the curve. We can construct a set of vectors $\mathbf{v}(t)$ along the curve that are arrived at by parallel transport if, for every t,

$$0 = \lim_{\Delta t \to 0} \frac{\mathbf{v}(t + \Delta t) - \mathbf{v}(t)}{\Delta t} = \frac{d\mathbf{v}}{dt} \tag{2.21}$$

or, in terms of the components in some coordinate system,

$$0 = \lim_{\Delta t \to 0} \frac{v^\alpha(t + \Delta t) - v^\alpha(t)}{\Delta t} = \frac{dv^\alpha}{dt} =: u^\mu \nabla_\mu v^\alpha \,. \tag{2.22}$$

This defines the *connection* ∇_α, also known as the *covariant derivative*, which is not always the ordinary derivative operator $\partial/\partial x^\alpha$ – though we can always find coordinates in which, locally, it is. Suppose we find a transformation $x'^\beta(x^\alpha)$ to a new set of coordinates in which the connection *is* simply an ordinary derivative operator. Recall that the components of the vector $\mathbf{u}$ in the new coordinates are related to the component in the old coordinates by $u'^\beta = u^\alpha \partial x'^\beta/\partial x^\alpha$, and similarly $v'^\beta = v^\alpha \partial x'^\beta/\partial x^\alpha$. It then follows that

$$\begin{aligned} 0 = u'^\nu \frac{\partial}{\partial x'^\nu} v'^\mu &= \left(u^\alpha \frac{\partial x'^\nu}{\partial x^\alpha}\right) \frac{\partial}{\partial x'^\nu} \left(v^\beta \frac{\partial x'^\mu}{\partial x^\beta}\right) = u^\alpha \frac{\partial}{\partial x^\alpha} \left(v^\beta \frac{\partial x'^\mu}{\partial x^\beta}\right) \\ &= u^\alpha \frac{\partial x'^\mu}{\partial x^\beta} \frac{\partial v^\beta}{\partial x^\alpha} + u^\alpha \frac{\partial^2 x'^\mu}{\partial x^\alpha \partial x^\beta} v^\beta \,. \end{aligned}$$

Now we multiply both sides by $\partial x^\gamma/\partial x'^\mu$ and use the fact that $\partial x^\gamma/\partial x^\beta = \delta^\gamma_\beta$ (the Kronecker delta) to obtain

$$0 = u^\alpha \left(\frac{\partial v^\gamma}{\partial x^\alpha} + \frac{\partial x^\gamma}{\partial x'^\mu} \frac{\partial^2 x'^\mu}{\partial x^\alpha \partial x^\beta} v^\beta \right) = u^\alpha \nabla_\alpha v^\gamma \,. \tag{2.23}$$

Because we can choose any curve to parallel transport $\mathbf{v}$ along, the choice of $\mathbf{u}$ is arbitrary. Thus,

$$\nabla_\alpha v^\gamma = \frac{\partial v^\gamma}{\partial x^\alpha} + \Gamma^\gamma_{\alpha\beta} v^\beta \,, \tag{2.24}$$

where

$$\Gamma^\gamma_{\alpha\beta} := \frac{\partial x^\gamma}{\partial x'^\mu} \frac{\partial^2 x'^\mu}{\partial x^\alpha \partial x^\beta} \tag{2.25}$$

are the *connection coefficients*. Note that these coefficients are symmetric in α and β.

While the covariant derivative of a vector is different from the ordinary derivative for a vector, the two types of derivatives must agree for a scalar. That is,

$$\nabla_\alpha \Phi(\mathbf{x}) = \frac{\partial \Phi(\mathbf{x})}{\partial x^\alpha} \,. \tag{2.26}$$

Using this identity, we can obtain expressions for the covariant derivative operating on arbitrary tensors in terms of the connection coefficients. For example, consider the scalar $u^\mu v_\mu$. We have:

$$\nabla_\alpha (u^\mu v_\mu) = \frac{\partial}{\partial x^\alpha}(u^\mu v_\mu) = v_\mu \frac{\partial u^\mu}{\partial x^\alpha} + u^\mu \frac{\partial v_\mu}{\partial x^\alpha} \,. \tag{2.27}$$

However, we also have

$$\nabla_\alpha(u^\mu v_\mu) = u^\mu \nabla_\alpha v_\mu + v_\mu \nabla_\alpha u^\mu = u^\mu \nabla_\alpha v_\mu + v_\mu \frac{\partial u^\mu}{\partial x^\alpha} + v_\mu \Gamma^\mu_{\alpha\nu} u^\nu \tag{2.28}$$

and combining these two equations we see that

$$u^\mu \frac{\partial v_\mu}{\partial x^\alpha} = u^\mu \nabla_\alpha v_\mu + v_\mu \Gamma^\mu_{\alpha\nu} u^\nu \tag{2.29}$$

and therefore, since u^μ is arbitrary,

$$\nabla_\alpha v_\beta = \frac{\partial v_\beta}{\partial x^\alpha} - \Gamma^\mu_{\alpha\beta} v_\mu \,. \tag{2.30}$$

For a general tensor, $T^{\alpha\cdots\beta}_{\gamma\cdots\delta}$ we have

$$\begin{aligned}\nabla_\lambda T^{\alpha\cdots\beta}_{\gamma\cdots\delta} = \frac{\partial}{\partial x^\lambda} T^{\alpha\cdots\beta}_{\gamma\cdots\delta} &+ \Gamma^\alpha_{\lambda\mu} T^{\mu\cdots\beta}_{\gamma\cdots\delta} + \cdots + \Gamma^\beta_{\lambda\mu} T^{\alpha\cdots\mu}_{\gamma\cdots\delta} \\ &- \Gamma^\mu_{\lambda\gamma} T^{\alpha\cdots\beta}_{\mu\cdots\delta} - \cdots - \Gamma^\mu_{\lambda\delta} T^{\alpha\cdots\beta}_{\gamma\cdots\mu} \,. \end{aligned} \tag{2.31}$$

Example 2.4 Flat-space connection in polar coordinates

Even in a two-dimensional (flat) plane the connection ∇_α is not generally an ordinary derivative. For example, consider two unit radial vectors $\mathbf{e}_{r,\mathcal{P}}$ and $\mathbf{e}_{r,\mathcal{Q}}$ at points $\mathcal{P} = (r, \phi)$ and $\mathcal{Q} = (r, \phi + \Delta\phi)$. These are clearly not parallel. (Translate both radially back to the origin and the first points in direction ϕ while the other points in direction $\phi + \Delta\phi$!) But the components (in polar coordinates) of the unit radial vector do not depend on ϕ, so, along the curve $r =$ const the sequence of unit radial vectors $\mathbf{e}_r(\phi)$ has components satisfying $\partial e^\alpha_r/\partial\phi = 0$. This shows that $\nabla_\alpha \neq \partial/\partial x^\alpha$ in polar coordinates.

To compute the quantity $\Gamma^\gamma_{\alpha\beta}$ we need to find the transformation $x'^\beta(x^\alpha)$ in which the connection is an ordinary derivative operator. We know that Cartesian coordinates are the coordinates that define parallel transport so the transformation is $x = r\cos\phi$ and $y = r\sin\phi$. Therefore, using $\partial r/\partial x = x/r$, $\partial r/\partial y = y/r$, $\partial\phi/\partial x = -y/r^2$, and $\partial\phi/\partial y = x/r^2$, as well as $\partial^2 x/\partial r^2 = \partial^2 y/\partial r^2 = 0$, $\partial^2 x/\partial r\partial\phi = -y/r$, $\partial^2 y/\partial r\partial\phi = x/r$, $\partial^2 x/\partial\phi^2 = -x$, and $\partial^2 y/\partial\phi^2 = -y$, we find

$$\begin{aligned}\Gamma^r_{rr} &= \frac{\partial r}{\partial x}\frac{\partial^2 x}{\partial r^2} + \frac{\partial r}{\partial y}\frac{\partial^2 y}{\partial r^2} \\ &= 0 \\ \Gamma^r_{r\phi} = \Gamma^r_{\phi r} &= \frac{\partial r}{\partial x}\frac{\partial^2 x}{\partial r\partial\phi} + \frac{\partial r}{\partial y}\frac{\partial^2 y}{\partial r\partial\phi} = \cos\phi(-\sin\phi) + \sin\phi\cos\phi \\ &= 0 \\ \Gamma^r_{\phi\phi} &= \frac{\partial r}{\partial x}\frac{\partial^2 x}{\partial\phi^2} + \frac{\partial r}{\partial y}\frac{\partial^2 y}{\partial\phi^2} = \cos\phi(-r\cos\phi) + \sin\phi(-r\sin\phi) \\ &= -r\end{aligned}$$

$$
\begin{aligned}
\Gamma^{\phi}_{rr} &= \frac{\partial \phi}{\partial x}\frac{\partial^2 x}{\partial r^2} + \frac{\partial \phi}{\partial y}\frac{\partial^2 y}{\partial r^2} \\
&= 0 \\
\Gamma^{\phi}_{r\phi} &= \Gamma^{\phi}_{\phi r} = \frac{\partial \phi}{\partial x}\frac{\partial^2 x}{\partial r\,\partial \phi} + \frac{\partial \phi}{\partial y}\frac{\partial^2 y}{\partial r\,\partial \phi} = \frac{-\sin\phi}{r}(-\sin\phi) + \frac{\cos\phi}{r}\cos\phi \\
&= \frac{1}{r} \\
\Gamma^{\phi}_{\phi\phi} &= \frac{\partial \phi}{\partial x}\frac{\partial^2 x}{\partial \phi^2} + \frac{\partial \phi}{\partial y}\frac{\partial^2 y}{\partial \phi^2} = \frac{-\sin\phi}{r}(-r\cos\phi) + \frac{\cos\phi}{r}(-r\sin\phi) \\
&= 0\,.
\end{aligned}
\tag{2.32}
$$

The equations of parallel transport for the components v^α of a vector $\mathbf{v}$ along a curve with tangent vector $\mathbf{u}$,

$$
u^\mu \nabla_\mu v^\alpha = u^\mu \frac{\partial v^\alpha}{\partial x^\mu} + \Gamma^\alpha_{\mu\nu} u^\mu v^\nu = 0
\tag{2.33}
$$

form a system of four first-order ordinary differential equations for these components. Given values $v^\alpha_{\mathcal{P}}$ at some point $\mathcal{P}$ on the curve, the values at any other point on the curve can be computed by integration.

Recall that the metric $g_{\mu\nu}$ gives us our natural definition for the inner product of two vectors: $\mathbf{v}\cdot\mathbf{w} = g_{\mu\nu}v^\mu w^\nu$. We know from flat space that the inner product of two vectors will remain the same if they are parallel transported together, so we demand that the inner product remains constant under parallel transport generally. That is, given a curve parameterized by t with tangent vector $\mathbf{u} = d/dt$, we demand that $d(\mathbf{v}\cdot\mathbf{w})/dt = 0$ when $\mathbf{v}$ and $\mathbf{w}$ are parallel transported along the curve:

$$
\begin{aligned}
0 &= \frac{d}{dt}(\mathbf{v}\cdot\mathbf{w}) = u^\rho \nabla_\rho (g_{\mu\nu} v^\mu w^\nu) \\
&= g_{\mu\nu} w^\nu u^\rho \nabla_\rho v^\mu + g_{\mu\nu} v^\mu u^\rho \nabla_\rho w^\nu + u^\rho v^\mu w^\nu \nabla_\rho g_{\mu\nu} \\
&= u^\rho v^\mu w^\nu \nabla_\rho g_{\mu\nu}\,,
\end{aligned}
\tag{2.34}
$$

where we have used $u^\rho \nabla_\rho v^\mu = 0$ and $u^\rho \nabla_\rho w^\nu = 0$ since these vectors are being parallel transported. Since the vectors $\mathbf{u}$, $\mathbf{v}$ and $\mathbf{w}$ are arbitrary, we conclude that the connection must satisfy

$$
\nabla_\gamma g_{\alpha\beta} = 0\,.
\tag{2.35}
$$

We can now express the connection coefficients in terms of the metric. We again compute the covariant derivative of the inner product $\mathbf{v}\cdot\mathbf{w}$:

$$
\begin{aligned}
\nabla_\delta (g_{\alpha\beta} v^\alpha w^\beta) &= g_{\alpha\beta} w^\beta \nabla_\delta v^\alpha + g_{\alpha\beta} v^\alpha \nabla_\delta w^\beta \\
&= g_{\alpha\beta} w^\beta \frac{\partial}{\partial x^\delta} v^\alpha + g_{\alpha\beta} w^\beta \Gamma^\alpha_{\delta\mu} v^\mu \\
&\quad + g_{\alpha\beta} v^\alpha \frac{\partial}{\partial x^\delta} w^\beta + g_{\alpha\beta} v^\alpha \Gamma^\beta_{\delta\mu} w^\mu
\end{aligned}
\tag{2.36}
$$

But the covariant derivative of a scalar function is the same as an ordinary derivative of that function so the left-hand-side is

$$\nabla_\delta \left(g_{\alpha\beta} v^\alpha w^\beta\right) = \frac{\partial}{\partial x_\delta}\left(g_{\alpha\beta} v^\alpha w^\beta\right)$$
$$= v^\alpha w^\beta \frac{\partial}{\partial x^\delta} g_{\alpha\beta} + g_{\alpha\beta} w^\beta \frac{\partial}{\partial x^\delta} v^\alpha + g_{\alpha\beta} v^\alpha \frac{\partial}{\partial x^\delta} w^\beta\,. \tag{2.37}$$

Combining these two equations, and using the fact that **v** and **w** are arbitrary vectors, we see that

$$\frac{\partial}{\partial x^\delta} g_{\alpha\beta} = g_{\mu\beta}\Gamma^\mu_{\delta\alpha} + g_{\alpha\mu}\Gamma^\mu_{\delta\beta} \tag{2.38a}$$

and, permuting the indices, we have

$$\frac{\partial}{\partial x^\alpha} g_{\beta\delta} = g_{\mu\delta}\Gamma^\mu_{\alpha\beta} + g_{\beta\mu}\Gamma^\mu_{\alpha\delta} \tag{2.38b}$$

$$\frac{\partial}{\partial x^\beta} g_{\delta\alpha} = g_{\mu\alpha}\Gamma^\mu_{\beta\delta} + g_{\delta\mu}\Gamma^\mu_{\beta\alpha}\,. \tag{2.38c}$$

Now we add Eqs. (2.38b) and (2.38c) and subtract Eq. (2.38a) to obtain

$$\frac{\partial}{\partial x^\alpha} g_{\beta\delta} + \frac{\partial}{\partial x^\beta} g_{\delta\alpha} - \frac{\partial}{\partial x^\delta} g_{\alpha\beta} = 2 g_{\delta\mu}\Gamma^\mu_{\alpha\beta} = 2\Gamma_{\delta\alpha\beta}\,, \tag{2.39}$$

where $\Gamma_{\delta\alpha\beta}$ are the connection coefficients with one index lowered with the metric. This system of equations can be solved for the connection coefficients (multiply by the inverse metric $g^{\gamma\delta}$):

$$\Gamma^\gamma_{\alpha\beta} = \frac{1}{2} g^{\gamma\delta}\left(\frac{\partial}{\partial x^\alpha} g_{\beta\delta} + \frac{\partial}{\partial x^\beta} g_{\delta\alpha} - \frac{\partial}{\partial x^\delta} g_{\alpha\beta}\right). \tag{2.40}$$

Example 2.5 Flat-space connection in polar coordinates (again)

We compute the connection coefficients for flat two-dimensional space in polar coordinates from the metric using Eq. (2.40). Actually, we will use Eq. (2.39) to compute $\Gamma_{\alpha\mu\nu}$ and then use the inverse-metric to obtain $\Gamma^\alpha_{\mu\nu}$. The line element in polar coordinates is $ds^2 = dr^2 + r^2 d\phi^2$ so the metric is $g_{\mu\nu} = \mathrm{diag}[1, r^2]$ and the inverse metric is $g^{\mu\nu} = \mathrm{diag}[1, 1/r^2]$. Clearly the only derivative of a metric component that does not vanish is $\partial g_{\phi\phi}/\partial r = 2r$ and so the only non-vanishing components of $\Gamma_{\alpha\mu\nu}$ are $\Gamma_{r\phi\phi} = -r$, and $\Gamma_{\phi r\phi} = \Gamma_{\phi\phi r} = r$. Now, using $\Gamma^\alpha_{\mu\nu} = g^{\alpha\beta}\Gamma_{\beta\mu\nu}$, we find that the only non-vanishing connection coefficients are $\Gamma^r_{\phi\phi} = -r$ and $\Gamma^\phi_{r\phi} = \Gamma^\phi_{\phi r} = 1/r$. This is the same result as was found in Example 2.4.

Example 2.6 Equation of continuity

The *equation of continuity* for some conserved quantity ρ with current $\boldsymbol{j}$ is

$$\nabla_\mu J^\mu = 0\,, \tag{2.41}$$

where $J^\mu := [\rho, \boldsymbol{j}]$. In flat spacetime this reduces to its usual form

$$\frac{\partial \rho}{\partial t} + \boldsymbol{\nabla}\cdot\boldsymbol{j} = 0 \quad \text{(flat spacetime).} \tag{2.42}$$

In curved spacetime, we have

$$\nabla_\mu J^\mu = \frac{\partial J^\mu}{\partial x^\mu} + \Gamma^\mu_{\mu\nu} J^\nu\,. \tag{2.43}$$

Now $\Gamma^\mu_{\mu\alpha}$ can be written in the useful form

$$\Gamma^\mu_{\mu\alpha} = \frac{1}{2} g^{\mu\nu}\left(\frac{\partial g_{\alpha\nu}}{\partial x^\mu} + \frac{\partial g_{\mu\nu}}{\partial x^\alpha} - \frac{\partial g_{\alpha\mu}}{\partial x^\nu}\right) = \frac{1}{2} g^{\mu\nu}\frac{\partial g_{\mu\nu}}{\partial x^\alpha} \tag{2.44}$$

and, using the identity $\det\mathbf{g} = \exp(\mathrm{Tr}\ln\mathbf{g})$, we note that $\partial\det\mathbf{g}/\partial x^\alpha = (\det\mathbf{g})\mathrm{Tr}(\mathbf{g}^{-1}\cdot\partial\mathbf{g}/\partial x^\alpha)$, so

$$\Gamma^\mu_{\mu\alpha} = \frac{1}{2}(\det\mathbf{g})^{-1}\frac{\partial\det\mathbf{g}}{\partial x^\alpha} = \frac{\partial}{\partial x^\alpha}\ln|\det\mathbf{g}|^{1/2}\,. \tag{2.45}$$

We therefore have

$$\begin{aligned} 0 = |\det\mathbf{g}|^{1/2}\nabla_\mu J^\mu &= |\det\mathbf{g}|^{1/2}\frac{\partial J^\mu}{\partial x^\mu} + J^\mu\frac{\partial}{\partial x^\mu}|\det\mathbf{g}|^{1/2} \\ &= \frac{\partial}{\partial x^\mu}\left(|\det\mathbf{g}|^{1/2} J^\mu\right)\,. \end{aligned} \tag{2.46}$$

If we now integrate this over some volume Ω of spacetime we find

$$0 = \int_\Omega \frac{\partial}{\partial x^\mu}\left(|\det\mathbf{g}|^{1/2} J^\mu\right) d^4x = \int_{\partial\Omega} |\det\mathbf{g}|^{1/2} J^\mu\, dS_\mu\,, \tag{2.47}$$

where we have used Stokes's law to convert the integral over the spacetime region Ω into an integral over the boundary of the spacetime region $\partial\Omega$. Here, dS_μ is the area element on the boundary. If we take the boundary $\partial\Omega$ to be two spatial surfaces V_1 and V_2 at times t_1 and t_2 connected by the evolution of a two-dimensional surface S between these two times, $S\times(t_1, t_2)$, then on the two surfaces V_1 and V_2 we have $|\det\mathbf{g}|^{1/2} J^\mu dS_\mu = \rho\, dV$ where dV is the spatial volume element on those surfaces, while on $S\times(t_1,t_2)$ we have $|\det\mathbf{g}|^{1/2} J^\mu dS_\mu = \boldsymbol{j}\cdot\hat{\boldsymbol{n}}\, d\tau\, dS$ where the surface S has an outward spatial normal vector $\hat{\boldsymbol{n}}$ and volume element dS. We find

$$\left[\int_V \rho\, dV\right]_{t_1}^{t_2} = -\int_{t_1}^{t_2} \mathcal{F}(t)\, d\tau \tag{2.48}$$

with

$$\mathcal{F}(t) := \int_S j \cdot \hat{n} \, dS \tag{2.49}$$

being the flux through the surface S at time t. This shows that the change in the amount of quantity contained within the set of surfaces S is the negative of the flux of quantity through the surfaces – that is, the quantity is conserved.

A *vector field*, $\mathbf{v}(\mathcal{P})$ is a smoothly varying collection of vectors defined at all points $\mathcal{P}$ in some region of spacetime. The vector at one point in the region is not necessarily the same as a vector at a nearby point that is parallel transported to the first point along some curve. They might be different because the spacetime is curved, but they also might be different because of how the vector field is defined. As an illustration consider the unit radial vector field $\mathbf{e}_r(r, \phi)$ in a polar coordinate system: the vector $\mathbf{e}_r(1, \pi/2)$, parallel transported to the point $(1, 0)$, is orthogonal to the vector $\mathbf{e}_r(1, 0)$.

Even in flat space, two vector fields $\mathbf{u}$ and $\mathbf{v}$ do not necessarily *commute* in the sense that when travelling a small distance Δt along the curve with tangent vector $\mathbf{u}$ and then a small distance Δs along the curve with tangent vector $\mathbf{v}$ you do not necessarily arrive at the same point as when the two steps are reversed: this time travelling first a distance Δs along the curve with tangent vector $\mathbf{v}$ and then a distance Δt along the curve with tangent vector $\mathbf{u}$. If the two vector fields *do* commute, which is symbolically expressed as $[\mathbf{u}, \mathbf{v}] = 0$, then the distances Δs and Δt uniquely identify the point that you arrive at, and thus the parameters s and t can be used as coordinates in the surface spanned by the two vectors.

For small values of Δs and Δt we can work in a local inertial frame and we can treat the vector fields $\mathbf{u}$ and $\mathbf{v}$ as displacement vectors. Start at point $\mathcal{P}$ and travel a distance Δs along vector $\mathbf{v}(\mathcal{P})$ to point $\mathcal{Q}$ and then a distance Δt along vector $\mathbf{u}(\mathcal{Q})$ to point $\mathcal{A}$. Now, starting again at point $\mathcal{P}$ travel first the distance Δt along vector $\mathbf{u}(\mathcal{P})$ to point $\mathcal{R}$ and then a distance Δs along vector $\mathbf{v}(\mathcal{R})$ to point $\mathcal{B}$. The difference between these endpoints is the displacement vector $\mathbf{w}\Delta s \Delta t = \mathcal{B} - \mathcal{A}$ (see Figure 2.1):

$$\begin{aligned}
\mathbf{w} &= \lim_{\Delta s \to 0} \lim_{\Delta t \to 0} \frac{1}{\Delta s \Delta t} \left\{ \left[\mathbf{u}(\mathcal{P}) \Delta t + \mathbf{v}(\mathcal{R}) \Delta s \right] - \left[\mathbf{v}(\mathcal{P}) \Delta s + \mathbf{u}(\mathcal{Q}) \Delta t \right] \right\} \\
&= \lim_{\Delta s \to 0} \lim_{\Delta t \to 0} \frac{1}{\Delta s \Delta t} \left\{ \left[\mathbf{v}(\mathcal{R}) - \mathbf{v}(\mathcal{P}) \right] \Delta s - \left[\mathbf{u}(\mathcal{Q}) - \mathbf{u}(\mathcal{P}) \right] \Delta t \right\} \\
&= \lim_{\Delta t \to 0} \frac{\mathbf{v}(\mathcal{R}) - \mathbf{v}(\mathcal{P})}{\Delta t} - \lim_{\Delta s \to 0} \frac{\mathbf{u}(\mathcal{Q}) - \mathbf{u}(\mathcal{P})}{\Delta s} \, .
\end{aligned} \tag{2.50}$$

The first term is the covariant derivative of vector $\mathbf{v}$ along the curve with tangent $\mathbf{u}$ while the second term is the covariant derivative of vector $\mathbf{u}$ along the curve with tangent $\mathbf{v}$. Thus,

$$w^\alpha = u^\mu \nabla_\mu v^\alpha - v^\mu \nabla_\mu u^\alpha \tag{2.51}$$

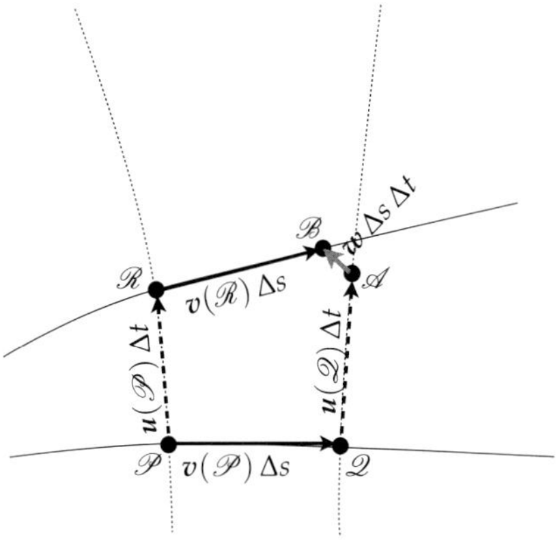

Figure 2.1 Illustration of the vector commutator $\mathbf{w} = [\mathbf{u}, \mathbf{v}]$. Dotted lines are the integral curves of $\mathbf{u}$ while solid lines are the integral curves of $\mathbf{v}$. The failure of points $\mathcal{A}$ and $\mathcal{B}$ to coincide indicates that the two vectors $\mathbf{u}$ and $\mathbf{v}$ do not commute.

is the commutator of vectors $\mathbf{u}$ and $\mathbf{v}$, which we write $\mathbf{w} = [\mathbf{u}, \mathbf{v}]$. It is easy to see, due to the fact that the connection coefficients $\Gamma^{\gamma}_{\alpha\beta}$ are symmetric in α and β, that

$$w^{\alpha} = u^{\mu}\frac{\partial}{\partial x^{\mu}}v^{\alpha} - v^{\mu}\frac{\partial}{\partial x^{\mu}}u^{\alpha}\,. \tag{2.52}$$

Example 2.7 Vector commutation

Consider again the two-dimensional plane in polar coordinates. There are two important sets of vector fields. The first set comprises the coordinate vectors $\mathbf{r}$ and $\boldsymbol{\phi}$ which have components $r^{\alpha} = [1, 0]$ and $\phi^{\alpha} = [0, 1]$. Note that these vectors are not orthonormal; in particular, $\|\boldsymbol{\phi}\|^2 = g_{\mu\nu}\phi^{\mu}\phi^{\nu} = r^2$. Instead, the orthonormal basis vectors $\mathbf{e}_r$ and $\mathbf{e}_{\phi}$ have components $(e_r)^{\alpha} = [1, 0]$ and $(e_{\phi})^{\alpha} = [0, 1/r]$.

Imagine travelling along the integral curve of vector $\mathbf{r}(\mathcal{P})$ (i.e. a radial line) a coordinate distance ϵ starting at point $\mathcal{P} = (r, \phi)$ and ending at point $\mathcal{Q} = (r + \epsilon, \phi)$; then travelling along the integral curve of vector $\boldsymbol{\phi}(\mathcal{Q})$ (i.e. along a circle of radius $r + \epsilon$) a coordinate distance ϵ, which takes you to point $\mathcal{A} = (r + \epsilon, \phi + \epsilon)$. Now imagine reversing the order, first travelling from $\mathcal{P}$ along a coordinate distance ϵ along the integral curve of vector $\boldsymbol{\phi}(\mathcal{R})$, that is the circle of radius r, to point $\mathcal{R} = (r, \phi + \epsilon)$, and then radially from this point along vector $\mathbf{r}(\mathcal{R})$ a coordinate distance ϵ; clearly this also arrives at point $\mathcal{A}$.

Now imagine doing the same thing with the two vectors $\mathbf{e}_r$ and $\mathbf{e}_{\phi}$. When traversing the "arc" paths (the integral curves of $\mathbf{e}_{\phi}$), the coordinate distance ϵ now corresponds to angular distances that depend on the radius of the circle. The first path, along $\mathbf{e}_r$ and then along $\mathbf{e}_{\phi}$, goes first from point $\mathcal{P}$ to point $\mathcal{Q}' = (r + \epsilon, \phi) = \mathcal{Q}$ and then to point $\mathcal{A}' = (r + \epsilon, \phi + \epsilon/(r + \epsilon))$. The second path, along $\mathbf{e}_{\phi}$ and then along $\mathbf{e}_r$, goes first from point $\mathcal{P}$ to point $\mathcal{R}' = (r, \phi + \epsilon/r)$ and then to point $\mathcal{B}' = (r+\epsilon, \phi+\epsilon/r)$. Note that $\mathcal{B}' \neq \mathcal{A}'$. The orthonormal vectors do not commute.

Example 2.8 Lie derivative

Suppose we choose a coordinate system in which the vector **u** is parameterized by the coordinate t, that is, $u^\mu(\partial/\partial x^\mu) = \partial/\partial t$, so $u^\alpha = [1,0,0,0]$. The vector commutator, $\mathbf{w} = u^\mu(\partial \mathbf{v}/\partial x^\mu) - v^\mu(\partial \mathbf{u}/\partial x^\mu)$, then has the particularly simple form $\mathbf{w} = \partial \mathbf{v}/\partial t$. That is, **w** describes the change in the vector **v** along the integral curve of **u**. Since vector commutation is covariant, this statement is true for all coordinate systems, not just the one adapted to the vector **u**. The change in the vector **v** along the integral curve of $\mathbf{u} = d/dt$ is known as the *Lie derivative* of **v** along **u**:

$$\begin{aligned}\Delta_{\mathbf{u}}\mathbf{v} = [\mathbf{u},\mathbf{v}] &:= u^\mu \frac{\partial \mathbf{v}}{\partial x^\mu} - v^\mu \frac{\partial \mathbf{u}}{\partial x^\mu} \\ &= \frac{\partial \mathbf{v}}{\partial t} \quad \text{(coordinate system where } \mathbf{u} = d/dt\text{)} .\end{aligned} \tag{2.53}$$

To extend the definition of the Lie derivative to tensors of arbitrary rank, we require the following two properties: (i) the Lie derivative is the same as the ordinary derivative when operating on scalars $\Delta_{\mathbf{u}} f = u^\mu(\partial f/\partial x^\mu)$ for any function f; and (ii) the Lie derivative obeys the Leibniz rule, for example $\Delta_{\mathbf{u}}(A^{\alpha\cdots\mu} B_{\beta\cdots\mu}) = A^{\alpha\cdots\mu}\Delta_{\mathbf{u}} B_{\beta\cdots\mu} + B_{\beta\cdots\mu}\Delta_{\mathbf{u}} A^{\alpha\cdots\mu}$. The Lie derivative of tensors of any order is then found to be

$$\begin{aligned}\Delta_{\mathbf{u}} T^{\alpha\cdots\beta}{}_{\gamma\cdots\delta} = {}& u^\mu \frac{\partial}{\partial x^\mu} T^{\alpha\cdots\beta}{}_{\gamma\cdots\delta} - T^{\mu\cdots\beta}{}_{\gamma\cdots\delta}\frac{\partial}{\partial x^\mu} u^\alpha - \cdots \\ & - T^{\alpha\cdots\mu}{}_{\gamma\cdots\delta}\frac{\partial}{\partial x^\mu} u^\beta + T^{\alpha\cdots\beta}{}_{\mu\cdots\delta}\frac{\partial}{\partial x^\gamma} u^\mu + \cdots \\ & + T^{\alpha\cdots\beta}{}_{\gamma\cdots\mu}\frac{\partial}{\partial x^\delta} u^\mu .\end{aligned} \tag{2.54}$$

A vector **v** is said to be *Lie dragged* along **u** if it has vanishing Lie derivative, $\Delta_{\mathbf{u}}\mathbf{v} = 0$, that is the two vectors commute. This can be understood physically as follows: suppose there are two nearby particles separated by some parameter amount Δs along the integral curve of **v**, that is, $\mathbf{v}\Delta s$ is initially the displacement vector between the particles. If **u** is the four-velocity of these particles and if **v** is Lie-dragged along **u**, then at future times the displacement vector between the two particles continues to be $\mathbf{v}\Delta s$.

2.1.4
Geodesics

In Newtonian mechanics, a particle will move in a straight line in the absence of any external force, and, due to the equivalence principle, this must also be true of a freely falling particle. By a "straight line" we mean straight in terms of a rectilinear coordinate system in a locally inertial frame. In curved spacetime such straight lines are known as *geodesics*. A geodesic is a curve for which the tangent vector to the curve is parallel transported along itself (the straightest possible curve in a curved spacetime). That is, a curve $x^\alpha(t)$ with tangent vector $\mathbf{u} = d/dt$ is a geodesic if

$$u^\mu \nabla_\mu u^\alpha = 0 \,. \tag{2.55}$$

This is known as the *geodesic equation*. Since $u^\alpha = dx^\alpha/dt$, the geodesic equation – that is, the trajectory of a freely falling particle – satisfies

$$0 = \frac{dx^\mu}{dt}\frac{\partial}{\partial x^\mu}\left(\frac{dx^\alpha}{dt}\right) + \Gamma^\alpha_{\mu\nu}\frac{dx^\mu}{dt}\frac{dx^\nu}{dt} \tag{2.56}$$

or

$$\frac{d^2x^\alpha}{dt^2} = -\Gamma^\alpha_{\mu\nu}\frac{dx^\mu}{dt}\frac{dx^\nu}{dt} \,. \tag{2.57}$$

The parameter t is the *affine* parameter of the geodesic (see Problem 2.1). In rectilinear coordinates in a locally inertial frame, the connection coefficients can be made to vanish at a point and the geodesic equations become simply $d^2x^\alpha/dt^2 = 0$, that is the equation of a straight line. The right hand side of Eq. (2.57) therefore encodes the acceleration of the particle with respect to a coordinate system (either because the system is not rectilinear or because the coordinate system is not inertial, e.g. not freely falling); in this sense, the connection coefficients $\Gamma^\gamma_{\alpha\beta}$ can be interpreted as providing the gravitational force (though in general relativity such forces are akin to the fictitious forces, such as centrifugal and Coriolis forces, of Newtonian mechanics in that such forces arise simply because one is not in a freely falling frame).

In fact, normal coordinates can be constructed using geodesics. Imagine an observer in a freely falling elevator sets up coordinates in the freely falling frame as follows: The proper time τ of the observer is used to measure time. A spatial coordinate grid is constructed by launching spatial geodesics from the observer's location (the origin of the coordinates) in three orthogonal directions $\boldsymbol{e}_1$, $\boldsymbol{e}_2$ and $\boldsymbol{e}_3$ and the affine distance (that is, the span of the affine parameter) along these three geodesics determines the coordinate value along that axis. Suppose, then, at time $\tau = x^0$, a point $\mathcal{P}$ is reached by travelling along the geodesic with tangent vector $\boldsymbol{v} = x^1\boldsymbol{e}_1 + x^2\boldsymbol{e}_2 + x^3\boldsymbol{e}_3$ an affine distance of 1; then the coordinates of $\mathcal{P}$ are (x^0, x^1, x^2, x^3). This procedure can be used to label all points within a *normal neighbourhood* of the observer, that is, a region small enough that there is a unique geodesic connecting every point in the neighbourhood to the origin. Such coordinates are known as *Riemann normal coordinates*. In these coordinates, the connection coefficients $\Gamma^\gamma_{\alpha\beta}$ vanish at the origin, though their derivatives do not. The metric at the origin in Riemann normal coordinates is the Minkowski metric $\eta_{\alpha\beta}$. First derivatives of the metric vanish at the origin but second derivatives do not.

2.1.5 Curvature

Our normal conception of a curved surface is that it appears bent when it is viewed from the outside. Such an *extrinsic* notion of curvature implicitly requires that the surface exist in some larger-dimensional space from which we can view the surface. But since we exist within spacetime, we seek an *intrinsic curvature* that can be determined from the properties of the spacetime itself without recourse to some higher-dimensional space in which spacetime exists. The intrinsic curvature of a surface is defined in terms of parallel transport of vectors: if a vector that is parallel transported around some closed path returns to its original point altered, then spacetime is curved in the vicinity of that point (see Example 2.9).

For simplicity, imagine two *commuting* vector fields, $\mathbf{u} = d/dt$ and $\mathbf{v} = d/ds$, and use these to construct a quadrilateral $\mathcal{PQRS}$ formed by starting at point $\mathcal{P}$, travelling along the integral curve of vector $\mathbf{u}$ a distance Δt to point $\mathcal{Q}$, then travelling along the integral curve of vector $\mathbf{v}$ a distance Δs to point $\mathcal{R}$, then travelling along the integral curve of vector $-\mathbf{u}$ a distance Δt to point $\mathcal{S}$, and finally travelling along the integral curve of vector $-\mathbf{v}$ a distance Δs to return to point $\mathcal{P}$. Now consider an arbitrary vector $\mathbf{w}_{\mathcal{P}}$ at point $\mathcal{P}$ and parallel transport it to the intermediate point $\mathcal{R}$ via point $\mathcal{Q}$, $\mathbf{w}_{\mathcal{P}\to\mathcal{Q}\to\mathcal{R}}$, and compare this vector to the same vector $\mathbf{w}_{\mathcal{P}}$ at point $\mathcal{P}$ and parallel transported to point $\mathcal{R}$ this time via point $\mathcal{S}$, $\mathbf{w}_{\mathcal{P}\to\mathcal{S}\to\mathcal{R}}$. The difference between these two vectors is what we wish to determine. It helps to consider, in addition, a smooth vector field $\mathbf{w}$ defined so that $\mathbf{w}(\mathcal{P}) = \mathbf{w}_{\mathcal{P}}$. We have

$$(\delta w^\delta)_{\mathcal{P}\to\mathcal{Q}} = w^\delta(\mathcal{Q}) - w^\delta_{\mathcal{P}\to\mathcal{Q}} = (\Delta t)\; u^\alpha \nabla_\alpha w^\delta \Big|_{\mathcal{P}}, \tag{2.58}$$

and

$$\begin{aligned}(\delta w^\delta)_{\mathcal{P}\to\mathcal{Q}\to\mathcal{R}} &= w^\delta(\mathcal{R}) - w^\delta_{\mathcal{P}\to\mathcal{Q}\to\mathcal{R}} = (\Delta s)\; v^\beta \nabla_\beta w^\delta \Big|_{\mathcal{Q}} \\ &= (\Delta s \Delta t) v^\beta \nabla_\beta \left(u^\alpha \nabla_\alpha w^\delta\right);\end{aligned} \tag{2.59}$$

similarly,

$$(\delta w^\delta)_{\mathcal{P}\to\mathcal{S}\to\mathcal{R}} = w^\delta(\mathcal{R}) - w^\delta_{\mathcal{P}\to\mathcal{S}\to\mathcal{R}} = (\Delta s \Delta t) u^\alpha \nabla_\alpha \left(v^\beta \nabla_\beta w^\delta\right). \tag{2.60}$$

Therefore,

$$\begin{aligned}\left(\delta w^\delta\right)_{\mathcal{P}\to\mathcal{Q}\to\mathcal{R}} - \left(\delta w^\delta\right)_{\mathcal{P}\to\mathcal{S}\to\mathcal{R}} &= w^\delta_{\mathcal{P}\to\mathcal{S}\to\mathcal{R}} - w^\delta_{\mathcal{P}\to\mathcal{Q}\to\mathcal{R}} \\ &= (\Delta s \Delta t)\, v^\beta \nabla_\beta \left(u^\alpha \nabla_\alpha w^\delta\right) - (\Delta s\, \Delta t)\, u^\alpha \nabla_\alpha \left(v^\beta \nabla_\beta w^\delta\right) \\ &= (\Delta s\, \Delta t) \Big\{\left(v^\beta \nabla_\beta u^\alpha\right) \nabla_\alpha w^\delta + u^\alpha v^\beta \nabla_\beta \nabla_\alpha w^\delta \\ &\qquad - \left(u^\alpha \nabla_\alpha v^\beta\right) \nabla_\beta w^\delta - u^\alpha v^\beta \nabla_\alpha \nabla_\beta w^\delta\Big\} \\ &= -(\Delta s\, \Delta t)\, u^\alpha v^\beta \left(\nabla_\alpha \nabla_\beta - \nabla_\beta \nabla_\alpha\right) w^\delta \\ &=: (\Delta s\, \Delta t)\, R_{\alpha\beta\gamma}{}^\delta u^\alpha v^\beta w^\gamma,\end{aligned} \tag{2.61}$$

where we have used the fact that $\mathbf{u}$ and $\mathbf{v}$ commute to obtain the second last line and in the last line we introduce the *Riemann curvature tensor*, which is defined by

$$R_{\alpha\beta\gamma}{}^\delta w^\gamma := -(\nabla_\alpha \nabla_\beta - \nabla_\beta \nabla_\alpha) w^\delta \tag{2.62}$$

for arbitrary $\mathbf{w}$.

Example 2.9 Curvature

On a two-dimensional flat plane, vectors that are parallel transported around any closed path return unchanged. This is depicted in Figure 2.2. This implies that the Riemann curvature tensor $R_{\alpha\beta\gamma}{}^{\delta}$ is zero and therefore the flat plane has no curvature.

On the two-dimensional surface of a sphere, however, a vector that is parallel transported from the North pole down to the equator along a line of longitude, along the equator to a new longitude, and then back up to the North pole along the new line of longitude, will return rotated compared to the initial vector. This is depicted in Figure 2.3. This demonstrates that the surface of a sphere has curvature: there are non-vanishing components of the Riemann curvature tensor $R_{\alpha\beta\gamma}{}^{\delta}$.

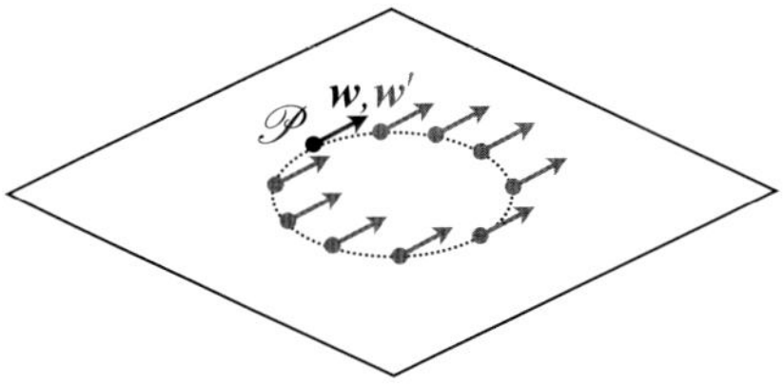

Figure 2.2 Parallel transport of a vector **w** around a closed curve on a two-dimensional flat plane. The resulting vector **w**′ is the same as the original vector **w**. The plane has no curvature.

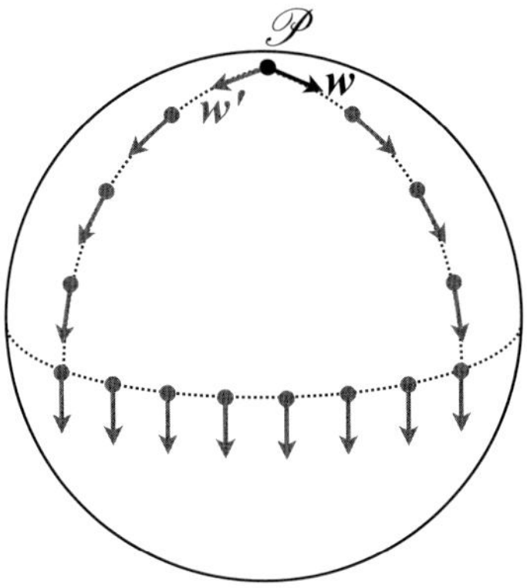

Figure 2.3 Parallel transport of a vector **w** around a closed curve on the two-dimensional surface of a sphere. The resulting vector **w**′ is rotated relative to the original vector **w**, which shows that the sphere is curved.

An explicit form of the Riemann tensor can be obtained from its definition given by Eq. (2.62):

$$\begin{aligned}
R_{\alpha\beta\gamma}{}^{\delta} w^{\gamma} &= -\left(\nabla_\alpha \nabla_\beta - \nabla_\beta \nabla_\alpha\right) w^{\delta} \\
&= -\frac{\partial}{\partial x^\alpha}\left(\nabla_\beta w^\delta\right) - \Gamma^\delta_{\alpha\gamma}\left(\nabla_\beta w^\gamma\right) + \Gamma^\mu_{\alpha\beta}\left(\nabla_\mu w^\delta\right) \\
&\quad + \frac{\partial}{\partial x^\beta}\left(\nabla_\alpha w^\delta\right) + \Gamma^\delta_{\beta\gamma}\left(\nabla_\alpha w^\gamma\right) - \Gamma^\mu_{\alpha\beta}\left(\nabla_\mu w^\delta\right) \\
&= -\frac{\partial}{\partial x^\alpha}\left(\frac{\partial}{\partial x^\beta} w^\delta + \Gamma^\delta_{\beta\gamma} w^\gamma\right) - \Gamma^\delta_{\alpha\gamma}\left(\frac{\partial}{\partial x^\beta} w^\gamma + \Gamma^\gamma_{\beta\mu} w^\mu\right) \\
&\quad + \frac{\partial}{\partial x^\beta}\left(\frac{\partial}{\partial x^\alpha} w^\delta + \Gamma^\delta_{\alpha\gamma} w^\gamma\right) + \Gamma^\delta_{\beta\gamma}\left(\frac{\partial}{\partial x^\alpha} w^\gamma + \Gamma^\gamma_{\alpha\mu} w^\mu\right) \\
&= -w^\gamma \frac{\partial}{\partial x^\alpha}\Gamma^\delta_{\beta\gamma} - \Gamma^\delta_{\beta\gamma}\frac{\partial}{\partial x^\alpha} w^\gamma - \Gamma^\delta_{\alpha\gamma}\frac{\partial}{\partial x^\beta} w^\gamma - \Gamma^\delta_{\alpha\gamma}\Gamma^\gamma_{\beta\mu} w^\mu \\
&\quad + w^\gamma \frac{\partial}{\partial x^\beta}\Gamma^\delta_{\alpha\gamma} + \Gamma^\delta_{\alpha\gamma}\frac{\partial}{\partial x^\beta} w^\gamma + \Gamma^\delta_{\beta\gamma}\frac{\partial}{\partial x^\alpha} w^\gamma + \Gamma^\delta_{\beta\gamma}\Gamma^\gamma_{\alpha\mu} w^\mu \\
&= \left(-\frac{\partial}{\partial x^\alpha}\Gamma^\delta_{\beta\gamma} + \frac{\partial}{\partial x^\beta}\Gamma^\delta_{\alpha\gamma} - \Gamma^\delta_{\alpha\mu}\Gamma^\mu_{\beta\gamma} + \Gamma^\delta_{\beta\mu}\Gamma^\mu_{\alpha\gamma}\right) w^\gamma
\end{aligned} \tag{2.63}$$

and, therefore,

$$R_{\alpha\beta\gamma}{}^{\delta} = -\frac{\partial}{\partial x^\alpha}\Gamma^\delta_{\beta\gamma} + \frac{\partial}{\partial x^\beta}\Gamma^\delta_{\alpha\gamma} - \Gamma^\delta_{\alpha\mu}\Gamma^\mu_{\beta\gamma} + \Gamma^\delta_{\beta\mu}\Gamma^\mu_{\alpha\gamma}\,. \tag{2.64}$$

Example 2.10 Riemann tensor in a locally inertial frame

The Riemann tensor takes a simple form in a locally inertial frame. Recall that a locally inertial frame has vanishing connection coefficients at the origin ($\Gamma^\gamma_{\alpha\beta} = 0$), though derivatives of the connection coefficients are not vanishing; equivalently, the first derivative of the metric vanishes but the second derivative of the metric does not. Therefore,

$$\begin{aligned}
R_{\alpha\beta\gamma\delta} &= -\frac{\partial \Gamma_{\delta\beta\gamma}}{\partial x^\alpha} + \frac{\partial \Gamma_{\delta\alpha\gamma}}{\partial x^\beta} \\
&= -\frac{1}{2}\frac{\partial}{\partial x^\alpha}\left(\frac{\partial g_{\gamma\delta}}{\partial x^\beta} + \frac{\partial g_{\beta\delta}}{\partial x^\gamma} - \frac{\partial g_{\beta\gamma}}{\partial x^\delta}\right) \\
&\quad + \frac{1}{2}\frac{\partial}{\partial x^\beta}\left(\frac{\partial g_{\gamma\delta}}{\partial x^\alpha} + \frac{\partial g_{\alpha\delta}}{\partial x^\gamma} - \frac{\partial g_{\alpha\gamma}}{\partial x^\delta}\right) \\
&= \frac{1}{2}\left(-\frac{\partial^2 g_{\beta\delta}}{\partial x^\alpha \partial x^\gamma} + \frac{\partial^2 g_{\beta\gamma}}{\partial x^\alpha \partial x^\delta} + \frac{\partial^2 g_{\alpha\delta}}{\partial x^\beta \partial x^\gamma} - \frac{\partial^2 g_{\alpha\gamma}}{\partial x^\beta \partial x^\delta}\right).
\end{aligned} \tag{2.65}$$

(Recall that $\Gamma_{\delta\alpha\beta} = g_{\gamma\delta}\Gamma^\gamma_{\alpha\beta}$.)

There are several important properties of the Riemann curvature tensor. First, the tensor is clearly antisymmetric in its first two indices:

$$R_{\alpha\beta\gamma}{}^{\delta} = -R_{\beta\alpha\gamma}{}^{\delta}\,. \tag{2.66a}$$

Second, it is straightforward to show that it is antisymmetric in the first *three* indices:

$$R_{\alpha\beta\gamma}{}^{\delta} + R_{\beta\gamma\alpha}{}^{\delta} + R_{\gamma\alpha\beta}{}^{\delta} = 0\,. \tag{2.66b}$$

A third property can be obtained by noting that the covariant derivative is compatible with the connection. Consider the scalar $g_{\mu\nu}u^{\mu}v^{\nu}$. Since the covariant derivative operators commute when acting on a scalar,

$$\begin{aligned}
0 &= (\nabla_\alpha\nabla_\beta - \nabla_\beta\nabla_\alpha)(g_{\mu\nu}u^{\mu}v^{\nu}) \\
&= g_{\mu\nu}u^{\mu}(\nabla_\alpha\nabla_\beta - \nabla_\beta\nabla_\alpha)v^{\nu} + g_{\mu\nu}v^{\nu}(\nabla_\alpha\nabla_\beta - \nabla_\beta\nabla_\alpha)u^{\mu} \\
&= -g_{\mu\nu}u^{\mu}R_{\alpha\beta\gamma}{}^{\nu}v^{\gamma} - g_{\mu\nu}v^{\nu}R_{\alpha\beta\gamma}{}^{\mu}u^{\gamma} \\
&= -u^{\delta}v^{\gamma}(R_{\alpha\beta\gamma\delta} + R_{\alpha\beta\delta\gamma})
\end{aligned} \tag{2.66c}$$

therefore

$$R_{\alpha\beta\gamma\delta} = -R_{\alpha\beta\delta\gamma}\,. \tag{2.66d}$$

From Eqs. (2.66a), (2.66b) and (2.66d), it can be shown that

$$R_{\alpha\beta\gamma\delta} = R_{\gamma\delta\alpha\beta} \tag{2.66e}$$

(Problem 2.2). A final property, known as the *Bianchi identity*, is

$$\nabla_\alpha R_{\beta\gamma\kappa}{}^{\lambda} + \nabla_\beta R_{\gamma\alpha\kappa}{}^{\lambda} + \nabla_\gamma R_{\alpha\beta\kappa}{}^{\lambda} = 0. \tag{2.66f}$$

To show that the Bianchi identity is true, consider parallel transporting an arbitrary vector **a** about the six faces of a cube where the path goes around opposite faces in opposite ways. Figure 2.4 shows a traversal of the top and bottom sides of the cube starting at point $\mathcal{P}$; note that the trip from $\mathcal{P}$ to $\mathcal{Q}$ is cancelled by the return trip from $\mathcal{Q}$ to $\mathcal{P}$, so Figure 2.4 essentially involves a clockwise traversal of the bottom side and a counterclockwise traversal of the top side of the cube. Similar paths will traverse the front and back sides and the right and left sides, and the resulting traversal of all sides is seen in Figure 2.5. From Figure 2.5 we notice that the vector is parallel transported along each edge twice in opposite directions; therefore the net shift after the vector is transported around the entire cube will be zero.

It is helpful to specialize to Riemann normal coordinates, in which $w^{\alpha}\nabla_\alpha = w^{\alpha}(\partial/\partial x^{\alpha})$. Let the sides of the cube all be small with length ϵ and suppose that **u**, **v** and **w** are the tangent vectors to the edges of the cube. Consider first going around the bottom face of the cube and then around the top face as shown in Figure 2.4. The shift in the vector **a** after traversing the bottom face (the black path shown in Figure 2.4) is, by the definition of the Riemann tensor,

$$(\delta a^{\alpha})_{\text{bot}} = -\epsilon^2 R_{\mu\nu\rho}{}^{\lambda}(\mathcal{P})u^{\mu}v^{\nu}a^{\rho}\,. \tag{2.67}$$

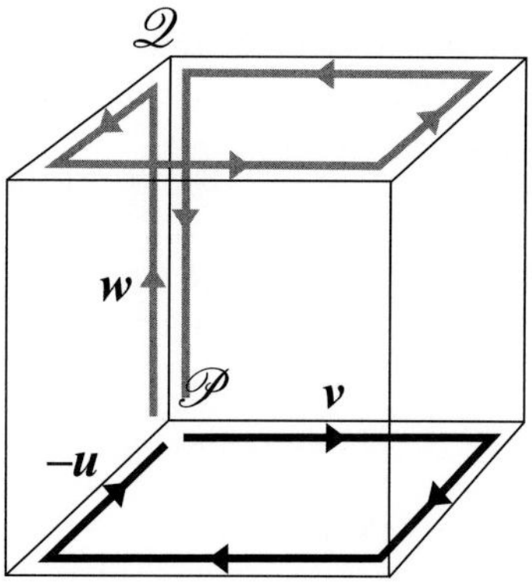

Figure 2.4 The path of parallel transport of some vector, first, starting at point $\mathcal{P}$ and traversing the bottom face of a cube (shown in black), and second, from point $\mathcal{P}$ to $\mathcal{Q}$, around the top face of the cube in the opposite sense, and then back down from point $\mathcal{Q}$ to point $\mathcal{P}$ (shown in grey). The total change in the vector from the legs $\mathcal{P} \rightarrow \mathcal{Q}$ and $\mathcal{Q} \rightarrow \mathcal{P}$ cancel, but the transports around the faces do *not* cancel, even though they are in the opposite sense because they are performed on different faces.

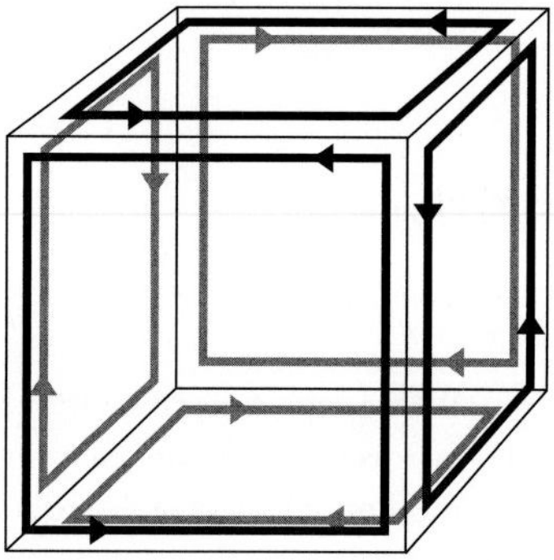

Figure 2.5 Same as Figure 2.4 but now showing transport around all six faces of the cube, with opposite faces traversed with cycles of opposite senses. Notice that each edge of the cube is traversed twice in opposite directions; therefore, the net shift in any transported vector will be zero after completing all cycles.

Similarly, the shift in the vector after traversing the top face (the grey path shown in Figure 2.4) is

$$(\delta a^{\alpha})_{\text{top}} = \epsilon^2 R_{\mu\nu\rho}{}^{\nu}(\mathcal{Q}) u^{\mu} v^{\nu} a^{\rho} \, . \tag{2.68}$$

Note that there is no contribution from the segment of the path between points $\mathcal{P}$ and $\mathcal{Q}$ because this edge is traversed twice in opposite directions. The overall shift in vector **a** after parallel transporting it about both the grey and the black paths is

$$\begin{aligned}(\delta a^{\alpha})_{\text{top+bot}} &= \epsilon^2 \left[R_{\mu\nu\rho}{}^{\alpha}(\mathcal{Q}) - R_{\mu\nu\rho}{}^{\alpha}(\mathcal{P}) \right] u^{\mu} v^{\nu} a^{\rho} \\ &= \epsilon^3 \left(w^{\sigma} \nabla_{\sigma} R_{\mu\nu\rho}{}^{\alpha} \right) u^{\mu} v^{\nu} a^{\rho} \, ,\end{aligned} \tag{2.69}$$

where the second equality holds in the limit $\epsilon \to 0$ because the covariant and ordinary derivatives are the same in Riemann normal coordinates. Now transport **a** similarly about the front and back faces as well as the left and right faces; as stated above, the net shift in **a** must be zero, so

$$0 = (\delta a^{\alpha})_{\text{all faces}} = \epsilon^3 \left(\nabla_\sigma R_{\mu\nu\rho}{}^{\alpha} + \nabla_\mu R_{\nu\sigma\rho}{}^{\alpha} + \nabla_\nu R_{\sigma\mu\rho}{}^{\alpha}\right) u^\mu v^\nu w^\sigma a^\rho \,. \tag{2.70}$$

This establishes the Bianchi identity of Eq. (2.66f) since the factor in parentheses must vanish.

2.1.6 Geodesic Deviation

Geometrically, the Riemann tensor encodes the information about the curvature of spacetime. Physically, as we have stressed, the coordinate invariant notion of gravity is the tidal acceleration of two nearby free-falling objects. This is known as *geodesic deviation* since the two objects are following geodesics that will experience a relative acceleration due to the effect of gravity. We therefore expect to see a relationship between the tidal tensor field and the Riemann tensor.

Consider two freely falling particles initially at points $\mathcal{P}$ and $\mathcal{Q}$ having four-velocity vectors (i.e. tangent vectors to their geodesics) $\mathbf{u}(\mathcal{P})$ and $\mathbf{u}(\mathcal{Q})$. Let $\boldsymbol{\zeta} = d/dx$ be the separation vector between the two geodesics. The rate of change of the separation vector between the two geodesics is the relative velocity

$$v^\alpha = \frac{d\zeta^\alpha}{dt} = u^\mu \nabla_\mu \zeta^\alpha \,. \tag{2.71}$$

Because **u** and $\boldsymbol{\zeta}$ are coordinate basis vectors, they commute and so we can write

$$v^\alpha = \zeta^\mu \nabla_\mu u^\alpha \,. \tag{2.72}$$

The relative acceleration is

$$\begin{aligned}
a^\alpha &= \frac{dv^\alpha}{dt} = u^\rho \nabla_\rho (u^\mu \nabla_\mu \zeta^\alpha) = u^\rho \nabla_\rho (\zeta^\mu \nabla_\mu u^\alpha) \\
&= (u^\rho \nabla_\rho \zeta^\mu)(\nabla_\mu u^\alpha) + \zeta^\mu u^\rho \nabla_\rho \nabla_\mu u^\alpha \\
&= (\zeta^\rho \nabla_\rho u^\mu)(\nabla_\mu u^\alpha) + \zeta^\mu u^\rho \nabla_\mu \nabla_\rho u^\alpha + \zeta^\mu u^\rho u^\nu R_{\mu\rho\nu}{}^{\alpha} \\
&= (\zeta^\rho \nabla_\rho u^\mu)(\nabla_\mu u^\alpha) + \zeta^\mu \nabla_\mu (u^\rho \nabla_\rho u^\alpha) \\
&\quad - (\zeta^\mu \nabla_\mu u^\rho)(\nabla_\rho u^\alpha) + \zeta^\mu u^\rho u^\nu R_{\mu\rho\nu}{}^{\alpha} \,.
\end{aligned} \tag{2.73}$$

Notice that the first and third terms cancel while the second term vanishes because of the geodesic equation $u^\rho \nabla_\rho u^\alpha = 0$. Therefore, after re-labelling the indices and using the symmetries of the Riemann tensor, we have

$$a^\alpha = -R_{\mu\rho\nu}{}^{\alpha} u^\mu \zeta^\rho u^\nu \,. \tag{2.74}$$

Recall from Newtonian gravity that the relative tidal acceleration was $-\mathcal{E}_{ij}\zeta^j$; therefore we see that

$$\mathcal{E}_{ij} = R_{\mu i \nu j} u^\mu u^\nu \tag{2.75}$$

or, in a coordinate system in which $x^0 = t$,

$$\mathcal{E}_{ij} = R_{0i0j} \,. \tag{2.76}$$

2.1.7
Ricci and Einstein Tensors

Three useful quantities constructed from the Riemann curvature tensor are the *Ricci tensor*, the *Ricci scalar* and the *Einstein tensor*.

The Ricci tensor and scalar are formed by contracting the indices of Riemann tensor; they are, respectively,

$$R_{\alpha\beta} := R_{\alpha\mu\beta}{}^{\mu} \tag{2.77}$$

and

$$R := g^{\mu\nu} R_{\mu\nu} \,. \tag{2.78}$$

It is straightforward to show that the Ricci tensor is symmetric in its indices. Another important identity follows from the contracted Bianchi identity:

$$\begin{aligned} 0 &= \nabla_\alpha R_{\beta\gamma\delta}{}^{\alpha} + \nabla_\beta R_{\gamma\alpha\delta}{}^{\alpha} + \nabla_\gamma R_{\alpha\beta\delta}{}^{\alpha} \\ &= -\nabla^\alpha R_{\beta\gamma\alpha\delta} + \nabla_\beta R_{\gamma\delta} - \nabla_\gamma R_{\beta\delta} \end{aligned} \tag{2.79}$$

now contract with $g^{\gamma\delta}$ to obtain

$$\begin{aligned} 0 &= -\nabla^\alpha R_{\beta\alpha} + \nabla_\beta R - \nabla^\delta R_{\beta\delta} \\ &= -2\nabla^\alpha \left(R_{\alpha\beta} - \frac{1}{2} g_{\alpha\beta} R \right) . \end{aligned} \tag{2.80}$$

The combination in parentheses is *divergenceless*; it is known as the Einstein tensor:

$$G_{\alpha\beta} := R_{\alpha\beta} - \frac{1}{2} g_{\alpha\beta} R \,, \tag{2.81}$$

which is symmetric in its indices and

$$\nabla^\alpha G_{\alpha\beta} = 0 \,. \tag{2.82}$$

2.2
Slow Motion in Weak Gravitational Fields

When gravity is weak, the spacetime metric is very close to the flat Minkowski spacetime:

$$g_{\alpha\beta} = \eta_{\alpha\beta} + h_{\alpha\beta} \,, \tag{2.83}$$

where $\eta_{\alpha\beta}$ is the Minkowski metric and $h_{\alpha\beta} := g_{\alpha\beta} - \eta_{\alpha\beta}$ is a small perturbation. In rectilinear coordinates the Minkowski metric is $\eta_{\alpha\beta} = \mathrm{diag}[-c^2, 1, 1, 1]$; then the perturbation is small if $|h_{ij}| \ll 1$, $|h_{0i}| \ll c$, and $|h_{00}| \ll c^2$. We will see that the scale of the components of the perturbations is $\sim \Phi/c^2$ where Φ is the Newtonian potential. For objects in the Solar System this number is $\lesssim G M_\odot/(c^2 R_\odot) \sim 10^{-6}$, so the weak field approximation is good.

To compute the motion of particles in this rectilinear coordinate system, the geodesic equations must be solved to leading order in the perturbation. It is straightforward to show that the connection coefficients are

$$\Gamma^\gamma_{\alpha\beta} = \frac{1}{2}\eta^{\gamma\delta}\left(\frac{\partial h_{\beta\delta}}{\partial x^\alpha} + \frac{\partial h_{\alpha\delta}}{\partial x^\beta} - \frac{\partial h_{\alpha\beta}}{\partial x^\delta}\right) + O(h^2)\,. \tag{2.84}$$

Here $\eta^{\gamma\delta}$ is the inverse of $\eta_{\gamma\delta}$ so that $\eta^{\alpha\mu}\eta_{\mu\beta} = \delta^\alpha_\beta$. Suppose that the worldline of a particle is given by $x^\alpha(\tau)$ where τ is the proper time for the particle (the affine parameter for the particle's geodesic). Then the worldline will satisfy the geodesic equation

$$u^\mu \nabla_\mu u^\alpha = 0\,, \tag{2.85}$$

where $u^\alpha = dx^\alpha/d\tau$ is the particle's four-velocity vector. This equation can be rewritten as (cf. Eq. (2.57))

$$\frac{d^2 x^\alpha}{d\tau^2} = -\Gamma^\alpha_{\mu\nu}\frac{dx^\mu}{d\tau}\frac{dx^\nu}{d\tau}\,. \tag{2.86}$$

If the particle is moving with spatial velocities much less than the speed of light, $v \ll c$, then the dominant component of the four-velocity is the x^0 component,

$$u^\alpha = \frac{dx^\alpha}{d\tau} = [1, 0, 0, 0] + O(v/c)\,, \tag{2.87}$$

so the spatial acceleration of the particle with respect to the coordinate system is

$$a^i = \frac{d^2 x^i}{d\tau^2} = -\Gamma^i_{00} + O(v/c)\,. \tag{2.88}$$

Now if the spacetime is approximately stationary (i.e. it varies in space, but only very slowly in time – which is consistent with a notion of slow motion of both the components of the gravitational source and the falling particle), then we ignore time derivatives of the metric perturbation. It follows that

$$a_i \approx \frac{1}{2}\frac{\partial}{\partial x^i} h_{00}\,. \tag{2.89}$$

Compare this to the Newtonian formula $\boldsymbol{a} = -\boldsymbol{\nabla}\Phi$, where Φ is the Newtonian potential, and we see that the metric perturbation is related to the Newtonian potential by $h_{00} = -2\Phi$. Recall that this acceleration is a *coordinate acceleration*, which is different from the relative tidal acceleration between two freely falling bodies that is given by the equation of geodesic deviation.

Example 2.11 Geodesic deviation in the weak-field slow-motion limit

We can also compute the *physical* tidal acceleration (as opposed to the coordinate acceleration we just described) in the weak-field slow-motion limit using the equation of geodesic deviation

$$a^\alpha = -R_{\mu\rho\nu}{}^{\alpha} u^\mu \zeta^\rho u^\nu \,, \tag{2.90}$$

where $\mathbf{u}$ is the four-velocity of one of the particles and $\boldsymbol{\zeta}$ is the separation vector between the two particles. For slow motion of the particles, the spatial components of the tidal acceleration are

$$a_j = -R_{0i0j}\zeta^i + O(v/c) \,. \tag{2.91}$$

Now we must compute the Riemann tensor to linear order in h. Recall that the connection coefficients are $O(h)$, and so the $\Gamma\Gamma$ terms in the Riemann tensor are $O(h^2)$; these are ignored. The calculation proceeds almost identically as in Example 2.10 and we obtain

$$R_{\alpha\beta\gamma\delta} = \frac{1}{2}\left(-\frac{\partial^2 h_{\beta\delta}}{\partial x^\alpha \partial x^\gamma} + \frac{\partial^2 h_{\beta\gamma}}{\partial x^\alpha \partial x^\delta} + \frac{\partial^2 h_{\alpha\delta}}{\partial x^\beta \partial x^\gamma} - \frac{\partial^2 h_{\alpha\gamma}}{\partial x^\beta \partial x^\delta}\right) + O(h^2) \,. \tag{2.92}$$

If we disregard the time derivatives of the metric perturbation (slow motion of the sources of the gravitational field) then we find

$$R_{0i0j} = -\frac{1}{2}\frac{\partial}{\partial x^i}\frac{\partial}{\partial x^j} h_{00} + O(h^2) \,. \tag{2.93}$$

Recall now that we have identified this component of the metric perturbation as being related to the Newtonian potential by $h_{00} = -2\Phi$, so the Newtonian result $a_j = -\mathcal{E}_{ij}\zeta^i + O(v/c)$ is recovered where

$$\mathcal{E}_{ij} = R_{0i0j} = -\frac{1}{2}\frac{\partial}{\partial x^i}\frac{\partial}{\partial x^j} h_{00} + O(h^2) = \frac{\partial^2 \Phi}{\partial x^i \partial x^j} + O(\Phi^2/c^4) \,. \tag{2.94}$$

While h_{00} is the only component of the metric that is important for computing the coordinate acceleration of a freely falling particle under the assumptions of slow motion (for the particle and slow variation for the gravitational field) and weak gravitational field, it is not the only component of the metric perturbation of order Φ.

2.3 Stress-Energy Tensor

Newtonian mechanics describes not only how matter moves within a gravitational field but also how the mass of matter generates the gravitational field. The Newtonian potential Φ contained all the information about the gravitational field, so it describes the motion of matter (for a fluid, $\rho d\boldsymbol{v}/dt + \nabla p = -\rho \nabla \Phi$), and the source of this field was the mass density of matter (via the Poisson equation $\nabla^2 \Phi = 4\pi G\rho$). In general relativity, the single potential Φ is replaced with the ten potentials $g_{\alpha\beta}$ (recall that $g_{\alpha\beta} = g_{\beta\alpha}$) which describe the spacetime geometry that manifests itself as gravitation. The Poisson equation is replaced by ten second-order partial differential equations for the ten components of the metric, and is sourced by a tensor that describes not only the mass density of matter (which is the dominant component) but also the current and internal stresses of the matter.

In classical mechanics, the *stress tensor* S determines the amount of force $d\boldsymbol{F}$ acting on an element of area $d\boldsymbol{A}$ of a material body,

$$d\boldsymbol{F} = \mathsf{S} \cdot d\boldsymbol{A} \quad \text{or} \quad dF_i = S_{ij}\hat{n}^j dA \,, \tag{2.95}$$

where $\hat{\boldsymbol{n}}$ is the unit vector normal to the area element. As an example, a fluid always produces a force per unit area (pressure) that is orthogonal to the element of area, and this is given by its isotropic pressure p: $S_{ij} = p\delta_{ij}$. For an electromagnetic field, the stress tensor is known as the Maxwell stress tensor and has the form

$$S_{ij} = \epsilon_0 E_i E_j + \frac{1}{\mu_0} B_i B_j - \frac{1}{2}\delta_{ij}\left(\epsilon_0 \boldsymbol{E}\cdot\boldsymbol{E} + \frac{1}{\mu_0}\boldsymbol{B}\cdot\boldsymbol{B}\right). \tag{2.96}$$

The energy density of a material body is ρc^2 where ρ is the mass density of the material. If there is a flow of matter in a continuous distribution then there is energy flux $c\boldsymbol{j}$ (or, equivalently, the material has a momentum density $\rho\boldsymbol{v} = \boldsymbol{j}$). Continuity of matter suggests that $\nabla_\mu J^\mu = 0$ where $J_\alpha = [-\rho c^2, j_i]$ is the energy-momentum vector (see Example 2.6). In gravitating systems, this does not generally hold because there can be an exchange between matter energy and gravitational field energy. In classical electrodynamics, the energy density of an electromagnetic field is $\frac{1}{2}(\epsilon_0 \boldsymbol{E}\cdot\boldsymbol{E} + \boldsymbol{B}\cdot\boldsymbol{B}/\mu_0)$ and the energy flux is given by the Poynting vector $\boldsymbol{E}\times\boldsymbol{B}/\mu_0$.

The energy density, the energy flux or momentum density and the stress tensor can be combined to form the *stress-energy tensor*,

$$T^{\alpha\beta} := \left[\begin{array}{c|c} \begin{pmatrix}\text{mass density}\\ \rho\end{pmatrix} & \begin{pmatrix}\text{momentum density}\\ \boldsymbol{j}\end{pmatrix} \\ \hline \begin{pmatrix}\text{momentum density}\\ \boldsymbol{j}\end{pmatrix} & \begin{pmatrix}\text{stress tensor}\\ \mathsf{S}\end{pmatrix}\end{array}\right]. \tag{2.97}$$

This tensor will play the same role in the Einstein field equations of General Relativity that mass density played in the Poisson equation for Newtonian gravity.

The stress-energy tensor is also involved in the equations of motion for matter. For a small element of fluid we can construct a normal neighbourhood. Then, in Riemann normal coordinates, we can write the local conservation of energy-momentum as $0 = \partial J^\mu/\partial x^\mu = \partial T^{\mu 0}/\partial x^\mu$ since $T^{\alpha 0} = J^\alpha$. From the definition of the stress tensor, $d\boldsymbol{F} - \mathsf{S}\cdot d\boldsymbol{A}$ we see that the momentum density is

$\partial j^i/\partial t = dF^i/dV = \partial S^{ij}/\partial x^j$ and therefore $\partial T^{\mu i}/\partial x^\mu = 0$. The general equations of motion for matter are

$$\nabla_\mu T^{\mu\alpha} = 0\,, \tag{2.98}$$

which holds in any coordinate system.

2.3.1
Perfect Fluid

A *perfect fluid* is a fluid that can be described completely in terms of the four-velocity $\mathbf{u}$ of the fluid elements, its mass-density ρ and its isotropic (in its rest frame) pressure p. The stress-energy tensor for a perfect fluid is

$$T^{\alpha\beta} = (\rho + p/c^2)u^\alpha u^\beta + p g^{\alpha\beta}\,. \tag{2.99}$$

In the local rest frame of the fluid, $u^\alpha = [1, 0, 0, 0]$ and $g^{\alpha\beta} = \eta^{\alpha\beta}$ (at least locally), the components of the stress-energy tensor take the values

$$T^{00} = \rho\,, \quad T^{0i} = 0\,, \quad T^{ij} = p\delta^{ij} \quad \text{(local rest frame)}. \tag{2.100}$$

The *equation of state* of a perfect fluid describes the relationship between the pressure and the density, which is given by the function $p(\rho)$. A pressureless perfect fluid, $p = 0$, is known as *dust*:

$$T^{\alpha\beta} = \rho u^\alpha u^\beta \quad \text{(dust)}. \tag{2.101}$$

A perfect fluid with $p = \frac{1}{3}\rho c^2$ describes an isotropic *radiation* fluid:

$$T^{\alpha\beta} = p\left(\frac{u^\alpha u^\beta}{c^2} + g^{\alpha\beta}\right) \quad \text{(radiation)}. \tag{2.102}$$

The equations of conservation of energy and momentum are $\nabla_\mu T^{\mu\alpha} = 0$. We can decompose the vector $\nabla_\mu T^{\mu\alpha}$ into a vector proportional to $\mathbf{u}$, $-c^{-2}u^\alpha u_\nu \nabla_\mu T^{\mu\nu}$, and a vector orthogonal to $\mathbf{u}$, $\nabla_\mu T^{\mu\alpha} + c^{-2}u^\alpha u_\nu \nabla_\mu T^{\mu\nu}$.[1] The component of the vector parallel to $\mathbf{u}$ therefore satisfies $u_\nu \nabla_\mu T^{\mu\nu} = 0$; since $g_{\mu\nu}u^\mu u^\nu = -c^2$ we have $0 = \nabla_\alpha(u_\mu u^\mu) = 2u_\mu \nabla_\alpha u^\mu$ and so

$$\begin{aligned} 0 &= u_\nu \nabla_\mu T^{\mu\nu} \\ &= u_\nu \nabla_\mu \left[\left(\rho + p/c^2\right) u^\mu u^\nu + p g^{\mu\nu}\right] \\ &= -c^2 \nabla_\mu \left[\left(\rho + p/c^2\right) u^\mu\right] + u^\mu \nabla_\mu p\,, \end{aligned} \tag{2.103}$$

which gives us the equation for the continuity of mass:

$$u^\mu \nabla_\mu \rho + \left(\rho + p/c^2\right) \nabla_\mu u^\mu = 0\,. \tag{2.104}$$

1) Note that a vector v^α can be split as follows: $v^\alpha = -c^{-2}u^\alpha u_\nu v^\nu + (v^\alpha + c^{-2}u^\alpha u_\nu v^\nu)$ where, by construction, the first term is parallel to u^α and the second term is orthogonal to u^α. Note that $u_\mu u^\mu = -c^2$.

The portion of the vector $\nabla_\mu T^{\mu\alpha}$ that is orthogonal to $\mathbf{u}$, which is $\nabla_\mu T^{\mu\alpha} + c^{-2}u^\alpha u_\nu \nabla_\mu T^{\mu\nu} = (\delta^\alpha_\nu + c^{-2}u^\alpha u_\nu)\nabla_\mu T^{\mu\nu}$, also vanishes, which results in the relativistic version of the *Euler equation*,

$$\left(\rho + p/c^2\right) u^\mu \nabla_\mu u^\alpha + \left(g^{\alpha\mu} + \frac{u^\alpha u^\mu}{c^2}\right) \nabla_\mu p = 0 \,. \tag{2.105}$$

For dust, these equations take a simple form. The Euler equation becomes simply $u^\mu \nabla_\mu u^\alpha = 0$ which is the geodesic equation for the dust particles: the dust is freely falling. The other equation is,

$$0 = u^\mu \nabla_\mu \rho + \rho \nabla_\mu u^\mu = \nabla_\mu \left(\rho u^\mu\right) , \tag{2.106}$$

which shows that ρu^μ is divergenceless. This is the equation of conservation of momentum.

Example 2.12 The Euler equations

In the non-relativistic, flat-spacetime limit, we write $\nabla_\alpha \simeq \partial/\partial x^\alpha$, $u^\alpha \simeq [1, v^i]$, and formally let $c \to \infty$. Equation (2.104) becomes

$$\frac{\partial \rho}{\partial t} + \boldsymbol{\nabla} \cdot (\rho \boldsymbol{v}) = 0 \,, \tag{2.107}$$

which we recognize as the equation of conservation of mass, while Eq. (2.105) becomes

$$\rho \frac{\partial \boldsymbol{v}}{\partial t} + \rho \boldsymbol{v} \cdot \boldsymbol{\nabla} \boldsymbol{v} = -\boldsymbol{\nabla} p \,, \tag{2.108}$$

which is the normal non-gravitational non-relativistic form of the Euler equation.

A more useful form of Eq. (2.105) can be obtained by introducing the *enthalpy* of the fluid. From classical thermodynamics, the enthalpy H is defined by

$$H := E + pV = \left(\rho c^2 + p\right) V \,, \tag{2.109}$$

where E is the energy of the system and V is the volume of the element of fluid. From the first law of thermodynamics,

$$dH = dE + p dV + V dp = đQ + V dp \,, \tag{2.110}$$

where $đQ$ is the heat transfer to the element of fluid. If we assume that the fluid flow is *adiabatic*, so that $đQ = 0$, then $dH = V dp$. In this case, we can multiply Eq. (2.105) by Vc^2 and using $V\nabla_\alpha p = \nabla_\alpha H$ we find

$$H u^\mu \nabla_\mu u_\alpha + c^2 \nabla_\alpha H + u_\alpha u^\mu \nabla_\mu H = 0 \,. \tag{2.111}$$

The first and the third terms can be combined to form $u^\mu \nabla_\mu (H u_\alpha)$ while the second term can be written as $c^2 \nabla_\alpha H = -u^\mu \nabla_\alpha (H u_\mu)$ since $u^\mu u_\mu = -c^2$ and $u^\mu \nabla_\alpha u_\mu = 0$. We therefore find an alternate form of the Euler equation,

$$u^\mu \left[\nabla_\mu (H u_\alpha) - \nabla_\alpha (H u_\mu)\right] = 0 \,. \tag{2.112}$$

For a constant enthalpy system, the fluid elements obey the geodesic equation.

2.3.2
Electromagnetism

In electrodynamics, the electric and magnetic fields can be written in terms of a scalar potential ϕ and a vector potential $\boldsymbol{A}$ as $\boldsymbol{B} = \boldsymbol{\nabla} \times \boldsymbol{A}$ and $\boldsymbol{E} = -\boldsymbol{\nabla}\phi - \partial \boldsymbol{A}/\partial t$; these can be combined to form a four-potential $A_\alpha = [-\phi, A_i]$ which can in turn be used to define the *Faraday tensor*

$$F_{\alpha\beta} := \frac{\partial A_\beta}{\partial x^\alpha} - \frac{\partial A_\alpha}{\partial x^\beta} \tag{2.113}$$

and we see that $E_i = F_{i0}$ and $B_i = \frac{1}{2}\varepsilon_{ijk}F^{jk}$, where ε_{ijk} is the three-dimensional *Levi-Civita symbol*,

$$\varepsilon_{ijk} := \begin{cases} +1 & \text{if } (i, j, k) \text{ is } (1, 2, 3), (2, 3, 1), \text{ or } (3, 1, 2) \\ -1 & \text{if } (i, j, k) \text{ is } (3, 2, 1), (2, 1, 3), \text{ or } (1, 3, 2) \\ 0 & \text{if } i = j, i = k, \text{ or } j = k\,. \end{cases} \tag{2.114}$$

Maxwell's equations can be written succinctly as

$$\frac{\partial F_{\alpha\beta}}{\partial x^\gamma} + \frac{\partial F_{\beta\gamma}}{\partial x^\alpha} + \frac{\partial F_{\gamma\alpha}}{\partial x^\beta} = 0 \tag{2.115}$$

$$\nabla_\mu F^{\alpha\mu} = J^\alpha\,, \tag{2.116}$$

where J^α is the charge-current four-vector. The electromagnetic stress-energy tensor is

$$T_{\alpha\beta} = F_{\alpha\mu}F^\mu_\beta - \frac{1}{4}g_{\alpha\beta}F_{\mu\nu}F^{\mu\nu}\,. \tag{2.117}$$

2.4
Einstein's Field Equations

Conservation of matter and its stress-energy implies that the stress-energy tensor, $T_{\alpha\beta}$, is divergenceless. By constructing field equations in which the geometry is related to the matter via the Einstein tensor, this conservation of matter-energy is ensured. The *Einstein field equations* have the form

$$G_{\alpha\beta} = \frac{8\pi G}{c^4}T_{\alpha\beta}\,. \tag{2.118}$$

The factor $8\pi G/c^4$ is found by requiring the equations to have the correct form when going to the Newtonian limit (slow motion and weak gravitational fields). The Einstein field equations describe how the gravitational field is generated by the matter. In addition, since the Einstein tensor is divergenceless, the field equations enforce the equations of motion for the matter:

$$\nabla_\mu T^{\mu\alpha} = 0\,. \tag{2.119}$$

The effects of the curved spacetime (the gravitational effects) are contained in the connection. That is, the equations of motion can be written as

$$\frac{\partial}{\partial x^\mu} T^{\mu\alpha} = -\Gamma^\mu_{\mu\nu} T^{\nu\alpha} - \Gamma^\alpha_{\mu\nu} T^{\mu\nu} , \tag{2.120}$$

where the left-hand side is the normal flat-space expression for the divergence of the matter stress-energy and the right-hand side contains the gravitational terms.

Example 2.13 Equations of motion for a point particle

The stress energy tensor for a point particle of mass m has the form

$$T^{\alpha\beta}(\mathbf{x}') = \rho(\mathbf{x}') u^\alpha u^\beta \tag{2.121}$$

where

$$\rho(\mathbf{x}') = m \frac{\delta^4 (\mathbf{x}' - \mathbf{x}(\tau))}{\sqrt{-\det \mathbf{g}}} \tag{2.122}$$

is the density (a delta function for a point particle). Here $\mathbf{x}(\tau)$ is the worldline of the particle (the thing we wish to obtain!) and $\mathbf{u} = d\mathbf{x}/d\tau$ is the particle's four-velocity. The equation of motion is

$$0 = \nabla_\mu T^{\mu\nu} = \nabla_\mu (\rho u^\mu u^\nu) = u^\nu \nabla_\mu (\rho u^\mu) + \rho u^\mu \nabla_\mu u^\nu . \tag{2.123}$$

Now contract Eq. (2.123) with u_ν; since the four-velocity is normalized, $u_\nu u^\nu = -c^2$ and

$$0 = u_\nu \nabla_\mu T^{\mu\nu} = -c^2 \nabla_\mu (\rho u^\mu) + \rho u^\mu u_\nu \nabla_\mu u^\nu . \tag{2.124}$$

The second term vanishes (since $u_\nu \nabla_\mu u^\nu = \frac{1}{2} \nabla_\mu (u_\nu u^\nu) = 0$) so one component of the equations of motion is

$$\nabla_\mu (\rho u^\mu) = 0 . \tag{2.125}$$

This is the equation of conservation of four-momentum. Now we substitute Eq. (2.125) into Eq. (2.123) and obtain

$$\rho u^\mu \nabla_\mu u^\nu = 0 . \tag{2.126}$$

That is, wherever the density is non-vanishing – at the location of the particle – the geodesic equations $u^\mu \nabla_\mu u^\nu = 0$ hold. We have therefore recovered the geodesic equations from the conservation of stress-energy for a point particle.

2.5
Newtonian Limit of General Relativity

We have already derived the slow-motion limit of the geodesic equation in a nearly flat, slowly changing spacetime in Section 2.2. Now we wish to write the Einstein field equations in a similar limit. First we will write the field equations in the weak-field approximation, which is called the *linearized gravity* approximation, then we will impose the additional assumption of slow motion which will give us the *Newtonian limit* of General Relativity.

2.5.1
Linearized Gravity

As in Section 2.2, we take the spacetime metric to be that of flat Minkowski spacetime, $\eta_{\alpha\beta}$, plus a small perturbation $h_{\alpha\beta}$: $g_{\alpha\beta} = \eta_{\alpha\beta} + h_{\alpha\beta}$. We adopt the following convention: *tensor indices are raised and lowered using the Minkowski metric, $\eta_{\alpha\beta}$ and its inverse $\eta^{\alpha\beta}$ rather than the actual metric $g_{\alpha\beta}$ and $g^{\alpha\beta}$*. That is, we have $h^{\alpha\beta} = \eta^{\mu\alpha}\eta^{\nu\beta}h_{\mu\nu}$ rather than $g^{\mu\alpha}g^{\nu\beta}h_{\mu\nu}$. The single exception that we make is with the spacetime metric $g^{\alpha\beta}$ itself: it is

$$\begin{aligned} g^{\alpha\beta} &= (g_{\alpha\beta})^{-1} = (\eta_{\alpha\beta} + h_{\alpha\beta})^{-1} \\ &= \eta^{\alpha\beta} - h^{\alpha\beta} + O(h^2) . \end{aligned} \tag{2.127}$$

We have already computed the linearized Riemann tensor, Eq. (2.92). From this it is possible to compute the linearized Ricci tensor, Ricci scalar and then the linearized Einstein tensor.

The linearized Ricci tensor is

$$\begin{aligned} R_{\alpha\beta} &= R_{\alpha\mu\beta}{}^{\mu} \\ &= \frac{1}{2}\left(-\frac{\partial^2 h}{\partial x^\alpha \partial x^\beta} + \frac{\partial^2 h^\mu{}_\beta}{\partial x^\alpha \partial x^\mu} + \frac{\partial^2 h_\alpha{}^\mu}{\partial x^\mu \partial x^\beta} - \eta^{\mu\nu}\frac{\partial^2 h_{\alpha\beta}}{\partial x^\mu \partial x^\nu}\right) + O(h^2) , \end{aligned} \tag{2.128}$$

where we write $h = h_\mu{}^\mu$ as the trace of $h_\mu{}^\nu$, and the linearized Ricci scalar is

$$\begin{aligned} R &= g^{\alpha\beta}R_{\alpha\beta} = \eta^{\alpha\beta}R_{\alpha\beta} + O(h^2) \\ &= \frac{\partial^2 h^{\mu\nu}}{\partial x^\mu \partial x^\nu} - \eta^{\mu\nu}\frac{\partial^2 h}{\partial x^\mu \partial x^\nu} + O(h^2) . \end{aligned} \tag{2.129}$$

Therefore the Einstein tensor is

$$\begin{aligned} G_{\alpha\beta} &= R_{\alpha\beta} - \frac{1}{2}g_{\alpha\beta}R = R_{\alpha\beta} - \frac{1}{2}\eta_{\alpha\beta}R + O(h^2) \\ &= \frac{1}{2}\left(-\frac{\partial^2 h}{\partial x^\alpha \partial x^\beta} + \frac{\partial^2 h^\mu{}_\beta}{\partial x^\alpha \partial x^\mu} + \frac{\partial^2 h_\alpha{}^\mu}{\partial x^\mu \partial x^\beta} - \eta^{\mu\nu}\frac{\partial^2 h_{\alpha\beta}}{\partial x^\mu \partial x^\nu}\right) \\ &\quad - \frac{1}{2}\eta_{\alpha\beta}\left(\frac{\partial^2 h^{\mu\nu}}{\partial x^\mu \partial x^\nu} - \eta^{\mu\nu}\frac{\partial^2 h}{\partial x^\mu \partial x^\nu}\right) + O(h^2) . \end{aligned} \tag{2.130}$$

The number of terms in the Einstein tensor can be reduced by expressing it in terms of the *trace-reversed metric perturbation*, $\bar{h}_{\alpha\beta}$, instead of the actual metric perturbation, $h_{\alpha\beta}$. The trace-reversed metric perturbation is defined by

$$\bar{h}_{\alpha\beta} := h_{\alpha\beta} - \frac{1}{2}\eta_{\alpha\beta}h \,. \tag{2.131}$$

Note that $\bar{h} = \eta^{\alpha\beta}\bar{h}_{\alpha\beta} = -h$, which is why it is called trace-reversed. The actual metric perturbation is recovered by trace-reversing the trace-reversed metric perturbation:

$$h_{\alpha\beta} = \bar{h}_{\alpha\beta} - \frac{1}{2}\eta_{\alpha\beta}\bar{h} \,. \tag{2.132}$$

Now we substitute this equation into the equation for the linearized Einstein tensor and we obtain (after a number of terms cancel)

$$G_{\alpha\beta} = \frac{1}{2}\left(\frac{\partial^2 \bar{h}^{\mu}_{\beta}}{\partial x^{\alpha}\,\partial x^{\mu}} + \frac{\partial^2 \bar{h}^{\mu}_{\alpha}}{\partial x^{\mu}\partial x^{\beta}} - \eta^{\mu\nu}\frac{\partial^2 \bar{h}_{\alpha\beta}}{\partial x^{\mu}\partial x^{\nu}} - \eta_{\alpha\beta}\frac{\partial^2 \bar{h}^{\mu\nu}}{\partial x^{\mu}\partial x^{\nu}}\right) + O(h^2) \,. \tag{2.133}$$

Therefore, the linearized Einstein field equations are

$$-\eta^{\mu\nu}\frac{\partial^2 \bar{h}_{\alpha\beta}}{\partial x^{\mu}\partial x^{\nu}} - \eta_{\alpha\beta}\frac{\partial^2 \bar{h}^{\mu\nu}}{\partial x^{\mu}\partial x^{\nu}} + \frac{\partial^2 \bar{h}^{\mu}_{\beta}}{\partial x^{\alpha}\partial x^{\mu}} + \frac{\partial^2 \bar{h}^{\mu}_{\alpha}}{\partial x^{\mu}\partial x^{\beta}} + O(h^2) = \frac{16\pi G}{c^4}T_{\alpha\beta} \,. \tag{2.134}$$

The first term on the left-hand side is simply $-\Box\bar{h}_{\alpha\beta}$, where $\Box$ is the d'Alembertian operator, the flat-spacetime wave operator. Is it possible to simplify the linearized Einstein field equations to a simple flat-spacetime wave equation?

The way to obtain a further simplification of the form of the field equations is to choose a specialized coordinate system, that is, to make an appropriate gauge choice. The terms that we wish to discard are all divergences operating on the trace-reversed metric perturbation so we wish to find a gauge, akin to the Lorenz gauge in electromagnetism, in which the divergence of the trace-reversed metric is zero: $\partial\bar{h}^{\mu\alpha}/\partial x^{\mu} = 0$. Since this gauge condition is very similar to the Lorenz gauge condition in electromagnetism, we shall call it the *Lorenz gauge* for the metric perturbation. It remains to be shown that this condition can be generically achieved.

We seek an infinitesimal coordinate transformation $\mathbf{x} \to \mathbf{x}' = \mathbf{x} + \boldsymbol{\xi}$, which transforms the metric according to (cf. Eq. (2.8))

$$g_{\alpha\beta} \to g'_{\alpha\beta} = g_{\alpha\beta} - \frac{\partial\xi_{\beta}}{\partial x^{\alpha}} - \frac{\partial\xi_{\alpha}}{\partial x^{\beta}} + O(h^2) \,, \tag{2.135}$$

where we will see below that ξ is $O(h)$, so the $O(\xi^2)$ corrections are written as $O(h^2)$. Here, $g_{\alpha\beta}(\mathbf{x}) = \eta_{\alpha\beta} + h_{\alpha\beta}(\mathbf{x})$ and $g'_{\alpha\beta}(\mathbf{x}') = \eta_{\alpha\beta} + h'_{\alpha\beta}(\mathbf{x}')$; therefore

$$h_{\alpha\beta} \to h'_{\alpha\beta} = h_{\alpha\beta} - \frac{\partial\xi_{\beta}}{\partial x^{\alpha}} - \frac{\partial\xi_{\alpha}}{\partial x^{\beta}} + O(h^2) \,. \tag{2.136}$$

Under such a coordinate transformation, the change to the Riemann tensor is second order, that is of order $O(h^2)$, so we say that the linearized Riemann tensor is invariant under infinitesimal coordinate transformations (see Problem 2.6). In terms of the trace-reversed metric perturbation, the coordinate transformation is

$$\bar{h}_{\alpha\beta} \to \bar{h}'_{\alpha\beta} = \bar{h}_{\alpha\beta} - \frac{\partial \xi_\beta}{\partial x^\alpha} - \frac{\partial \xi_\alpha}{\partial x^\beta} + \eta_{\alpha\beta}\eta^{\mu\nu}\frac{\partial \xi_\nu}{\partial x^\mu} + O(h^2) . \tag{2.137}$$

Now we require the Lorenz gauge condition in the new gauge,

$$0 = \frac{\partial}{\partial x'^\mu}\bar{h}'^\mu{}_\beta = \frac{\partial}{\partial x^\mu}\bar{h}^\mu{}_\beta - \Box\xi_\beta + O(h^2) , \tag{2.138}$$

where we see that two terms have cancelled. Therefore, given an existing trace-reversed metric perturbation, $\bar{h}_{\alpha\beta}$, we need to find a coordinate transformation generated by a vector $\boldsymbol{\xi}$ that solves the equation

$$\Box\xi_\beta = \frac{\partial \bar{h}^\mu{}_\beta}{\partial x^\mu} . \tag{2.139}$$

A solution to this equation can always be found (and furthermore we see that the solution will be $O(h)$ as anticipated), so we are always free to choose the Lorenz gauge. In fact, the Lorenz gauge is not unique since a homogeneous solution of the form

$$\Box\xi_\beta = 0 \tag{2.140}$$

can always be added; that is, if you are in a Lorenz gauge, you are free to go to a different Lorenz gauge via a gauge transformation generated by a $\boldsymbol{\xi}$ which satisfies $\Box\boldsymbol{\xi} = 0$.

Having now chosen a Lorenz gauge, the linearized Einstein field equations take the very simple form

$$-\Box\bar{h}_{\alpha\beta} = \frac{16\pi G}{c^4} T_{\alpha\beta} \quad \text{(Lorenz gauge).} \tag{2.141}$$

Example 2.14 Harmonic coordinates

The coordinate system associated with the Lorenz gauge is known as *harmonic coordinates*. This is because the coordinates themselves are harmonic functions on spacetime. A harmonic function is a scalar field that is a homogeneous solution to the *curved spacetime* wave operator. The four harmonic coordinates $x^\alpha = \{x^0, x^1, x^2, x^3\}$ therefore satisfy four separate *scalar* equations, that is, we treat x^α not as a component of a vector but as one of the four scalars, and, in particular, we therefore have $\nabla_\beta x^\alpha = \partial x^\alpha/\partial x^\beta = \delta^\alpha_\beta$. The harmonic coordinates thus satisfy

$$\begin{aligned} 0 &= g^{\mu\nu}\nabla_\mu\nabla_\nu x^\alpha \\ &= g^{\mu\nu}\frac{\partial}{\partial x^\mu}(\nabla_\nu x^\alpha) - g^{\mu\nu}\Gamma^\gamma_{\mu\nu}(\nabla_\gamma x^\alpha) \\ &= -g^{\mu\nu}\Gamma^\alpha_{\mu\nu} = -\Gamma^\alpha . \end{aligned} \tag{2.142}$$

Therefore, harmonic coordinates are coordinates in which $\Gamma^\alpha = g^{\mu\nu}\Gamma^\alpha_{\mu\nu} = 0$.

In harmonic coordinates, then, the following identity holds

$$\begin{aligned}0 = g^{\mu\nu}\Gamma^{\alpha}_{\mu\nu} &= \frac{1}{2}g^{\mu\nu}g^{\alpha\rho}\left(\frac{\partial g_{\rho\nu}}{\partial x^{\mu}} + \frac{\partial g_{\mu\rho}}{\partial x^{\nu}} - \frac{\partial g_{\mu\nu}}{\partial x^{\rho}}\right)\\ &= g^{\mu\nu}g^{\alpha\rho}\frac{\partial g_{\rho\nu}}{\partial x^{\mu}} - \frac{1}{2}g^{\mu\nu}g^{\alpha\rho}\frac{\partial g_{\mu\nu}}{\partial x^{\rho}}\\ &= g^{\mu\nu}\frac{\partial}{\partial x^{\mu}}(g^{\alpha\rho}g_{\rho\nu}) - g^{\mu\nu}g_{\rho\nu}\frac{\partial g^{\alpha\rho}}{\partial x^{\mu}} - \frac{1}{2}g^{\mu\nu}g^{\alpha\rho}\frac{\partial g_{\mu\nu}}{\partial x^{\rho}}\\ &= -\frac{\partial g^{\alpha\rho}}{\partial x^{\rho}} - \frac{1}{2}g^{\mu\nu}g^{\alpha\rho}\frac{\partial g_{\mu\nu}}{\partial x^{\rho}}\,. \end{aligned} \tag{2.143}$$

In linearized gravity, this identity becomes

$$0 = \frac{\partial}{\partial x^{\rho}}h^{\alpha\rho} - \frac{1}{2}\frac{\partial}{\partial x^{\rho}}(\eta^{\alpha\rho}h) + O(h^2) = \frac{\partial}{\partial x^{\rho}}\bar{h}^{\alpha\rho} + O(h^2)\,, \tag{2.144}$$

that is, the Lorenz gauge conditions. Notice that the switching of the sign in the first term arises from the definition of the inverse metric $g^{\alpha\beta} = \eta^{\alpha\beta} - h^{\alpha\beta} + O(h^2)$.

2.5.2
Newtonian Limit

In addition to a weak gravitational field, the Newtonian limit requires slowly moving mass distributions with small internal stresses. That is, it must be possible to find some coordinate system in which

$$T_{00}/c^4 = \rho \quad \text{(mass energy density)} \tag{2.145a}$$

$$|T_{0i}|/c^3 \sim \rho(v/c) \ll T_{00}/c^4 \quad \text{(slow motion } v \ll c) \tag{2.145b}$$

$$|T_{ij}|/c^2 \sim p/c^2 \;\&\; \rho(v/c)^2 \ll T_{00}/c^4 \quad \text{(small internal stresses)} \tag{2.145c}$$

(or else the Newtonian limit is not valid). Since the matter is slowly moving, the metric perturbation should also be slowly moving (since the matter generates the gravitational field). Therefore we replace the d'Alembertian operator with a spatial Laplace operator, $\Box \to \nabla^2$, in the field equations. The resulting Newtonian limit is the system of equations

$$\nabla^2\bar{h}_{00} = -16\pi G\rho \tag{2.146a}$$

$$\nabla^2\bar{h}_{0i} = 0 \tag{2.146b}$$

$$\nabla^2\bar{h}_{ij} = 0\,, \tag{2.146c}$$

where the post-Newtonian terms have all been dropped. The solution to the last two sets of equations with correct boundary conditions (vanishing perturbation at large distances from the source) are the trivial solutions $\bar{h}_{ij} = 0$ and $\bar{h}_{0i} = 0$. Recall from our consideration of the motion of slowly moving particles in weak gravitational fields that the metric perturbation is related to the Newtonian potential by $h_{00} = -2\Phi$; since $h_{00} = \bar{h}_{00} + \frac{1}{2}c^2\bar{h}$ and $\bar{h} = \eta^{\mu\nu}\bar{h}_{\mu\nu} = -c^{-2}\bar{h}_{00}$ (all other components are zero), we see that $\bar{h}_{00} = 2h_{00} = -4\Phi$. Therefore, the only non-trivial field equation can be written as

$$\nabla^2\Phi = 4\pi G\rho\,, \tag{2.147}$$

which is precisely the Poisson equation for the potential in Newtonian gravity. (This essentially identifies the factor $8\pi G/c^4$ in the Einstein field equations.)

The metric perturbation in the Newtonian limit is $h_{\mu\nu} = \text{diag}[-2\Phi, -2\Phi/c^2, -2\Phi/c^2, -2\Phi/c^2]$, so the Newtonian metric is

$$g_{\mu\nu} = \begin{bmatrix} -c^2 - 2\Phi & 0 & 0 & 0 \\ 0 & 1 - 2\Phi/c^2 & 0 & 0 \\ 0 & 0 & 1 - 2\Phi/c^2 & 0 \\ 0 & 0 & 0 & 1 - 2\Phi/c^2 \end{bmatrix} + O(\Phi^2/c^4)\,. \tag{2.148}$$

As was shown earlier, the Newtonian equation of motion $\boldsymbol{a} = -\nabla\Phi$ is recovered in the slow-motion, weak-gravity limit. We now consider the case of weak gravity without the slow-motion assumption.

2.5.3
Fast Motion

If we can continue to neglect the internal stress of the matter present then we write

$$T^{00} = \rho\,, \quad T^{0i} = j^i\,, \quad T^{ij} = 0\,, \tag{2.149}$$

where ρ is the mass density of the matter and $\boldsymbol{j}$ is the current density of the matter. We can also express the time-time and the time-space components of the trace-reversed metric perturbation in terms of a scalar potential Φ and a vector potential $\boldsymbol{A}$,

$$\bar{h}_{00} = -4\Phi \quad \text{and} \quad \bar{h}_{0i} = A_i \tag{2.150}$$

and the Lorenz gauge condition $\partial\bar{h}^{0\mu}/\partial x^\mu = 0$ enforces

$$\frac{\partial\Phi}{\partial t} = -\frac{1}{4}c^2\,\nabla\cdot\boldsymbol{A}\,. \tag{2.151}$$

Then the field equations for these two potentials are

$$\Box\Phi = 4\pi G\rho \quad \text{and} \quad \Box\boldsymbol{A} = \frac{16\pi G}{c^2}\boldsymbol{j}\,. \tag{2.152}$$

If we assume a solution $\bar{h}_{ij} = 0$ to the field equations $\Box \bar{h}_{ij} = 0$, then the geodesic equations can be used to compute the gravitational force on an object (a fictional force). An object of mass m moving with three-velocity $\boldsymbol{v} = d\boldsymbol{x}/dt$ has a four-velocity $u^\alpha = \gamma[1, v^i]$ and four-momentum $\mathbf{p} = m\mathbf{u}$ that is governed by the fictitious force

$$\frac{d\boldsymbol{p}}{dt} = \gamma m \left\{ -\boldsymbol{\nabla}\Phi - \frac{\partial \boldsymbol{A}}{\partial t} + \boldsymbol{v} \times (\boldsymbol{\nabla} \times \boldsymbol{A}) + \frac{1}{c^2}\left[4\frac{\partial \Phi}{\partial t}\boldsymbol{v} + 2(\boldsymbol{v} \cdot \boldsymbol{\nabla}\Phi)\boldsymbol{v} - v^2 \boldsymbol{\nabla}\Phi \right] \right\} . \tag{2.153}$$

The terms in square brackets are all $O(v^2/c^2)$; if these are ignored then the equation has the same form as the electromagnetic Lorentz force. In particular, this suggests that general relativity contains a magnetic-like gravitational effect that is produced by moving masses.

2.6 Problems

Problem 2.1

Technically, any curve whose tangent vector $\mathbf{u} = d/d\lambda$ satisfies

$$u^\alpha \nabla_\alpha u^\mu = \kappa u^\mu ,$$

where κ is a constant is also a geodesic. Show that such a curve can be reparameterized so that the new tangent vector $\hat{\mathbf{u}} = d/dt$ with $t = t(\lambda)$ satisfies the usual geodesic equation

$$\hat{u}^\alpha \nabla_\alpha \hat{u}^\mu = 0 .$$

Such a parameterization is called *affine* and the new parameter t is called an *affine parameter.* Show that the affine parameter is unique up to scaling by a constant factor and a shift by a constant value.

Problem 2.2

Use Eqs. (2.66a), (2.66b) and (2.66d) to show that $R_{\alpha\beta\gamma\delta} = R_{\gamma\delta\alpha\beta}$.

Problem 2.3

The motion of a fluid can be found from an action principle.

a) Consider a fluid of dust particles. A particular particle will travel from its location $\mathcal{P}$ at one time to a point $\mathcal{Q}$ in the future along a path that *maximizes* the proper time elapsed:

$$\Delta\tau = \int_{\gamma} d\tau = \int_{0}^{1} \sqrt{g_{\mu\nu}\frac{dx^\mu}{d\sigma}\frac{dx^\nu}{d\sigma}}\, d\sigma\,, \tag{2.154}$$

where σ is some parameterization of an arbitrary curve $\gamma(\sigma)$ beginning at point $\mathcal{P}$ at $\sigma = 0$ and ending at $\mathcal{Q}$ at $\sigma = 1$. The proper time of the trajectory $\Delta\tau$ is maximized when the Euler–Lagrange equations

$$\frac{d}{d\sigma}\frac{\partial \mathcal{L}}{\partial(dx^\alpha/d\sigma)} = \frac{\partial \mathcal{L}}{\partial x^\alpha} \tag{2.155}$$

hold, where the Lagrangian $\mathcal{L}$ is

$$\mathcal{L} = -mc^2\sqrt{g_{\mu\nu}(\mathbf{x})\frac{dx^\mu}{d\sigma}\frac{dx^\nu}{d\sigma}}\,, \tag{2.156}$$

where m is the mass of the dust particle and $-mc^2 d\tau = \mathcal{L} d\sigma$. Show that the Euler–Lagrange equations yield the geodesic equations, $u^\mu \nabla_\mu u^\alpha$ where $u^\alpha = dx^\alpha/d\tau$.

b) Now consider a fluid with enthalpy H undergoing adiabatic flow. Show that extremization of the action

$$S = -\int_{\gamma} H d\tau \tag{2.157}$$

yields the Euler equations $u^\mu[\nabla_\mu(Hu_\alpha) - \nabla_\alpha(Hu_\mu)] = 0$ for the fluid.

Problem 2.4

Consider a perfect fluid undergoing a non-adiabatic process.

a) Starting with Eq. (2.104), show that

$$u^\mu \nabla_\mu S = 0\,, \tag{2.158}$$

where S is the entropy of a fluid element. That is, the entropy of a fluid element is conserved along its flow line. Hint: you will need to use the first law of thermodynamics, $TdS = c^2 d(\rho V) + p dV$ and $\nabla_\mu u^\mu = u^\mu \nabla_\mu \ln V$ where V is the volume of the fluid element.

b) Next use this identity and the Euler equation (2.105) to show that

$$u^\mu\left[\nabla_\mu(Hu_\alpha) - \nabla_\alpha(Hu_\mu)\right] = T\nabla_\alpha S\,. \tag{2.159}$$

Problem 2.5

The *Schwarzschild spacetime* describes an isolated, non-rotating black hole. It has the line element

$$ds^2 = -\left(c^2 - \frac{2GM}{r}\right) dt^2 + \left(1 - \frac{2GM}{c^2 r}\right)^{-1} dr^2 + r^2 d\theta^2 + r^2 \sin^2\theta \, d\phi^2 \,, \tag{2.160}$$

where M is the mass of the black hole.

a) Compute the connection coefficients for the Schwarzschild metric.
b) Compute the equations of motion, $d^2t/d\tau^2$ and $d^2\phi/d\tau^2$, for circular, equatorial orbits $r = a$ (where a is constant) and $\theta = \pi/2$. Obtain Kepler's law, $a^3\omega^2 = GM$ where $\omega = d\phi/dt$.
c) Compute the Riemann tensor and show that the Ricci tensor (and hence the Einstein tensor) vanishes. This shows that the Schwarzschild metric is a vacuum solution to Einstein's equations.

Problem 2.6

If $g_{\alpha\beta} = \eta_{\alpha\beta} + h_{\alpha\beta}$ where $h_{\alpha\beta}$ is a small perturbation to flat spacetime, show that the Riemann tensor, given by Eq. (2.92), is invariant (at first order) under the gauge transformation

$$h_{\alpha\beta} \to h'_{\alpha\beta} = h_{\alpha\beta} - \frac{\partial \xi_\beta}{\partial x^\alpha} - \frac{\partial \xi_\alpha}{\partial x^\beta} \,.$$

Problem 2.7

Use the geodesic equation to derive Eq. (2.153) for linearized gravity with a trace-reversed metric perturbation $\bar{h}_{00} = -4\Phi$, $\bar{h}_{0i} = A_i$ and $\bar{h}_{ij} = 0$.

References

Hartle, J.B. (2003) *Gravity: An Introduction to Einstein's General Relativity*, Benjamin Cummings.

Misner, C.W., Thorne, K.S. and Wheeler, J.A. (1973) *Gravitation*, Freeman, San Francisco.

Schutz, B. (2009) *A First Course in General Relativity*, 2nd edn, Cambridge University Press.

Wald, R.M. (1984) *General Relativity*, University of Chicago Press.

Weinberg, S. (1972) *Gravitation and Cosmology: Principles and Applications of the General Theory of Relativity*, John Wiley & Sons.

3
Gravitational Waves

Equation (2.141) shows that the equation of motion for linear perturbations to flat spacetime in the Lorenz gauge is a wave equation, with the matter stress-energy tensor acting as a source term. In vacuum, the metric perturbation solutions will be waves, and where there is matter present it can generate the gravitational waves. We will now explore the basic properties of gravitational waves and describe how they are generated and how they interact with matter. Other treatments of these topics can be found in Misner *et al.* (1973, Part VIII), Hartle (2003, Chap. 16), Schutz (2009, Chap. 9), Wald (1984, Sec. 4.4b) and Weinberg (1972, Chap. 10), and a more extensive discussion of gravitational waves can also be found in Maggiore (2007, Part I).

3.1
Description of Gravitational Waves

We have seen that the linearized Einstein equation can be expressed as a wave equation, in the Lorenz gauge, for the trace-reversed metric perturbation, and that the matter stress-energy tensor provides a source term for this wave equation. The linearized vacuum Einstein equations are

$$\Box \bar{h}_{\alpha\beta} = 0 \quad \text{(Lorenz gauge)} \tag{3.1}$$

and the Lorenz gauge condition is

$$\frac{\partial \bar{h}^{\mu\alpha}}{\partial x^{\mu}} = 0 \,. \tag{3.2}$$

An important solution is the plane-wave solution, which we choose as travelling in the $z = x^3$-direction. Equation (3.1) implies that the components of the metric perturbation (both the actual perturbation and the trace-reversed version) must all be functions of the retarded time $t - z/c$. The Lorenz gauge condition then requires $\partial \bar{h}^{0\alpha}/\partial t = 0$ and $\partial \bar{h}^{3\alpha}/\partial z = 0$, so $\bar{h}_{0\alpha}(t - z/c)$ and $\bar{h}_{3\alpha}(t - z/c)$ are constant, and we are free to choose the constant to be zero: $\bar{h}_{0\alpha} = 0$ and $\bar{h}_{3\alpha} = 0$. Therefore, the

Gravitational-Wave Physics and Astronomy, First Edition. Jolien D. E. Creighton, Warren G. Anderson.

only non-vanishing components of the trace-reversed metric perturbation are

$$\bar{h}_{11} = \bar{h}_{11}(t - z/c) \tag{3.3a}$$

$$\bar{h}_{22} = \bar{h}_{22}(t - z/c) \tag{3.3b}$$

$$\bar{h}_{12} = \bar{h}_{21} = \bar{h}_{12}(t - z/c)\ . \tag{3.3c}$$

In terms of the actual perturbation, $h_{\alpha\beta} = \bar{h}_{\alpha\beta} - \frac{1}{2}\eta_{\alpha\beta}\bar{h}$, the non-vanishing components are

$$h_{00} = -c^2 h_{33} = \frac{1}{2}c^2\left(\bar{h}_{11} + \bar{h}_{22}\right) \tag{3.4a}$$

$$h_{11} = -h_{22} = \frac{1}{2}\left(\bar{h}_{11} - \bar{h}_{22}\right) \tag{3.4b}$$

$$h_{12} = h_{21} = \bar{h}_{12}\ . \tag{3.4c}$$

Clearly the metric perturbation is travelling along the z-axis at the speed of light. We call such a solution a *gravitational wave.*

There are three independent functions of $t - z/c$, $\bar{h}_{11}$, $\bar{h}_{22}$, and $\bar{h}_{12} = \bar{h}_{21}$, or, equivalently, $h_{00} = -c^2 h_{33}$, $h_{11} = -h_{22}$, and $h_{12} = h_{21}$. Which of these, if any, are physical, and how many are artifacts of the choice of gauge (or can be removed by a different gauge choice)? To answer this, recall that the Riemann tensor is gauge invariant to linear order (Problem 2.6). Therefore, the independent components of the Riemann tensor must represent the true degrees of freedom of the gravitational-wave solution, and, additionally, we can compute the components of the linearized Riemann tensor in any gauge we choose. Recall Eq. (2.92) for the Riemann tensor,

$$R_{\alpha\beta\gamma\delta} = \frac{1}{2}\left(-\frac{\partial^2 h_{\beta\delta}}{\partial x^\alpha \partial x^\gamma} + \frac{\partial^2 h_{\beta\gamma}}{\partial x^\alpha \partial x^\delta} + \frac{\partial^2 h_{\alpha\delta}}{\partial x^\beta \partial x^\gamma} - \frac{\partial^2 h_{\alpha\gamma}}{\partial x^\beta \partial x^\delta}\right), \tag{3.5}$$

where we have kept only the terms linear in the metric perturbation. Since this plane-wave perturbation is a function only of t and z, the derivatives $\partial/\partial t$ and $\partial/\partial z$ alone will be non-vanishing – and since the metric components are functions of t and z only in the form $t - z/c$, $\partial/\partial z = -c^{-1}\partial/\partial t$. So we need consider solely the derivative $\partial/\partial t$. There are two derivatives of the metric, so at least two indices of the Riemann tensor will be 0; by the (anti-)symmetries of the Riemann tensor, along with Eq. (3.4) relating the non-vanishing metric components in the Lorenz gauge, it can be seen that the other two components must be 1 or 2 (since an index values of 3 are interchangeable with 0, up to an overall factor).[1] Thus, we see that only $h_{11} = -h_{22}$, and h_{12} will appear in the result. Depending on where we choose to put the 0s in the components of the Riemann tensor, we will select just one of the four terms in the equation for the linearized Riemann tensor; the

1) For example, $R_{0303} = -\frac{1}{2}(\partial^2 h_{33}/\partial t^2 + \partial^2 h_{00}/\partial z^2)$ which vanishes because $h_{00} = -c^2 h_{33}$ in the Lorenz gauge and $\partial/\partial z = -c^{-1}\partial/\partial t$.

other components of Riemann can then be deduced by its symmetries. Therefore, without loss of generality, we conclude that the only two independent components of the Riemann tensor are

$$R_{0101} = -\frac{1}{2}\frac{\partial^2}{\partial t^2} h_{11} = -\frac{1}{4}\frac{\partial^2}{\partial t^2}\left(\bar{h}_{11} - \bar{h}_{22}\right) \tag{3.6a}$$

$$R_{0102} = -\frac{1}{2}\frac{\partial^2}{\partial t^2} h_{12} = -\frac{1}{2}\frac{\partial^2}{\partial t^2}\bar{h}_{12} \,. \tag{3.6b}$$

In addition, we have $R_{0202} = -R_{0101}$ and $R_{0201} = R_{0102}$; replacing 0 by 3 simply introduces a factor of $-c^{-1}$ (so that, for example, $R_{3101} = -c^{-1}R_{0101}$); the rest of the non-vanishing components of Riemann follow from the symmetries of the Riemann tensor.

Since *some* components of the Riemann tensor are non-vanishing, the gravitational wave must be physical – it cannot be purely an artifact of a peculiar gauge choice. The Riemann tensor contains *two* independent functions of $t - z/c$ which we will call the two independent degrees of freedom h_+ and $h_\times$ (for reasons that will become clear in the next section) where $h_+ = h_{11} = -h_{22}$ and $h_\times = h_{12} = h_{21}$:

$$R_{0101} = -R_{0202} = -\frac{1}{2}\ddot{h}_+ \tag{3.7a}$$

$$R_{0102} = R_{0201} = -\frac{1}{2}\ddot{h}_\times \,. \tag{3.7b}$$

The components of the Riemann tensor depend on only the two independent functions of $t - z/c$, $h_+ = \frac{1}{2}(\bar{h}_{11} - \bar{h}_{22})$ and $h_\times = \bar{h}_{12} = \bar{h}_{21}$, rather than the three implied by the field equations, in which it appears that $\bar{h}_{11}$, $\bar{h}_{22}$ and $\bar{h}_{12}$ (or h_{00}, h_{11} and h_{12}) would all be independent functions. There must be more gauge freedom within the Lorenz gauge that is responsible for the extra (non-physical) degree of freedom (specifically, for $h_{00} = -h_{33} = \frac{1}{2}(\bar{h}_{11} + \bar{h}_{22})$, which does not appear in the Riemann tensor).

To understand a bit better the nature of the gauge freedom, consider a monochromatic plane wave:

$$\bar{h}_{\alpha\beta} = A_{\alpha\beta}\cos\left(k_\mu x^\mu\right) , \tag{3.8}$$

where $A_{\alpha\beta}$ is a constant (and symmetric) tensor and $k_\alpha = [-\omega, \boldsymbol{k}]$ with $\omega = c\|\boldsymbol{k}\|$ is a constant null-vector (i.e. $k_\mu k^\mu = 0$). The Lorenz gauge condition is

$$0 = \frac{\partial \bar{h}^{\mu\alpha}}{\partial x^\mu} = -k_\mu A^{\mu\alpha}\sin\left(k_\nu x^\nu\right) = -k_\mu A^{\mu\alpha}\sin(\boldsymbol{k}\cdot\boldsymbol{x} - \omega t) \,, \tag{3.9}$$

which is satisfied if $k_\mu A^{\mu\alpha} = 0$, that is the wave is *transverse*. The plane wave has a frequency $\omega = c\|\boldsymbol{k}\| = c(k_1^2 + k_2^2 + k_3^2)^{1/2}$ and propagates in the direction specified by the spatial part of the vector k_α: $\hat{\boldsymbol{n}} = c\boldsymbol{k}/\omega$ is the unit vector in the direction of propagation. Once k_α is chosen, there are 10 independent components of the metric perturbation $\bar{h}_{\alpha\beta}$ specified by the ten independent components of the symmetric matrix $A_{\alpha\beta}$. However, the Lorenz condition $k_\mu A^{\mu\alpha} = 0$ imposes four conditions on these 10 components, which leaves six independent components. But we saw that there are really only two independent fields in the Riemann tensor. The extra degrees of freedom that we see are related to the remaining gauge freedom within the Lorenz gauge.

Recall that there is the freedom within the Lorenz gauge to make an additional infinitesimal gauge transformation generated by a vector ξ_α that satisfies the harmonic condition $\Box\xi_\alpha = 0$. A solution to this equation that is compatible with the monochromatic plane-wave metric perturbation is

$$\xi_\alpha = -C_\alpha \sin\left(k_\mu x^\mu\right), \tag{3.10}$$

where C_α are 4 arbitrary constants. After accounting for these four degrees of gauge freedom we see that the metric perturbation has only 2 independent degrees of freedom remaining.

The Lorenz gauge choice makes the metric perturbation look like a *transverse* wave; the remaining freedom within the Lorenz gauge can be used to choose a gauge in which the perturbation is *traceless* and also purely spatial. Given the trace-reversed metric perturbation $\bar{h}_{\alpha\beta}$ in an arbitrary Lorenz gauge, the corresponding *transverse traceless gauge* metric perturbation is

$$\begin{aligned} h^{\mathrm{TT}}_{\alpha\beta} &= \bar{h}_{\alpha\beta} - \frac{\partial\xi_\beta}{\partial x^\alpha} - \frac{\partial\xi_\alpha}{\partial x^\beta} + \eta_{\alpha\beta}\eta^{\mu\nu}\frac{\partial\xi_\mu}{\partial x^\nu} \\ &= A_{\alpha\beta}\cos\left(k_\mu x^\mu\right) + C_\alpha k_\beta \cos\left(k_\mu x^\mu\right) + C_\beta k_\alpha \cos\left(k_\mu x^\mu\right) \\ &\quad - \eta_{\alpha\beta} C_\nu k^\nu \cos\left(k_\mu x^\mu\right). \end{aligned} \tag{3.11}$$

So, identifying $h^{\mathrm{TT}}_{\alpha\beta} = A^{\mathrm{TT}}_{\alpha\beta}\cos(k_\mu x^\mu)$, we see that

$$A^{\mathrm{TT}}_{\alpha\beta} = A_{\alpha\beta} + C_\alpha k_\beta + C_\beta k_\alpha - \eta_{\alpha\beta} C_\nu k^\nu . \tag{3.12}$$

Note that we write $h^{\mathrm{TT}}_{\alpha\beta}$ rather than $\bar{h}^{\mathrm{TT}}_{\alpha\beta}$ (where the over-bar represents trace-reversal) because, as we noted earlier, this gauge has a trace-free metric perturbation so the trace-reversal has no effect. We require first that $A^{\mathrm{TT}}_{\alpha\beta}$ is spatial, that is, if $\mathbf{u} = d/dt$ is a time-like vector, then we require $A^{\mathrm{TT}}_{\alpha\nu} u^\nu = 0$. Although this appears to specify all four components of C_α, it only in fact specifies three because of the identity

$$k^\mu A^{\mathrm{TT}}_{\mu\nu} u^\nu = k^\mu A_{\mu\nu} u^\nu + k^\mu C_\mu k_\nu u^\nu + k^\mu k_\mu C_\nu u^\nu - k_\mu u^\mu C_\nu k^\nu \equiv 0\,, \tag{3.13}$$

where we see that the first term vanishes due to the transverse condition, the third term vanishes since $\boldsymbol{k}$ is null, and the second and fourth terms cancel. Therefore the component $k^\mu C_\mu$ is unspecified in demanding the spatial requirement. This additional component can be specified in such a manner as to make the perturbation trace-free:

$$0 = \eta^{\mu\nu} A^{\mathrm{TT}}_{\mu\nu} = \eta^{\mu\nu} A_{\mu\nu} + 2C_\mu k^\mu - 4C_\nu k^\nu \tag{3.14}$$

so

$$C_\mu k^\mu = \frac{1}{2}\eta^{\mu\nu} A_{\mu\nu} . \tag{3.15}$$

This now specifies the last component of C_α and so uses up the last of the gauge freedom.

It is natural to write the time-like vector **u** in a frame where $u^0 = 1$ and $u^i = 0$. Then we have the spatial condition

$$h^{\mathrm{TT}}_{\alpha 0} = 0 , \tag{3.16a}$$

which means that only h_{ij} can be non-zero (i.e. purely spatial), which leaves only six independent components of this 3×3 symmetric matrix; we also have

$$\delta^{ij} h^{\mathrm{TT}}_{ij} = 0 , \tag{3.16b}$$

which is a single equation that enforces the traceless condition and leaves five independent components in the 3×3 symmetric and traceless matrix h_{ij}; and we have

$$\delta^{ik} \frac{\partial h^{\mathrm{TT}}_{ij}}{\partial x^k} = 0 , \tag{3.16c}$$

which are 3 equations that impose the transverse condition, and leave just two degrees of freedom.

This procedure for going to the TT-gauge works for an arbitrary monochromatic plane wave. Since any plane wave is a Fourier superposition of monochromatic waves, and since the gauge conditions $h^{\mathrm{TT}}_{\alpha 0} = 0$, $\delta^{ij} h^{\mathrm{TT}}_{ij} = 0$, and $\delta^{ik} \partial h^{\mathrm{TT}}_{ij} / \partial x^k = 0$ are all linear, we can enforce the TT-gauge for any *radiative* gravitational-wave perturbation by decomposing it into its Fourier components and applying the above procedure to determine the corresponding Fourier component of the necessary gauge transformation. Note that this procedure does not work for all metric perturbations, but solely for those radiative ones that can be written as a Fourier superposition of monochromatic null plane waves. The spatial, transverse and traceless gauge is sometimes called the *radiative gauge*.

Example 3.1 Transformation from TT coordinates to a locally inertial frame

For concreteness, consider the metric of a purely plus-polarized plane gravitational wave propagating in the z-direction in a spatial, transverse and traceless (TT) gauge:

$$ds^2 = -c^2 dt^2 + \left[1 + h_+(t - z/c)\right] dx^2 + \left[1 - h_+(t - z/c)\right] dy^2 + dz^2 . \tag{3.17}$$

We want to write this in a locally inertial frame where the metric is as close to the Minkowski metric as possible, up to corrections that are proportional to the curvature:

$$g'_{\alpha\beta}(x^\mu) = \eta_{\alpha\beta} + O\left(R_{\alpha\mu\beta\nu} x^\mu x^\nu \text{ , and so on}\right). \tag{3.18}$$

Recall that all the components of the Riemann tensor are proportional to $\ddot{h}_+$ and so the corrections are proportional to $\ddot{h}_+$ times terms that are quadratic in the coordinates $\{t, x, y, z\}$ which are measured relative to the centre of the inertial frame.

First, take the metric in TT-gauge and make a coordinate transformation to the null coordinates $u = t - z/c$ and $v = t + z/c$. Then $dt = \frac{1}{2}(du + dv)$ and $dz = \frac{1}{2}c(du - dv)$ and

$$ds^2 = -c^2 du dv + \left[1 + h_+(u)\right] dx^2 + \left[1 - h_+(u)\right] dy^2 . \tag{3.19}$$

Next change from unprimed coordinates (in the TT-gauge) to primed coordinates (in the locally inertial frame) via the relations

$$\begin{aligned}
u &= u' \\
v &= v' - \frac{1}{2}\dot{h}_+(u')\left(x'^2 - y'^2\right)/c^2 \\
x &= \left[1 - \frac{1}{2}h_+(u')\right] x' \\
y &= \left[1 + \frac{1}{2}h_+(u')\right] y' ,
\end{aligned} \tag{3.20}$$

which yield

$$\begin{aligned}
du &= du' \\
dv &= dv' - \dot{h}_+(u')\left(x'dx' - y'dy'\right)/c^2 - \frac{1}{2}\ddot{h}_+(u')(x'^2 - y'^2)du'/c^2 \\
dx^2 &= \left[1 - h_+(u')\right] dx'^2 - x'\dot{h}_+(u')du'dx' + O\left(h^2\right) \\
dy^2 &= \left[1 + h_+(u')\right] dx'^2 + y'\dot{h}_+(u')du'dy' + O\left(h^2\right)
\end{aligned} \tag{3.21}$$

and therefore

$$ds^2 = -c^2 du'dv' + \frac{1}{2}\left(x'^2 - y'^2\right)\ddot{h}_+(u')du'^2 + dx'^2 + dy'^2 , \tag{3.22}$$

where we drop the $O(h^2)$ terms. Finally, let $u' = t' - z'/c$ and $v' = t' + z'/c$ to obtain

$$\begin{aligned}
ds^2 &= -c^2 dt'^2 + dx'^2 + dy'^2 + dz'^2 \\
&\quad + \frac{1}{2}\left(x'^2 - y'^2\right)\ddot{h}_+(t' - z'/c)(dt' - dz'/c)^2 .
\end{aligned} \tag{3.23}$$

This is the desired form of a locally inertial frame.

Example 3.2 Wave equation for the Riemann tensor

Rather than construct a wave equation for the gauge-dependent metric perturbation, we can use the Bianchi identity as well as the vacuum field equations to show that the gauge-invariant linear Riemann tensor obeys a wave equation. Recall the Bianchi identity:

$$0 = \frac{\partial}{\partial x^\alpha} R_{\beta\gamma\delta}{}^\epsilon + \frac{\partial}{\partial x^\beta} R_{\gamma\alpha\delta}{}^\epsilon + \frac{\partial}{\partial x^\gamma} R_{\alpha\beta\delta}{}^\epsilon \,. \tag{3.24}$$

Contracting the index α with ϵ and using the vacuum field equation $R_{\alpha\beta} = 0$ we find $\partial R_{\beta\gamma\delta}{}^\mu / \partial x^\mu = 0$; in fact, due to the symmetries of the Riemann tensor, the Riemann tensor is divergenceless on all indices in a vacuum spacetime. Now, apply $\eta^{\alpha\nu}\partial/\partial x^\nu$ to the Bianchi identity and use this divergenceless property to obtain $\Box R_{\beta\gamma\delta}{}^\epsilon = 0$. Hence the components of the linear Riemann tensor obey the wave equation

$$\Box R_{\alpha\beta\gamma\delta} = 0 \tag{3.25}$$

in a vacuum spacetime.

3.1.1
Propagation of Gravitational Waves

We have seen that the plane-wave gravitational-wave perturbations to flat spacetime propagate linearly at the speed of light. What happens to gravitational waves in curved spacetime? In order to address this problem we need to distinguish between what part of the spacetime curvature is the background and what part is the gravitational-wave perturbation; to uniquely separate these two we describe a perturbation as a gravitational wave if (i) the amplitude of the perturbation is small, and (ii) the characteristic length scale over which the perturbation varies, the wavelength λ, is much smaller than the length scale of the background curvature $\mathcal{R}$ in a locally inertial frame. This means that our description of a gravitational wave in flat-spacetime is also a valid description of a gravitational wave in the locally inertial frame. We now want to find out how that wave propagates across spacetime as it leaves the neighbourhood in which the locally inertial frame was constructed. The limit that we are exploring here is the *short-wavelength approximation* or the *geometric optics* limit of the propagation equation for metric perturbations.

Let us define the trace-reversed metric perturbation to the curved spacetime as

$$\bar{h}_{\alpha\beta} := h_{\alpha\beta} - \frac{1}{2}\overset{0}{g}_{\alpha\beta} h \,, \tag{3.26}$$

where $h = \overset{0}{g}{}^{\mu\nu} h_{\mu\nu}$ and $\overset{0}{g}_{\alpha\beta}$ is the background metric. Because Eqs. (3.1) and (3.2) must hold in a neighbourhood of the origin of a locally inertial frame, we see that this metric perturbation must satisfy the curved-space wave equation up to corrections that depend on the curvature scale of the background spacetime:

$$0 = \overset{0}{g}{}^{\mu\nu}\overset{0}{\nabla}_\mu \overset{0}{\nabla}_\nu \bar{h}_{\alpha\beta} + O\left(h\mathcal{R}^{-2}\right) \quad \text{(Lorenz gauge)}, \tag{3.27}$$

where the amplitude of the perturbation is $O(h)$, and we assume it is in the curved-space Lorenz gauge

$$0 = \overset{0}{\nabla}_\mu \bar{h}^{\mu\alpha} . \tag{3.28}$$

Let us write the solution to the wave equation in the form

$$\bar{h}_{\alpha\beta}(\mathbf{x}) = A_{\alpha\beta}(\mathbf{x}) \cos\varphi(\mathbf{x}) , \tag{3.29}$$

where the amplitude $A_{\alpha\beta}(\mathbf{x})$ is slowly varying (on the length scale $\mathcal{R}$) while the phase is rapidly varying (on the length scale λ); the wave vector is $k_\alpha = \partial\varphi/\partial x^\alpha$, which is $O(\lambda^{-1})$.

From the Lorenz condition we see that

$$0 = \overset{0}{\nabla}_\mu \bar{h}^{\mu\alpha} = -k_\mu A^{\mu\alpha} \sin\varphi + \left(\overset{0}{\nabla}_\mu A^{\mu\alpha}\right) \cos\varphi . \tag{3.30}$$

The first term on the right-hand side is $O(h\lambda^{-1})$ while the second term is $O(h\mathcal{R}^{-1})$, which can be neglected. We find, therefore,

$$k_\mu A^{\mu\alpha} = 0 , \tag{3.31}$$

that is the waves are transverse. From the curved-space wave equation we find

$$0 = -k_\mu k^\mu A_{\alpha\beta} \cos\varphi - \left[A_{\alpha\beta} \overset{0}{\nabla}_\mu k^\mu + 2k^\mu \overset{0}{\nabla}_\mu A_{\alpha\beta}\right] \sin\varphi + O\left(h\mathcal{R}^{-2}\right) . \tag{3.32}$$

The first term on the right-hand side is $O(h\lambda^{-2})$, which dominates over the terms in brackets, which are $O(h\lambda^{-1}\mathcal{R}^{-1})$, and these in turn dominate over the neglected $O(h\mathcal{R}^{-2})$; each one of these must individually vanish, so

$$0 = k_\mu k^\mu \tag{3.33}$$

(the waves are null), and

$$0 = A_{\alpha\beta} \overset{0}{\nabla}_\mu k^\mu + 2k^\mu \overset{0}{\nabla}_\mu A_{\alpha\beta} . \tag{3.34}$$

Equation (3.33), along with the fact that k_α is the derivative of a scalar $\varphi(\mathbf{x})$, implies that the wave vector **k** satisfies the geodesic equation:

$$\begin{aligned} 0 &= \frac{1}{2}\overset{0}{\nabla}_\alpha (k^\mu k_\mu) = k^\mu \overset{0}{\nabla}_\alpha k_\mu = k^\mu \overset{0}{\nabla}_\alpha \overset{0}{\nabla}_\mu \varphi = k^\mu \overset{0}{\nabla}_\mu \overset{0}{\nabla}_\alpha \varphi \\ &= k^\mu \overset{0}{\nabla}_\mu k_\alpha . \end{aligned} \tag{3.35}$$

The waves follow geodesics of the background spacetime. Equation (3.34) can be interpreted by writing $A_{\alpha\beta} = \mathcal{A} e_{\alpha\beta}$, that is by dividing it into a scalar amplitude, $\mathcal{A}$, and a polarization tensor $e_{\alpha\beta}$ that is normalized so that $e^{\mu\nu} e_{\mu\nu} = 2$. Then Eq. (3.34) becomes

$$0 = \frac{1}{2} \mathcal{A} e_{\alpha\beta} \overset{0}{\nabla}_\mu k^\mu + e_{\alpha\beta} k^\mu \overset{0}{\nabla}_\mu \mathcal{A} + \mathcal{A} k^\mu \overset{0}{\nabla}_\mu e_{\alpha\beta} \,. \tag{3.36}$$

If we contract this equation with $A^{\alpha\beta} = \mathcal{A} e^{\alpha\beta}$ and note that $0 = \overset{0}{\nabla}_\mu (e^{\alpha\beta} e_{\alpha\beta}) = 2 e^{\alpha\beta} \overset{0}{\nabla}_\mu e_{\alpha\beta}$ then

$$0 = \mathcal{A}^2 \overset{0}{\nabla}_\mu k^\mu + 2\mathcal{A} k^\mu \overset{0}{\nabla}_\mu \mathcal{A} \quad = \overset{0}{\nabla}_\mu \left(\mathcal{A}^2 k^\mu \right) , \tag{3.37}$$

and, using this relation in Eq. (3.36), we obtain

$$k^\mu \overset{0}{\nabla}_\mu e_{\alpha\beta} = 0 \,. \tag{3.38}$$

Equation (3.37) is a conservation equation for the current $\mathcal{A}^2 k^\alpha$ (see Problem 3.2), while Eq. (3.38) shows that the polarization tensor is parallel transported along the null geodesics.

Example 3.3 Attenuation of gravitational waves

In Problem 3.3 it is shown that the length scale of attenuation of gravitational waves in a viscous medium is

$$\ell = \frac{c^3}{8\pi G \eta} \,, \tag{3.39}$$

where η is the coefficient of viscosity of the medium. For a gas containing molecules of mass m moving with speed v that have an interaction cross-section σ, the coefficient of viscosity is $\eta \sim m v/\sigma$. The mean free path of the molecules in the gas is $d = 1/(n\sigma)$ where n is the number density of molecules in the gas, so $\eta \sim \rho d v$ where $\rho = nm$ is the mass density of the gas. We have

$$\ell \sim \frac{c^2}{8\pi G \rho} \frac{1}{d} \frac{c}{v} = \mathcal{R}^2 \frac{1}{d} \frac{c}{v} \,, \tag{3.40}$$

where $\mathcal{R} = c/\sqrt{8\pi G \rho}$ is the spacetime curvature scale produced by the mass of the gas. Within our short-wavelength approximation we have $\mathcal{R} \gg \lambda$, and we have also required $\lambda \gg d$ in order for there to be viscous damping of the gravitational wave. Finally, since $v \leq c$, we find

$$\ell \gg \mathcal{R} \,, \tag{3.41}$$

that is, the length scale over which the gravitational wave is attenuated by a medium is much greater than the length scale of spacetime curvature produced by the medium. This means that we can effectively ignore the process of attenuation of gravitational waves in almost all cases of interest.

3.2
Physical Properties of Gravitational Waves

Having now shown that gravitational waves can be described by waves having two polarization states and travelling at the speed of light, we now explore the effects these waves have on observers (and how they may be detected), and show that the waves carry energy.

3.2.1
Effects of Gravitational Waves

Let us continue with the example of a generic plane gravitational wave and examine the physical effects and properties of the gravitational wave. Specifically, we will use the equation of geodesic deviation to examine the effect of the gravitational wave on the relative motions of two nearby freely falling particles.

Let the two particles be separated by the (spatial) vector $\boldsymbol{\zeta} = \zeta(\sin\theta\cos\phi\,\boldsymbol{e}_1 + \sin\theta\sin\phi\,\boldsymbol{e}_2 + \cos\theta\,\boldsymbol{e}_3)$. The relative acceleration of the two particles is $a_j = -R_{0i0j}\zeta^j$ so

$$a_1 = -R_{0101}\zeta^1 - R_{0102}\zeta^2 = \frac{1}{2}\ddot{h}_+\zeta\sin\theta\cos\phi + \frac{1}{2}\ddot{h}_\times\zeta\sin\theta\sin\phi \quad (3.42a)$$

$$a_2 = -R_{0202}\zeta^2 - R_{0201}\zeta^1 = -\frac{1}{2}\ddot{h}_+\zeta\sin\theta\sin\phi + \frac{1}{2}\ddot{h}_\times\zeta\sin\theta\cos\phi \quad (3.42b)$$

$$a_3 = 0\,. \quad (3.42c)$$

Note that the acceleration is *transverse* to the direction of propagation of the wave. We can depict the acceleration as a function of the positions of the two particles with a lines-of-force diagram. Consider the case of particles separated on the transverse plane so $\theta = \pi/2$. For a pure plus-polarized wave (with $\ddot{h}_\times = 0$ and $\ddot{h}_+ > 0$) we see that $a_1 \propto x^1$ and $a_2 \propto -x^2$ where $(x^1, x^2) = (\zeta\cos\phi, \zeta\sin\phi)$ are the coordinates on the transverse plane of the second particle relative to the reference particle; the lines-of-force diagram is shown in Figure 3.1. For a pure cross-polarized wave (with $\ddot{h}_+ = 0$ and $\ddot{h}_\times > 0$) we see that $a_1 \propto x^2$ and $a_2 \propto x^1$; Figure 3.2 shows this lines-of-force diagram.

We can deduce how the separation vector evolves in time. Consider the magnitude of the separation vector, which is a measure of the distance between two nearby freely falling particles. The unit vector along the separation vector is $\boldsymbol{e}_\zeta = \boldsymbol{\zeta}/\zeta$ and the component of geodesic deviation acceleration in this direction is

$$\begin{aligned} \boldsymbol{a}_\zeta = \boldsymbol{e}_\zeta \cdot \boldsymbol{a} &= -\frac{1}{\zeta}R_{0i0j}\zeta^i\zeta^j \\ &= a_1\sin\theta\cos\phi + a_2\sin\theta\sin\phi \\ &= \zeta\left(\frac{1}{2}\ddot{h}_+\sin^2\theta\cos 2\phi + \frac{1}{2}\ddot{h}_\times\sin^2\theta\sin 2\phi\right). \end{aligned} \quad (3.43)$$

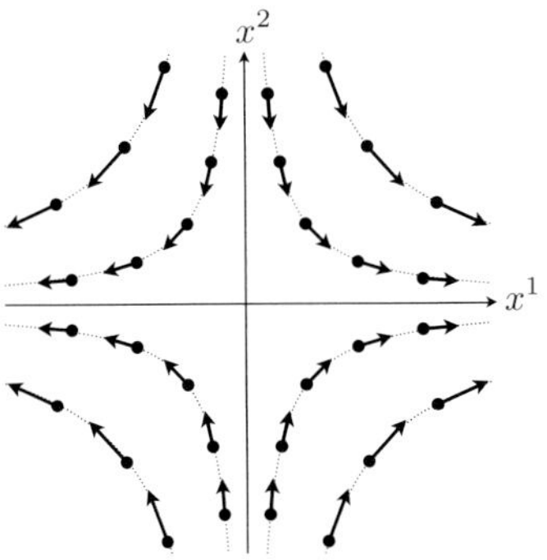

Figure 3.1 Lines-of-force diagram in the transverse plane of a purely plus-polarized gravitational wave. The name "plus" arises from the shape of these lines of force.

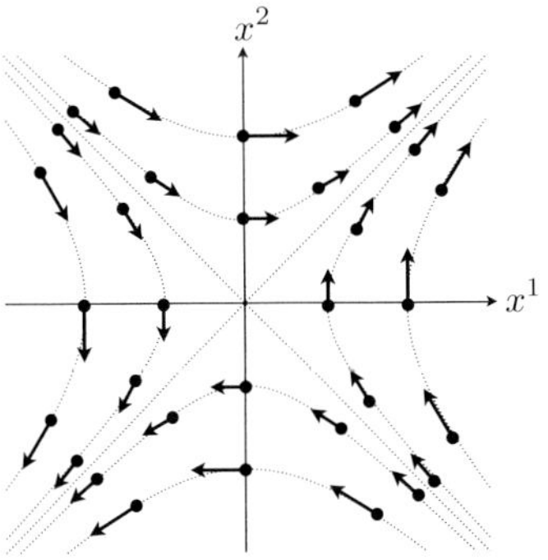

Figure 3.2 Lines-of-force diagram in the transverse plane of a purely cross-polarized gravitational wave. The name "cross" arises from the shape of these lines of force, cf. Figure 3.1.

Integrate this twice (assuming that the particles are initially at rest with respect to each other) to obtain

$$\zeta(t) = \zeta(0)\left(1 + \frac{1}{2}h_+ \sin^2\theta \cos 2\phi + \frac{1}{2}h_\times \sin^2\theta \sin 2\phi\right). \tag{3.44}$$

When $\theta = 0$ or $\theta = \pi$, the separation vector does not change with time. This is because the gravitational wave is transverse and does not affect the distance between freely falling particles initially separated along the direction of propagation. The largest change in ζ will occur when $\theta = \pi/2$ – that is, when the separation vector lies on the transverse plane. The rotation between h_+ and $h_\times$ depends on *twice* the azimuthal angle, 2ϕ, so these are quadrupolar fields. This can be seen by analyzing the effect of each polarization, h_+ and $h_\times$, on a hoop of particles. As a gravitational wave passes, an initially circular hoop is distorted into an ellipse, the orientation of which depends on the polarization state of the wave. See Figure 3.3a,b.

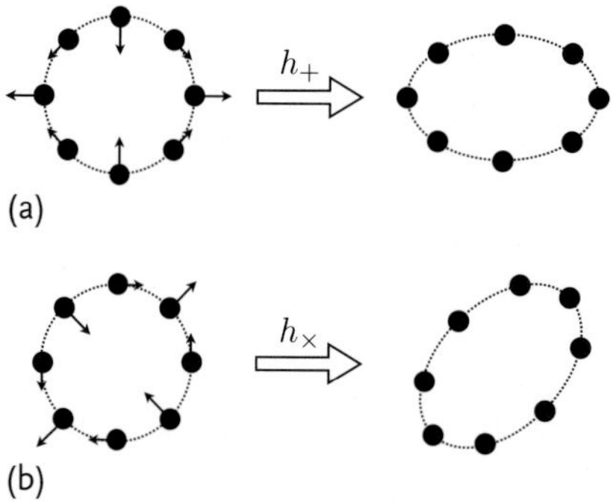

Figure 3.3 The distortion of a hoop of particles lying in the transverse plane of a passing gravitational wave; (a) the effect of a purely plus-polarized gravitational wave and (b) the effect of a purely cross-polarized gravitational wave.

Example 3.4 Degrees of freedom of a plane gravitational wave

It has been argued that there are only two independent degrees of freedom of a plane gravitational wave. It is instructive to see how this depends on the specific theory of gravity – General Relativity – that we have restricted our attention to, and what other degrees of freedom there could be in other theories of gravity. Recall that the physical manifestation of a gravitational wave results in the perturbation to the curvature tensor $R_{\alpha\beta\gamma\delta}$. If the gravitational wave is a plane wave travelling along the $z = x^3$-direction at the speed of light then

$$R_{\alpha\beta\gamma\delta} = R_{\alpha\beta\gamma\delta}(t - z/c) . \tag{3.45}$$

The Bianchi identity

$$0 = \frac{\partial}{\partial x^\alpha} R_{\beta\gamma\delta\epsilon} + \frac{\partial}{\partial x^\beta} R_{\gamma\alpha\delta\epsilon} + \frac{\partial}{\partial x^\gamma} R_{\alpha\beta\delta\epsilon} \tag{3.46}$$

with $\alpha = 0$, $\beta = 1$, $\gamma = 2$ shows that $R_{12\delta\epsilon}$ must be a constant, which we take to be zero. Due to the antisymmetry of the Riemann tensor in the first two indices we have

$$R_{11\delta\epsilon} = R_{22\delta\epsilon} = R_{12\delta\epsilon} = R_{\delta\epsilon 11} = R_{\delta\epsilon 22} = R_{\delta\epsilon 12} = 0 . \tag{3.47}$$

Therefore, for a non-vanishing component, one of the first pair of indices and one of the second pair of indices must be a 0 or a 3, though $R_{00\delta\epsilon} = 0$ and $R_{33\delta\epsilon} = 0$, again due to the antisymmetry of the Riemann tensor in the first two indices. Returning to the Bianchi identity and choosing $\alpha = 0$, $\beta = 1$ and $\gamma = 3$ we find

$$R_{13\delta\epsilon} = -c^{-1} R_{10\delta\epsilon} \tag{3.48}$$

(plus an arbitrary constant which we take to be zero) and similarly, with the choice $\beta = 2$, we find

$$R_{23\delta\epsilon} = -c^{-1} R_{20\delta\epsilon} , \tag{3.49}$$

where we have used the fact that $\partial R_{\alpha\beta\gamma\delta}/\partial z = -c^{-1}\partial R_{\alpha\beta\gamma\delta}/\partial t$. With these relations it is clear that the only independent components of the Riemann tensor are $R_{0i0j} = \mathcal{E}_{ij}$.

So far we have not imposed the Einstein field equations (though we have assumed a plane-wave solution to whatever field equations hold). The symmetric 3×3 tidal tensor $\mathcal{E}_{ij}$ has six independent components. It is illustrative to see the physical effects of such a tidal tensor on a sphere of particles. The six polarizations are (i) the spin-0 longitudinal polarization, given when only $\mathcal{E}_{33}$ is non-vanishing; (ii) and (iii) the two spin-1 longitudinal-transverse polarizations when only $\mathcal{E}_{31} = \mathcal{E}_{13}$ or $\mathcal{E}_{32} = \mathcal{E}_{23}$ are non-vanishing; (iv) the transverse spin-0 "breathing mode" when $\mathcal{E}_{11} = \mathcal{E}_{22}$ and all other components are zero; (v) the transverse spin-2 "plus" polarization when $\mathcal{E}_{11} = -\mathcal{E}_{22}$ and all other components are zero; and (vi) the transverse spin-2 "cross" polarization when $\mathcal{E}_{12} = \mathcal{E}_{21}$ and all other components are zero. These modes are depicted in terms of their effects on a sphere of freely falling particles in Figure 3.4a–e.

The vacuum field equations, $R_{\alpha\beta} = R_{\alpha\mu\beta}{}^{\mu} = 0$, impose additional constraints. In particular, $R_{31} = 0$ implies $R_{3010} = 0$ and $R_{32} = 0$ implies $R_{3020} = 0$; so the longitudinal-transverse modes do not exist. Also, $R_{00} = 0$ implies $R_{0101} + R_{0202} + R_{0303} = 0$ while $R_{33} = 0$ implies $R_{0101} + R_{0202} - R_{0303} = 0$; together these show that $R_{0303} = 0$ which excludes the longitudinal spin-0 mode, and $R_{0101} = -R_{0202}$ which excludes the transverse spin-0 breathing mode. All that remains are the two transverse spin-2 modes, the plus- and cross-modes.

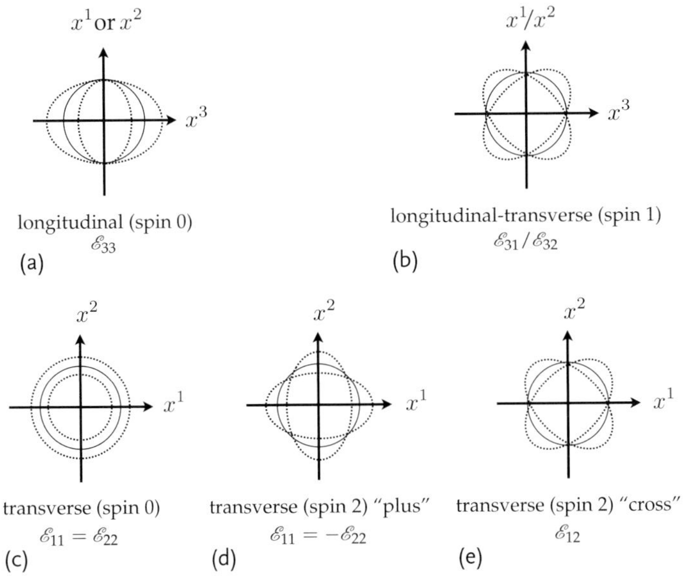

Figure 3.4 The six polarization states found in a generalized gravitational wave. These diagrams show the effect of each mode of the gravitational wave, which is propagating in the x^3-direction, on a sphere of particles: (a) the spin-0 longitudinal mode; (b) the two spin-1 longitudinal-transverse modes; (c) the purely transverse modes known as the spin-0 "breathing mode", and the two traceless (d) "plus" and (e) "cross" modes.

Example 3.5 Plus- and cross-polarization tensors

Gravitational radiation produces tides (geodesic deviation) with a tidal field $\mathcal{E}_{ij} = R_{0i0j}$, which can be expressed as a linear combination of two polarization tensors, e^+_{ij} and $e^\times_{ij}$: $\mathcal{E}_{ij} = -\frac{1}{2}\ddot{h}_+ e^+_{ij} - \frac{1}{2}\ddot{h}_\times e^\times_{ij}$. For a gravitational wave travelling in the z-direction, these polarization tensors have the form

$$\mathbf{e}_+ = \begin{bmatrix} 1 & 0 & 0 \\ 0 & -1 & 0 \\ 0 & 0 & 0 \end{bmatrix} \quad \text{and} \quad \mathbf{e}_\times = \begin{bmatrix} 0 & 1 & 0 \\ 1 & 0 & 0 \\ 0 & 0 & 0 \end{bmatrix} . \tag{3.50}$$

Suppose we choose a new coordinate system, $\{x', y', z'\}$, obtained by a rotation by the angle ϕ in the transverse plane:

$$\begin{bmatrix} x' \\ y' \\ z' \end{bmatrix} = \mathbf{R} \begin{bmatrix} x \\ y \\ z \end{bmatrix} \quad \text{and} \quad \mathbf{R} = \begin{bmatrix} \cos\phi & \sin\phi & 0 \\ -\sin\phi & \cos\phi & 0 \\ 0 & 0 & 1 \end{bmatrix} . \tag{3.51}$$

Under such a coordinate transformation we find

$$\begin{aligned} \mathbf{e}'_+ &= \mathbf{R}\mathbf{e}_+\mathbf{R}^{-1} = \begin{bmatrix} 1 & 0 & 0 \\ 0 & -1 & 0 \\ 0 & 0 & 0 \end{bmatrix} \cos 2\phi - \begin{bmatrix} 0 & 1 & 0 \\ 1 & 0 & 0 \\ 0 & 0 & 0 \end{bmatrix} \sin 2\phi \\ \mathbf{e}'_\times &= \mathbf{R}\mathbf{e}_\times\mathbf{R}^{-1} = \begin{bmatrix} 1 & 0 & 0 \\ 0 & -1 & 0 \\ 0 & 0 & 0 \end{bmatrix} \sin 2\phi + \begin{bmatrix} 0 & 1 & 0 \\ 1 & 0 & 0 \\ 0 & 0 & 0 \end{bmatrix} \cos 2\phi \end{aligned} \tag{3.52}$$

and hence

$$\begin{aligned} \mathbf{e}_+ \to \mathbf{e}'_+ &= \mathbf{e}_+ \cos 2\phi - \mathbf{e}_\times \sin 2\phi \\ \mathbf{e}_\times \to \mathbf{e}'_\times &= \mathbf{e}_\times \cos 2\phi + \mathbf{e}_+ \sin 2\phi . \end{aligned} \tag{3.53}$$

Note that the polarization tensors are rotated by 2ϕ rather than ϕ which shows that they represent a quadrupolar perturbation.

3.2.1.1 Detection of Gravitational Waves

A gravitational-wave detector exploits the physical effects of a gravitational wave on bodies to detect the waves. A gravitational wave detector may, for example, monitor the separation between two freely falling bodies; as a gravitational wave passes, the two bodies will experience a tidal acceleration relative to each other. A common question arises: "How does one measure this relative acceleration?" For large gravitational-wave metric perturbations, the distance between freely falling bodies could be measured using a rigid object such as a ruler. The rigidity of the ruler would resist (somewhat) the tidal force while the freely falling bodies would experience the full tidal acceleration. (In a similar way, earth tides are much smaller than ocean tides because of the capacity of the solid earth to resist tidal deformations.)

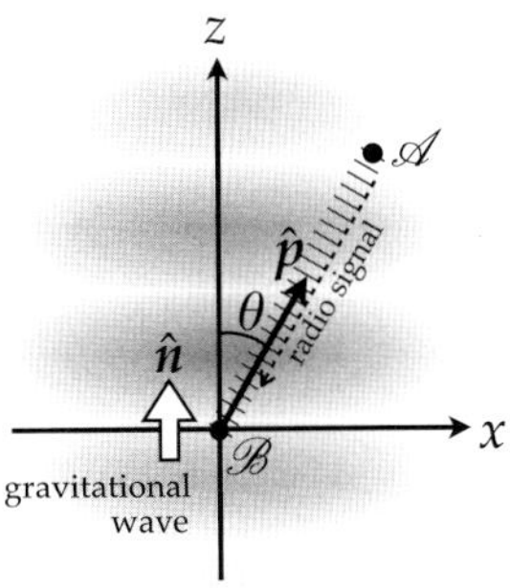

Figure 3.5 A radio signal from a source at point $\mathcal{A}$ travels to the receiver at point $\mathcal{B}$ in the direction $-\hat{\boldsymbol{p}}$. A gravitational wave is also present and travelling in direction $\hat{\boldsymbol{n}} = \boldsymbol{e}_3$. The angle θ is the angle between the vectors $\hat{\boldsymbol{n}}$ and $\hat{\boldsymbol{p}}$ so that $\hat{\boldsymbol{n}} \cdot \hat{\boldsymbol{p}} = \cos\theta$.

But for very weak gravitational waves, it is not clear how to make a rigid body: the atoms that make up an object are essentially in free fall with respect to each other on the tiny scales of gravitational-wave distortions.

The distance between two objects could be monitored by radar ranging or some similar technique. Then the question arises of whether electromagnetic waves that are being used to measure the distance between the objects are affected by the gravitational wave. Another method of sensing the changing distance between two objects caused by a gravitational wave would be to look for a Doppler shift in a signal sent from one object to the other. Here we will consider this problem.

For a specific example of a gravitational-wave detector, consider two objects, $\mathcal{A}$ and $\mathcal{B}$, separated by some distance. Object $\mathcal{A}$ emits a radio signal with frequency $\nu_{\mathcal{A}}$, which is received by object $\mathcal{B}$ with an observed frequency $\nu_{\mathcal{B}}$, which may be different due to relative motion of the two objects (the Doppler effect) and due to a gravitational wave. The redshift of the radio signal is $\mathfrak{z} = (\nu_{\mathcal{A}} - \nu_{\mathcal{B}})/\nu_{\mathcal{B}}$ and this redshift is affected by a gravitational wave.[2)]

Construct a coordinate system in which object $\mathcal{B}$ is at the origin and object $\mathcal{A}$ is on the x–z plane so that the direction $\hat{\boldsymbol{p}}$ from $\mathcal{B}$ to $\mathcal{A}$ makes an angle θ to the $\boldsymbol{e}_3$-axis (see Figure 3.5). For simplicity, we consider a purely plus-polarized plane gravitational wave propagating in the $\hat{\boldsymbol{n}} = \boldsymbol{e}_3$ direction, in a spatial, transverse and traceless (TT) gauge so that $h_{11} = -h_{22} = h_+(t - z/c)$. The radio signal follows a null geodesic with a tangent vector[3)]

$$\hat{k}^\alpha = \frac{dx^\alpha}{d\lambda} = \nu\left[1, -\left(1 - \frac{1}{2}h_+\right)c\sin\theta, 0, -c\cos\theta\right]. \tag{3.54}$$

It is straightforward to show that this vector is null to first order in the gravitational-wave amplitude h. The vector $\hat{k}^\alpha$ is an affine vector, with affine parameter λ, so $\hat{k}^0 = \nu = dt/d\lambda$ is the frequency of the radio signal and the geodesic equation

2) Be careful to distinguish between the redshift, $\mathfrak{z}$ and the coordinate, z.

3) In the unperturbed spacetime, the null geodesic is $\overset{0}{k}{}^\alpha \propto [1, -c\sin\theta, 0, -c\cos\theta]$. When a perturbation given by $h_{\alpha\beta}$ is induced, the null geodesic becomes $k^\alpha = \overset{0}{k}{}^\alpha - \frac{1}{2}\eta^{\alpha\mu}h_{\mu\nu}\overset{0}{k}{}^\nu$. The normalization constant of $\hat{k}^\alpha$ is chosen so that $\hat{k}^\alpha = \nu$ is the frequency of the radio signal.

requires

$$\frac{d\hat{k}^\alpha}{d\lambda} = -\Gamma^\alpha_{\mu\nu}\hat{k}^\mu\hat{k}^\nu \,. \tag{3.55}$$

To determine the measured Doppler shift, we must compute $d\hat{k}^0/d\lambda = d\nu/d\lambda$ and integrate this over the geodesic from point $\mathcal{A}$ with $\lambda = \lambda_\mathcal{A}$, to point $\mathcal{B}$ with $\lambda = \lambda_\mathcal{B}$. Given the form of the metric perturbation, the geodesic equation simplifies to

$$\begin{aligned}\frac{d\nu}{d\lambda} &= -\Gamma^0_{11}\hat{k}^1\hat{k}^1 = -\frac{1}{2}\frac{1}{c^2}\frac{\partial h_+}{\partial t}\left[\nu\left(1-\frac{1}{2}h_+\right)c\sin\theta\right]^2 \\ &= -\frac{1}{2}\frac{\partial h_+}{\partial t}\nu^2\sin^2\theta + O\left(h^2\right),\end{aligned} \tag{3.56}$$

where $\Gamma^0_{11} = \frac{1}{2}c^{-2}(\partial h_+/\partial t)$. The partial derivative of $h_+(t - z/c)$ with respect to t can be expressed in terms of a total derivative of h_+ with respect to λ:

$$\begin{aligned}\frac{dh_+}{d\lambda} &= \frac{dt}{d\lambda}\frac{\partial h_+}{\partial t} + \frac{dz}{d\lambda}\frac{\partial h_+}{\partial z} = \frac{dt}{d\lambda}\frac{\partial h_+}{\partial t} - \frac{1}{c}\frac{dz}{d\lambda}\frac{\partial h_+}{\partial t} \\ &= \left(\hat{k}^0 - \hat{k}^3/c\right)\frac{\partial h_+}{\partial t} = \nu(1+\cos\theta)\frac{\partial h_+}{\partial t}\,.\end{aligned} \tag{3.57}$$

Therefore, to first order in h, the quantity $d\ln\nu/d\lambda$ is a total derivative

$$\frac{d\ln\nu}{d\lambda} = -\frac{d}{d\lambda}\left[\frac{1}{2}(1-\cos\theta)h_+\right] + O\left(h^2\right) \tag{3.58}$$

since ν and θ are constant up to $O(h)$ corrections; thus the change in frequency depends only on the values of the metric perturbation when the radio signal is emitted and when it is received,

$$\begin{aligned}\log(\nu_\mathcal{A}/\nu_\mathcal{B}) &= -\int_{\lambda_\mathcal{A}}^{\lambda_\mathcal{B}}\frac{d\ln\nu}{d\lambda}d\lambda = \int_{\lambda_\mathcal{A}}^{\lambda_\mathcal{B}}\frac{d}{d\lambda}\left[\frac{1}{2}(1-\cos\theta)h_+\right]d\lambda + O\left(h^2\right) \\ &= \frac{1}{2}(1-\cos\theta)\left[h_+\left(x^\alpha_\mathcal{B}\right) - h_+(x^\alpha_\mathcal{A})\right] + O\left(h^2\right),\end{aligned} \tag{3.59}$$

and, since $\log(\nu_\mathcal{A}/\nu_\mathcal{B}) = \log(1+\mathfrak{z}) \simeq \mathfrak{z}$, the Doppler shift is

$$\mathfrak{z} = \frac{\nu_\mathcal{A} - \nu_\mathcal{B}}{\nu_\mathcal{B}} = \frac{1}{2}(1-\cos\theta)\left[h_+\left(x^\alpha_\mathcal{B}\right) - h_+\left(x^\alpha_\mathcal{A}\right)\right] + O\left(h^2\right). \tag{3.60}$$

Note that the redshift depends on the value of the metric perturbation at the time and place at which the radio signal is emitted and the value of the metric perturbation at the time and place where it is received: for a plane gravitational wave, only the values of the metric perturbation on the *endpoints* of the null geodesic connecting points $\mathcal{A}$ and $\mathcal{B}$ count toward the observed redshift.

For an arbitrary polarization gravitational wave travelling in the direction $\hat{\boldsymbol{n}}$, the general result is

$$\mathfrak{z} = \frac{\nu_{\rm em} - \nu_{\rm rec}}{\nu_{\rm rec}} = \frac{1}{2}\frac{\hat{p}^i \hat{p}^j}{1 + \hat{\boldsymbol{n}} \cdot \hat{\boldsymbol{p}}}\left[h_{ij}^{\rm TT}(t_{\rm rec}, \boldsymbol{x}_{\rm rec}) - h_{ij}^{\rm TT}(t_{\rm em}, \boldsymbol{x}_{\rm em})\right] . \tag{3.61}$$

A gravitational wave can therefore be directly detected by seeking a Doppler modulation of a signal sent from one body to another.

Example 3.6 A resonant mass detector

The first gravitational-wave detectors built were *resonant mass detectors* or *bar detectors*. These were large, heavy, metal bars. The detector would absorb gravitational waves and be set into oscillation, and such oscillations would hopefully be detectable.

We can model a bar detector as a simple, damped spring. Suppose that two masses, m_1 and m_2, are placed on the x-axis and connected by a spring of spring constant k which is in equilibrium when the masses are separated by length L. Let x measure the displacement of the masses with respect to this equilibrium. If a plane gravitational wave, travelling along the z-axis and polarized along the x-axis so that $h_{11} = -h_{22} = h_+ = h\cos\omega t$, is incident on the oscillator then the oscillations are driven and the equation of motion of these oscillations is

$$\frac{d^2x}{dt^2} + 2\beta\frac{dx}{dt} + \omega_0^2 x = -R_{0101}L = -\frac{1}{2}hL\omega^2\cos\omega t\, . \tag{3.62}$$

Here $\omega_0 = \sqrt{k/\mu}$ is the characteristic frequency of the oscillator, $\mu := m_1 m_2/(m_1 + m_2)$ is the reduced mass of the system, and $\beta = b/2\mu$ is the damping parameter where the dissipative force is $F_{\rm diss} = -b(dx/dt)$. We have assumed that $L \ll \lambda$ where $\lambda = 2\pi c/\omega$ is the wavelength of the gravitational wave.

After initial transients, the equilibrium state of motion of the oscillator will be

$$x(t) = x_{\max}\cos(\omega t + \delta)\, , \tag{3.63}$$

where

$$x_{\max} = \frac{1}{2}hL\frac{\omega^2}{\sqrt{\left(\omega_0^2 - \omega^2\right)^2 + 4\omega^2\beta^2}} \tag{3.64}$$

and

$$\tan\delta = \frac{2\omega\beta}{\omega^2 - \omega_0^2}\, . \tag{3.65}$$

Resonant excitation of the oscillation occurs when the gravitational-wave frequency ω is near the characteristic frequency of the oscillator ω_0. Specifically, amplitude resonance (when driving frequency excites the largest amplitude oscillations) occurs when $\omega = \bar{\omega} = \sqrt{\omega_0^2 - 2\beta^2}$, while the kinetic energy resonance (when the driving frequency produces oscillations with the greatest average kinetic energy) occurs when $\omega = \omega_0$. When $\omega = \omega_0$, the amplitude of the oscillations is

$$x_{\text{max,res}} = \frac{1}{2} h L Q \,, \tag{3.66}$$

where

$$Q := \frac{\omega_0}{2\beta} \tag{3.67}$$

is the *quality factor* of the oscillator.

The kinetic energy of the oscillations is $E_{\text{kin}} = \frac{1}{2}\mu(dx/dt)^2$, the potential energy of the oscillator is $E_{\text{pot}} = \frac{1}{2}kx^2$, the work done on the oscillator by the gravitational wave is $W = \frac{1}{2}\mu(dx/dt)Lh\cos\omega t$, and the rate of energy dissipated is $(dE_{\text{tot}}/dt)_{\text{diss}} = -(dx/dt)F_{\text{diss}}$; averaging over a cycle of oscillation we obtain

$$\langle E_{\text{kin}} \rangle = \frac{1}{4}\mu x_{\text{max}}^2 \omega^2 \tag{3.68}$$

$$\langle E_{\text{pot}} \rangle = \frac{1}{4}\mu x_{\text{max}}^2 \omega_0^2 \tag{3.69}$$

$$\langle W \rangle = -\left\langle \frac{dE_{\text{tot}}}{dt} \right\rangle_{\text{diss}} = \beta\mu x_{\text{max}}^2 \omega^2 \,. \tag{3.70}$$

On resonance, the *decay rate* of the oscillations is

$$\Gamma = \frac{-\langle dE_{\text{tot}}/dt \rangle_{\text{diss}}}{\langle E_{\text{tot}} \rangle} = 2\beta \,, \tag{3.71}$$

where $E_{\text{tot}} = E_{\text{kin}} + E_{\text{pot}}$ is the total energy in the oscillation.

If a gravitational wave with $h \sim 10^{-21}$ is incident on a resonant mass detector having $L \sim 1\,\text{m}$, $\mu \sim 1000\,\text{kg}$, and $Q \sim 10^6$ with frequency $f = \omega_0/2\pi \sim 1\,\text{kHz}$ over some long duration, then the oscillations of amplitude $x_{\text{max,res}} \sim 10^{-15}\,\text{m}$ will be excited. The energy in the oscillator will be $\langle E_{\text{tot}} \rangle \sim 10^{-21}\,\text{J}$. The thermal energy in the oscillator will be $k_B T$, so if $T = 300\,\text{K}$, the thermal energy will also be $\sim 10^{-21}\,\text{J}$, so it will not be possible to distinguish the oscillations due to the gravitational wave from the thermal oscillations if the detector is at room temperature.

3.2.2
Energy Carried by a Gravitational Wave

Since gravitational waves produce a real, physical effect on nearby particles – they accelerate toward or away from each other as a wave passes – it is clear that the wave must be carrying energy. Imagine two beads on a rough rigid rod. As a gravitational wave passes, the beads will slide toward and away from each other, sliding across the rod. Friction will heat the rod and this energy must have originated from the gravitational wave.

We wish to calculate the energy associated with gravitational radiation. We cannot define a unique local energy of a gravitational wave because the notion of local gravitational energy does not exist in general relativity: one can always transform to a locally inertial frame of reference in which there is no gravitational field, and thus no local energy. Instead we will calculate the amount of energy associated with gravitational radiation in a region of spacetime that is large enough to contain many wavelengths of the radiation but is small compared to any background curvature scale. In essence, we wish to perform an integral average of an effective stress-energy tensor over a volume large enough that the bulk contributions are greater than the boundary contributions (which implies many wavelengths of radiation). The volume must be small enough, however, to remain a normal neighbourhood so that integrals of tensors can be defined.

The Einstein field equations in vacuum are

$$0 = G_{\alpha\beta} = \overset{0}{G}_{\alpha\beta} + \overset{1}{G}_{\alpha\beta}[h_{\mu\nu}] + \overset{2}{G}_{\alpha\beta}[h_{\mu\nu}] + \cdots , \tag{3.72}$$

where $G_{\alpha\beta}$ is the full Einstein tensor of the metric $g_{\alpha\beta} = \eta_{\alpha\beta} + h_{\alpha\beta}$, $\overset{0}{G}_{\alpha\beta} = 0$ is the Einstein tensor of the flat-space background metric $\eta_{\alpha\beta}$ (which vanishes, of course), and

$$\overset{1}{G}_{\alpha\beta}, \overset{2}{G}_{\alpha\beta} ,$$

and so on, are the $O(h)$, $O(h^2)$, and so on, parts of the series expansion of $G_{\alpha\beta}$ in powers of the perturbation $h_{\alpha\beta}$. If we now write

$$g_{\alpha\beta} = \eta_{\alpha\beta} + h_{\alpha\beta} = \eta_{\alpha\beta} + \lambda \overset{1}{h}_{\alpha\beta} + \lambda^2 \overset{2}{h}_{\alpha\beta} + \cdots , \tag{3.73}$$

where λ is a formal order parameter, then we have

$$0 = G_{\alpha\beta} = \lambda \overset{1}{G}_{\alpha\beta}[\overset{1}{h}_{\mu\nu}] + \lambda^2 \left(\overset{1}{G}_{\alpha\beta}[\overset{2}{h}_{\mu\nu}] + \overset{2}{G}_{\alpha\beta}[\overset{1}{h}_{\mu\nu}] \right) + O\left(\lambda^3\right), \tag{3.74}$$

and since this equation must hold order-by-order in λ, we find

$$\overset{1}{G}_{\alpha\beta}[\overset{1}{h}_{\mu\nu}] = 0 \tag{3.75a}$$

$$\overset{1}{G}_{\alpha\beta}[\overset{2}{h}_{\mu\nu}] = -\overset{2}{G}_{\alpha\beta}[\overset{1}{h}_{\mu\nu}] \tag{3.75b}$$

and so on.

These equations give us the first-order ($\overset{1}{h}_{\alpha\beta}$) and second-order ($\overset{2}{h}_{\alpha\beta}$) corrections to the background metric. Until now we have solved Eq. (3.75a) for the first-order metric perturbation, and we have ignored the fact that the field equations only hold up to $O(h^2)$ terms; now we note that the $O(h^2)$ terms that have been dropped in fact form an effective stress-energy tensor that is a source term for the second-order corrections to the metric by virtue of Eq. (3.75b). That is, we can rewrite Eq. (3.75b) as

$$\overset{1}{G}_{\alpha\beta}[\overset{2}{h}_{\mu\nu}] = \frac{8\pi G}{c^4} T^{\rm GW}_{\alpha\beta} \,, \tag{3.76a}$$

where

$$T^{\rm GW}_{\alpha\beta} = -\frac{c^4}{8\pi G} \overset{2}{G}_{\alpha\beta}[\overset{1}{h}_{\mu\nu}] \tag{3.76b}$$

is the effective stress-energy tensor created by the first-order gravitational-wave perturbation. To make $T^{\rm GW}_{\alpha\beta}$ gauge invariant we must perform an integral average over a region of spacetime large enough to contain several gravitational-wave oscillations. Our goal is therefore to calculate

$$T^{\rm GW}_{\alpha\beta} = -\frac{c^4}{8\pi G} \left\langle \overset{2}{R}_{\alpha\beta} - \frac{1}{2}\eta_{\alpha\beta} \overset{2}{R} \right\rangle , \tag{3.77}$$

where $\langle\cdot\rangle$ represents the integral average.

To perform this calculation, we use

$$R_{\alpha\beta} = \frac{\partial \Gamma^{\mu}_{\alpha\beta}}{\partial x^{\mu}} - \frac{\partial \Gamma^{\mu}_{\mu\beta}}{\partial x^{\alpha}} + \Gamma^{\mu}_{\alpha\beta}\Gamma^{\nu}_{\mu\nu} - \Gamma^{\mu}_{\nu\beta}\Gamma^{\nu}_{\alpha\mu} \tag{3.78}$$

and retain the terms quadratic in $h_{\alpha\beta}$ while applying the following tricks: (i) we adopt harmonic coordinates in the TT-gauge so that $h^{\rm TT}_{0\mu} = 0$, $\delta^{ik}\partial h^{\rm TT}_{ij}/\partial x^k = 0$, and $\delta^{ij} h^{\rm TT}_{ij} = 0$; (ii) we assume that under the integral average, all terms of the form $\langle \partial T_{\beta\cdots\gamma}/\partial x^{\alpha}\rangle$ can be neglected since such terms contribute only to the boundary of the region and can be made arbitrarily small compared to the bulk by expanding the region in which the integral average is performed (these are effectively higher-order contributions); and (iii) we use the field equations $\Box h^{\rm TT}_{ij} = 0$. The result is (Isaacson, 1968a, 1968b)

$$T^{\rm GW}_{\alpha\beta} = \frac{c^4}{32\pi G} \left\langle \frac{\partial h^{ij}_{\rm TT}}{\partial x^{\alpha}} \frac{\partial h^{\rm TT}_{ij}}{\partial x^{\beta}} \right\rangle \tag{3.79}$$

(see Problem 3.1).

It is useful to obtain an expression of Eq. (3.79) for a gravitational plane wave. Consider a plane wave travelling in the $z = x^3$-direction so that

$$h^{\rm TT}_{ij} = h^{\rm TT}_{ij}(t - z/c) = h_{+}(t - z/c) e^{+}_{ij} + h_{\times}(t - z/c) e^{\times}_{ij} \,. \tag{3.80}$$

Since $e^{ij}_{+} e^{+}_{ij} = e^{ij}_{\times} e^{\times}_{ij} = 2$ and $e^{ij}_{+} e^{\times}_{ij} = 0$, we find

$$T^{\rm GW}_{00} = -c T^{\rm GW}_{03} = -c T^{\rm GW}_{30} = c^2 T^{\rm GW}_{33} = \frac{c^4}{16\pi G} \left\langle \dot{h}^2_{+} + \dot{h}^2_{\times} \right\rangle \tag{3.81}$$

and all other components vanish.

For a monochromatic plane wave with frequency ω travelling in the $z = x^3$-direction

$$h^{\rm TT}_{ij} = \mathcal{A}_{+} \cos[\omega(t - z/c) + \delta_{+}] e^{+}_{ij} + \mathcal{A}_{\times} \cos[\omega(t - z/c) + \delta_{\times}] e^{\times}_{ij} \tag{3.82}$$

the spacetime average can be taken to be a time average over many cycles, $\langle\cos^2\omega(t - z/c)\rangle = 1/2$, and the gravitational-wave stress-energy tensor takes a particularly simple form:

$$T_{00}^{\text{GW}} = -cT_{03}^{\text{GW}} = -cT_{30}^{\text{GW}} = c^2T_{33}^{\text{GW}} = \frac{c^4}{32\pi G}\omega^2\left(\mathcal{A}_+^2 + \mathcal{A}_\times^2\right) \tag{3.83}$$

with all other components vanishing.

3.3
Production of Gravitational Radiation

Gravitational waves result from the acceleration of masses. Because the waves carry energy and momentum, the bodies that generate the waves experience a back-reaction from the waves. To compute the production of gravitational waves by a dynamical system we need to obtain a far-field expression for the metric perturbation that depends on the motion of the bodies, and we need to obtain the equations of motion for the bodies themself by examining the near-field equations of motion, including the near-field metric perturbation (which might not be small at all!). However, in this chapter, we will presume that the Newtonian equations of motion are good enough for describing the near-zone dynamics, and incorporate the radiation reaction as a secular back-reaction that enforces energy and angular momentum balance. (In the next chapter, we will consider the post-Newtonian equations of motion.)

3.3.1
Far- and Near-Zone Solutions

Until now we have considered the nature of gravitational radiation in the far-field zone, well away from the source of radiation. In the far-field the metric perturbation is radiative and the fields are solutions to the vacuum Einstein equations. Now we turn to the problem of connecting the far-field solution to the near-field solution to Einstein's equations in a non-vacuum spacetime. This will allow us to analyze how gravitational radiation is produced.

As before, we choose harmonic coordinates in which the trace-reversed metric perturbation satisfies $\partial\bar{h}^{\mu\alpha}/\partial x^\mu = 0$. The Einstein field equations are

$$\Box\bar{h}^{\alpha\beta} = -\frac{16\pi G}{c^4}T^{\alpha\beta} + O\left(h^2\right). \tag{3.84}$$

Here $\Box$ is the flat-space d'Alembertian operator and the $O(h^2)$ terms, which are the terms in the Einstein tensor that are second order in the metric perturbation, can be thought of as an additional effective source for the linear metric perturbation – a contribution of the gravitational radiation itself to the overall stress-energy tensor. We define the *effective stress-energy tensor*, $\tau^{\alpha\beta}$, to include both the stress-energy tensor from matter, $T^{\alpha\beta}$, and the $O(h^2)$ terms. The *exact* field equations then read

$$\Box \bar{h}^{\alpha\beta} = -\frac{16\pi G}{c^4} \tau^{\alpha\beta} . \tag{3.85}$$

Also, due to the Lorenz gauge condition, we have

$$0 = \Box \frac{\partial \bar{h}^{\mu\alpha}}{\partial x^\mu} = \frac{\partial}{\partial x^\mu} \Box \bar{h}^{\mu\alpha} = -\frac{16\pi G}{c^4} \frac{\partial \tau^{\mu\alpha}}{\partial x^\mu} \tag{3.86}$$

so

$$\frac{\partial \tau^{\mu\alpha}}{\partial x^\mu} = 0 \tag{3.87}$$

are the *exact* conservation laws in this coordinate system.

Since $\Box$ is the flat-space wave operator, we know the solution to the field equations. It is

$$\bar{h}^{\alpha\beta}(t, \boldsymbol{x}) = \frac{4G}{c^4} \int \frac{\tau^{\alpha\beta}(t - \|\boldsymbol{x} - \boldsymbol{x}'\|/c, \boldsymbol{x}')}{\|\boldsymbol{x} - \boldsymbol{x}'\|} d^3 \boldsymbol{x}' . \tag{3.88}$$

We will solve this equation in both the far-field and near-field zone. We will also assume slow motion of the source.

3.3.1.1 Far Zone

The far-field solution allows us to connect the radiative gravitational field to the dynamics of the source that generated it. The far-field zone is where the distance from the source to the field point r is much greater than the gravitational-wave wavelength λ, which in turn is much greater than the size of the source R:

$$\underset{R}{\text{size of source}} \ll \underset{\lambda}{\text{gravitational-wave wavelength}} \ll \underset{r}{\text{distance to source}} .$$

The far-field approximation means that $\|\boldsymbol{x} - \boldsymbol{x}'\| \simeq r$ is approximately constant over the entire source. The slow motion of the source means that we can approximate $t - \|\boldsymbol{x} - \boldsymbol{x}'\|/c \approx t - r/c$ over the entire source, that is, we neglect the relative retardation effects of one region of the source relative to another. With these approximations, we have

$$\bar{h}^{\alpha\beta}(t, \boldsymbol{x}) \simeq \frac{4G}{c^4 r} \int \tau^{\alpha\beta}(t - r/c, \boldsymbol{x}') \, d^3 \boldsymbol{x}' . \tag{3.89}$$

We will be interested in computing the far-field solution in the spatial, transverse and traceless gauge or TT-gauge. We therefore only need to compute the spatial components of the trace-reversed metric perturbation $\bar{h}^{ij}$. An identity that relates the spatial components τ^{ij} of the effective stress-energy tensor to the component τ^{00},

$$\tau^{ij} = \frac{1}{2} \frac{\partial^2}{\partial t^2} \left(x^i x^j \tau^{00}\right) + \frac{\partial}{\partial x^k} \left(x^i \tau^{jk} + x^j \tau^{ki}\right) - \frac{1}{2} \frac{\partial^2}{\partial x^k \partial x^l} \left(x^i x^j \tau^{kl}\right), \tag{3.90}$$

follows from the conservation laws (see Problem 3.4). With this identity we see that

$$\bar{h}^{ij}(t, \boldsymbol{x}) \simeq \frac{2G}{c^4 r}\frac{\partial^2}{\partial t^2}\int x'^i x'^j \tau^{00}\left(t - r/c, \boldsymbol{x}'\right) d^3\boldsymbol{x}' , \tag{3.91}$$

where the boundary terms generated by the terms in τ^{ij} that involve spatial derivatives are dropped because the boundary can be taken outside of any source matter (so $T^{00} = 0$) and the radiation field is weak (so the $O(h^2)$ contributions to τ^{00} are vanishingly small). We define the quadrupole tensor

$$I^{ij}(t) = \int x^i x^j \tau^{00}(t - r/c, \boldsymbol{x})\, d^3\boldsymbol{x} \tag{3.92}$$

so our solution is

$$\bar{h}^{ij}(t, \boldsymbol{x}) \simeq \frac{2G}{c^4 r}\ddot{I}^{ij}(t - r/c) . \tag{3.93}$$

If we include only the matter component of the effective stress-energy tensor then $\tau^{00} = \rho$ and the quadrupole tensor is the same as the one we introduced earlier, but this definition now includes the contribution from the gravitational field within the source.

To finish the development of our solution, we now want to project into the TT-gauge (for radially travelling waves) using the transverse projection operator

$$P_{ij} = \delta_{ij} - \hat{n}_i \hat{n}_j , \tag{3.94}$$

where $\hat{n}^i = x^i/r$ is the unit vector in the propagation direction (normal to the transverse plane). The solution in the TT-gauge is then

$$h^{\mathrm{TT}}_{ij}(t) \simeq \frac{2G}{c^4 r}\ddot{I}^{\mathrm{TT}}_{ij}(t - r/c) \tag{3.95}$$

with

$$I^{\mathrm{TT}}_{ij} = P_{ik} I^{kl} P_{lj} - \frac{1}{2} P_{ij} P_{kl} I^{kl} . \tag{3.96}$$

(The second term on the right-hand side makes I^{TT}_{ij} trace-free.)

3.3.1.2 **Near Zone**

The near-field solution is needed in order to compute the equations of motion that describe the source dynamics. The near-field zone is where the distance from the source to the field point, r is much smaller than the gravitational-wave wavelength λ but is much larger than the size of the components of the system R:

$$\underset{R}{\text{size of source}} \ll \underset{r}{\text{distance to source}} \ll \underset{\lambda}{\text{gravitational-wave wavelength}} .$$

For example, for two stars of size R orbiting each other with angular velocity ω, the near-field solution is valid when $R \ll r \ll c/\omega$. In the Newtonian limit, we need only compute the Newtonian potential in order to determine the dynamics of the system, hence we seek a solution for

$$\Phi = -\frac{1}{2}c^4 h^{00} = -\frac{1}{4}c^4\left(\bar{h}^{00} + c^{-2}\delta_{ij}\bar{h}^{ij}\right), \tag{3.97}$$

which will be, if we ignore retardation effects within the near-zone,

$$\Phi(t, x) = -G\int \frac{\tau^{00}(t, x') + c^{-2}\delta_{ij}\tau^{ij}(t, x')}{\|x - x'\|} d^3x' . \tag{3.98}$$

Also, since we are working in the Newtonian limit, we will ignore the term involving the internal stresses, $\delta_{ij}\tau^{ij}$, since these will be much smaller than the effective mass energy density $c^2\tau^{00}$. Now we expand $\|x - x'\|^{-1}$ in powers of $1/r$ and find

$$\Phi(t, x) = -G\left[\frac{M}{r} + \frac{D_i x^i}{r^3} + \frac{3}{2}\frac{ƚ_{ij}x^i x^j}{r^5} + \cdots\right], \tag{3.99}$$

where

$$M := \int \tau^{00}(x)d^3x$$
$$D^i := \int x^i\tau^{00}(x)d^3x$$
$$ƚ^{ij} := \int \left(x^i x^j - \frac{1}{3}r^2\delta^{ij}\right)\tau^{00}(x)d^3x . \tag{3.100}$$

For nearly Newtonian gravity we can always choose the origin of coordinates to be at the centre of mass, which does not change (for nearly Newtonian motion), so $D_i = 0$ for all time in centre of mass coordinates. The quadrupolar term, the $1/r^5$ term, can be read off the Newtonian potential in the near zone. Because $ƚ_{ij}^{TT} = I_{ij}^{TT}$, since $ƚ_{ij}$ and I_{ij} only differ by a trace, the far-field solution can be related to this term in the near-zone Newtonian potential:

$$h_{ij}^{TT}(t) \simeq \frac{2G}{c^4 r}\ddot{ƚ}_{ij}^{TT}(t - r/c) \quad \text{(wave zone)}. \tag{3.101}$$

Example 3.7 Order of magnitude estimates of gravitational-wave amplitude

Let M be the mass of a system, R be the size of the system and r be the distance from the system to an observer. The amplitude of gravitational waves is

$$h \sim \frac{G}{c^4}\frac{\ddot{I}}{r} . \tag{3.102}$$

Here $I \sim MR^2$ so $\ddot{I} \sim Mv_{NS}^2 \sim E_{kin}^{NS}$, where v_{NS} is the velocity of the non-spherically symmetric motion of the source and E_{kin}^{NS} is the kinetic energy associated with such motion. This shows that an order of magnitude estimate for the gravitational-wave amplitude can be expressed in terms of the non-spherically symmetric kinetic energy in a system by

$$h \sim \frac{G\left(E_{\text{kin}}^{\text{NS}}/c^2\right)}{c^2 r} . \tag{3.103}$$

Alternatively, if the motion within the body obeys the virial theorem, then $v_{\text{NS}}^2 \sim GM/R \sim -\Phi_{\text{int}}$, the scale of the Newtonian potential within the system, and

$$h \sim \frac{1}{c^4}\Phi_{\text{ext}}\Phi_{\text{int}} , \tag{3.104}$$

where $\Phi_{\text{ext}} = GM/r$ is the external Newtonian potential associated with the source at the location of the observer.

For example, a bar with mass M, length ℓ, spinning with angular frequency ω has $E_{\text{kin}}^{\text{NS}} \sim M\ell^2\omega^2$, so an observer at a distance r would receive a gravitational wave with amplitude

$$h \sim \frac{GM\ell^2\omega^2}{c^4 r} . \tag{3.105}$$

To understand the order of magnitude, suppose typical laboratory measurements are used: let $M = 1\,\text{kg}$, $\ell = 1\,\text{m}$, $\omega = 1\,\text{s}^{-1}$. The observer must be in the wave zone to detect gravitational waves so $r \gg c/\omega$. Therefore

$$h \ll \frac{ML^2\omega^3}{c^5/G} = \frac{1\,\text{W}}{3.63\times 10^{52}\,\text{W}} \sim 10^{-53} . \tag{3.106}$$

This is a very small gravitational perturbation.

Example 3.8 Fourier solution for the gravitational wave

We can solve Eq. (3.88) via a Fourier decomposition of the source term $\tau^{\alpha\beta}$. We begin by taking the Fourier transform of Eq. (3.88):

$$\begin{aligned}\bar{h}^{\alpha\beta}(\omega, \boldsymbol{x}) &= \int_{-\infty}^{\infty} \bar{h}^{\alpha\beta}(t, \boldsymbol{x})\mathrm{e}^{-i\omega t} dt \\ &= \frac{4G}{c^4}\int d^3x' \int_{-\infty}^{\infty} dt\,\mathrm{e}^{-i\omega t}\frac{\tau^{\alpha\beta}(t - \|\boldsymbol{x}-\boldsymbol{x}'\|/c, \boldsymbol{x}')}{\|\boldsymbol{x}-\boldsymbol{x}'\|} \\ &= \frac{4G}{c^4}\int \frac{\tau^{\alpha\beta}(\omega, \boldsymbol{x}')}{\|\boldsymbol{x}-\boldsymbol{x}'\|}\mathrm{e}^{-i\omega\|x-x'\|/c} d^3x' ,\end{aligned} \tag{3.107}$$

where

$$\tau^{\alpha\beta}(\omega, \boldsymbol{x}) := \int_{-\infty}^{\infty} \tau^{\alpha\beta}(t, \boldsymbol{x})\mathrm{e}^{-i\omega t} dt \tag{3.108}$$

is the Fourier transform of the effective stress energy tensor. If we are in the wave-zone then $r = \|\boldsymbol{x}\|$ is much larger than the domain of the spatial integral (the size of the source) so we can make the approximation $\|\boldsymbol{x} - \boldsymbol{x}'\| \simeq r - \hat{\boldsymbol{n}} \cdot \boldsymbol{x}'$ where $\hat{\boldsymbol{n}} = \boldsymbol{x}/r$ is the direction to the observer. Then we have

$$\bar{h}^{\alpha\beta}(\omega, \boldsymbol{x}) = \frac{4G}{c^4 r} e^{-i\omega r/c} \int \tau^{\alpha\beta}(\omega, \boldsymbol{x}') e^{i\omega\hat{\boldsymbol{n}}\cdot\boldsymbol{x}'/c} d^3\boldsymbol{x}' \,. \tag{3.109}$$

The inverse Fourier transform of this equation gives us a formula for the trace-reversed metric perturbation. Using $\omega\hat{\boldsymbol{n}} \cdot \boldsymbol{x}'/c = \boldsymbol{k} \cdot \boldsymbol{x}'$, we obtain

$$\bar{h}^{\alpha\beta}(t, \boldsymbol{x}) = \frac{4G}{c^4 r} \int_{-\infty}^{\infty} \tilde{\tau}^{\alpha\beta}(\omega, \omega\boldsymbol{n}/c) e^{i\omega(t-r/c)} \frac{d\omega}{2\pi} \tag{3.110}$$

with

$$\tilde{\tau}^{\alpha\beta}(\omega, \boldsymbol{k}) := \int \tau^{\alpha\beta}(\omega, \boldsymbol{x}') e^{i\boldsymbol{k}\cdot\boldsymbol{x}'} d^3\boldsymbol{x}' = \int \tau^{\alpha\beta}(t, \boldsymbol{x}') e^{i(\boldsymbol{k}\cdot\boldsymbol{x}'-\omega t)} dt\, d^3\boldsymbol{x}' \tag{3.111}$$

(the Fourier transform of the effective stress energy tensor both in the time coordinate and in the spatial coordinates). Here, Eq. (3.110) gives the expression for the trace-reversed metric perturbation at large r that is travelling in the direction $\hat{\boldsymbol{n}}$.

3.3.2
Gravitational Radiation Luminosity

The amount of gravitational radiation energy dE passing through an area element dA of a sphere of radius r surrounding the gravitational-wave source in a time dt is known as the gravitational-wave *flux*

$$\frac{dE}{dt\,dA} = T_{03}^{\mathrm{GW}} = -c^{-1} T_{00}^{\mathrm{GW}} = -\frac{c^3}{32\pi G} \left\langle \dot{h}_{\mathrm{TT}}^{ij} \dot{h}_{ij}^{\mathrm{TT}} \right\rangle = -\frac{c^3}{16\pi G} \left\langle \dot{h}_+^2 + \dot{h}_\times^2 \right\rangle, \tag{3.112}$$

which can be re-written in terms of the quadrupole tensor using Eq. (3.95) as

$$\frac{dE}{dt\,dA} = -\frac{G}{8\pi c^5 r^2} \left\langle \dddot{I\!\!\!-}_{\mathrm{TT}}^{ij} \dddot{I\!\!\!-}_{ij}^{\mathrm{TT}} \right\rangle, \tag{3.113}$$

where

$$I\!\!\!-_{ij}^{\mathrm{TT}} = \left(P_{ik} P_{jl} - \frac{1}{2} P_{ij} P_{kl} \right) I\!\!\!-^{kl} \,. \tag{3.114}$$

At the source, the energy radiated per unit time into solid angle $d\Omega$ is therefore

$$-\frac{dE}{dt\,d\Omega} = \frac{G}{8\pi c^5}\left\langle\left(P_{ik}P_{jl} - \frac{1}{2}P_{ij}P_{kl}\right)\left(P^i_m P^j_n - \frac{1}{2}P^{ij}P_{mn}\right)\dddot{\mathcal{I}}^{kl}\dddot{\mathcal{I}}^{mn}\right\rangle$$
$$= \frac{1}{4\pi}\frac{G}{c^5}\left\langle\frac{1}{2}\dddot{\mathcal{I}}_{ij}\dddot{\mathcal{I}}^{ij} - \hat{n}_i\hat{n}_j\dddot{\mathcal{I}}^{ik}\dddot{\mathcal{I}}^{j}_{k} + \frac{1}{4}\hat{n}_i\hat{n}_j\hat{n}_k\hat{n}_l\dddot{\mathcal{I}}^{ij}\dddot{\mathcal{I}}^{kl}\right\rangle, \tag{3.115}$$

where we note that $P_{ij}\dddot{\mathcal{I}}^{ij} = -\hat{n}_i\hat{n}_j\dddot{\mathcal{I}}^{ij}$ since $\mathcal{I}^{ij}$ is a trace-free tensor. Now to obtain the gravitational-wave luminosity, we integrate over all solid angles using the identities

$$\frac{1}{4\pi}\int d\Omega = 1 \tag{3.116}$$

$$\frac{1}{4\pi}\int \hat{n}_i\hat{n}_j\,d\Omega = \frac{1}{3}\delta_{ij} \tag{3.117}$$

$$\frac{1}{4\pi}\int \hat{n}_i\hat{n}_j\hat{n}_k\hat{n}_l\,d\Omega = \frac{1}{15}\left(\delta_{ij}\delta_{kl} + \delta_{ik}\delta_{jl} + \delta_{il}\delta_{jk}\right) \tag{3.118}$$

the result is the gravitational-wave luminosity

$$L_{\mathrm{GW}} = -\frac{dE}{dt} = \frac{1}{5}\frac{G}{c^5}\left\langle\dddot{\mathcal{I}}_{ij}\dddot{\mathcal{I}}^{ij}\right\rangle. \tag{3.119}$$

Example 3.9 Order of magnitude estimates of gravitational-wave luminosity

Let M be the mass of a system, R be the size of the system and T be the time-scale of motion within the system. Recall from Example 3.7 that $\ddot{I} \sim E^{\mathrm{NS}}_{\mathrm{kin}}$, the kinetic energy associated with the non-spherically symmetric dynamics. Then $\dddot{I} \sim E^{\mathrm{NS}}_{\mathrm{kin}}/T$ is the power flowing from one side of the system to the other. For some violent burst of energy the associated gravitational-wave luminosity would be

$$L_{\mathrm{GW}} \sim \frac{G}{c^5}\dddot{I}^2 \sim \frac{\left(E^{\mathrm{NS}}_{\mathrm{kin}}/T\right)^2}{c^5/G}\,. \tag{3.120}$$

If the dynamics are quasi-stationary and gravitationally dominated then the luminosity can be estimated as follows. In terms of v_{NS} we have $T \sim R/v_{\mathrm{NS}}$ and so $\dddot{I} \sim (Mv^2_{\mathrm{NS}})/(R/v_{\mathrm{NS}})$. The luminosity is then

$$L_{\mathrm{GW}} \sim \frac{G}{c^5}\dddot{I}^2 \sim \frac{c^5}{G}\left(\frac{GM}{c^2R}\right)^2\left(\frac{v_{\mathrm{NS}}}{c}\right)^6. \tag{3.121}$$

For the one-metre long one-kilogram rotating at one radian per second example in Example 3.7, we would have $v_{\mathrm{NS}} \sim 1\,\mathrm{m\,s^{-1}}$ and $M/R \sim M/\ell = 1\,\mathrm{kg\,m^{-1}}$; therefore the luminosity would be a paltry $L_{\mathrm{GW}} \sim 10^{-53}$ W. The largest value of the luminosity would occur for highly relativistic ($v_{\mathrm{NS}} \sim c$) systems with sizes compared to their Schwarzschild radius for their mass ($R \sim GM/c^2$); such systems would have luminosities approaching the bound c^5/G. It is quite large:

$$\frac{c^5}{G} = 3.63 \times 10^{52}\,\mathrm{W} = 3.63 \times 10^{59}\,\mathrm{erg\,s^{-1}} = 2.03 \times 10^5\, M_\odot c^2\,\mathrm{s^{-1}}\,.$$

If the system is in virial equilibrium then $GM/R \sim v_{\mathrm{NS}}^2$ and

$$L_{\mathrm{GW}} \sim \frac{c^5}{G}\left(\frac{v_{\mathrm{NS}}}{c}\right)^{10}. \qquad (3.122)$$

For example, if a binary neutron star system contains two neutron stars orbiting each other at 10% of the speed of light then the gravitational-wave luminosity from this motion is $L_{\mathrm{GW}} \sim 10^{-10} c^5/G \sim 10^{42}$ W.

Example 3.10 Gravitational-wave spectrum

Recall Example 3.8 where we obtained a solution for the gravitational-wave perturbation in the radiation zone in terms of the Fourier transform of the effective stress tensor. This solution, Eqs. (3.110) and (3.111), was found without assuming that the wavelength λ is much larger than the size R of the radiating system. We can use this formula to express the gravitational-wave flux in terms of the Fourier transform of the effective stress-energy tensor.

We begin by inserting the form of the metric perturbation given by Eq. (3.110) into the formula for the gravitational-wave flux, Eq. (3.112). We obtain

$$\left|\frac{dE}{dt\,dA}\right| = \frac{G}{2\pi c^5}\left\langle \frac{1}{r^2}\int_{-\infty}^{\infty}\frac{d\omega}{2\pi}\int_{-\infty}^{\infty}\frac{d\omega'}{2\pi}\,\omega\omega' \times \tilde{\tau}^{ij*}_{\mathrm{TT}}(\omega,\omega\hat{\boldsymbol{n}}/c)\,\tilde{\tau}^{\mathrm{TT}}_{ij}(\omega',\omega'\hat{\boldsymbol{n}}/c)\,\mathrm{e}^{-i(\omega-\omega')t}\mathrm{e}^{i(\omega-\omega')r/c}\right\rangle. \qquad (3.123)$$

The averaging over several wavelengths depicted by $\langle\cdot\rangle$ can be replaced by a time integral over a duration larger than $1/\omega$; when this is performed, the factor $\exp[-i(\omega-\omega')t]$ becomes $2\pi\delta(\omega-\omega')$, and the integral over ω' becomes trivial. We then have an expression for the energy radiated into solid angle $d\Omega$ in direction $\boldsymbol{n}$:

$$\left|\frac{dE}{d\Omega}\right| = \frac{G}{c^5}\frac{1}{\pi}\int_0^{\infty}\omega^2\tilde{\tau}^{ij*}_{\mathrm{TT}}(\omega,\omega\hat{\boldsymbol{n}}/c)\,\tilde{\tau}^{\mathrm{TT}}_{ij}(\omega,\omega\hat{\boldsymbol{n}}/c)\,\frac{d\omega}{2\pi}\,. \qquad (3.124)$$

Note that we have used the reality of $\tau^{ij}(t,\boldsymbol{x})$ to map the negative frequency components into positive frequency components, so our integral is now over positive frequencies only. Then we can write the gravitational-wave spectrum as

$$\left|\frac{dE}{d\omega\,d\Omega}\right| = \left(p^{ik}p^{jl} - \frac{1}{2}p^{ij}p^{kl}\right)\frac{G}{4\pi^2c^5}\,\omega^2\tilde{\tau}^*_{ij}(\omega,\omega\hat{\boldsymbol{n}}/c)\,\tilde{\tau}_{kl}(\omega,\omega\hat{\boldsymbol{n}}/c)\,, \qquad (3.125)$$

where the spectrum is now over positive and negative frequencies and we have used $\tilde{\tau}^{\mathrm{TT}}_{kl} = (P_{ik}P_{jl} - \frac{1}{2}P_{ij}P_{kl})\tilde{\tau}^{ij}$ to explicitly express the TT-gauge; recall that $P_{ij} = \delta_{ij} - \hat{n}_i\hat{n}_j$.

Example 3.11 Cross-section of a resonant mass detector

Recall Example 3.6 where we discussed a resonant mass detector comprised of a spring connecting two masses. The total cross-section of interaction with a gravitational wave is the ratio of the rate of energy absorbed from the wave into the oscillator to the energy flux in the wave. Recall the gravitational wave was of the form $h_{11} = -h_{22} = h_+ = h\cos\omega t$, so the gravitational-wave flux that is incident on the oscillator is

$$\left|\frac{dE}{dt\,dA}\right| = \frac{c^3}{16\pi G}\langle \dot{h}_+^2\rangle = \frac{c^3}{32\pi G}h^2\omega^2\,, \qquad (3.126)$$

while the wave performs work on the oscillator, $\langle W\rangle = \beta\mu x_{\max}^2\omega^2$ where $x_{\max}$ is the amplitude of the induced oscillations. The total cross-section is the ratio of the work done on the oscillator to the incident flux,

$$\begin{aligned}\sigma &:= \frac{\langle W\rangle}{|dE/dt\,dA|} = \frac{32\pi G}{c^3}\beta\mu(x_{\max}/h)^2\\ &= L^2\frac{8\pi G\mu\beta}{c^3}\frac{\omega^4}{\left(\omega_0^2-\omega^2\right)^2+4\omega^2\beta^2}\,.\end{aligned} \qquad (3.127)$$

The cross-section is the greatest if the incident waves are near the resonance frequency of the oscillator. If $\omega = \omega_0$ then $x_{\max,\mathrm{res}} = \frac{1}{2}hLQ$ (recall $Q := \omega_0/2\beta$) and

$$\sigma_{\mathrm{res}} = L^2\frac{8\pi G\mu\beta}{c^3}Q^2 = L^2\frac{4\pi G\mu\omega_0}{c^3}Q = L^2\frac{8\pi^2 G\mu}{c^2\lambda_0}Q\,, \qquad (3.128)$$

where λ_0 is the wavelength of the gravitational wave. Note that the curvature scale outside a body of size L and mass μ is $\mathcal{R} \sim c/\sqrt{8\pi G(\mu/L^3)}$, so $\sigma_{\mathrm{res}} \sim \sigma_{\mathrm{geom}}(L/\mathcal{R})^2(L/\lambda_0)Q$ where $\sigma_{\mathrm{geom}} \sim \pi L^2$ is the geometric cross-section of the system. If we are in the long-wavelength limit $\lambda_0 \gg L$ and the geometric optics approximation for the gravitational wave $\lambda_0 \ll \mathcal{R}$ is valid then $L \ll \lambda_0 \ll \mathcal{R}$ and we have $\sigma_{\mathrm{res}} \ll \sigma_{\mathrm{geom}}$ unless Q is *extremely* large. That is, we expect that the cross-section σ is much less than the geometric cross-section of the system.

If, however, the quality factor is extremely large, one might be able to have a large cross-section. The quality factor is determined by the damping parameter β, so we would like to reduce the dissipative effects as much as possible. There is a fundamental lower bound to the level of dissipation: the energy loss of the oscillator due to the gravitational radiation it itself produces! The oscillations are in the x-direction only and the only non-vanishing component of the quadrupole tensor I^{ij} is

$$I_{11} = \mu\left[L + x_{\max}\cos(\omega t - \delta)\right]^2 \qquad (3.129)$$

so we find

$$\dddot{I}_{11} = 2\mu L x_{\max}\omega^3 \sin(\omega t - \delta) + O\left(x_{\max}^2\right). \tag{3.130}$$

From Eq. (3.119) we find

$$L_{\mathrm{GW}} = -\frac{dE}{dt} = \frac{4}{15}\frac{G}{c^5}\mu^2 L^2 x_{\max}^2 \omega^6 . \tag{3.131}$$

The decay rate due to gravitational radiation by the oscillator is

$$\Gamma_{\mathrm{GW}} = \frac{L_{\mathrm{GW}}}{\langle E\rangle} = \frac{8}{15}\frac{G}{c^5}\mu L^2 \omega^4 \tag{3.132}$$

near resonance. Since Γ_{GW} is the smallest attainable value of the decay rate $\Gamma = 2\beta$, we express Γ in terms of Γ_{GW} via the quantity $\eta = \Gamma_{\mathrm{GW}}/\Gamma$, which is the fraction of the dissipation that is from energy loss to gravitational radiation. Note that $\eta \leq 1$ (and typically $\eta \ll 1$). In terms of η and Γ we find the total cross-section is

$$\sigma = \eta \frac{15\pi c^2}{2} \frac{\Gamma^2}{\left(\omega_0^2 - \omega^2\right)^2 + \Gamma^2\omega^2} \simeq \eta\lambda_0^2 \frac{15}{8\pi} \frac{\Gamma^2}{4\left(\omega_0 - \omega\right)^2 + \Gamma^2} , \tag{3.133}$$

where the second expression holds for gravitational-wave frequencies near resonance. The cross-section is peaked in the narrow frequency range $\omega_0 - \frac{1}{2}\Gamma \lesssim \omega \lesssim \omega_0 + \frac{1}{2}\Gamma$ so if we average the cross-section over gravitational-wave frequencies in this interval near the resonant frequency we have

$$\bar{\sigma} = \Gamma^{-1} \int_{\omega_0 - \frac{1}{2}\Gamma}^{\omega_0 + \frac{1}{2}\Gamma} \sigma\, d\omega \approx \frac{15}{32}\eta\lambda_0^2 . \tag{3.134}$$

If gravitational radiation is the only dissipation mechanism then $\eta = 1$ and the cross-section is quite large with $\bar{\sigma} \sim \lambda_0^2$. However, η is typically very small, and the cross-section is therefore very small. For example, for the resonant detector described in Example 3.6 with $L \sim 1\,\mathrm{m}$, $\mu \sim 1000\,\mathrm{kg}$, $Q \sim 10^6$, and $\omega_0/2\pi \sim 1\,\mathrm{kHz}$, $\Gamma = \omega_0/Q \sim 10^{-3}\,\mathrm{s}^{-1}$ while $\Gamma_{\mathrm{GW}} \sim 10^{-35}\,\mathrm{s}^{-1}$, so $\eta \sim 10^{-32}$.

3.3.3 Radiation Reaction

Because moving bodies generate gravitational waves that carry away energy, there must be a gravitational *radiation reaction* acting back on the bodies, enforcing energy conservation. The force that a body encounters as a result of its own gravitational field is called its *self-force*, and it turns out that the self-force not only contains a non-conservative piece (which is responsible for the radiation of energy), but also a conservative piece (which changes the motion of the body though it does not produce any radiation). The form of the radiation reaction force is known (Mino *et al.*, 1997; Quinn and Wald, 1997), though its computation is quite difficult.

The existence of a self-force is paradoxical in that the force must accelerate heavier bodies more than lighter bodies (since heavier bodies will produce stronger gravitational radiation), which seems contrary to the equivalence principle (that the gravitational acceleration is independent of a body's mass). In fact, the effect of the self-force ensures that bodies *do* move on geodesics of spacetime – it is simply that they move on geodesics of the spacetime *which includes their own gravitational field*; therefore the equivalence principle does hold.[4)]

Rather than addressing the general problem of finding a radiation reaction force, we restrict attention here to finding a radiation reaction potential that is appropriate to enforce the conservation of energy in weakly gravitating and slowly moving (i.e. nearly Newtonian) systems.

A radiation reaction force $\boldsymbol{F}^{\mathrm{RR}}$ should perform work on a moving body that equals the negative of the power that is radiated from the body in gravitational waves. Therefore, we aim to find a force for which

$$\int \boldsymbol{F}^{\mathrm{RR}} \cdot \boldsymbol{v}\, dt = -\frac{1}{5}\frac{G}{c^5}\int \dddot{\mathcal{I}}_{ij}\dddot{\mathcal{I}}^{ij}\, dt\,, \tag{3.135}$$

where the integral is over several cycles of motion or several gravitational-wave cycles (or, in the case of a single gravitational encounter, for example gravitational radiation from a particle in a hyperbolic orbit about a black hole, then the integral is over the duration of the encounter). We integrate by parts twice and discard the boundary terms to obtain

$$\int \boldsymbol{F}^{\mathrm{RR}} \cdot \boldsymbol{v}\, dt = -\frac{1}{5}\frac{G}{c^5}\int \overset{.....}{\mathcal{I}}_{ij}\, \dot{I}^{ij}\, dt \tag{3.136}$$

(because the first factor in the integrand, $\overset{.....}{\mathcal{I}}_{ij}$, is trace-free, the second factor $\dot{I}^{ij}$ need not be). Now, for a point particle, the quadrupole tensor is $I^{ij} = m x^i x^j$, so $\dot{I}^{ij} = m(x^i v^j + x^j v^i)$. We now have

$$\int F_j^{\mathrm{RR}} v^j\, dt = -\frac{2}{5}\frac{G}{c^5}\int \left(m\, \overset{.....}{\mathcal{I}}_{ij} x^i\right) v^j\, dt \tag{3.137}$$

so we identify

$$F_j^{\mathrm{RR}} = -\frac{2}{5}\frac{G}{c^5} m x^i \frac{d^5 \mathcal{I}_{ij}}{dt^5} \tag{3.138}$$

as our radiation reaction force. This radiation reaction is derived from a radiation reaction potential, Φ^{RR}, as $\boldsymbol{F}^{\mathrm{RR}} = -\boldsymbol{\nabla}\Phi^{\mathrm{RR}}$, with

$$\Phi^{\mathrm{RR}} = \frac{1}{5}\frac{G}{c^5} x^i x^j \frac{d^5 \mathcal{I}_{ij}}{dt^5}\,. \tag{3.139}$$

4) A light body falling into a black hole will have a different trajectory than a heavy body falling into the same black hole, but the two spacetimes are not the same because the heavy body creates a greater perturbation of the black hole spacetime.

This radiation reaction potential is added to the Newtonian potential to obtain the equations of motion for a system which include the effects of energy loss to gravitational waves. Beyond the Newtonian limit, additional potentials must be introduced to describe the radiation reaction (see Blanchet, 1997).

3.3.4 Angular Momentum Carried by Gravitational Radiation

A particle moving under the influence of a radiation reaction potential will lose angular momentum as well as energy, and so the radiation reaction force can be used to obtain a formula for the angular momentum carried by a gravitational wave.

The rate of change of angular momentum $\boldsymbol{J}$ of a particle acted upon by a radiation reaction force $\boldsymbol{F}^{\mathrm{RR}}$ is

$$\frac{d\boldsymbol{J}}{dt} = \boldsymbol{x} \times \boldsymbol{F}^{\mathrm{RR}} , \tag{3.140}$$

where $\boldsymbol{x}$ is the position of the particle. Using our expression for the radiation reaction force, Eq. (3.138), we find that the angular momentum radiated over some period of time containing several gravitational-wave cycles is

$$\Delta J_i = -\frac{2}{5}\frac{G}{c^5}\varepsilon_{ijk}\int m x^j x^l \frac{d^5 \text{I\!\!-}^k_l}{dt^5} dt , \tag{3.141}$$

where ε_{ijk} is the Levi-Civita symbol defined in Eq. (2.114). We note that $m x^j x^l = I^{jl}$ and integrate by parts twice (discarding the boundary terms) to obtain an expression for the rate of angular momentum radiated by a system:

$$\frac{dJ_i}{dt} = -\frac{2}{5}\frac{G}{c^5}\varepsilon_{ijk}\left\langle \ddot{\text{I\!\!-}}^{jl}\dddot{\text{I\!\!-}}^k_l \right\rangle . \tag{3.142}$$

3.4 Demonstration: Rotating Triaxial Ellipsoid

We now illustrate the results of Section 3.3 by applying these results to two important models of sources of gravitational waves. The first demonstration is that of a rotating triaxial ellipsoid, which could model, for example, a rapidly rotating and distorted neutron star. In the next section we will discuss the important example of gravitational-wave production by a binary system of objects.

Consider a triaxial ellipsoidal bar spinning about its $z = x^3$-axis which corresponds to one of the principle axes of inertia. Suppose at $t = 0$ the quadrupole tensor is diagonal, that is, the $x = x^1$-axis and the $y = x^2$-axis also correspond to principle axes of inertia:

$$\mathbf{I} = \begin{bmatrix} I_1 & 0 & 0 \\ 0 & I_2 & 0 \\ 0 & 0 & I_3 \end{bmatrix} . \tag{3.143}$$

Now suppose the bar is spinning with angular velocity ω. Then, at time t the quadrupole tensor will be

$$\begin{aligned}
\mathbf{I}(t) &= \mathbf{R}_3(\omega t)\mathbf{I}_0\mathbf{R}_3^{-1}(\omega t) \\
&= \begin{bmatrix} \cos\omega t & \sin\omega t & 0 \\ -\sin\omega t & \cos\omega t & 0 \\ 0 & 0 & 1 \end{bmatrix}
\begin{bmatrix} I_1 & 0 & 0 \\ 0 & I_2 & 0 \\ 0 & 0 & I_3 \end{bmatrix}
\begin{bmatrix} \cos\omega t & -\sin\omega t & 0 \\ \sin\omega t & \cos\omega t & 0 \\ 0 & 0 & 1 \end{bmatrix} \\
&= \begin{bmatrix} \frac{1}{2}I + \frac{1}{2}\varepsilon I_3\cos 2\omega t & -\frac{1}{2}\varepsilon I_3 \sin 2\omega t & 0 \\ -\frac{1}{2}\varepsilon I_3 \sin 2\omega t & \frac{1}{2}I - \frac{1}{2}\varepsilon I_3 \cos 2\omega t & 0 \\ 0 & 0 & I_3 \end{bmatrix} ,
\end{aligned} \tag{3.144}$$

where $I = I_1 + I_2$ and $\varepsilon = (I_1 - I_2)/I_3$ is known as the *ellipticity*; the matrix $\mathbf{R}_3(\phi)$ is the rotation matrix around the z-axis:

$$\mathbf{R}_3(\phi) := \begin{bmatrix} \cos\phi & \sin\phi & 0 \\ -\sin\phi & \cos\phi & 0 \\ 0 & 0 & 1 \end{bmatrix} . \tag{3.145}$$

Taking two time derivatives of $\mathbf{I}$ we find

$$\ddot{\mathbf{I}}(t) = 2\varepsilon I_3\omega^2 \begin{bmatrix} -\cos 2\omega t & \sin 2\omega t & 0 \\ \sin 2\omega t & \cos 2\omega t & 0 \\ 0 & 0 & 0 \end{bmatrix} . \tag{3.146}$$

Notice that this tensor is already trace-free and transverse with respect to the z-axis. Therefore, for an observer at a distance r from the ellipsoid along the z-axis, the gravitational-wave perturbation has the form

$$h_{ij}^{\mathrm{TT}} = \frac{2G}{c^4 r}\ddot{I}_{ij} = \frac{4G\varepsilon I_3\omega^2}{c^4 r}\begin{bmatrix} -\cos 2\omega t & \sin 2\omega t & 0 \\ \sin 2\omega t & \cos 2\omega t & 0 \\ 0 & 0 & 0 \end{bmatrix} \tag{3.147}$$

from which we see that the two polarizations of radiation are

$$h_+ = -\frac{4G\varepsilon I_3\omega^2}{c^4 r}\cos 2\omega t \quad \text{and} \quad h_\times = \frac{4G\varepsilon I_3\omega^2}{c^4 r}\sin 2\omega t . \tag{3.148}$$

Next we will compute the power radiated by the system. We first take one more time derivative of the quadrupole tensor:

$$\dddot{\mathbf{I}}(t) = 4\varepsilon I_3\omega^3 \begin{bmatrix} \sin 2\omega t & \cos 2\omega t & 0 \\ \cos 2\omega t & -\sin 2\omega t & 0 \\ 0 & 0 & 0 \end{bmatrix} . \tag{3.149}$$

This tensor is already traceless, so

$$\dddot{\bar{I}}_{ij}\dddot{\bar{I}}^{ij} = \dddot{I}_{ij}\dddot{I}^{ij} = 32\varepsilon^2 I_3^2\omega^6 . \tag{3.150}$$

Note that this is constant so the spacetime average in computing the luminosity will be trivial: the luminosity is

$$L_{\mathrm{GW}} = \frac{1}{5}\frac{G}{c^5}\left\langle \dddot{I\!\!\!-}_{ij} \dddot{I\!\!\!-}^{ij} \right\rangle = \frac{32}{5}\frac{G}{c^5}\varepsilon^2 I_3^2 \omega^6 \,. \tag{3.151}$$

Now we compute the angular momentum radiated by the system,

$$\begin{aligned}\frac{dJ_3}{dt} &= -\frac{2}{5}\frac{G}{c^5}\epsilon_{3ij}\left\langle \ddot{I\!\!\!-}^{ik} \dddot{I\!\!\!-}^{j}{}_{k} \right\rangle = -\frac{2}{5}\frac{G}{c^5}\epsilon_{3ij}\left\langle \dddot{I}^{ik} \ddot{I}^{j}{}_{k} \right\rangle \\ &= -\frac{2}{5}\frac{G}{c^5}\left\langle \dddot{I}^{1k}\ddot{I}^{2}{}_{k} - \dddot{I}^{2k}\ddot{I}^{1}{}_{k} \right\rangle \\ &= -\frac{32}{5}\frac{G}{c^5}\varepsilon^2 I_3^2 \omega^5 \\ &= -\frac{L_{\mathrm{GW}}}{\omega} \,.\end{aligned} \tag{3.152}$$

Note that $dE_{\mathrm{GW}} = \omega\, dJ_{\mathrm{GW}}$.

Now we will analyze the polarization of the wave for observers not on the z-axis. For an observer at inclination ι we need to rotate the tensor $\ddot{\mathbf{I}}$ by an angle ι about the x-axis (the *line of nodes*); then we take the transverse-trace-free projection with respect to the new z' axis. We find

$$\begin{aligned}\ddot{\mathbf{I}}' &= \mathbf{R}_1(\iota)\ddot{\mathbf{I}}\mathbf{R}_1^{-1}(\iota) \\ &= 2\varepsilon I_3\omega^2 \begin{bmatrix} 1 & 0 & 0 \\ 0 & \cos\iota & \sin\iota \\ 0 & -\sin\iota & \cos\iota \end{bmatrix} \begin{bmatrix} -\cos 2\omega t & \sin 2\omega t & 0 \\ \sin 2\omega t & \cos 2\omega t & 0 \\ 0 & 0 & 0 \end{bmatrix} \\ &\quad \cdot \begin{bmatrix} 1 & 0 & 0 \\ 0 & \cos\iota & -\sin\iota \\ 0 & \sin\iota & \cos\iota \end{bmatrix} \\ &= 2\varepsilon I_3\omega^2 \begin{bmatrix} -\cos 2\omega t & \cos\iota\sin 2\omega t & -\sin\iota\sin 2\omega t \\ \cos\iota\sin 2\omega t & \cos^2\iota\cos 2\omega t & -\sin\iota\cos\iota\cos 2\omega t \\ -\sin\iota\sin 2\omega t & -\sin\iota\cos\iota\cos 2\omega t & \sin^2\iota\cos 2\omega t \end{bmatrix},\end{aligned} \tag{3.153}$$

where

$$\mathbf{R}_1(\iota) := \begin{bmatrix} 1 & 0 & 0 \\ 0 & \cos\iota & \sin\iota \\ 0 & -\sin\iota & \cos\iota \end{bmatrix}. \tag{3.154}$$

To take the transverse projection, we simply zero-out all entries in the third row as well as all entries in the third column:

$$\ddot{\mathbf{I}}'_{\mathrm{transverse}} = 2\varepsilon I_3\omega^2 \begin{bmatrix} -\cos 2\omega t & \cos\iota\sin 2\omega t & 0 \\ \cos\iota\sin 2\omega t & \cos^2\iota\cos 2\omega t & 0 \\ 0 & 0 & 0 \end{bmatrix}. \tag{3.155}$$

Now we compute the trace of this matrix and subtract it to obtain the transverse-traceless matrix. The trace is

$$\mathrm{Tr}(\ddot{\mathbf{I}}'_{\text{transverse}}) = -2\varepsilon\, I_3\omega^2(1-\cos^2\iota)\cos 2\omega t\ , \tag{3.156}$$

and the transverse traceless quadrupole-moment tensor is

$$\ddot{\mathbf{I}}'^{\mathrm{TT}} = 2\varepsilon\, I_3\omega^2 \begin{bmatrix} -\frac{1}{2}(1+\cos^2\iota)\cos 2\omega t & \cos\iota\sin 2\omega t & 0 \\ \cos\iota\sin 2\omega t & \frac{1}{2}(1+\cos^2\iota)\cos 2\omega t & 0 \\ 0 & 0 & 0 \end{bmatrix}. \tag{3.157}$$

The metric perturbation is

$$\begin{aligned} h'^{\mathrm{TT}}_{ij} &= \frac{2G}{c^4 r}\ddot{I}'^{\mathrm{TT}}_{ij} \\ &= \frac{4G\varepsilon\, I_3\omega^2}{c^4 r} \begin{bmatrix} -\frac{1}{2}(1+\cos^2\iota)\cos 2\omega t & \cos\iota\sin 2\omega t & 0 \\ \cos\iota\sin 2\omega t & \frac{1}{2}(1+\cos^2\iota)\cos 2\omega t & 0 \\ 0 & 0 & 0 \end{bmatrix} \end{aligned} \tag{3.158}$$

from which we see that the two polarizations of radiation are

$$h_+ = -\frac{4G\varepsilon\, I_3\omega^2}{c^4 r}\frac{1+\cos^2\iota}{2}\cos 2\omega t \tag{3.159a}$$

$$h_\times = \frac{4G\varepsilon\, I_3\omega^2}{c^4 r}\cos\iota\sin 2\omega t\ . \tag{3.159b}$$

Note that when $\iota = 0$ or $\iota = \pi$, the gravitational waveform is circularly polarized. When $\iota = \pi/2$ then only the plus-polarization remains and the wave is linearly polarized. Also, if $I_1 = I_2$ then $\varepsilon = 0$ and there is no radiation. This arises when there is axial symmetry about the rotation axis.

Example 3.12 Point particle in rotating reference frame

For a point particle of mass m fixed at a distance a from the origin in a frame of reference that is rotating with angular velocity ω, $I_1 = ma^2$, $I_2 = I_3 = 0$, with $\varepsilon I_3 = I_1 = ma^2$. The luminosity is

$$L_{\mathrm{GW}} = \frac{32}{5}\frac{G}{c^5}m^2a^4\omega^6\ . \tag{3.160}$$

The Earth orbits the Sun at a distance of $a = 1.496\times10^{11}$ m with an angular velocity of 1.991×10^{-7} s. The mass of the Earth is 5.974×10^{24} kg. The gravitational-wave luminosity from the Earth's orbital motion is 196 W.

Example 3.13 The Crab pulsar

The Crab pulsar has the following (approximate) observed physical parameters:

Rotation period	$P = 0.0333\,\mathrm{s}$
Spin-down rate	$\dot{P} = 4.21 \times 10^{-13}$
Mass	$M = 1.4\, M_\odot$
Radius	$R = 10\,\mathrm{km}$
Mass quadrupole	$I_3 \approx \frac{2}{5} M R^2 = 1.1 \times 10^{38}\,\mathrm{m}^2\,\mathrm{kg}$
Inclination of spin axis	$\iota = 62°$
Distance	$r = 2.5\,\mathrm{kpc}$

If the observed spin-down is entirely due to gravitational radiation (which we know is not the case since there is substantial electromagnetic breaking) then what ellipticity is required?

The torque is

$$I_3 \dot{\omega} = \frac{d J_3}{dt} = -\frac{32 G \varepsilon^2 I_3^2 \omega^5}{5 c^5} . \tag{3.161}$$

The angular velocity is $\omega = 2\pi/P$ so $\dot{\omega} = -(2\pi/P^2)\dot{P}$ and

$$\dot{P} = \frac{512\pi^4}{5} \frac{G}{c^5} \varepsilon^2 I_3 P^{-3} . \tag{3.162}$$

Solving for ε we find

$$\varepsilon = \sqrt{\frac{5}{512\pi^4} \frac{c^5}{G} \frac{\dot{P} P^3}{I_3}} \simeq 7.2 \times 10^{-4} . \tag{3.163}$$

If such an ellipticity existed then we would be receiving gravitational waves from the Crab pulsar with amplitudes

$$A_+ = \frac{4 G \varepsilon I_3 \omega^2}{c^4 r} \frac{1 + \cos^2 \iota}{2} = 7.3 \times 10^{-25} \tag{3.164a}$$

$$A_\times = \frac{4 G \varepsilon I_3 \omega^2}{c^4 r} \cos \iota = 5.6 \times 10^{-25} . \tag{3.164b}$$

3.5 Demonstration: Orbiting Binary System

Consider a binary system consisting of two point particles with masses m_1 and m_2 in a circular orbit on the x^1–x^2 plane so that the orbital angular momentum

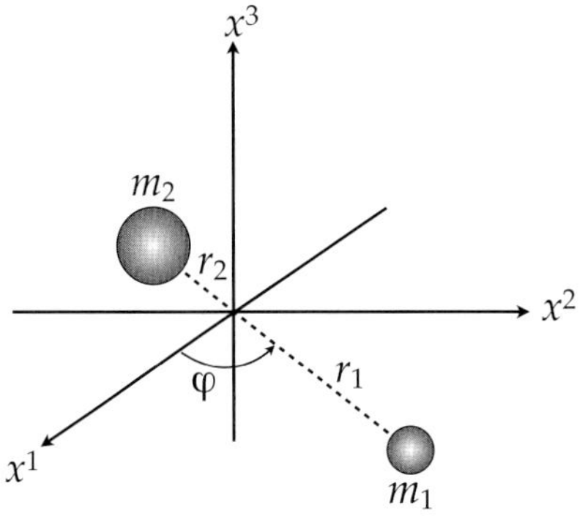

Figure 3.6 A binary system orbits on the x^1–x^2 plane.

vector points along the x^3-axis. In centre-of-mass coordinates, these bodies are at a constant distance r_1 and r_2 from the origin of the coordinates. (See Figure 3.6.) Define the orbital separation $a = r_1 + r_2$, the total mass of the system $M := m_1 + m_2$, and the *reduced mass* of the system $\mu := m_1 m_2/M$. Then, $r_1 = a m_2/M$ and $r_2 = a m_1/M$. The non-vanishing components of the quadrupole tensor I_{ij} are

$$\begin{aligned} I_{11} &= m_1(r_1\cos\varphi)^2 + m_2\left[r_2\cos(\varphi+\pi)\right]^2 \\ &= \mu a^2\cos^2\varphi = \frac{1}{2}\mu a^2(1+\cos 2\varphi) \end{aligned} \tag{3.165a}$$

$$\begin{aligned} I_{22} &= m_1(r_1\sin\varphi)^2 + m_2\left[r_2\sin(\varphi+\pi)\right]^2 \\ &= \mu a^2\sin^2\varphi = \frac{1}{2}\mu a^2(1-\cos 2\varphi) \end{aligned} \tag{3.165b}$$

$$\begin{aligned} I_{12} = I_{21} &= m_1(r_1\cos\varphi)(r_1\sin\varphi) + m_2\left[r_2\cos(\varphi+\pi)\right]\left[r_2\sin(\varphi+\pi)\right] \\ &= \mu a^2\sin\varphi\cos\varphi = \frac{1}{2}\mu a^2\sin 2\varphi \end{aligned} \tag{3.165c}$$

where the orbital phase $\varphi = \omega t$ uniformly increases with time (uniform circular motion).

To compute the metric perturbation, we take two time derivatives:

$$\ddot{I}_{11} = -2\mu a^2\omega^2\cos 2\varphi \tag{3.166a}$$

$$\ddot{I}_{22} = 2\mu a^2\omega^2\cos 2\varphi \tag{3.166b}$$

$$\ddot{I}_{12} = \ddot{I}_{21} = -2\mu a^2\omega^2\sin 2\varphi\,. \tag{3.166c}$$

Then, for an observer at a distance of r on the x^3-axis, this matrix is already in the TT-gauge and

$$h_{ij}^{\mathrm{TT}} = -\frac{4G\mu a^2\omega^2}{c^4 r}\begin{bmatrix}\cos 2\varphi & \sin 2\varphi & 0\\ \sin 2\varphi & -\cos 2\varphi & 0\\ 0 & 0 & 0\end{bmatrix}, \tag{3.167}$$

from which we can read off the two polarizations

$$h_+ = -\frac{4G\mu a^2\omega^2}{c^4 r}\cos 2\varphi \tag{3.168a}$$

$$h_\times = -\frac{4G\mu a^2\omega^2}{c^4 r}\sin 2\varphi\ . \tag{3.168b}$$

Note that we observe a gravitational wave at *twice* the orbital frequency, $f = 2f_{\text{orbital}} = \omega/\pi$. It is convenient to use $v = a\omega$ to eliminate a and ω. From Kepler's third law, $GM = a^3\omega^2$, we have $GM\omega = v^3$. Therefore, v is a surrogate variable for the gravitational-wave frequency,

$$v = (\pi GMf)^{1/3} \tag{3.169a}$$

or the orbital period P,

$$v = \left(\frac{2\pi GM}{P}\right)^{1/3}, \tag{3.169b}$$

or the orbital separation,

$$v = \sqrt{\frac{GM}{a}}\ . \tag{3.169c}$$

In terms of v, an observer on the axis of the binary will see the gravitational-wave polarizations

$$h_+ = -\frac{4G\mu}{c^2 r}\left(\frac{v}{c}\right)^2\cos 2\varphi \tag{3.170a}$$

$$h_\times = -\frac{4G\mu}{c^2 r}\left(\frac{v}{c}\right)^2\sin 2\varphi\ . \tag{3.170b}$$

Note that the orbital phase can also be expressed in terms of v and the time t:

$$\varphi = \omega t = \left(\frac{v}{c}\right)^3\frac{c^3 t}{GM}\ . \tag{3.171}$$

For an observer not on the orbital axis, but at an angle of inclination ι from the axis, the gravitational waveform is

$$h_+ = -\frac{2G\mu}{c^2 r}(1+\cos^2\iota)\left(\frac{v}{c}\right)^2\cos 2\varphi \tag{3.172a}$$

$$h_\times = -\frac{4G\mu}{c^2 r}\cos\iota\left(\frac{v}{c}\right)^2\sin 2\varphi\ . \tag{3.172b}$$

The gravitational wave is monochromatic at twice the orbital frequency – except that the loss of energy from the system will cause the orbit to decay, the frequency to sweep up and the amplitude of the radiation to sweep up too. To describe this we need to compute the energy loss from the system. We need the third time derivative of the quadrupole tensor:

$$\dddot{I}_{11} = -\dddot{I}_{22} = 4\frac{G}{c^5}\frac{\mu}{M}\left(\frac{v}{c}\right)^5 \sin 2\phi \tag{3.173a}$$

$$\dddot{I}_{12} = \dddot{I}_{21} = -4\frac{G}{c^5}\frac{\mu}{M}\left(\frac{v}{c}\right)^5 \cos 2\phi\ . \tag{3.173b}$$

Because the matrix I_{ij} is trace-free, $I_{ij} = \bar{I}_{ij}$ and the gravitational-wave luminosity is

$$L_{\text{GW}} = \frac{1}{5}\frac{G}{c^5}\left\langle \dddot{I}_{11}^2 + \dddot{I}_{22}^2 + 2\dddot{I}_{12}^2 \right\rangle = \frac{32}{5}\frac{c^5}{G}\eta^2\left(\frac{v}{c}\right)^{10}, \tag{3.174}$$

where $\eta = \mu/M$ is the *symmetric mass ratio*.

The energy lost from the system comes from the orbital energy. The (Newtonian) energy of the system is

$$\begin{aligned} E &= \frac{1}{2}m_1 v_1^2 + \frac{1}{2}m_2 v_2^2 - \frac{G m_1 m_2}{a} \\ &= \frac{1}{2}m_1 (r_1\omega)^2 + \frac{1}{2}m_2 (r_2\omega)^2 - \frac{G M\mu}{a} \\ &= \frac{1}{2}m_1 (m_2 a\omega/M)^2 + \frac{1}{2}m_2 (m_1 a\omega/M)^2 - \frac{(a^3\omega^2)\mu}{a} \\ &= \frac{1}{2}\mu a^2\omega^2 - \mu a^2\omega^2 \\ &= -\frac{1}{2}\mu v^2\ . \end{aligned} \tag{3.175}$$

Since $L_{\text{GW}} = -dE/dt$, we find

$$\frac{d(v/c)}{dt} = \frac{32\eta}{5}\frac{c^3}{GM}\left(\frac{v}{c}\right)^9. \tag{3.176}$$

Time until coalescence The time until coalescence, τ_c, starting from orbital velocity $v_0 \ll c$ can be found by directly integrating Eq. (3.176):

$$\int_{v_0/c}^{\infty} \frac{d(v/c)}{(v/c)^9} = \frac{32\eta}{5}\frac{c^3}{GM}\int_0^{\tau_c} dt \tag{3.177}$$

(note that we integrate to $v \to \infty$ rather than some cutoff frequency less than the speed of light for simplicity: the correction is very small). We solve this equation for τ_c and obtain the following forms depending on whether we want to express the coalescence time in terms of the initial velocity v_0, orbital period P_0, orbital separation a_0, or gravitational-wave frequency f_0:

$$\tau_{\mathrm{c}} = \frac{5}{256\eta}\frac{GM}{c^3}\left(\frac{v_0}{c}\right)^{-8} \tag{3.178a}$$

$$\tau_{\mathrm{c}} = \frac{5}{256\eta}\frac{GM}{c^3}\left(\frac{c^3 P_0}{2\pi GM}\right)^{8/3} \tag{3.178b}$$

$$\tau_{\mathrm{c}} = \frac{5}{256\eta}\frac{GM}{c^3}\left(\frac{c^2 a_0}{GM}\right)^{4} \tag{3.178c}$$

$$\tau_{\mathrm{c}} = \frac{5}{256\eta}\frac{GM}{c^3}\left(\frac{\pi GM f_0}{c^3}\right)^{-8/3} . \tag{3.178d}$$

Phase evolution Because the orbit is decaying with time, the gravitational waveform will not be monochromatic and the phase will not increase uniformly with time. In computing the phase evolution it will be helpful to introduce two dimensionless functions, the *energy function* $\mathcal{E}(v)$, and the *flux function* $\mathcal{F}(v)$. The energy function is related to the energy by

$$E(v) - Mc^2 =: Mc^2\mathcal{E}(v) , \tag{3.179}$$

while the flux function is related to the luminosity by

$$L_{\mathrm{GW}}(v) := \frac{c^5}{G}\mathcal{F}(v) . \tag{3.180}$$

Since $L_{\mathrm{GW}} = -dE/dt$, we find

$$\frac{dt}{dv} = -\frac{GM}{c^3}\frac{1}{\mathcal{F}}\frac{d\mathcal{E}}{dv} \tag{3.181}$$

and the time at which the system reaches orbital velocity v can be related to the time of coalescence t_{c} by

$$t(v) = t_{\mathrm{c}} + \frac{GM}{c^3}\int_{v}^{v_{\mathrm{c}}} \frac{1}{\mathcal{F}}\frac{d\mathcal{E}}{dv} dv . \tag{3.182}$$

Here v_{c} is the velocity where coalescence is defined. The phase evolution can now be computed using

$$\frac{d\varphi}{dv} = \frac{d\varphi}{dt}\frac{dt}{dv} = -\omega\frac{GM}{c^3}\frac{1}{\mathcal{F}}\frac{d\mathcal{E}}{dv} = -\left(\frac{v}{c}\right)^3\frac{1}{\mathcal{F}}\frac{d\mathcal{E}}{dv} \tag{3.183}$$

which yields

$$\varphi(v) = \varphi_{\mathrm{c}} + \int_{v}^{v_{\mathrm{c}}}\left(\frac{v}{c}\right)^3\frac{1}{\mathcal{F}}\frac{d\mathcal{E}}{dv} dv . \tag{3.184}$$

We can thus construct the waveform parametrically in terms of $v = (\pi G M f)^{1/3}$ as

$$h_+(t(v)) = -\frac{2G\mu}{c^2 r}(1+\cos^2\iota)\left(\frac{v}{c}\right)^2 \cos 2\varphi(v) \tag{3.185a}$$

$$h_\times(t(v)) = -\frac{4G\mu}{c^2 r}\cos\iota\left(\frac{v}{c}\right)^2 \sin 2\varphi(v) \tag{3.185b}$$

along with Eqs. (3.182) and (3.184). As the orbit decays, the frequency of the gravitational waveform sweeps upwards as does its amplitude. This is known as a *chirp* and such an inspiral waveform is often referred to as a chirp waveform.

Example 3.14 Newtonian chirp

For the Newtonian case, the energy and flux functions are

$$\mathcal{E} = -\frac{1}{2}\eta\left(\frac{v}{c}\right)^2 \quad \text{and} \quad \mathcal{F} = \frac{32}{5}\eta^2\left(\frac{v}{c}\right)^{10}.$$

Using Eq. (3.182) and taking $v_c \to \infty$ we find

$$t(v) = t_c - \frac{5}{256\eta}\frac{GM}{c^3}\left(\frac{v}{c}\right)^{-8}, \tag{3.186}$$

and using Eq. (3.184) we find

$$\varphi(v) = \varphi_c - \frac{1}{32\eta}\left(\frac{v}{c}\right)^{-5}. \tag{3.187}$$

Since the gravitational-wave frequency is $f = v^3/\pi G M$,

$$\begin{aligned}\frac{df}{dt} &= \frac{df}{dv}\frac{dv}{dt} = \left(\frac{3v^2}{\pi G M}\right)\left(\frac{32\eta}{5}\frac{c^4}{GM}\left(\frac{v}{c}\right)^9\right)\\ &= \frac{96}{5}\pi^{8/3}\eta\left(\frac{GM}{c^3}\right)^{5/3} f^{11/3}.\end{aligned} \tag{3.188}$$

This equation depends on the masses of the two companions only in the combination known as the *chirp mass*, $\mathcal{M} = \eta^{3/5}M = \mu^{3/5}M^{2/5} = (m_1 m_2)^{3/5}(m_1+m_2)^{-1/5}$. The gravitational-wave frequency evolves according to

$$\frac{df}{dt} = \frac{96}{5}\pi^{8/3}\left(\frac{G\mathcal{M}}{c^3}\right)^{5/3} f^{11/3}. \tag{3.189}$$

The chirp mass is particularly important because not only does the frequency evolution depend on the companion masses only through the chirp mass, but also the entire gravitational waveforms depend on the chirp mass alone, and not through any other combination of the component masses (though as the binary orbit becomes relativistic, other mass combinations become important). The waveforms are

$$h_+(t) = -\frac{G\mathcal{M}}{c^2 r}\frac{1+\cos^2\iota}{2}\left(\frac{c^3(t_c - t)}{5G\mathcal{M}}\right)^{-1/4}\cos\left[2\varphi_c - 2\left(\frac{c^3(t_c - t)}{5G\mathcal{M}}\right)^{5/8}\right] \tag{3.190a}$$

$$h_\times(t) = -\frac{G\mathcal{M}}{c^2 r}\cos\iota\left(\frac{c^3(t_c - t)}{5G\mathcal{M}}\right)^{-1/4}\sin\left[2\varphi_c - 2\left(\frac{c^3(t_c - t)}{5G\mathcal{M}}\right)^{5/8}\right] . \tag{3.190b}$$

Example 3.15 The Hulse–Taylor binary pulsar

The first binary neutron star system discovered, PSR B1913+16, was discovered by Russell Hulse and Joseph Taylor in 1974. Very precise measurements of the masses of the two neutron stars and their orbital elements system have provided some of the strongest tests of General Relativity. Some of the parameters of PSR B1913+16 are given below:

Pulsar mass	$m_1 = 1.4414\ M_\odot$
Companion mass	$m_2 = 1.3867\ M_\odot$
Orbital period	$P = 0.322\,997\,448\,930$ days
Rate of decrease of orbital period	$-\dot{P} = 75.9\ \mu\text{s}\,\text{yr}^{-1} = 2.405 \times 10^{-12}$
Orbital eccentricity	$e = 0.617\,133\,8$

Under gravitational radiation, General Relativity predicts an orbital decay

$$\frac{dP}{dt} = -\frac{192\pi}{5}\frac{m_1 m_2}{(m_1+m_2)^2}\left(\frac{2\pi G(m_1+m_2)}{c^3 P}\right)^{5/3}\left(\frac{1+\frac{73}{24}e^2+\frac{37}{96}e^4}{(1-e^2)^{7/2}}\right) \tag{3.191}$$

(see Problems 3.5 and 3.9). Given the masses of the neutron stars, m_1 and m_2, the orbital period P and eccentricity e, the General Relativity prediction for the orbital decay is $\dot{P} = 2.402 \times 10^{-12}$, which is in excellent agreement with the observed orbital decay rate (Weisberg and Taylor, 2005).

3.6 Problems

Problem 3.1

Given

$$T_{\alpha\beta}^{\rm GW} = -\frac{c^4}{8\pi G}\left\langle R_{\alpha\beta}^{(2)} - \frac{1}{2}\eta_{\alpha\beta}R^{(2)}\right\rangle + O\left(h^3\right),$$

where $\langle\cdot\rangle$ means an integral average over a spacetime region containing many wavelengths (but small enough so that the region is a normal neighbourhood, that is there is a unique geodesic connecting any two points in the region so that an integral over a tensor can be defined) show that, in a transverse traceless gauge:

$$T_{\alpha\beta}^{\rm GW} = \frac{c^4}{32\pi G}\left\langle\frac{\partial h_{ij}^{\rm TT}}{\partial x^\alpha}\frac{\partial h_{\rm TT}^{ij}}{\partial x^\beta}\right\rangle$$

where the $O(h^3)$ terms are discarded. See Isaacson (1968a, 1968b). Hint: you can integrate by parts to get most terms in the form of divergences and then use the Lorenz gauge to zero these out.

Problem 3.2

In curved spacetime, in the short-wavelength approximation, the stress-energy tensor for gravitational waves, which in flat spacetime is given by Eq. (3.79), is

$$T_{\alpha\beta}^{\rm GW} = \frac{c^4}{32\pi G}\left\langle\left(\overset{0}{\nabla}_\alpha \bar{h}^{\mu\nu}\right)\left(\overset{0}{\nabla}_\beta \bar{h}_{\mu\nu}\right)\right\rangle$$

in a gauge where $\overset{0}{\nabla}_\mu \bar{h}^{\mu\alpha} = 0$ and $\bar{h} = 0$. For the perturbation given by Eq. (3.29), show that the stress-energy tensor is

$$T_{\alpha\beta}^{\rm GW} = \frac{c^4}{32\pi G}\mathcal{A}^2 k_\alpha k_\beta\,.$$

Using Eqs. (3.35) and (3.37), show that this stress-energy tensor is conserved,

$$\overset{0}{\nabla}_\mu T^{\mu\alpha} = 0\,.$$

Suppose an observer is in a locally inertial frame with four-velocity $u^\alpha = [1, 0, 0, 0]$ which satisfies

$$\overset{0}{\nabla}_\alpha u^\beta = 0$$

in a neighbourhood of the observer in this frame. Then show that Eq. (3.37) is a conservation law for the current $j_\alpha = T_{0\alpha}^{\rm GW}$.

Problem 3.3

A gravitational wave encounters a viscous fluid, which is initially at rest with fluid four-velocity given by $u^\alpha = [1, 0, 0, 0]$.

a) The shearing of the fluid is described by the *shear tensor*:

$$\sigma_{\alpha\beta} = \frac{1}{2}\nabla_\alpha u_\beta + \frac{1}{2}\nabla_\beta u_\alpha + \frac{1}{2}u_\alpha u^\mu \nabla_\mu u_\beta + \frac{1}{2}u_\beta u^\mu \nabla_\mu u_\alpha - \frac{1}{3}(g_{\alpha\beta} + u_\alpha u_\beta)\nabla_\mu u^\mu \ . \tag{3.192}$$

Show that the shear caused by the gravitational wave in the TT-gauge is purely spatial with

$$\sigma_{ij} = \frac{1}{2}\frac{\partial}{\partial t} h_{ij}^{\mathrm{TT}} \ . \tag{3.193}$$

b) The shearing of the viscous fluid generates a contribution to the stress-energy tensor of the form

$$T_{\alpha\beta}^{\text{viscosity}} = -2\eta\sigma_{\alpha\beta} \ , \tag{3.194}$$

where η is the *coefficient of viscosity*. The linearized field equations for the gravitational wave are therefore

$$\Box h_{ij}^{\mathrm{TT}} = \eta\frac{16\pi G}{c^4}\frac{\partial}{\partial t} h_{ij}^{\mathrm{TT}} \ . \tag{3.195}$$

Show that a plane wave travelling along the z-axis is attenuated by the fluid by an amount $\mathrm{e}^{-z/\ell}$ where ℓ is the attenuation length scale,

$$\ell = \frac{c^3}{8\pi G\eta} \ . \tag{3.196}$$

c) Ketchup has a coefficient of viscosity of $\eta = 50\,\mathrm{kg\,m^{-1}\,s^{-1}}$. Calculate the distance ℓ that a gravitational wave must travel through ketchup before it is attenuated by a factor of $1/\mathrm{e}$.

Problem 3.4

Beginning with the conservation law $\partial\tau^{\mu\alpha}/\partial x^\mu = 0$, derive the identities

$$\tau^{ij} = \frac{1}{2}\frac{\partial^2}{\partial t^2}\left(x^i x^j \tau^{00}\right) + \frac{\partial}{\partial x^k}\left(x^i \tau^{jk} + x^j \tau^{ki}\right) - \frac{1}{2}\frac{\partial^2}{\partial x^k \partial x^l}\left(x^i x^j \tau^{kl}\right)$$

and

$$\tau^{ij} x^k = \frac{1}{2}\frac{\partial}{\partial t}\left(\tau^{0i} x^j x^k + \tau^{0j} x^i x^k - \tau^{0k} x^i x^j\right) + \frac{1}{2}\frac{\partial}{\partial x^l}\left(\tau^{li} x^j x^k + \tau^{lj} x^i x^k - \tau^{lk} x^i x^j\right).$$

Problem 3.5

Consider a binary system composed of two 1.4 $M_\odot$ neutron stars in a circular orbit.

a) If the orbital period is 7.75 h, how long will it be until the neutron stars collide?
b) Show that the rate of change of the orbital period is

$$\frac{dP}{dt} = -\frac{192\pi}{5}\frac{m_1 m_2}{(m_1+m_2)^2}\left(\frac{2\pi G(m_1+m_2)}{c^3 P}\right)^{5/3}$$

and compute dP/dt in $\mu\mathrm{s\,yr^{-1}}$ when $P = 7.75$ h.
c) How far apart can the companions be if the time to collision is less than 10^{10} yr?
d) How much time remains before collision once the gravitational-wave emission frequency reaches 40 Hz? 100 Hz? (Note: the gravitational-wave frequency is twice the orbital frequency.)

Problem 3.6

For a binary neutron star system with companion masses $m_1 = m_2 = 1.4\,M_\odot$ at a distance $r = 1$ Mpc viewed with inclination $\theta = 0$ (i.e. face-on), plot the functions $h_+(t)$ and $f(t)$ for gravitational-wave frequencies between 100 and 300 Hz.

Problem 3.7

Consider the Newtonian chirp waveform

$$h(t) = \frac{G\mathcal{M}}{c^2 r}\left(\frac{c^3(t_c - t)}{5G\mathcal{M}}\right)^{-1/4}\cos\left[2\varphi_c - 2\left(\frac{c^3(t_c-t)}{5G\mathcal{M}}\right)^{5/8}\right].$$

Compute the Fourier transform of the chirp waveform,

$$\tilde{h}(f) = \int_{-\infty}^{\infty} e^{-2\pi i f t} h(t)\,dt\,,$$

under the stationary phase approximation:

a) Divide the integrand into a slowly varying amplitude factor and an exponential of i times a rapidly varying (except at one point) phase factor. That is, write the integrand in the form $A(t)\exp[i\Phi(t)]$ with $d\ln A/dt \ll d\Phi/dt$ and $d^2\Phi/dt^2 \ll (d\Phi/dt)^2$ (see Problem 3.8 for the validity of these restrictions).
b) Identify the value of t, t_{SP}, at which the phase function becomes stationary.
c) Expand the phase function to quadratic order about t_{SP} (the linear piece vanishes). Shift the integration contour off the real axis and determine the angle that must be traversed over the saddle point in order for the quadratic piece to be negative.
d) Evaluate the amplitude function at t_{SP} and perform the integral over the quadratic parameter in the phase function.

Show that the result is (for $f > 0$):

$$\tilde{h}(f) \approx \left(\frac{5\pi}{24}\right)^{1/2} \frac{G^2 \mathcal{M}^2}{c^5 r} \left(\pi G \mathcal{M} f/c^3\right)^{-7/6} e^{-2\pi i f t_c} e^{2i\varphi_c} e^{-i\Psi(f)}$$

with

$$\Psi(f) = -\frac{\pi}{4} + \frac{3}{128}\left(\pi G \mathcal{M} f/c^3\right)^{-5/3}.$$

Problem 3.8

In the test particle limit ($\mu \ll M$ where μ is the reduced mass and M is the total mass of the system) the inspiral waveform will cut off at a frequency that corresponds to the point at which the orbital separation becomes the *innermost stable circular orbit*, $r = 6GM/c^2$ in Schwarzschild coordinates; at this point, the smaller mass rapidly plunges onto the larger mass. At what gravitational-wave frequency does this occur? Compute the ratios $(d \ln A/dt)/(d\Phi/dt)$ and $(d^2\Phi/dt^2)/(d\Phi/dt)^2$ described in Problem 3.7 and show that these are small for the entire inspiral (so the stationary phase approximation is valid).

Problem 3.9

Consider a binary system with total mass M and reduced mass μ in an eccentric Keplerian orbit with semi-major axis a and orbital eccentricity e.

a) Show, for an observer located at a great distance r on the orbital angular momentum axis, that

$$h_+ = -\frac{G\mu}{c^2 r}\frac{GM}{c^2 a(1-e^2)}(4\cos 2\varphi + 5e\cos\varphi + e\cos 3\varphi)$$

$$h_\times = -\frac{G\mu}{c^2 r}\frac{GM}{c^2 a(1-e^2)}(4\sin 2\varphi + 5e\sin\varphi + e\sin 3\varphi).$$

b) Show that the orbital energy and angular momentum radiated per unit time are

$$-\frac{dE}{dt} = \frac{32}{5}\frac{G^4\mu^2 M^3}{c^5 a^5} f(e) \quad \text{and} \quad -\frac{dL}{dt} = \frac{32}{5}\frac{G^{7/2}\mu^2 M^{5/2}}{c^5 a^{7/2}} g(e),$$

where

$$f(e) = \frac{1 + \frac{73}{24}e^2 + \frac{37}{96}e^4}{(1-e^2)^{7/2}} \quad \text{and} \quad g(e) = \frac{1 + \frac{7}{8}e^2}{(1-e^2)^2}.$$

c) Compute da/dt and de/dt and show that the elliptical orbit is circularized by gravitational-wave emission.

The following relations for Keplerian orbits are useful:

- Equation of an ellipse:

$$r_{12} = \frac{a(1-e^2)}{1+e\cos\varphi} \, .$$

- Kepler's second law (area law):

$$r_{12}^2\dot{\varphi} = \sqrt{GMa(1-e^2)} = L/\mu$$

(the orbital angular momentum per reduced mass). See Peters (1964) and Peters and Mathews (1963).

References

Blanchet, L. (1997) Gravitational radiation reaction and balance equations to post-Newtonian order. *Phys. Rev.*, **D55**, 714–732, doi: 10.1103/PhysRevD.55.714.

Hartle, J.B. (2003) *Gravity: An Introduction to Einstein's General Relativity*, Benjamin Cummings.

Isaacson, R.A. (1968a) Gravitational radiation in the limit of high frequency. i. the linear approximation and geometrical optics. *Phys. Rev.*, **166**(5), 1263–1271, doi: 10.1103/PhysRev.166.1263.

Isaacson, R.A. (1968b) Gravitational radiation in the limit of high frequency. ii. nonlinear terms and the effective stress tensor. *Phys. Rev.*, **166**(5), 1272–1280, doi: 10.1103/PhysRev.166.1272.

Maggiore, M. (2007) *Gravitational Waves, Vol. 1: Theory and Experiments*, Oxford University Press.

Mino, Y., Sasaki, M. and Tanaka, T. (1997) Gravitational radiation reaction to a particle motion. *Phys. Rev.*, **D55**, 3457–3476, doi: 10.1103/PhysRevD.55.3457.

Misner, C.W., Thorne, K.S. and Wheeler, J.A. (1973) *Gravitation*, Freeman, San Francisco.

Peters, P.C. (1964) Gravitational radiation and the motion of two point masses. *Phys. Rev.*, **136**(4B), B1224–B1232, doi: 10.1103/PhysRev.136.B1224.

Peters, P.C. and Mathews, J. (1963) Gravitational radiation from point masses in a keplerian orbit. *Phys. Rev.*, **131**(1), 435–440, doi: 10.1103/PhysRev.131.435.

Quinn, T.C. and Wald, R.M. (1997) An axiomatic approach to electromagnetic and gravitational radiation reaction of particles in curved spacetime. *Phys. Rev.*, **D56**, 3381–3394, doi: 10.1103/PhysRevD.56.3381.

Schutz, B. (2009) *A First Course in General Relativity*, 2nd edn, Cambridge University Press.

Wald, R.M. (1984) *General Relativity*, University of Chicago Press.

Weinberg, S. (1972) *Gravitation and Cosmology: Principles and Applications of the General Theory of Relativity*, John Wiley & Sons.

Weisberg, J. and Taylor, J. (2005) The relativistic binary pulsar b1913+16: Thirty years of observations and analysis, in *Binary Radio Pulsars, ASP Conference Series*, Vol. 328 (eds F. Rasio and I. Stairs), Astronomical Society of the Pacific, San Francisco, ASP Conference Series, pp. 25–32.

4
Beyond the Newtonian Limit

So far we have considered only the Newtonian limit of General Relativity when analyzing the motion of bodies, and only the leading order radiative effects from General Relativity. However, when we study systems that are highly relativistic or strongly gravitating we are required to go beyond this quasi-Newtonian limit. In this chapter we will give an introduction to some of the methods that are used. Post-Newtonian theory is an expansion of the equations of motion of bodies in General Relativity in orders of the velocity of the bodies, and it introduces relativistic corrections to the Newtonian results. Perturbation theory can be used to study gravitational waves in curved spacetimes, such as cosmological or black hole spacetimes, and can allow us to model highly relativistic systems, though only when the perturbation to the background spacetime is small. To fully model complex relativistic systems requires the computational machinery of numerical relativity.

4.1
Post-Newtonian

In linearized gravity, we examined the weak gravitational limit in which the gravitational field is small, $GM/c^2R \ll 1$, where M is the mass of the system and R is its size. Only the first order was kept. This can be thought of as the first term in an expansion in powers of G.

In post-Newtonian theory, the internal motions within the system are taken to be small. If the system dynamics are governed by self-gravitation, the virial theorem suggests that $GM/c^2R \sim v^2/c^2$ so we are interested in an expansion in powers of $1/c^2$. Formally, we will introduce the post-Newtonian order parameter ϵ^2 for our expansion in $1/c^2$.

Our treatment of post-Newtonian theory closely follows Epstein and Wagoner (1975), Weinberg (1972, Chap. 9) and Misner *et al.* (1973, Chap. 39). We will only consider the *first* post-Newtonian order of expansion, though post-Newtonian calculations have now been completed to higher orders. The second post-Newtonian calculation is presented by Will and Wiseman (1996) and a review of the state-of-the-art post-Newtonian calculations is given by Blanchet (2002).

Gravitational-Wave Physics and Astronomy, First Edition. Jolien D. E. Creighton, Warren G. Anderson.

Table 4.1 Post-Newtonian expansion of the metric components. In the Newtonian limit, components up to and including ϵ^2 are required. Post-Newtonian adds terms up to and including ϵ^4; higher-order terms are post-post-Newtonian. The no-gravity (Minkowski spacetime) limit is indicated by the terms highlighted in grey. Linearized gravity introduces the additional terms indicated by boxes; these were computed previously. To obtain the post-Newtonian metric, we must compute the single additional $O(\epsilon^4)$ term that will appear in g_{00}.

Order	Newtonian			Post-Newtonian		Post-post-Newtonian	
	ϵ^0	ϵ^1	ϵ^2	ϵ^3	ϵ^4	ϵ^5	ϵ^6
$g_{00}/c^2 =$	-1		$-2\Phi/c^2$		?		$+O(\epsilon^6)$
$g_{0i}/c =$				A_i/c		$+O(\epsilon^5)$	
$g_{ij} =$			δ_{ij}		$-(2\Phi/c^2)\delta_{ij}$		$+O(\epsilon^6)$

We begin by expanding the metric in the post-Newtonian order parameter to identify the terms that we need to calculate. Consider the line element

$$ds^2 = g_{00}dt^2 + 2\epsilon g_{0i}dt dx^i + \epsilon^2 g_{ij}dx^i dx^j \,. \tag{4.1}$$

Compared to the g_{00} metric component, g_{0i} is $O(1/c)$ and g_{ij} is $O(1/c^2)$, so we have included factors of ϵ to keep count of the relative order of each term. In the Newtonian limit of General Relativity we found $g_{00} = -c^2 - 2\Phi$, $g_{0i} = 0$ and $g_{ij} = \delta_{ij}$, where Φ is the Newtonian gravitational potential. In linear gravity, we obtained corrections to g_{0i} and g_{ij}: $g_{0i} = A_i$ and $g_{ij} = -(2\Phi/c^2)\delta_{ij}$. These terms are formally $O(\epsilon^3)$ and $O(\epsilon^4)$ in the line element. The post-Newtonian metric therefore requires us to obtain the $O(\epsilon^4)$ correction to g_{00}. See Table 4.1.

Now consider the matter and assume, for definiteness, that we have a perfect fluid

$$T^{\alpha\beta} = (\rho + p/c^2)u^\alpha u^\beta + p\, g^{\alpha\beta} \,, \tag{4.2}$$

where $\mathbf{u}$ is the fluid four-velocity, $\mathbf{u} = u^0[1, \boldsymbol{v}]$, which satisfies the normalization

$$-c^2 = g_{\mu\nu}u^\mu u^\nu = (u^0)^2 \left\{\left[-c^2 - 2\Phi + O\left(\epsilon^4\right)\right] + v^2 + O\left(\epsilon^4\right)\right\} \tag{4.3}$$

and so

$$u^0 = 1 - \frac{\Phi}{c^2} + \frac{1}{2}\frac{v^2}{c^2} + O\left(\epsilon^4\right). \tag{4.4}$$

The density ρ contains not only the baryon rest-mass density of the fluid, ρ_0 (i.e. number of baryons in the fluid times the mass per baryon) but also the specific internal energy $\rho_0\Pi$, which includes the energy of compression, the thermal energy, and so on, per unit rest mass: $\rho c^2 := \rho_0(c^2 + \Pi)$. Now, ρ_0 is effectively $O(\epsilon^2)$ since $\nabla^2\Phi = 4\pi G\rho_0$ (from the Newtonian limit) and $\Phi \sim O(\epsilon^2)$. Therefore we

Table 4.2 Post-Newtonian expansion of the stress-energy components. In the Newtonian limit, components up to and including ϵ^2 are required. Notice that we are taking ρ_0 to be $O(\epsilon^2)$ since this is what appears in Newtonian gravitational field equations. Post-Newtonian adds terms up to and including ϵ^4; higher-order terms are post-post-Newtonian. Linearized gravity includes the terms indicated by boxes. For post-Newtonian calculations, the additional $O(\epsilon^4)$ term of T^{00} must be included.

Order	Newtonian	Post-Newtonian		Post-post-Newtonian	
	ϵ^2	ϵ^3	ϵ^4	ϵ^5	ϵ^6
$T^{00} =$	$\boxed{\rho_0}$		$+\rho_0(\Pi + v^2 - 2\Phi)/c^2$		$+O(\epsilon^6)$
$T^{0i} =$		$\boxed{\rho_0 v^i}$		$+O(\epsilon^5)$	
$T_{ij} =$			$\boxed{\rho_0 v^i v^j + p\delta^{ij}}$		$+O(\epsilon^6)$

have the components of the stress-energy, expanded in the post-Newtonian order parameter:

$$T^{00} = \rho(u^0)^2$$
$$= \rho_0\left[1 + \left(\frac{\Pi}{c^2} - \frac{2\Phi}{c^2} + \frac{v^2}{c^2}\right) + O\left(\epsilon^4\right)\right] \tag{4.5a}$$

$$T^{0i} = \rho c (u^0)^2 (v^i/c)\left[1 + p/(\rho c^2)\right]$$
$$= \rho_0 c\left[\frac{v^i}{c} + O(\epsilon^3)\right] \tag{4.5b}$$

$$T^{ij} = \rho c^2 (u^0)^2 (v^i v^j)/c^2\left[1 + p/(\rho c^2)\right] + p g^{ij}$$
$$= \rho_0 c^2\left[\left(\frac{v^i}{c}\frac{v^j}{c} + \frac{p}{\rho_0 c^2}\delta^{ij}\right) + O\left(\epsilon^4\right)\right]. \tag{4.5c}$$

Table 4.2 illustrates the expansion of the components of the stress-energy tensor in terms of the post-Newtonian order parameter.

We wish to (i) compute the post-Newtonian equations of motion for the body, which will describe the dynamics of the system, and (ii) compute the post-Newtonian radiation in the wave zone. The starting point for both tasks are the *exact* field equations for the metric perturbation in harmonic coordinates

$$\Box \bar{h}^{\alpha\beta} = -\frac{16\pi G}{c^4}\tau^{\alpha\beta}\,, \tag{4.6}$$

where we recall that $\Box$ is the flat-space d'Alembertian operator, $\tau^{\alpha\beta}$ is the effective stress-energy tensor which includes both the matter-stress-energy tensor, $T^{\alpha\beta}$, plus $O(h^2)$ corrections which give rise to a gravitational contribution to the stress-energy tensor $t^{\alpha\beta}$:

$$\tau^{\alpha\beta} := T^{\alpha\beta} + t^{\alpha\beta}\,. \tag{4.7}$$

Example 4.1 Effective stress-energy tensor

The effective stress-energy tensor contains the matter stress energy $T_{\alpha\beta}$ as well as the stress energy associated with the gravitational field $t_{\alpha\beta}$, which is composed of terms that are $O(h^2)$. The latter is defined via Eq. (4.6) as

$$\frac{16\pi G}{c^4} t_{\alpha\beta} := 2G_{\alpha\beta} - \Box \bar{h}_{\alpha\beta} \tag{4.8}$$

in the Lorenz gauge, and can be obtained by computing the Einstein tensor $G_{\alpha\beta}$ in terms of the trace-reversed metric perturbation $\bar{h}_{\alpha\beta}$. The result is

$$\begin{aligned}
\frac{16\pi G}{c^4} t_{\alpha\beta} = & -\frac{1}{2}\frac{\partial \bar{h}_{\mu\nu}}{\partial x^\alpha}\frac{\partial \bar{h}^{\mu\nu}}{\partial x^\beta} \\
& - \bar{h}^{\mu\nu}\left(\frac{\partial^2 \bar{h}_{\mu\nu}}{\partial x^\alpha \partial x^\beta} + \frac{\partial^2 \bar{h}_{\alpha\beta}}{\partial x^\mu \partial x^\nu} - \frac{\partial^2 \bar{h}_{\mu\alpha}}{\partial x^\beta \partial x^\nu} - \frac{\partial^2 \bar{h}_{\mu\beta}}{\partial x^\alpha \partial x^\nu}\right) \\
& + \frac{1}{2}\bar{h}\frac{\partial^2 \bar{h}}{\partial x^\alpha \partial x^\beta} + \frac{1}{4}\frac{\partial \bar{h}}{\partial x^\alpha}\frac{\partial \bar{h}}{\partial x^\beta} + \frac{3}{2}\bar{h}\Box\bar{h}_{\alpha\beta} + \frac{1}{2}\bar{h}_{\alpha\beta}\Box\bar{h} \\
& + \eta^{\mu\nu}\left[\frac{\partial \bar{h}}{\partial x^\mu}\left(\frac{\partial \bar{h}_{\alpha\beta}}{\partial x^\nu} - \frac{1}{2}\frac{\partial \bar{h}_{\nu\alpha}}{\partial x^\beta} - \frac{1}{2}\frac{\partial \bar{h}_{\nu\beta}}{\partial x^\alpha}\right)\right. \\
& \left. - \frac{\partial \bar{h}^\sigma_\alpha}{\partial x^\mu}\frac{\partial \bar{h}_{\beta\sigma}}{\partial x^\nu} + \frac{\partial \bar{h}^\sigma_\alpha}{\partial x^\mu}\frac{\partial \bar{h}_{\beta\nu}}{\partial x^\sigma}\right] \\
& - \frac{1}{2}\bar{h}^\mu_\alpha\frac{\partial^2 \bar{h}}{\partial x^\beta \partial x^\mu} - \frac{1}{2}\bar{h}^\mu_\beta\frac{\partial^2 \bar{h}}{\partial x^\alpha \partial x^\mu} - \bar{h}^\mu_\alpha\Box\bar{h}_{\mu\beta} - \bar{h}^\mu_\beta\Box\bar{h}_{\mu\alpha} \\
& + \frac{1}{2}\eta_{\alpha\beta}\left(\frac{3}{2}\eta^{\mu\nu}\frac{\partial \bar{h}_{\rho\sigma}}{\partial x^\mu}\frac{\partial \bar{h}^{\rho\sigma}}{\partial x^\nu} - \frac{3}{4}\eta^{\mu\nu}\frac{\partial \bar{h}}{\partial x^\mu}\frac{\partial \bar{h}}{\partial x^\nu} + 2\bar{h}^{\mu\nu}\Box\bar{h}_{\mu\nu}\right. \\
& \left. - \bar{h}\Box\bar{h} - \frac{\partial \bar{h}^\nu_\mu}{\partial x^\sigma}\frac{\partial \bar{h}^{\mu\sigma}}{\partial x^\nu} + \bar{h}^{\mu\nu}\frac{\partial^2 \bar{h}}{\partial x^\mu \partial x^\nu}\right) + O\left(h^3\right).
\end{aligned} \tag{4.9}$$

The Lorenz gauge conditions have been used to simplify this result.

The near-zone solution to the field equations in linear theory, which neglected $t^{\alpha\beta}$, was

$$\bar{h}_{00}/c^2 = -\frac{4\Phi}{c^2} + O\left(\epsilon^4\right), \quad \bar{h}_{0i}/c = \frac{A_i}{c} + O\left(\epsilon^5\right), \quad \bar{h}_{ij} = O\left(\epsilon^4\right)$$

and recall that we need the $O(\epsilon^4)$ term of h^{00} in order to have a complete post-Newtonian metric (to compute the equations of motion). To obtain h^{00} we need to solve

$$\begin{aligned}
\Box h^{00} &= \Box\left(\bar{h}^{00} - \frac{1}{2}\eta^{00}\bar{h}\right) = \frac{1}{2}\Box\bar{h}^{00} + \frac{1}{2}c^{-2}\delta_{ij}\Box\bar{h}^{ij} \\
&= -\frac{8\pi G}{c^4}\left(\tau^{00} + c^{-2}\delta_{ij}\tau^{ij}\right).
\end{aligned} \tag{4.10}$$

We already know the required terms of the matter-stress-energy tensor T^{00} and T^{ij} to first post-Newtonian order; we now need to compute the gravitational contribution to the effective stress-energy tensor. It is

$$16\pi G t_{\alpha\beta} = -4\frac{\partial\Phi}{\partial x^\alpha}\frac{\partial\Phi}{\partial x^\beta} - 8\Phi\frac{\partial^2\Phi}{\partial x^\alpha \partial x^\beta} + \eta_{\mu\nu}\left(8\Phi\nabla^2\Phi + 6(\nabla\Phi)\cdot(\nabla\Phi)\right) \tag{4.11}$$

and so we have

$$\tau^{00} = \rho_0\left(1 + \frac{\Pi}{c^2} + \frac{v^2}{c^2} - \frac{4\Phi}{c^2}\right) - \frac{3}{8\pi G c^2}(\nabla\Phi)\cdot(\nabla\Phi) \tag{4.12}$$

and

$$\begin{aligned}\tau^{ij} &= \rho_0 v^i v^j + p\delta^{ij} \\ &\quad - \frac{1}{4\pi G}\frac{\partial\Phi}{\partial x_i}\frac{\partial\Phi}{\partial x_j} - \frac{1}{2\pi G}\Phi\frac{\partial^2\Phi}{\partial x_i \partial x_j} \\ &\quad + \frac{1}{8\pi G}\delta^{ij}\left(4\Phi\nabla^2\Phi + 3(\nabla\Phi)\cdot(\nabla\Phi)\right).\end{aligned} \tag{4.13}$$

We then have

$$\begin{aligned}\Box h^{00} &= -\frac{8\pi G}{c^4}\rho_0\left(1 + \frac{\Pi}{c^2} + 2\frac{v^2}{c^2} - \frac{4\Phi}{c^2} + 3\frac{p}{\rho_0 c^2}\right) \\ &\quad - \frac{1}{c^6}\left(8\Phi\nabla^2\Phi + 4(\nabla\Phi)\cdot(\nabla\Phi)\right)\end{aligned} \tag{4.14}$$

and, since the Newtonian potential satisfies $\nabla^2\Phi = 4\pi G\rho_0$,

$$\begin{aligned}\Box h^{00} &= -\frac{1}{c^2}\frac{\partial^2 h^{00}}{\partial t^2} + \nabla^2 h^{00} \\ &= -\frac{8\pi G}{c^4}\rho_0\left(1 + \frac{\Pi}{c^2} + 2\frac{v^2}{c^2} + 3\frac{p}{\rho_0 c^2}\right) - \frac{4}{c^6}(\nabla\Phi)\cdot(\nabla\Phi).\end{aligned} \tag{4.15}$$

This equation can be written in terms of three potentials, the Newtonian potential Φ, a post-Newtonian potential Ψ, and a *super-potential* χ:

$$h^{00} = \frac{1}{c^2}\left\{-\frac{2\Phi}{c^2} - 2\left(\frac{\Phi}{c^2}\right)^2 + \frac{4\Psi}{c^2} - \frac{1}{c^4}\frac{\partial^2\chi}{\partial t^2}\right\}, \tag{4.16}$$

where the potentials satisfy

$$\nabla^2\Phi = 4\pi G\rho_0 \tag{4.17a}$$

$$\nabla^2\Psi = -4\pi G\rho_0\left(\frac{1}{2}\frac{\Pi}{c^2} + \frac{v^2}{c^2} - \frac{\Phi}{c^2} + \frac{3}{2}\frac{p}{\rho_0 c^2}\right) \tag{4.17b}$$

$$\nabla^2\chi = 2\Phi \tag{4.17c}$$

and have solutions

$$\Phi(\boldsymbol{x}, t) = -G \int \frac{\rho_0(\boldsymbol{x}', t)}{\|\boldsymbol{x} - \boldsymbol{x}'\|} d^3\boldsymbol{x}' \tag{4.18a}$$

$$\Psi(\boldsymbol{x}, t) = +G \int \frac{\rho_0(\boldsymbol{x}', t)}{\|\boldsymbol{x} - \boldsymbol{x}'\|} \times \left(\frac{1}{2} \frac{\Pi(\boldsymbol{x}', t)}{c^2} + \frac{v^2}{c^2} - \frac{\Phi(\boldsymbol{x}', t)}{c^2} + \frac{3}{2} \frac{p(\boldsymbol{x}', t)}{\rho_0(\boldsymbol{x}', t) c^2} \right) d^3\boldsymbol{x}' \tag{4.18b}$$

$$\chi(\boldsymbol{x}, t) = -G \int \rho_0(\boldsymbol{x}', t) \|\boldsymbol{x} - \boldsymbol{x}'\| \, d^3\boldsymbol{x}' . \tag{4.18c}$$

At this stage, the near-zone metric $g_{\alpha\beta}$ is known to first post-Newtonian order and the equations of motion $\nabla_\mu T^{\mu\nu} = 0$ can be found.

We next consider the wave-zone (far-field) metric perturbation. As with our Newtonian calculation in the previous chapter, we have

$$\bar{h}^{\alpha\beta}(t, \boldsymbol{x}) = \frac{4G}{c^4} \int \frac{\tau^{\alpha\beta}(t - \|\boldsymbol{x} - \boldsymbol{x}'\|/c), \boldsymbol{x}')}{\|\boldsymbol{x} - \boldsymbol{x}'\|} d^3\boldsymbol{x}' . \tag{4.19}$$

As before, we can express $\|\boldsymbol{x} - \boldsymbol{x}'\|^{-1} \approx r^{-1}$, but now we need to consider the retardation effects within the source integration. We specialize to the TT-gauge where we must compute

$$h^{ij}_{\mathrm{TT}}(t, \boldsymbol{x}) = \frac{4G}{c^4 r} \int \tau^{ij}_{\mathrm{TT}} \left(t - r/c + \hat{\boldsymbol{n}} \cdot \boldsymbol{x}'/c, \boldsymbol{x}'\right) d^3\boldsymbol{x}' \tag{4.20}$$

which we then expand in multipoles

$$h^{ij}_{\mathrm{TT}}(t, \boldsymbol{x}) = \frac{4G}{c^4 r} \sum_{m=0}^{\infty} \frac{1}{m!} \frac{\partial^m}{\partial t^m} \int \tau^{ij}_{\mathrm{TT}} \left(t - r/c, \boldsymbol{x}'\right) \left(\hat{\boldsymbol{n}} \cdot \boldsymbol{x}'/c\right)^m d^3\boldsymbol{x}' , \tag{4.21}$$

where $\hat{\boldsymbol{n}} = \boldsymbol{x}/r$. Notice that each higher multiple moment introduces an additional factor of $1/c$. The quadrupole and octupole moments can be simplified using the two identities

$$\tau^{ij} = \frac{1}{2} \frac{\partial^2}{\partial t^2} \left(x^i x^j \tau^{00}\right) + \frac{\partial}{\partial x^k} \left(x^i \tau^{jk} + x^j \tau^{ki}\right) - \frac{1}{2} \frac{\partial^2}{\partial x^k \partial x^l} \left(x^i x^j \tau^{kl}\right) \tag{4.22a}$$

and

$$\tau^{ij} x^k = \frac{1}{2} \frac{\partial}{\partial t} \left(\tau^{0i} x^j x^k + \tau^{0j} x^i x^k - \tau^{0k} x^i x^j\right) + \frac{1}{2} \frac{\partial}{\partial x^l} \left(\tau^{li} x^j x^k + \tau^{lj} x^i x^k - \tau^{lk} x^i x^j\right) \tag{4.22b}$$

which both follow from the conservation of the effective stress-energy tensor

$$\frac{\partial \tau^{\mu\alpha}}{\partial x^\mu} = 0 \tag{4.23}$$

(see Problem 3.4) and we obtain

$$h_{\mathrm{TT}}^{ij} = \frac{2G}{c^4 r}\left[\frac{\partial^2}{\partial t^2}\sum_{m=0}^{\infty}\hat{n}_{k_1}\hat{n}_{k_2}\cdots\hat{n}_{k_m} I^{ijk_1k_2\cdots k_m}(t - r/c)\right]_{\mathrm{TT}}, \tag{4.24}$$

where

$$I^{ij}(t) := \int \tau^{00}(t, \boldsymbol{x}) x^i x^j \, d^3\boldsymbol{x} \tag{4.25a}$$

$$I^{ijk}(t) := \int \left[\tau^{0i}(t, \boldsymbol{x}) x^j x^k + \tau^{0j}(t, \boldsymbol{x}) x^i x^k - \tau^{0k}(t, \boldsymbol{x}) x^i x^j\right] d^3\boldsymbol{x} \tag{4.25b}$$

$$I^{ijk_1k_2\cdots k_m}(t) := \frac{2}{m!}\frac{\partial^{m-2}}{\partial t^{m-2}}\int \tau^{ij}(t, \boldsymbol{x}) x^{k_1} x^{k_2}\cdots x^{k_m} \, d^3\boldsymbol{x} \quad (m \geq 2) . \tag{4.25c}$$

(We have ignored surface terms of the integral since the wave zone is in a region with no source matter and very weak fields.) This gives us the metric perturbation in the wave zone.

The energy loss can also be computed:

$$\begin{aligned} -\frac{dE}{dt\,d\Omega} &= \frac{c^3 r^2}{32\pi G}\left\langle \dot{h}_{\mathrm{TT}}^{ij}\dot{h}_{ij}^{\mathrm{TT}}\right\rangle \\ &= \frac{G}{8\pi c^5}\left\langle \sum_{p=0}^{\infty}\sum_{q=0}^{\infty}\hat{n}_{k_1}\cdots\hat{n}_{k_p}\hat{n}^{l_1}\cdots\hat{n}^{l_q} \right. \\ &\quad \times\left(\dddot{I\!\!\!-}^{ijk_1\cdots k_p}\dddot{I\!\!\!-}_{ijl_1\cdots l_q} - 2\hat{n}_i\hat{n}^j\dddot{I\!\!\!-}^{imk_1\cdots k_p}\dddot{I\!\!\!-}_{jml_1\cdots l_q}\right. \\ &\quad \left.\left. + \frac{1}{2}\hat{n}_i\hat{n}_j\hat{n}^m\hat{n}^n\dddot{I\!\!\!-}^{ijk_1\cdots k_p}\dddot{I\!\!\!-}_{mnl_1\cdots l_q}\right)\right\rangle, \end{aligned} \tag{4.26}$$

where

$$I\!\!\!-^{ijk_1\cdots k_p} := I^{ijk_1\cdots k_p} - \frac{1}{3}\delta^{ij}\delta_{mn}I^{mnk_1\cdots k_p} . \tag{4.27}$$

Next we integrate over the solid angle and use the identity

$$\frac{1}{4\pi}\int n_{k_1}\cdots n_{k_m}\,d\Omega = \begin{cases} \frac{(\delta_{k_1k_2}\cdots\delta_{k_{m-1}k_m} + \text{permutations})}{(m+1)!!} & m \text{ even} \\ 0 & m \text{ odd.} \end{cases} \tag{4.28}$$

The result is

$$\begin{aligned} L_{\mathrm{GW}} &= \frac{1}{5}\frac{G}{c^5}\left\langle \dddot{I\!\!\!-}^{ij}\dddot{I\!\!\!-}_{ij}\right\rangle \\ &+ \frac{1}{105}\frac{G}{c^5}\left\langle 11\ddddot{I\!\!\!-}^{ijk}\ddddot{I\!\!\!-}_{ijk} - 6\delta_{jk}\delta^{lm}\ddddot{I\!\!\!-}^{ijk}\ddddot{I\!\!\!-}_{ilm} - 6\ddddot{I\!\!\!-}^{ijk}\ddddot{I\!\!\!-}_{ikj}\right. \\ &\left. + 22\delta^{kl}\ddddot{I\!\!\!-}^{ij}\ddddot{I\!\!\!-}_{ijkl} - 24\delta^{kl}\ddddot{I\!\!\!-}^{ij}\ddddot{I\!\!\!-}_{iklj}\right\rangle + \text{higher-order terms.} \end{aligned} \tag{4.29}$$

4.1.1
System of Point Particles

We now consider a specific application of the post-Newtonian formalism to a gravitating system of N particles (e.g. the planets in the Solar System or two orbiting black holes). In this section we follow Wagoner and Will (1976) (see also Misner *et al.*, 1973, Chap. 39).

For a system of point particles the matter-stress-energy tensor is distributional:

$$T^{00} = \sum_A m_A \left\{1 + \frac{1}{2}\frac{v_A^2}{c^2} - \sum_{B \neq A} \frac{G m_B}{c^2 r_{AB}}\right\} \delta^3\left(\boldsymbol{x} - \boldsymbol{x}_A(t)\right) \tag{4.30a}$$

$$T^{0j} = \sum_A m_A v_A^j \delta^3\left(\boldsymbol{x} - \boldsymbol{x}_A(t)\right) \tag{4.30b}$$

$$T^{ij} = \sum_A m_A v_A^i v_A^j \delta^3\left(\boldsymbol{x} - \boldsymbol{x}_A(t)\right), \tag{4.30c}$$

where m_A is the mass of particle A, $\boldsymbol{x}_A$ is its position, $\boldsymbol{v}_A$ is its velocity, and $r_{AB} := \|\boldsymbol{x}_A - \boldsymbol{x}_B\|$ is the distance between particles A and B. The effective stress-energy tensor is

$$\tau^{00} = \sum_A m_A \left\{1 + \frac{1}{2}\frac{v_A^2}{c^2} - \sum_{B \neq A} \frac{G m_B}{c^2 r_{AB}}\right\} \delta^3\left(\boldsymbol{x} - \boldsymbol{x}_A(t)\right) - \frac{1}{8\pi G c^2}\left[4\Phi \nabla^2 \Phi + 3(\nabla\Phi)\cdot(\nabla\Phi)\right] \tag{4.31a}$$

$$\tau^{0j} = \sum_A m_A v_A^j \delta^3\left(\boldsymbol{x} - \boldsymbol{x}_A(t)\right) \tag{4.31b}$$

$$\tau^{ij} = \sum_A m_A v_A^i v_A^j \delta^3\left(\boldsymbol{x} - \boldsymbol{x}_A(t)\right) + \frac{1}{8\pi G}\left\{-2\frac{\partial\Phi}{\partial x_i}\frac{\partial\Phi}{\partial x_j} - 4\Phi\frac{\partial^2\Phi}{\partial x_i \partial x_j} + \delta^{ij}\left[4\Phi\nabla^2\Phi + 3(\nabla\Phi)\cdot(\nabla\Phi)\right]\right\}, \tag{4.31c}$$

where

$$\Phi(\boldsymbol{x}, t) = -\sum_B \frac{G m_B}{\|\boldsymbol{x} - \boldsymbol{x}_B(t)\|}\,. \tag{4.32}$$

4.1.1.1 Post-Newtonian Metric

To compute the post-Newtonian equations of motion for the system of particles, we need to compute the additional potentials Ψ and χ. These are

$$\Psi(\boldsymbol{x}, t) = \frac{1}{4}\sum_A \frac{G m_A}{\|\boldsymbol{x} - \boldsymbol{x}_A(t)\|}\left(3\frac{v_A^2}{c^2} - 2\sum_{B \neq A}\frac{G m_B}{c^2 r_{AB}}\right) \tag{4.33}$$

and

$$\chi(\boldsymbol{x}, t) = -G \sum_A m_A \|\boldsymbol{x} - \boldsymbol{x}_A(t)\| . \tag{4.34}$$

The metric is

$$\begin{aligned} g_{00} &= -c^2 + h_{00} = -c^2 - 2\Phi - 2\frac{\Phi^2}{c^2} + \frac{4\Psi}{c^2} - \frac{1}{c^2}\frac{\partial^2 \chi}{\partial t^2} + O\left(\epsilon^6\right) \\ &= -c^2 + 2\sum_A \frac{G m_A}{r_A} - 2\frac{1}{c^2}\left(\sum_A \frac{G m_A}{r_A}\right)^2 \\ &\quad + 3\sum_A \frac{G m_A}{r_A}\frac{v_A^2}{c^2} - 2\sum_A \sum_{B\neq A} \frac{G m_A}{r_A}\frac{G m_B}{c^2 r_{AB}} - \frac{1}{c^2}\frac{\partial^2 \chi}{\partial t^2} + O\left(\epsilon^6\right) \end{aligned} \tag{4.35a}$$

$$g_{0i} = -4\sum_A \frac{G m_A}{c^2 r_A}(v_A)_i + O\left(\epsilon^5\right) \tag{4.35b}$$

$$g_{ij} = \delta_{ij}\left(1 + 2\sum_A \frac{G m_A}{c^2 r_A}\right) + O\left(\epsilon^6\right). \tag{4.35c}$$

Here we use the notation $\boldsymbol{r}_A := \boldsymbol{x} - \boldsymbol{x}_A(t)$.

Notice that the super-potential describes the centre of mass of the system of particles. Its second derivative therefore describes the acceleration of the coordinate system with respect to the centre of mass of the system. We can perform a coordinate transformation into the *standard post-Newtonian gauge* that eliminates the super-potential from the g_{00} metric component. An infinitesimal coordinate transformation $\boldsymbol{x} \to \boldsymbol{x}' = \boldsymbol{x} + \boldsymbol{\xi}$ induces a change in the metric

$$g_{\alpha\beta} \to g'_{\alpha\beta} = g_{\alpha\beta} - \frac{\partial \xi_\beta}{\partial x^\alpha} - \frac{\partial \xi_\alpha}{\partial x^\beta} \tag{4.36}$$

so we choose

$$\xi_0 = -\frac{1}{2}\frac{1}{c^2}\frac{\partial \chi}{\partial t}, \quad \xi_i = 0 , \tag{4.37}$$

which yields

$$g_{00} \to g'_{00} = g_{00} + \frac{1}{c^2}\frac{\partial^2 \chi}{\partial t^2} \tag{4.38a}$$

$$g_{0i} \to g'_{0i} = g_{0i} + \frac{1}{2}\frac{1}{c^2}\frac{\partial}{\partial x^i}\frac{\partial \chi}{\partial t} \tag{4.38b}$$

$$g_{ij} \to g'_{ij} = g_{ij} . \tag{4.38c}$$

Since

$$\frac{\partial}{\partial x^i}\frac{\partial \chi}{\partial t} = \sum_A \frac{G m_A}{r_A}\left\{\boldsymbol{v}_A - \frac{(\boldsymbol{v}_A \cdot \boldsymbol{r}_A)}{r_A^2}\boldsymbol{r}_A\right\} \tag{4.39}$$

we find that the metric to first post-Newtonian order in the standard gauge is

$$g_{00} = -c^2 + 2\sum_A \frac{G m_A}{r_A} - 2\frac{1}{c^2}\left(\sum_A \frac{G m_A}{r_A}\right)^2 + 3\sum_A \frac{G m_A}{r_A}\frac{v_A^2}{c^2} - 2\sum_A\sum_{B\neq A}\frac{G m_A}{r_A}\frac{G m_B}{c^2 r_{AB}} + O\left(\epsilon^6\right) \tag{4.40a}$$

$$g_{0i} = -\frac{7}{2}\sum_A \frac{G m_A}{c^2 r_A}(v_A)_i - \frac{1}{2}\sum_A \frac{G m_A (\boldsymbol{v}_A\cdot\boldsymbol{r}_A)}{c^2 r_A^3}(r_A)_i + O(\epsilon^5) \tag{4.40b}$$

$$g_{ij} = \delta_{ij}\left(1 + 2\sum_A \frac{G m_A}{c^2 r_A}\right) + O\left(\epsilon^6\right). \tag{4.40c}$$

4.1.1.2 **Equations of Motion**

Given the post-Newtonian metric, we can use the geodesic equation to obtain the equations of motion for all bodies in the system. It is somewhat easier to obtain the equations of motion from the Lagrangian, and, in addition, once the Lagrangian is computed, the Hamiltonian for the system can be constructed, which yields the energy of the system.

The Lagrangian is

$$\begin{aligned}\mathcal{L} := -\sum_A m_A c^2 \frac{d\tau}{dt} &= -\sum_A m_A c\sqrt{-g_{00} - 2 g_{0i} v_A^i - g_{ij} v_A^i v_A^j} \\ &= -\sum_A m_A\left(c^2 - \frac{1}{2}v_A^2 - \frac{1}{8}\frac{v_A^4}{c^2}\right) \\ &\quad + \frac{1}{2}\sum_A\sum_{B\neq A}\frac{G m_A m_B}{r_{AB}}\left[1 + 3\frac{v_A^2}{c^2} - \sum_{C\neq A}\frac{G m_C}{c^2 r_{AC}} - \frac{7}{2}\frac{\boldsymbol{v}_A\cdot\boldsymbol{v}_B}{c^2}\right. \\ &\quad \left. - \frac{1}{2}\frac{(\boldsymbol{v}_A\cdot\boldsymbol{r}_{AB})(\boldsymbol{v}_B\cdot\boldsymbol{r}_{AB})}{c^2 r_{AB}^2}\right]. \end{aligned} \tag{4.41}$$

It follows from the Euler–Lagrange equations,

$$\frac{d}{dt}\frac{\partial\mathcal{L}}{\partial v_A^i} = \frac{\partial\mathcal{L}}{\partial x_A^i}, \quad \text{where} \quad \frac{d}{dt} = \frac{\partial}{\partial t} + v_A^j\frac{\partial}{\partial x_A^j} \tag{4.42}$$

that the equations of motion are

$$\begin{aligned}\frac{d\boldsymbol{v}_A}{dt} &= -\sum_{B\neq A}\frac{Gm_B\boldsymbol{r}_{AB}}{r_{AB}^3}\left[1-4\sum_{C\neq A}\frac{Gm_C}{c^2r_{AC}}\right.\\ &\quad+\sum_{C\neq A,B}\left(-\frac{Gm_C}{c^2r_{BC}}+\frac{1}{2}\frac{Gm_C(\boldsymbol{r}_{AB}\cdot\boldsymbol{r}_{BC})}{c^2r_{BC}^3}\right)\\ &\quad\left.-5\frac{Gm_A}{c^2r_{AB}}+\frac{v_A^2}{c^2}-4\frac{\boldsymbol{v}_A\cdot\boldsymbol{v}_B}{c^2}+2\frac{v_B^2}{c^2}-\frac{3}{2}\left(\frac{\boldsymbol{v}_B\cdot\boldsymbol{r}_{AB}}{cr_{AB}}\right)^2\right]\\ &\quad-\frac{7}{2}\sum_{B\neq A}\frac{Gm_B}{r_{AB}}\sum_{C\neq A,B}\frac{Gm_C\boldsymbol{r}_{BC}}{c^2r_{BC}^3}\\ &\quad+\sum_{B\neq A}\frac{Gm_B}{r_{AB}^3}\frac{\boldsymbol{r}_{AB}\cdot(4\boldsymbol{v}_A-3\boldsymbol{v}_B)}{c}\frac{\boldsymbol{v}_A-\boldsymbol{v}_B}{c}\,. \end{aligned}\tag{4.43}$$

These are called the *Einstein–Infeld–Hoffman equations.*

The total energy of the system is found by computing the Hamiltonian

$$\mathcal{H} := \sum_A \boldsymbol{p}_A\cdot\boldsymbol{v}_A - \mathcal{L}\,,\quad \text{where}\quad (p_A)_j := \frac{\partial\mathcal{L}}{\partial v_A^j}\,. \tag{4.44}$$

4.1.1.3 **Gravitational Radiation**

In the wave zone, the metric perturbation in the TT-gauge is

$$h_{\mathrm{TT}}^{ij} = \frac{2G}{c^2r}\frac{\partial^2}{\partial t^2}\left[I^{ij}(t-r/c)+\hat{n}_k I^{ijk}(t-r/c)+\hat{n}_k\hat{n}_l I^{ijkl}(t-r/c)\right]_{\mathrm{TT}}\,. \tag{4.45}$$

We need to calculate the transverse-traceless projections of I^{ij}, I^{ijk} and I^{ijkl} to obtain the metric perturbation to the required order. First, it can be shown

$$\begin{aligned} I^{ij} &:= \int \tau^{00}x^ix^j\,d^3\boldsymbol{x}\\ &= \sum_A m_A x_A^i x_A^j\left(1+\frac{1}{2}\frac{v_A^2}{c^2}-\frac{1}{2}\sum_{B\neq A}\frac{Gm_B}{c^2r_{AB}}\right)\\ &\qquad + \text{terms that vanish under TT projection}\,, \end{aligned}\tag{4.46}$$

where several integrations-by-parts are used and the surface terms discarded. The second term in the metric perturbation is obtained directly:

$$\begin{aligned} I^{ijk} &:= \int\left(\tau^{0i}x^jx^k+\tau^{0j}x^ix^k-\tau^{0k}x^ix^j\right)d^3\boldsymbol{x}\\ &= \sum_A m_A\left(v_A^ix_A^jx_A^k+v_A^jx_A^ix_A^k-v_A^kx_A^ix_A^j\right). \end{aligned}\tag{4.47}$$

The third term is somewhat more complex:

$$
\begin{aligned}
I^{ijkl} &:= \int \tau^{ij} x^k x^l d^3\boldsymbol{x} \\
&= \sum_A m_A v_A^i v_A^j x_A^k x_A^l - \frac{1}{4\pi G} \int \Phi \frac{\partial^2 \Phi}{\partial x_i \partial x_j} x^k x^l d^3\boldsymbol{x} \\
&\quad + \text{terms that vanish under TT projection,}
\end{aligned}
\tag{4.48}
$$

where again integration by parts has been performed and the surface term discarded. The calculation of the integral is involved, but the result is

$$
\begin{aligned}
I^{ijkl} &= \sum_A m_A v_A^i v_A^j x_A^k x_A^l \\
&\quad - \frac{1}{12} \sum_A \sum_{B \neq A} \frac{G m_A m_B r_{AB}^i r_{AB}^j}{r_{AB}} \\
&\quad \times \left[\delta^{kl} + \frac{1}{r_{AB}^2} \left(2 r_{AB}^k r_{AB}^l - 3 r_{AB}^k x_A^l - 3 r_{AB}^l x_A^k + 6 x_A^k x_A^l \right) \right] \\
&\quad + \text{terms that vanish under TT projection.}
\end{aligned}
\tag{4.49}
$$

Finally, we find

$$
\begin{aligned}
h_{\mathrm{TT}}^{ij} = \frac{2G}{c^4 r} \frac{\partial^2}{\partial t^2} \sum_A m_A \Bigg\{ & \left(1 - \frac{\hat{\boldsymbol{n}} \cdot \boldsymbol{v}_A}{c} + \frac{1}{2} \frac{v_A^2}{c^2} \right) x_A^i x_A^j \\
& - \frac{1}{2} \sum_{B \neq A} \frac{G m_B}{c^2 r_{AB}} x_A^i x_A^j \\
& + \hat{\boldsymbol{n}} \cdot \boldsymbol{x}_A \frac{\left(v_A^i x_A^j + v_A^j x_A^i \right)}{c} + (\hat{\boldsymbol{n}} \cdot \boldsymbol{x})^2 \frac{v_A^i}{c} \frac{v_A^j}{c} \\
& - \frac{1}{12} \sum_{B \neq A} \frac{G m_B}{c^2 r_{AB}} r_{AB}^i r_{AB}^j \left[1 - \left(\frac{\hat{\boldsymbol{n}} \cdot \boldsymbol{r}_{AB}}{r_{AB}} \right)^2 + 6 \left(\frac{\hat{\boldsymbol{n}} \cdot \boldsymbol{x}_A}{r_{AB}} \right)^2 \right] \Bigg\}_{\mathrm{TT}} .
\end{aligned}
\tag{4.50}
$$

The gravitational-wave luminosity is found from

$$
L_{\mathrm{GW}} = \frac{c^3 r^2}{32\pi G} \int \left\langle \left(\frac{\partial h_{ij}^{\mathrm{TT}}}{\partial t} \right) \left(\frac{\partial h_{\mathrm{TT}}^{ij}}{\partial t} \right) \right\rangle d\Omega \,. \tag{4.51}
$$

4.1.2
Two-Body Post-Newtonian Motion

We now specialize to the case of two particles in orbit since such systems are of particular interest as gravitational-wave sources. The Lagrangian reduces to

$$\mathcal{L} = -(m_1 + m_2)c^2 + \frac{1}{2}m_1 v_1^2 + \frac{1}{2}m_2 v_2^2 + \frac{1}{8}m_1\frac{v_1^4}{c^2} + \frac{1}{8}m_2\frac{v_2^4}{c^2}$$
$$+ \frac{G m_1 m_2}{r_{12}}\left[1 + \frac{3}{2}\frac{v_1^2 + v_2^2}{c^2} - \frac{1}{2}\frac{G(m_1+m_2)}{c^2 r_{12}} - \frac{7}{2}\frac{\boldsymbol{v}_1\cdot\boldsymbol{v}_2}{c^2}\right.$$
$$\left. - \frac{1}{2}\frac{(\boldsymbol{v}_1\cdot\boldsymbol{r}_{12})(\boldsymbol{v}_1\cdot\boldsymbol{r}_{12})}{c^2 r_{12}^2}\right] \tag{4.52}$$

and the Hamiltonian (written in terms of v rather than p) is

$$\mathcal{H} = (m_1 + m_2)c^2 + \frac{1}{2}m_1 v_1^2 + \frac{1}{2}m_2 v_2^2 + \frac{3}{8}m_1\frac{v_1^4}{c^2} + \frac{3}{8}m_2\frac{v_2^4}{c^2}$$
$$- \frac{G m_1 m_2}{r_{12}}\left[1 - \frac{3}{2}\frac{v_1^2 + v_2^2}{c^2} - \frac{1}{2}\frac{G(m_1+m_2)}{c^2 r_{12}} + \frac{7}{2}\frac{\boldsymbol{v}_1\cdot\boldsymbol{v}_2}{c^2}\right.$$
$$\left. + \frac{1}{2}\frac{(\boldsymbol{v}_1\cdot\boldsymbol{r}_{12})(\boldsymbol{v}_1\cdot\boldsymbol{r}_{12})}{c^2 r_{12}^2}\right], \tag{4.53}$$

where

$$\boldsymbol{p}_1 = m_1\boldsymbol{v}_1 + \frac{1}{2}\frac{v_1^2}{c^2}m_1\boldsymbol{v}_1 + \frac{G m_1 m_2}{c^2 r_{12}}\left[3\boldsymbol{v}_1 - \frac{7}{2}\boldsymbol{v}_2 - \frac{1}{2}\frac{\boldsymbol{v}_2\cdot\boldsymbol{r}_{12}}{r_{12}^2}\boldsymbol{r}_{12}\right] \tag{4.54}$$

and similarly for $\boldsymbol{p}_2$.

In centre-of-mass coordinates,

$$\boldsymbol{r}_1 = -\frac{m_2}{M}\boldsymbol{a}\,, \quad \boldsymbol{r}_2 = \frac{m_1}{M}\boldsymbol{a}\,, \quad \boldsymbol{v}_1 = -\frac{m_2}{M}\boldsymbol{v}\,, \quad \boldsymbol{v}_2 = \frac{m_1}{M}\boldsymbol{v}\,,$$

with $M := m_1 + m_2$ and $\boldsymbol{a} := \boldsymbol{r}_{12}$, the energy of the system (the Hamiltonian) is

$$E := \mathcal{H} = Mc^2 + \mu\left\{\frac{1}{2}v^2 - \frac{GM}{a} + \frac{3}{8}(1-3\eta)\frac{v^4}{c^2}\right.$$
$$\left. + \frac{1}{2}\frac{GM}{c^2 a}\left(\frac{GM}{a} + (3+\eta)v^2 + \frac{(\boldsymbol{v}\cdot\boldsymbol{a})^2}{a^2}\right)\right\}, \tag{4.55}$$

where $\mu := m_1 m_2/M$ is the reduced mass, and $\eta := \mu/M$ is the symmetric mass ratio. The equation of motion in the centre of mass coordinates is

$$\frac{d\boldsymbol{v}}{dt} = -\frac{GM}{a^3}\boldsymbol{a}\left[1 - (4+2\eta)\frac{GM}{c^2 a} + (1+3\eta)\frac{v^2}{c^2} - \frac{3}{2}\eta\frac{(\boldsymbol{v}\cdot\boldsymbol{a})^2}{c^2 a^2}\right]$$
$$+ (4-2\eta)\frac{GM}{c^2 a^3}(\boldsymbol{v}\cdot\boldsymbol{a})\boldsymbol{v}\,. \tag{4.56}$$

If the two bodies are in a circular orbit then $\boldsymbol{v} \cdot \boldsymbol{a} = 0$ and the equation of motion reduces to

$$\frac{d\boldsymbol{v}}{dt} = -\frac{GM}{a^3}\boldsymbol{a}\left[1 - (4 + 2\eta)\frac{GM}{c^2 a} + (1 + 3\eta)\frac{v^2}{c^2}\right]. \tag{4.57}$$

Since $d\boldsymbol{v}/dt = -\omega^2 \boldsymbol{a}$ we find that Kepler's law is modified at first post-Newtonian order

$$\omega^2 = \frac{GM}{a^3}\left[1 + \frac{GM}{c^2 a}(\eta - 3)\right]. \tag{4.58}$$

Since $v = a\omega$, this implies

$$v^2 = \frac{GM}{a}\left[1 - (3 - \eta)\frac{GM}{c^2 a}\right] \tag{4.59}$$

or

$$\frac{GM}{a} = v^2\left[1 + (3 - \eta)\frac{v^2}{c^2}\right]. \tag{4.60}$$

The energy and luminosity can also be computed; they are

$$E = Mc^2 - \frac{1}{2}\mu v^2 - \frac{1}{8}\mu\frac{v^4}{c^2}(5 - 3\eta) \tag{4.61}$$

and

$$L_{\mathrm{GW}} = \frac{32c^5}{5G}\eta^2\left(\frac{GM}{c^2 a}\right)^5\left[1 - \frac{GM}{c^2 a}\left(\frac{2927}{336} + \frac{5}{4}\eta\right)\right]. \tag{4.62}$$

With these expressions, we can use energy balance to compute the orbital evolution to first post-Newtonian order. There is a subtlety, however, as the modification of Kepler's law has an important consequence: when we earlier derived the evolution of a binary orbit under gravitational radiation, we used the fact that $v = a\omega = (GM\omega)^{1/3}$, thereby relating the orbital velocity (which is a coordinate-dependent quantity) with the quantity $GM\omega$, which is the observable at infinite distance, that describes the gravitational-wave radiation. However, this relationship no longer holds. Therefore we continue to use $v = a\omega$ to be the orbital velocity in our coordinate system, and we defined a new variable $x := (GM\omega/c^3)^{2/3}$ as our post-Newtonian parameter with

$$x^{3/2} = \frac{GM\omega}{c^3} = \frac{GM}{c^2 a}\frac{v}{c} = v^3\left[1 + (3 - \eta)v^2\right] \tag{4.63}$$

so

$$x = \frac{v^2}{c^2}\left[1 + \frac{2}{3}(3 - \eta)\frac{v^2}{c^2}\right] \tag{4.64a}$$

or

$$v^2 = x c^2 \left[1 - \frac{2}{3}(3 - \eta) x \right] . \tag{4.64b}$$

With this substitution, we find the energy function $\mathcal{E} := E - Mc^2$ and the flux function $\mathcal{F} := (G/c^5) L_{\mathrm{GW}}$ are

$$\mathcal{E}(x) = -\frac{1}{2} \eta x \left[1 - \frac{1}{12}(9 + \eta) x \right] \tag{4.65}$$

and

$$\mathcal{F}(x) = \frac{32}{5} \eta^2 x^5 \left[1 - \left(\frac{1247}{336} + \frac{35}{12} \eta \right) x \right] . \tag{4.66}$$

The time $t(x)$ in which the binary system reaches an orbital frequency set by x and the corresponding orbital phase $\varphi(x)$ are found, as in the Newtonian case, from the relationships

$$t(x) = t_{\mathrm{c}} + \frac{GM}{c^3} \int_x^{x_{\mathrm{c}}} \frac{1}{\mathcal{F}} \frac{d\mathcal{E}}{dx} dx \tag{4.67}$$

and

$$\varphi(x) = \varphi_{\mathrm{c}} + \int_x^{x_{\mathrm{c}}} x^{3/2} \frac{1}{\mathcal{F}} \frac{d\mathcal{E}}{dx} dx , \tag{4.68}$$

where x_{c} is set by the termination frequency (the "coalescence" frequency), and t_{c} and φ_{c} is the time and phase at that frequency. These result in (if we formally take $x_{\mathrm{c}} \to \infty$)

$$t(x) = t_{\mathrm{c}} - \frac{5}{256\eta} \frac{GM}{c^3} x^{-4} \left[1 + \left(\frac{743}{252} + \frac{11}{3} \eta \right) x \right] \tag{4.69}$$

$$\varphi(x) = \varphi_{\mathrm{c}} - \frac{1}{32\eta} x^{-5/2} \left[1 + \left(\frac{3715}{1008} + \frac{55}{12} \eta \right) x \right] . \tag{4.70}$$

This parametric solution to the phase, $\{t(x), \varphi(x)\}$, is known as a *TaylorT2* time-domain approximate. There are several other alternative evolution schemes: The *TaylorT3* method is obtained from the TaylorT2 by reversion of the power series for $t(x)$ to obtain $x(t)$; this is done in terms of a surrogate dimensionless time variable

$$\Theta := \frac{\eta}{5} \frac{c^3 (t_{\mathrm{c}} - t)}{GM} \tag{4.71}$$

and one obtains

$$x = \frac{1}{4} \Theta^{-1/4} \left[1 + \left(\frac{743}{4032} + \frac{11}{48} \eta \right) \Theta^{-1/4} + O\left(\epsilon^3\right) \right] . \tag{4.72}$$

Now the expression $\varphi(x)$ can be written in terms of the time variable Θ as

$$\varphi = \varphi_c - \frac{1}{\eta}\Theta^{5/8}\left[1 + \left(\frac{3715}{8064} + \frac{55}{96}\eta\right)\Theta^{-1/4} + O\left(\epsilon^3\right)\right]. \tag{4.73}$$

The *TaylorT1* method is to numerically integrate the coupled ODEs

$$\frac{dx}{dt} = -\frac{c^3}{GM}\frac{\mathcal{F}}{d\mathcal{E}/dx} \tag{4.74}$$

$$\frac{d\varphi}{dt} = \frac{c^3}{GM}x^{3/2}. \tag{4.75}$$

The *TaylorT4* method is the same as the TaylorT1 method except that the right-hand side of the expression for dx/dt is expanded in a power series to the required post-Newtonian order; the resulting coupled ODEs

$$\frac{dx}{dt} = \frac{64}{5}\eta\frac{c^3}{GM}x^5\left[1 - \left(\frac{743}{336} + \frac{11}{4}\eta\right)x + O(\epsilon^3)\right] \tag{4.76}$$

$$\frac{d\varphi}{dt} = \frac{c^3}{GM}x^{3/2} \tag{4.77}$$

are then integrated numerically. All of these methods, TaylorT1–T4, will give phase evolutions that are formally valid to first post-Newtonian order; the differences between the evolutions result from higher-order terms that are discarded by the various truncation schemes, and it is not possible (without doing a higher-order post-Newtonian calculation) to determine which scheme is "better".

The gravitational waveform h_{ij}^{TT} for an observer located at position (r, ι, ϕ) in polar coordinates relative to the coordinate system is written in terms of the two polarizations h_+ and $h_\times$. It is most convenient to write these in an expansion in terms of spin-2-weighted spherical harmonics,

$$h_+ - i h_\times = \sum_{\ell=2}^{\infty}\sum_{m=-\ell}^{\ell} {}_{-2}Y_{\ell m}(\iota, \phi) h_{\ell m}, \tag{4.78}$$

where the complex modes are

$$h_{\ell m} := \int {}_{-2}Y^*_{\ell m}(\iota, \phi)(h_+ - i h_\times) d\Omega \tag{4.79}$$

and satisfy

$$h_{\ell, m} = (-1)^\ell h^*_{\ell, -m}. \tag{4.80}$$

The *spin-weighted spherical harmonics* of spin-weight s are

$$\begin{aligned} {}_{-s}Y_{\ell m}(\theta, \phi) := (-1)^s \sqrt{\frac{2\ell+1}{4\pi}}\sqrt{\frac{(\ell+m)!(\ell-m)!}{(\ell+s)!(\ell-s)!}} e^{im\phi}\cos^{2\ell}\frac{\theta}{2} \\ \times \sum_{k=\max(0,m-s)}^{\min(\ell+m,\ell-s)} (-1)^k \binom{\ell-s}{k}\binom{\ell+s}{k+s-m}\tan^{2k+s-m}\frac{\theta}{2} \end{aligned} \tag{4.81}$$

and satisfy

$$ {}_{s}Y_{\ell,m}(\theta,\phi) = (-1)^{s+m}\,{}_{-s}Y^{*}_{\ell,-m}(\theta,\phi) \tag{4.82}$$

and

$$ {}_{s}Y_{\ell m}(\theta,\phi) = (-1)^{\ell}\,{}_{-s}Y_{\ell m}(\pi-\theta,\phi+\pi)\,. \tag{4.83}$$

Table B.1 provides the $s=-2$, $\ell=2,3,4$ spin-weighted spherical harmonics.

To first post-Newtonian order, the relevant modes are (Kidder, 2008)

$$h_{22} = -8\sqrt{\frac{\pi}{5}}\frac{G\mu}{c^2 r}\mathrm{e}^{-2i\varphi}x\left[1-\left(\frac{107}{42}-\frac{55}{42}\eta\right)x\right]+O\left(\epsilon^5\right) \tag{4.84a}$$

$$h_{21} = -i\frac{8}{3}\sqrt{\frac{\pi}{5}}\frac{G\mu}{c^2 r}\frac{\delta m}{M}\mathrm{e}^{-i\varphi}x^{3/2}+O\left(\epsilon^5\right) \tag{4.84b}$$

$$h_{33} = 3i\sqrt{\frac{6\pi}{7}}\frac{G\mu}{c^2 r}\frac{\delta m}{M}\mathrm{e}^{-3i\varphi}x^{3/2}+O\left(\epsilon^5\right) \tag{4.84c}$$

$$h_{32} = -\frac{8}{3}\sqrt{\frac{\pi}{7}}\frac{G\mu}{c^2 r}\mathrm{e}^{-2i\varphi}(1-3\eta)x^2+O(\epsilon^6) \tag{4.84d}$$

$$h_{31} = -\frac{i}{3}\sqrt{\frac{2\pi}{35}}\frac{G\mu}{c^2 r}\frac{\delta m}{M}\mathrm{e}^{-i\varphi}x^{3/2}+O\left(\epsilon^5\right) \tag{4.84e}$$

$$h_{44} = \frac{64}{9}\sqrt{\frac{\pi}{7}}\frac{G\mu}{c^2 r}\mathrm{e}^{-4i\varphi}(1-3\eta)x^2+O(\epsilon^6) \tag{4.84f}$$

$$h_{42} = -\frac{8}{63}\sqrt{\pi}\frac{G\mu}{c^2 r}\mathrm{e}^{-2i\varphi}(1-3\eta)x^2+O(\epsilon^6), \tag{4.84g}$$

where $\delta m := m_1 - m_2$. The waveform can then be written

$$h_{+,\times} = \frac{2G\mu}{c^2 r}x\left\{H^{(0)}_{+,\times}+x^{1/2}H^{(1/2)}_{+,\times}+xH^{(1)}_{+,\times}+O\left(\epsilon^3\right)\right\} \tag{4.85a}$$

with

$$H^{(0)}_{+} := -(1+\cos^2\iota)\cos 2\varphi \tag{4.85b}$$

$$H^{(0)}_{\times} := -2\cos\iota\sin 2\varphi \tag{4.85c}$$

$$H^{(1/2)}_{+} := -\frac{1}{8}\frac{\delta m}{M}\sin\iota\left[\left(5+\cos^2\iota\right)\cos\varphi-9\left(1+\cos^2\iota\right)\cos 3\varphi\right] \tag{4.85d}$$

$$H^{(1/2)}_{\times} := -\frac{3}{4}\frac{\delta m}{M}\sin\iota\cos\iota\left[\sin\varphi-3\sin 3\varphi\right] \tag{4.85e}$$

$$H^{(1)}_{+} := \frac{1}{6}\left[(19+9\cos^2\iota-2\cos^4\iota)-\eta(19-11\cos^2\iota-6\cos^4\iota)\right]\cos 2\varphi - \frac{4}{3}\sin^2\iota(1+\cos^2\iota)(1-3\eta)\cos 4\varphi \tag{4.85f}$$

$$H_{\times}^{(1)} := \frac{1}{3}\cos\iota\left[(17 - 4\cos\iota) - \eta(13 - 12\cos^2\iota)\right]\sin 2\varphi - \frac{8}{3}\cos\iota\sin^2\iota(1 - 3\eta)\sin 4\varphi\,. \tag{4.85g}$$

Here we have assumed that the observer lies along the x–z-plane of the system, that is, with $\phi = 0$ (which effectively absorbs ϕ into the definition of the constant φ_c).

4.1.3 Higher-Order Post-Newtonian Waveforms for Binary Inspiral

The gravitational waveform given by Eq. (4.85) contains terms up to and including $O(\epsilon^2)$ beyond the quadrupole formula, and so contains the first post-Newtonian corrections. However, higher-order post-Newtonian calculations are required to obtain gravitational waveforms that describe the late stages of binary inspiral. Post-Newtonian calculations up to third post-Newtonian order, that is $O(\epsilon^6)$, are now available (see Appendix B). These calculations also include the corrections to the phase arising due to spin of the binary components (spin-orbit and spin-spin coupling). If the component spins are not aligned with the orbital angular momentum, then the plane of the orbit will precess and the waveforms can become quite complicated. In addition, tidal deformation of finite-sized objects can affect the waveform, and corrections to the energy and flux equations to take into account such finite-size effects have been calculated. For binary neutron star systems, such finite-size effects may significantly alter the waveform for gravitational-wave frequencies $f \gtrsim 400$ Hz, depending on the structure of the neutron stars.

Post-Newtonian calculations will never be able to give a complete description of the full inspiral and collision of binary systems of compact objects such as neutron stars or black holes as the motion will eventually become highly relativistic and the post-Newtonian approximation will fail. To understand the final stages of inspiral, and the post collision remnants, requires numerical relativity.

4.2 Perturbation about Curved Backgrounds

In a curved spacetime, we can still obtain a wave equation for the Riemann tensor, but it will no longer have the simple form $\nabla^2 R_{\alpha\beta\gamma\delta} = 0$; instead, it will have a source term that is constructed from the curvature tensor itself.

To begin, return to the Bianchi identities

$$\nabla_\nu R_{\alpha\beta\gamma\delta} + \nabla_\alpha R_{\beta\nu\gamma\delta} + \nabla_\beta R_{\nu\alpha\gamma\delta} = 0 \tag{4.86}$$

and take a second covariant derivative to obtain

$$\nabla_\mu\nabla_\nu R_{\alpha\beta\gamma\delta} + \nabla_\mu\nabla_\alpha R_{\beta\nu\gamma\delta} + \nabla_\mu\nabla_\beta R_{\nu\alpha\gamma\delta} = 0\,. \tag{4.87}$$

Now subtract from this equation the same equation with the indices α and μ swapped, and subtract the equation again but with indices β and μ swapped. We then have

$$\begin{aligned}0 = {} & \nabla_\mu \nabla_\nu R_{\alpha\beta\gamma\delta} + (\nabla_\mu \nabla_\alpha - \nabla_\alpha \nabla_\mu) R_{\beta\nu\gamma\delta} - \nabla_\alpha \nabla_\nu R_{\mu\beta\gamma\delta} \\ & + \nabla_\alpha \nabla_\beta R_{\mu\nu\gamma\delta} - (\alpha \leftrightarrow \beta)\,. \end{aligned} \tag{4.88}$$

Now we express the commutator of the derivatives of the Riemann tensor as products of Riemann tensors,

$$\begin{aligned}\left(\nabla_\mu \nabla_\alpha - \nabla_\alpha \nabla_\mu\right) R_{\beta\nu\gamma\delta} = {} & R_{\mu\alpha\beta}{}^{\rho} R_{\rho\nu\gamma\delta} + R_{\mu\alpha\nu}{}^{\rho} R_{\beta\rho\gamma\delta} + R_{\mu\alpha\gamma}{}^{\rho} R_{\beta\nu\rho\delta} \\ & + R_{\mu\alpha\delta}{}^{\rho} R_{\beta\nu\gamma\rho} \end{aligned} \tag{4.89}$$

and we evoke the Bianchi identities again to re-express

$$-\nabla_\alpha \nabla_\nu R_{\mu\beta\gamma\delta} = -\nabla_\alpha \nabla_\nu R_{\gamma\delta\mu\beta} = \nabla_\alpha \nabla_\gamma R_{\delta\nu\mu\beta} + \nabla_\alpha \nabla_\delta R_{\nu\gamma\mu\beta}\,, \tag{4.90}$$

which gives us

$$\begin{aligned}0 = {} & \nabla_\mu \nabla_\nu R_{\alpha\beta\gamma\delta} + R_{\mu\alpha\beta}{}^{\rho} R_{\rho\nu\gamma\delta} + R_{\mu\alpha\nu}{}^{\rho} R_{\beta\rho\gamma\delta} + R_{\mu\alpha\gamma}{}^{\rho} R_{\beta\nu\rho\delta} \\ & + R_{\mu\alpha\delta}{}^{\rho} R_{\beta\nu\gamma\rho} + \nabla_\alpha \nabla_\gamma R_{\delta\nu\mu\beta} + \nabla_\alpha \nabla_\delta R_{\nu\gamma\mu\beta} + \nabla_\alpha \nabla_\beta R_{\mu\nu\gamma\delta} \\ & - (\alpha \leftrightarrow \beta)\,. \end{aligned} \tag{4.91}$$

Now we contract the indices μ and ν with the metric tensor $g^{\mu\nu}$ and use the definition of the Ricci tensor to obtain

$$\begin{aligned}0 = {} & \nabla^2 R_{\alpha\beta\gamma\delta} + R^{\mu}{}_{\alpha\beta}{}^{\rho} R_{\rho\mu\gamma\delta} + R_\alpha{}^{\rho} R_{\beta\rho\gamma\delta} + R^{\mu}{}_{\alpha\gamma}{}^{\rho} R_{\beta\mu\rho\delta} + R^{\mu}{}_{\alpha\delta}{}^{\rho} R_{\beta\mu\gamma\rho} \\ & - \nabla_\alpha \nabla_\gamma R_{\delta\beta} + \nabla_\alpha \nabla_\delta R_{\gamma\beta} \\ & - (\alpha \leftrightarrow \beta)\,. \end{aligned} \tag{4.92}$$

Finally, employing the identities of the Riemann tensor (and replacing the dummy index ρ with ν), we find

$$\begin{aligned} & \nabla^2 R_{\alpha\beta\gamma\delta} + R_{\alpha\beta}{}^{\mu\nu} R_{\mu\nu\gamma\delta} - 2 R_\alpha{}^{\mu}{}_\delta{}^{\nu} R_{\beta\mu\gamma\nu} + 2 R_\beta{}^{\mu}{}_\delta{}^{\nu} R_{\alpha\mu\gamma\nu} \\ & = -\nabla_\alpha \nabla_\gamma R_{\delta\beta} - \nabla_\alpha \nabla_\delta R_{\gamma\beta} + \nabla_\beta \nabla_\gamma R_{\delta\alpha} + \nabla_\beta \nabla_\delta R_{\gamma\alpha} \\ & \quad + R_\alpha{}^{\nu} R_{\beta\nu\gamma\delta} - R_\beta{}^{\nu} R_{\alpha\nu\gamma\delta}\,. \end{aligned} \tag{4.93}$$

In a vacuum spacetime, the right-hand side of this equation vanishes. If there is matter present, terms on the right-hand side become terms that are derivatives of the matter-stress-energy tensor or products of the stress-energy tensor and the Riemann tensor.

Consider a vacuum spacetime and suppose that the curvature consists of a background curvature

$$\overset{0}{R}_{\alpha\beta\gamma\delta}\,,$$

which has some long curvature scale $\mathcal{R}$, and a small gravitational-wave perturbation curvature

$$\overset{1}{R}_{\alpha\beta\gamma\delta}\,,$$

which varies over short length scales $\lambda \ll \mathcal{R}$. We write

$$R_{\alpha\beta\gamma\delta} = \overset{0}{R}_{\alpha\beta\gamma\delta} + \overset{1}{R}_{\alpha\beta\gamma\delta} \,. \tag{4.94}$$

We then find a wave equation for the gravitational-wave perturbation

$$\begin{aligned}\nabla^2 \overset{1}{R}_{\alpha\beta\gamma\delta} &= -\overset{0}{R}_{\alpha\beta}{}^{\mu\nu}\overset{1}{R}_{\mu\nu\gamma\delta} + 2\overset{0}{R}_{\alpha}{}^{\mu}{}_{\delta}{}^{\nu}\overset{1}{R}_{\beta\mu\gamma\nu} - 2\overset{0}{R}_{\beta}{}^{\mu}{}_{\delta}{}^{\nu}\overset{1}{R}_{\alpha\mu\gamma\nu} \\ &\quad - \overset{1}{R}_{\alpha\beta}{}^{\mu\nu}\overset{0}{R}_{\mu\nu\gamma\delta} + 2\overset{1}{R}_{\alpha}{}^{\mu}{}_{\delta}{}^{\nu}\overset{0}{R}_{\beta\mu\gamma\nu} - 2\overset{1}{R}_{\beta}{}^{\mu}{}_{\delta}{}^{\nu}\overset{0}{R}_{\alpha\mu\gamma\nu} \,. \end{aligned} \tag{4.95}$$

Note that the left-hand side has the form $\nabla^2 \overset{1}{R} \sim \overset{1}{R}/\lambda^2$ while the terms on the right hand are of the form

$$\overset{0}{R}\overset{1}{R} \sim \overset{1}{R}/\mathcal{R}^2$$

since

$$\overset{0}{R} \sim 1/\mathcal{R}^2 \,.$$

These terms are much smaller than the terms on the left-hand side so we have the approximate equation

$$\nabla^2 \overset{1}{R}_{\alpha\beta\gamma\delta} = 0 \tag{4.96}$$

for the gravitational-wave perturbation *if the wavelength of the gravitational wave is much smaller than the curvature length scale of the background spacetime.* If this is the case, then the gravitational wave described by the perturbation to the Riemann tensor satisfies the curved-space wave equation where the $\nabla^2 := g^{\mu\nu}\nabla_\mu\nabla_\nu$ operator is the curved-space wave operator, which can be approximated as the background curved-space wave operator if the gravitational-wave perturbation is small.

An expression for the metric perturbation of a curved spacetime can be constructed as follows. Let

$$g_{\alpha\beta} = \overset{0}{g}_{\alpha\beta} + h_{\alpha\beta}$$

be the spacetime metric where

$$\overset{0}{g}_{\alpha\beta}$$

is the metric of the background spacetime and $h_{\alpha\beta}$ is the perturbation.

The covariant derivative ∇_γ that is compatible with the spacetime metric $g_{\alpha\beta}$ (so $\nabla_\gamma g_{\alpha\beta} = 0$) can be related to the covariant derivative

$$\overset{0}{\nabla}_\gamma$$

compatible with the background metric

$$\overset{0}{g}_{\alpha\beta} \text{ (which satisfies } \overset{0}{\nabla}_\gamma \overset{0}{g}_{\alpha\beta}\text{):}$$

for an arbitrary vector **v**, both $\nabla_\alpha v^\gamma$ and $\overset{0}{\nabla}_\alpha v^\gamma$ must be tensors, so the difference between these tensors must be a linear map on the vector, which defines a new tensor $C^\gamma_{\alpha\beta}$ by

$$\nabla_\alpha v^\gamma - \overset{0}{\nabla}_\alpha v^\gamma =: C^\gamma_{\alpha\mu} v^\mu \,. \tag{4.97}$$

We can evaluate $C^\gamma_{\alpha\beta}$ in the locally inertial frame of the background metric: in such a frame,

$$\nabla_\alpha v^\gamma = \frac{\partial}{\partial x^\alpha} v^\gamma + \Gamma^\gamma_{\alpha\mu} v^\mu \tag{4.98}$$

and

$$\overset{0}{\nabla}_\alpha v^\gamma = \frac{\partial}{\partial x^\alpha} v^\gamma \quad \text{(locally inertial frame of background spacetime);} \tag{4.99}$$

therefore, since **v** was arbitrary,

$$C^\gamma_{\alpha\beta} = \Gamma^\gamma_{\alpha\beta} = \frac{1}{2} g^{\gamma\mu} \left(\frac{\partial}{\partial x^\alpha} g_{\mu\beta} + \frac{\partial}{\partial x^\beta} g_{\alpha\mu} - \frac{\partial}{\partial x^\mu} g_{\alpha\beta} \right)$$
$$\text{(locally inertial frame of background spacetime),} \tag{4.100}$$

where we have used Eq. (2.40) for $\Gamma^\gamma_{\alpha\beta}$. This formula for the components of $C^\gamma_{\alpha\beta}$ holds only in the locally inertial coordinate frame of the background spacetime; however, $C^\gamma_{\alpha\beta}$ is a tensor, so we wish to find a covariant expression that holds in any coordinates. Such an expression can be obtained by replacing the ordinary partial derivatives $\partial/\partial x^\alpha$ with the covariant derivatives

$$\overset{0}{\nabla}_\alpha$$

since these two derivatives agree in the locally inertial coordinates frame of the background spacetime. Thus we find

$$\begin{aligned} C^\gamma_{\alpha\beta} &= \frac{1}{2} g^{\gamma\mu} \left(\overset{0}{\nabla}_\alpha g_{\mu\beta} + \overset{0}{\nabla}_\beta g_{\alpha\mu} - \overset{0}{\nabla}_\mu g_{\alpha\beta} \right) \\ &= \frac{1}{2} g^{\gamma\mu} \left(\overset{0}{\nabla}_\alpha h_{\mu\beta} + \overset{0}{\nabla}_\beta h_{\alpha\mu} - \overset{0}{\nabla}_\mu h_{\alpha\beta} \right) . \end{aligned} \tag{4.101}$$

A similar argument shows that

$$R_{\alpha\beta\gamma}{}^\delta = \overset{0}{R}_{\alpha\beta\gamma}{}^\delta - \overset{0}{\nabla}_\alpha C^\delta_{\beta\gamma} + \overset{0}{\nabla}_\beta C^\delta_{\alpha\gamma} - C^\delta_{\alpha\mu} C^\mu_{\beta\gamma} + C^\delta_{\beta\mu} C^\mu_{\alpha\gamma} \,. \tag{4.102}$$

From this we obtain the linearized Ricci tensor (note that $C^\gamma_{\alpha\beta}$ is linear in the metric perturbation $h_{\alpha\beta}$ so we may discard the terms that are quadratic in $C^\gamma_{\alpha\beta}$):

$$\begin{aligned} R_{\alpha\beta} = \overset{0}{R}_{\alpha\beta} &+ \frac{1}{2} \left(-\overset{0}{\nabla}_\alpha \overset{0}{\nabla}_\beta h + \overset{0}{\nabla}_\mu \overset{0}{\nabla}_\alpha h^\mu{}_\beta + \overset{0}{\nabla}_\mu \overset{0}{\nabla}_\beta h_\alpha{}^\mu - \overset{0}{g}{}^{\mu\nu} \overset{0}{\nabla}_\mu \overset{0}{\nabla}_\nu h_{\alpha\beta} \right) \\ &+ O\left(h^2\right) \end{aligned} \tag{4.103}$$

(cf. Eq. (2.128)); here $h = h_\mu{}^\mu = \overset{0}{g}{}^{\mu\nu} h_{\mu\nu}$. Following the development presented in Section 2.5.1, we consider the trace-reversed metric perturbation $\bar{h}_{\alpha\beta}$, which is now

$$\bar{h}_{\alpha\beta} = h_{\alpha\beta} - \frac{1}{2}\overset{0}{g}_{\mu\nu} h\,, \tag{4.104}$$

and we compute the Einstein tensor,

$$\begin{aligned} G_{\alpha\beta} &= \overset{0}{G}_{\alpha\beta} \\ &\quad + \frac{1}{2}\left(-\overset{0}{g}{}^{\mu\nu}\overset{0}{\nabla}_\mu\overset{0}{\nabla}_\nu\bar{h}_{\alpha\beta} - \overset{0}{g}_{\alpha\beta}\overset{0}{\nabla}_\mu\overset{0}{\nabla}_\nu\bar{h}^{\mu\nu} + \overset{0}{\nabla}_\mu\overset{0}{\nabla}_\alpha\bar{h}^\mu{}_\beta + \overset{0}{\nabla}_\mu\overset{0}{\nabla}_\beta\bar{h}_\alpha{}^\mu\right) \\ &\quad + O\left(h^2\right). \end{aligned} \tag{4.105}$$

It is always possible to adopt a Lorenz-like gauge so that the metric perturbation in this particular coordinate system satisfies

$$\overset{0}{\nabla}_\mu \bar{h}^\mu{}_\alpha = 0 \quad \text{(Lorenz gauge)}; \tag{4.106}$$

also, the difference $G_{\alpha\beta} - \overset{0}{G}_{\alpha\beta}$ is proportional to the difference between the true stress-energy tensor $T_{\alpha\beta}$, and the stress-energy tensor of the background spacetime $\overset{0}{T}_{\alpha\beta}$: $\delta T_{\alpha\beta} := T_{\alpha\beta} - \overset{0}{T}_{\alpha\beta}$. Therefore, in the Lorenz gauge

$$\overset{0}{g}{}^{\mu\nu}\overset{0}{\nabla}_\mu\overset{0}{\nabla}_\nu\bar{h}_{\alpha\beta} - \overset{0}{\nabla}_\mu\overset{0}{\nabla}_\alpha\bar{h}_\beta{}^\mu - \overset{0}{\nabla}_\mu\overset{0}{\nabla}_\beta\bar{h}_\alpha{}^\mu = \frac{16\pi G}{c^4}\delta T_{\alpha\beta} \quad \text{(Lorenz gauge)}. \tag{4.107}$$

The perturbation to the stress-energy tensor contains terms that arise (i) because the stress-energy tensor depends on the metric, so perturbations of the metric also result in perturbations of the stress-energy tensor; (ii) due to an additional component of stress energy above the background, which is the source of the metric perturbations; (iii) from the metric perturbation itself, which are quadratic in the metric perturbation (and so can be ignored in a linear analysis, but ultimately describes the gravitational-wave stress-energy). Finally, the background Riemann tensor has the property

$$\left(\overset{0}{\nabla}_\alpha\overset{0}{\nabla}_\beta - \overset{0}{\nabla}_\beta\overset{0}{\nabla}_\alpha\right)\bar{h}_{\gamma\delta} = \overset{0}{R}_{\alpha\beta\gamma}{}^\mu\bar{h}_{\mu\delta} + \overset{0}{R}_{\alpha\beta\delta}{}^\mu\bar{h}_{\gamma\mu} \tag{4.108}$$

so we find

$$\overset{0}{g}{}^{\mu\nu}\overset{0}{\nabla}_\mu\overset{0}{\nabla}_\nu\bar{h}_{\alpha\beta} + 2\overset{0}{R}_\alpha{}^\mu{}_\beta{}^\nu\bar{h}_{\mu\nu} - \overset{0}{R}_\alpha{}^\mu\bar{h}_{\beta\mu} - \overset{0}{R}_\beta{}^\mu\bar{h}_{\alpha\mu} = \frac{16\pi G}{c^4}\delta T_{\alpha\beta} \quad \text{(Lorenz gauge)}. \tag{4.109}$$

In a vacuum spacetime, it possible to adopt a gauge in which the metric perturbation is transverse and trace-free; in such a gauge the wave equation for the metric perturbation becomes

$$g^{\mu\nu}\overset{0}{\nabla}_\mu\overset{0}{\nabla}_\nu h_{\alpha\beta} + 2\overset{0}{R}_\alpha{}^\mu{}_\beta{}^\nu h_{\mu\nu} = 0 \quad \text{(vacuum spacetime; Lorenz gauge)} \tag{4.110}$$

where we have ignored the terms that are quadratic in the metric perturbation. For waves that are much smaller than the curvature scale, the linearized gravity approximation will hold and we can also choose coordinates in which $h_{0\mu} = 0$ locally; however it is not generally possible to construct such coordinates globally.

4.2.1 Gravitational Waves in Cosmological Spacetimes

A homogeneous, isotropic universe is described by the *Robertson–Walker metric*

$$ds^2 = -c^2 dt^2 + a^2(t)\left[dx^2 + dy^2 + dz^2\right] \tag{4.111}$$

where $a(t)$ is the *scale factor* of the Universe, which is a function only of the *cosmological time*, *t*. (In fact, this is just one special case of a Robertson–Walker spacetime: one that is spatially flat.) This metric is a solution to the Einstein equations for a homogeneous perfect fluid stress-energy tensor,

$$T_{\alpha\beta} = (\rho + p/c)u_\alpha u_\beta + p g_{\alpha\beta}\,, \tag{4.112}$$

where p and ρ are functions of t alone, provided that the scale factor satisfies the *Friedmann equations*[1)]

$$\left(\frac{1}{a}\frac{da}{dt}\right)^2 = \frac{8\pi G}{3}\rho \tag{4.113}$$

$$\frac{1}{a}\frac{d^2a}{dt^2} = -\frac{4\pi G}{3}(\rho + 3p/c^2)\,. \tag{4.114}$$

The Robertson–Walker spacetime that is the solution to the Friedmann equations (for some specified perfect fluid stress-energy tensor) is known as a *Friedmann–Lemaître–Robertson–Walker cosmology*. Important solutions are presented in Table 4.3, and a more detailed description is given in Section 5.3.1.1.

Now consider a perturbation of the Robertson–Walker metric of the form

$$ds^2 = -c^2 dt^2 + a^2(t)\left[dx^2 + dy^2 + dz^2\right] + h_{\mu\nu}dx^\mu dx^\nu\,. \tag{4.115}$$

1) The Universe contains a dark energy, which can be incorporated into Einstein's equations as a *cosmological constant*, Λ, but can also be thought of as an additional contribution to the perfect fluid stress-energy tensor with a constant (over space and time) density ρ_{vac} and *negative* pressure $p_{\text{vac}} = -\rho_{\text{vac}}c^2$; the cosmological constant is then $\Lambda = 8\pi G\rho_{\text{vac}}/c^2$.

Table 4.3 Flat Friedmann–Lemaître–Robertson–Walker solutions for cosmologies that are matter dominated (dust), radiation dominated, and inflating cosmologies dominated by vacuum energy (e.g. a cosmological constant). The scale factor is given in terms of both cosmological time t and conformal time η where $dt = a d\eta$.

Cosmology type	Equation of state	Scale factor
Matter-dominated	Dust	$a(t) = (t/t_0)^{2/3}$
	$p = 0$	$a(\eta) = (\eta/\eta_0)^2$
Radiation-dominated	Radiation	$a(t) = (t/t_0)^{1/2}$
	$p = \frac{1}{3}\rho c^2$	$a(\eta) = \eta/\eta_0$
Inflationary	Cosmological constant	$a(t) = \exp[H(t - t_0)]$
	$p = -\rho c^2 = \text{const}$	$a(\eta) = [1 - H(\eta - \eta_0)]^{-1}$

We are interested in metric perturbations that correspond at present time to gravitational waves, so we consider only spatial, transverse, traceless perturbations: $h = 0$, $h_{0\mu} = 0$, $\nabla_i h^i{}_j = 0$. We call this the synchronous transverse traceless gauge. The equation for such a metric perturbation is given by Eq. (4.107). The stress-energy tensor contains both the background stress energy solution of the homogeneous perfect fluid plus an anisotropic stress term, $\pi_{\alpha\beta}$:

$$T_{\alpha\beta} = (\rho + p/c)u_\alpha u_\beta + p\overset{0}{g}_{\alpha\beta} + \pi_{\alpha\beta} \,. \tag{4.116}$$

The equation for the metric perturbation depends only on the spatial, transverse and traceless part of this stress-energy tensor, $\pi_{\alpha\beta}$, which is a purely spatial and traceless tensor: $\pi_{0\alpha} = \pi_\mu{}^\mu = 0$. The anisotropic stress tensor contains any anisotropic matter perturbations (including the perturbation $p h_{ij}$ generated by the metric perturbation of the background fluid), as well as the effective stress energy from the metric perturbation itself (second order in h) – the transverse-traceless projection of the anisotropic stress tensor π_{ij}^{TT} will provide the source of gravitational waves. We then obtain a wave equation for the spatial components of the metric perturbation:

$$\frac{1}{a^2}\nabla^2 h_{ij} + \frac{1}{c^2}\left[-\frac{\partial^2}{\partial t^2}h_{ij} + \frac{1}{a}\frac{da}{dt}\frac{\partial}{\partial t}h_{ij} + 2\frac{1}{a}\frac{d^2a}{dt^2}h_{ij}\right] = \frac{16\pi G}{c^4}\pi_{ij}^{\text{TT}} \,. \tag{4.117}$$

This equation is often written in terms of the conformal metric perturbation $\hat{h}_{ij}$ which is related to the physical metric perturbation by the scale factor: $\hat{h}_{ij} := a^{-2}h_{ij}$. The equation for the conformal metric perturbation is then

$$\nabla^2\hat{h}_{ij} - \frac{1}{c^2}\left[a^2\frac{\partial^2}{\partial t^2}\hat{h}_{ij} - 3a\frac{da}{dt}\frac{\partial}{\partial t}\hat{h}_{ij}\right] = \frac{16\pi G}{c^4}\pi_{ij}^{\text{TT}} \,. \tag{4.118}$$

Other useful forms of the wave equation for the metric perturbation can be obtained in terms of the *conformal time* η in terms of which we write the Robertson–

Walker metric in the form

$$ds^2 = \overset{0}{g}_{\mu\nu} dx^\mu dx^\nu = a^2(\eta)\left[-c^2 d\eta^2 + dx^2 + dy^2 + dz^2\right]. \tag{4.119}$$

In this form the metric is conformal to flat Minkowski spacetime, $\overset{0}{g}_{\alpha\beta} = a^2(\eta)\eta_{\alpha\beta}$, where η is the Minkowski time coordinate and $a(\eta)$ is the conformal factor. The conformal time coordinate is related to the cosmological time coordinate by $a(\eta)d\eta = dt$. If we include the metric perturbation, the metric becomes

$$ds^2 = g_{\mu\nu} dx^\mu dx^\nu = a^2(\eta)\left[-c^2 d\eta^2 + dx^2 + dy^2 + dz^2 + \hat{h}_{ij} dx^i dx^j\right]. \tag{4.120}$$

and the wave equation for the metric perturbation is particularly simple:

$$\nabla^2 \hat{h}_{ij} - \frac{1}{c^2}\frac{\partial^2}{\partial\eta^2}\hat{h}_{ij} - 2\frac{1}{c^2}\frac{1}{a}\frac{da}{d\eta}\frac{\partial}{\partial\eta}\hat{h}_{ij} = \frac{16\pi G}{c^4}\pi_{ij}^{\mathrm{TT}}. \tag{4.121}$$

In this form, the conformal metric perturbation obeys the wave equation (in conformal time) with one additional damping term; the damping term can be ignored when $\partial\hat{h}_{ij}/\partial\eta \gg d\ln a/d\eta$, or when the physical frequency of the wave is much greater than $H := d\ln a/dt$ (which is known as the *Hubble parameter*).

To see what happens to waves that are of comparable size to the curvature scale of the Universe, consider a single polarization of a plane wave travelling along the z-direction with wave number k in an unperturbed ($\pi_{ij} = 0$) cosmological spacetime. Let

$$h_{ij}(\eta, z) = a(\eta)u(\eta)\mathrm{e}^{ikz}e_{ij}^{+}, \tag{4.122}$$

where the time-dependent factor $u(\eta)$ obeys the equation

$$\frac{d^2u}{d\eta^2} + \left(\hat{\omega}^2 - \frac{1}{a}\frac{d^2a}{d\eta^2}\right)u = 0 \tag{4.123}$$

with $\hat{\omega} := ck$. Notice that when $\hat{\omega}^2$ is much greater than the effective potential $V_{\mathrm{eff}}(\eta) = a^{-1}d^2a/d\eta^2$ then the solutions for u are $u \sim \mathrm{e}^{i\hat{\omega}\eta}$ where $\hat{\omega}$ is the frequency in conformal time of the wave, which is related to the true frequency f (in Hz) of the wave by the scale factor: $2\pi f = \hat{\omega}/a$. Notice that the only effect of cosmology on such a wave is that the true frequency of the wave gets redshifted by cosmological expansion. On the other hand, when $\hat{\omega}^2$ is much smaller than the effective potential, the size of the perturbation grows with time (for positive V_{eff}). The wave frequencies that are comparable to the effective potential have $\hat{\omega}^2 \sim V_{\mathrm{eff}}$ where

$$\begin{aligned} V_{\mathrm{eff}} &:= \frac{1}{a}\frac{d^2a}{d\eta^2} = a^2\left[\left(\frac{1}{a}\frac{da}{dt}\right)^2 + \frac{1}{a}\frac{d^2a}{dt^2}\right] = \frac{1}{2}H^2a^2\left(1 - 3\frac{p}{\rho c^2}\right) \\ &= a^2\begin{cases} 0 & \text{radiation, } p = \frac{1}{3}\rho c^2 \\ \frac{1}{2}H^2 & \text{dust, } p = 0 \\ 2H^2 & \text{dark energy, } p = -\rho c^2 . \end{cases} \end{aligned} \tag{4.124}$$

Waves with frequencies much greater than the Hubble parameter (i.e. with wavelengths much smaller than the size of the Universe), $f \gg H$, are those that unaffected by the effective potential, while waves with frequencies much less than the Hubble parameter (i.e. with wavelengths much greater than the size of the Universe), are affected by the effective potential. The exceptional case is that of a radiation-dominated universe, where the effective potential vanishes.

An important case to consider is when an initially small effective potential grows for some time and then once again becomes small. When gravitational waves encounter the effective potential, they are amplified (for an expanding universe). Notice that when $V_{\text{eff}} \gg \hat{\omega}^2$, Eq. (4.123) has an approximate solution $u(\eta) \propto a(\eta)$, so if waves initially "inside the horizon" with $f > H$ encounter a growing effective potential so that they "exit the horizon" with $f < H$ when the Universe has a scale factor a_1, and then at some later time if the effective potential decays to the point that $f > H$ once again at a time when the Universe has a (larger) scale factor a_2, then $u_2/u_1 \approx a_2/a_1$.

Example 4.2 Amplification of gravitational waves by inflation

Consider a Universe having an epoch of dark-energy dominated inflation that begins when the Universe has size a_1 and ends when the Universe has size a_2, at which point the Universe suddenly becomes radiation dominated. During inflation, the Hubble parameter has a constant value, H_{inf}, and since $H = a^{-1}(da/dt) = a^{-2}(da/d\eta)$, we have $V_{\text{eff}} = a^{-1}(d^2a/d\eta^2) = 2(H_{\text{inf}}a)^2$. The size of the Universe when a mode having amplitude u_1 and frequency $\hat{\omega}$ encounters the effective potential is therefore determined by $\hat{\omega}^2 = 2(H_{\text{inf}}a_1)^2$. At the end of inflation, $u_2/u_1 \approx a_2/a_1$, at which point the Universe suddenly becomes radiation-dominated and *the effective potential suddenly vanishes.* At this point, the growing mode with $u \simeq u_1 a/a_1$ must be matched onto an oscillating mode with $u/u_1 \sim \alpha \sin \hat{\omega}(\eta - \eta_2) + \beta \cos \hat{\omega}(\eta - \eta_2)$. Continuity requires $\alpha \sim a_2/a_1$ while matching the first derivative requires $\beta \sim (\hat{\omega} a_1)^{-1}(da_2/d\eta) \sim (a_2/a_1)^2 \gg \alpha$ (the scale factor of the Universe is expected to increase by a factor of at least 10^{27} during inflation). We see that the gravitational-wave mode is launched by inflation into a sinusoid with almost exactly zero phase.

The quantity β describes the amplification factor. In terms of the present day scale factor a_0 and the present day frequency $\omega = \hat{\omega}/a_0$, we have $\beta \sim (H_{\text{inf}}/\omega)^2(a_2/a_0)^2$. During the radiation era, $a \sim t^{1/2}$ and so $H_{\text{inf}}a_2^2 = H_0 a_0^2$ where H_0 is the present-day (if we ignore the recent matter-dominated era) value of the Hubble parameter, the *Hubble constant*, so we find $\beta \sim H_{\text{inf}} H_0/\omega^2$.

Suppose the pre-inflation era had Planck-scale fluctuations in the metric perturbation. Then the energy density in the frequency interval between frequency ω and frequency $\omega + d\omega$ is given by $d\rho_{\text{GW}}/d\omega \sim (\hbar/c^5)\omega^3$; this primordial energy spectrum is amplified by a factor of β^2, which gives

$$\frac{d\rho_{\text{GW}}}{d\ln\omega} \sim \frac{\hbar}{c^5} H_{\text{inf}}^2 H_0^2 \,. \tag{4.125}$$

Notice that this energy density per logarithmic frequency interval is *flat*, that is independent of frequency. We say that inflation produces a flat gravitational-wave spectrum. From the first Friedmann equation $H = \frac{8}{3}\pi G\rho$ we see that $H_{\text{inf}} = \frac{8}{3}\pi G\rho_{\text{inf}}$ and $H_0 = \frac{8}{3}\pi G\rho_{\text{crit}}$ where ρ_{crit} is the critical density today required for the Universe to be closed. We then define the gravitational-wave energy spectrum to be the fraction of energy density of gravitational waves in a logarithmic frequency interval compared to the total energy density in the Universe:

$$\Omega_{\text{GW}}(f) := \frac{1}{\rho_{\text{crit}}}\frac{d\rho_{\text{GW}}}{d\ln f} \sim \frac{\hbar G^2}{c^5}\rho_{\text{inf}} \sim \left(\frac{E_{\text{inf}}}{E_{\text{Planck}}}\right)^4 , \tag{4.126}$$

where E_{inf} is the energy scale of inflation, at which time the density was $\rho_{\text{inf}} \sim E^4_{\text{GUT}}/(c^5\hbar^3)$, and $E_{\text{Planck}} = \sqrt{\hbar c^5/G} \approx 2 \times 10^{19}$ GeV is the Planck energy. If inflation occurs at an energy scale of grand unification of the electroweak and strong particle theories (the GUT scale), $E_{\text{inf}} \sim E_{\text{GUT}} \sim 10^{16}$ GeV, then $\Omega_{\text{GW}} \sim (10^{16}\,\text{GeV}/10^{19}\,\text{GeV})^4 \sim 10^{-12}$.

We have ignored the matter-dominated era that occurred at a redshift of $z_{\text{eq}} = a_0/a_{\text{eq}} \approx 3000$ when the density of matter and the density of radiation were equal (which happened at the scale factor a_{eq}). To account for this recent epoch we must multiply Eq. (4.126) by the factor $(1+z_{\text{eq}})^{-1}$. This means that the predicted value of Ω_{GW} would be $\Omega_{\text{GW}} \sim 10^{-15}$, but this value is highly sensitive to the exact energy scale at which inflation occurred.

4.2.2
Black Hole Perturbation

Perturbations about a black hole spacetime are useful for studying the effects of small masses on the black hole. Unlike the post-Newtonian approximation, such perturbation calculations will be fully relativistic. In addition, the modes of oscillation of a perturbed black hole can be analyzed. We consider first the perturbation of a spherically symmetric black hole spacetime, which is described by the *Schwarzschild spacetime*, and then we discuss briefly the generalization of this result to rotating black holes, which are described by the *Kerr spacetime*.

We take as our background spacetime the (vacuum) Schwarzschild metric, which is given by the line element (in spherical-polar coordinates):

$$ds^2 = -\left(1 - \frac{2GM}{c^2 r}\right)(c\,dt)^2 + \left(1 - \frac{2GM}{c^2 r}\right)^{-1} dr^2 + r^2 d\theta^2 + r^2 \sin^2\theta\, d\phi^2 . \tag{4.127}$$

From this metric one can compute the background curvature tensor $\overset{0}{R}_{\alpha\beta\gamma\delta}$. We consider a perturbation of this spacetime, so that

$$R_{\alpha\beta\gamma\delta} = \overset{0}{R}_{\alpha\beta\gamma\delta} + \overset{1}{R}_{\alpha\beta\gamma\delta} ,$$

where $\overset{1}{R}_{\alpha\beta\gamma\delta}$ represents an outgoing gravitational-wave perturbation. It is helpful to introduce the following four basis vectors

$$l^\alpha := \frac{1}{\sqrt{2}}\left[\frac{1}{c}\left(1-\frac{2GM}{c^2 r}\right)^{-1}, 1, 0, 0\right] \tag{4.128a}$$

$$k^\alpha := \frac{1}{\sqrt{2}}\left[\frac{1}{c}, -\left(1-\frac{2GM}{c^2 r}\right), 0, 0\right] \tag{4.128b}$$

$$m^\alpha := \frac{1}{\sqrt{2}}\left[0, 0, \frac{1}{r}, \frac{i}{r\sin\theta}\right], \tag{4.128c}$$

which have the property that $\mathbf{l}\cdot\mathbf{k} = -1$, $\mathbf{m}\cdot\mathbf{m}^* = 1$, and $\mathbf{l}\cdot\mathbf{l} = \mathbf{k}\cdot\mathbf{k} = \mathbf{m}\cdot\mathbf{m} = \mathbf{l}\cdot\mathbf{m} = \mathbf{k}\cdot\mathbf{m} = 0$. Such a null basis is useful because we will be able to identify the gravitational-wave content that is experienced by an observer at a large distance from the black hole as the single component of the curvature perturbation. In particular, the single component of the Riemann tensor,

$$\begin{aligned} \Psi_4 &:= -k^\mu m^{*\nu} k^\rho m^{*\sigma} R_{\mu\nu\rho\sigma} \\ &= -k^\mu m^{*\nu} k^\rho m^{*\sigma} \overset{1}{R}_{\mu\nu\rho\sigma}, \end{aligned} \tag{4.129}$$

contains a complete description of the outgoing gravitational-wave content.[2)] Note that only the perturbation contributes to this scalar since

$$k^\mu m^{*\nu} k^\rho m^{*\sigma} \overset{0}{R}_{\mu\nu\rho\sigma} = 0$$

for the background Schwarzschild solution. At large distances, the waves are travelling in the outward radial direction, which we now call the z-axis of a Cartesian system in which the vectors $\mathbf{l}$, $\mathbf{k}$ and $\mathbf{m}$ asymptotically become

$$l^\alpha = \frac{1}{\sqrt{2}}\left[c^{-1}, 0, 0, 1\right] \tag{4.130a}$$

$$k^\alpha = \frac{1}{\sqrt{2}}\left[c^{-1}, 0, 0, -1\right] \tag{4.130b}$$

$$m^\alpha = \frac{1}{\sqrt{2}}[0, 1, i, 0]. \tag{4.130c}$$

We know that the gravitational wave will be described by $R_{0101} = -cR_{0131} = -cR_{3101} = c^2 R_{3131} = -R_{0202} = cR_{0232} = cR_{3202} = -c^2 R_{3232} = -\frac{1}{2}\ddot{h}_+$ and $R_{0102} = -cR_{0132} = -cR_{3102} = c^2 R_{3132} = R_{0201} = -cR_{0231} = -cR_{3201} =$

2) In fact, Ψ_4 is normally defined in terms of the *Weyl tensor*, $C_{\alpha\beta\gamma\delta} = R_{\alpha\beta\gamma\delta} - \frac{1}{2}(g_{\alpha\gamma}R_{\beta\delta} - g_{\beta\gamma}R_{\alpha\delta} + g_{\alpha\delta}R_{\beta\gamma} - g_{\beta\delta}R_{\alpha\gamma}) + \frac{1}{6}R(g_{\alpha\gamma}g_{\beta\delta} - g_{\alpha\delta}g_{\beta\gamma})$, as $\Psi_4 := -k^\mu m^{*\nu} k^\rho m^{*\sigma} C_{\mu\nu\rho\sigma}$. In vacuum, $R_{\alpha\beta} = 0$ and $R = 0$ so the Riemann tensor and the Weyl tensor are the same. The *Weyl scalars*, including Ψ_4, are the components of the Weyl tensor in the null basis. Notice however that Ψ_4 can be defined either in terms of the Riemann tensor or the Weyl tensor, even for non-vacuum spacetimes.

$c^2 R_{3231} = -\frac{1}{2}\ddot{h}_\times$, and with $R_{0303} = 0$ (transverse wave), so we see that Ψ_4 can be written as

$$\Psi_4 = c^{-2}(\ddot{h}_+ - i\ddot{h}_\times) \qquad (r \to \infty) . \tag{4.131}$$

Thus, at large distances the complex scalar Ψ_4 contains the outgoing gravitational-wave content.

We seek a wave equation for this component of the curvature tensor. We begin with Eq. (4.95) by computing

$$k^\mu m^{*\nu} k^\rho m^{*\sigma} \nabla^2 R_{\mu\nu\rho\sigma} = k^\mu m^{*\nu} k^\rho m^{*\sigma} g^{\kappa\tau} g^{\lambda\phi} \times \left\{ -\overset{0}{R}_{\mu\nu\kappa\lambda} \overset{1}{R}_{\tau\phi\rho\sigma} + 2\overset{0}{R}_{\mu\kappa\sigma\lambda} \overset{1}{R}_{\nu\tau\rho\phi} - 2\overset{0}{R}_{\nu\kappa\sigma\lambda} \overset{1}{R}_{\mu\tau\rho\phi} \right\} . \tag{4.132}$$

The left-had side will have one term that is $-\nabla^2 \Psi_4$, but there will be many other terms arising from the wave operator acting on the other components of the Riemann curvature perturbation. Furthermore, the right hand side will depend on Ψ_4 but it too will also depend on other components of the Riemann perturbation. Remarkably, the whole combination only depends on the single curvature perturbation Ψ_4, that is, this component of the perturbation *decouples* from the other components of the perturbation (see Ryan, 1974). The resulting wave equation for Ψ_4 is the *Teukolsky equation* for a Schwarzschild black hole (Teukolsky, 1972, 1973):

$$\frac{r^4}{\Delta} \frac{\partial^2 \psi}{c^2 \partial t^2} - \frac{1}{\sin^2\theta} \frac{\partial^2 \psi}{\partial \phi^2} - \Delta^2 \frac{\partial}{\partial r}\left(\frac{1}{\Delta}\frac{\partial \psi}{\partial r}\right) - \frac{1}{\sin\theta}\frac{\partial}{\partial\theta}\left(\sin\theta \frac{\partial\psi}{\partial\theta}\right) + 4i\frac{\cos\theta}{\sin^2\theta}\frac{\partial\psi}{\partial\phi} + 4\left[\frac{GMr^2/c^2}{\Delta} - r\right]\frac{\partial\psi}{c\,\partial t} + 2(2\cot^2\theta + 1)\psi = 0 , \tag{4.133}$$

where $\psi := r^4 \Psi_4$ and $\Delta := r^2 - 2GMr/c^2$. If a matter source is generating the perturbation, then a source term related to the matter stress energy appears on the right hand side of this equation.

The Teukolsky equation is also separable: let

$$\psi(t, r, \theta, \phi) = \sum_{\ell=2}^{\infty} \sum_{m=-\ell}^{\ell} {}_{-2}Y_{\ell m}(\theta, \phi) \int_{-\infty}^{\infty} R_\ell(r, \omega) e^{i\omega t} \frac{d\omega}{2\pi} , \tag{4.134}$$

where ${}_{-2}Y_{\ell m}(\theta, \phi)$ are the spin-weighted spherical harmonics of spin-weight $s = -2$. Then the function $R_\ell(r, \omega)$ satisfies the ordinary differential equation

$$\Delta^2 \frac{d}{dr}\left(\frac{1}{\Delta}\frac{dR_\ell}{dr}\right) + \left[\frac{r^4(\omega/c)^2 - 4ir^2(r - 3GM/c^2)(\omega/c)}{\Delta} - \ell(\ell+1) + 2\right] R_\ell = 0 . \tag{4.135}$$

Again, if a matter source is present, the source term will appear on the right hand side. If there is *no* source, the homogeneous equation (with suitable boundary conditions corresponding to outward going gravitational waves as $r \to \infty$ and inward

Table 4.4 The first few (complex) quasi-normal mode frequencies $GM\omega_{n\ell}/c^3$ for a non-spinning Schwarzschild black hole of mass M. See Leaver (1985) (note that Leaver sets $GM/c^3 = 1/2$ and has the opposite sign convention for ω).

	$\ell = 2$	$\ell = 3$	$\ell \gg 1$
$n = 1$	$\pm 0.373\,672 + 0.088\,962 i$	$\pm 0.599\,443 + 0.092\,703 i$	
$n = 2$	$\pm 0.346\,711 + 0.273\,915 i$	$\pm 0.582\,644 + 0.281\,298 i$	$\pm \frac{\ell+\frac{1}{2}}{3\sqrt{3}} + \frac{n+\frac{1}{2}}{3\sqrt{3}} i$
$n = 3$	$\pm 0.301\,054 + 0.478\,272 i$	$\pm 0.551\,685 + 0.479\,093 i$	
$n \to \infty$	$\pm 0.08 + (\frac{1}{4} n - 0.1) i$	$\pm 0.08 + (\frac{1}{4} n - 0.32) i$	

going gravitational waves at the black hole horizon) can be solved only for certain complex values of the eigenfrequency ω. These values are known as the *quasi-normal modes* of the Schwarzschild black hole, and are given in Table 4.4. These are the natural frequencies of oscillation of the black hole. Notice that the quasi-normal mode frequencies are complex, and so they decay exponentially with time: the black hole radiates any perturbation away and settles down to its stationary state. From these quasi-normal mode frequencies $\omega_{n\ell}$ we obtain the spectrum of gravitational-wave frequencies $f_{n\ell} = 2\pi \operatorname{Re} \omega_{n\ell}$ and the exponential decay time $\tau_{n\ell} = (\operatorname{Im} \omega_{n\ell})^{-1}$. Notice that the $n = 1$, $\ell = 2$ mode has the longest decay time; this is typically the most strongly excited mode of oscillation of a black hole.

In fact, we can obtain a decoupled curvature perturbation Ψ_4 not only for the non-rotating Schwarzschild black hole, but also for the generic stationary rotating black hole, which is described by the Kerr metric

$$\begin{aligned} ds^2 = &-\left(1 - \frac{2GMr}{c^2\Sigma}\right)(c\,dt)^2 - \frac{4GMar\sin^2\theta}{c^2\Sigma}(c\,dt)d\phi + \frac{\Sigma}{\Delta}dr^2 \\ &+ \Sigma\, d\theta^2 + \left(r^2 + a^2 + \frac{2GMa^2r}{c^2\Sigma}\sin^2\theta\right)\sin^2\theta\, d\phi^2\,, \end{aligned} \tag{4.136}$$

where $\Delta := r^2 - 2GMr/c^2 + a^2$, $\Sigma := r^2 + a^2\cos^2\theta$, and the parameter a is related to the black hole's spin angular momentum $\boldsymbol{S}$ and mass M by $a := S/(Mc)$. The basis vectors that are analogous to Eq. (4.128) are

$$l^a := \frac{1}{\sqrt{2}}\left[\frac{r^2 + a^2}{c\Delta}, 1, 0, \frac{a}{\Delta}\right] \tag{4.137a}$$

$$k^a := \frac{1}{\sqrt{2}}\left[\frac{r^2 + a^2}{c\Sigma}, -\frac{\Delta}{\Sigma}, 0, \frac{a}{\Sigma}\right] \tag{4.137b}$$

$$m^a := \frac{1}{\sqrt{2}}\left[\frac{ia\sin\theta}{c(r^2 + ia\cos\theta)}, 0, \frac{1}{r + ia\cos\theta}, \frac{i}{(r + ia\cos\theta)\sin\theta}\right]. \tag{4.137c}$$

These vectors also satisfy $\mathbf{l}\cdot\mathbf{k} = -1$, $\mathbf{m}\cdot\mathbf{m}^* = 1$ and $\mathbf{l}\cdot\mathbf{l} = \mathbf{k}\cdot\mathbf{k} = \mathbf{m}\cdot\mathbf{m} = \mathbf{l}\cdot\mathbf{m} = \mathbf{k}\cdot\mathbf{m} = 0$, and $\Psi_4 := -k^\mu m^{*\nu} k^\rho m^{*\sigma} R_{\mu\nu\rho\sigma}$ continues to represent the outgoing gravitational waves from the curvature perturbation (Eq. (4.131)). The decoupled Teukolsky equation now has the form

$$\begin{aligned}
&\left[\frac{(r^2+a^2)^2}{\Delta} - a^2 \sin^2\theta\right]\frac{\partial^2\psi}{c^2\partial t^2} + \frac{4GMar}{c^3\Delta}\frac{\partial^2\psi}{\partial t\partial\phi} + \left[\frac{a^2}{\Delta} - \frac{1}{\sin^2\theta}\right]\frac{\partial^2\psi}{\partial\phi^2}\\
&-\Delta^2\frac{\partial}{\partial r}\left(\frac{1}{\Delta}\frac{\partial\psi}{\partial r}\right) - \frac{1}{\sin\theta}\frac{\partial}{\partial\theta}\left(\sin\theta\frac{\partial\psi}{\partial\theta}\right)\\
&+4\left[\frac{a(r-GM/c^2)}{\Delta} + i\frac{\cos\theta}{\sin^2\theta}\right]\frac{\partial\psi}{\partial\phi}\\
&+4\left[\frac{GM(r^2-a^2)/c^2}{\Delta} - r - ia\cos\theta\right]\frac{\partial\psi}{c\partial t} + 2(2\cot^2\theta + 1)\psi = 0\,,
\end{aligned} \tag{4.138}$$

where $\psi := (r - ia\cos\theta)^4\Psi_4$. The Teukolsky equation for the Kerr spacetime is again separable. Let

$$\psi(t, r, \theta, \phi) = \sum_{\ell=2}^{\infty}\sum_{m=-\ell}^{\ell} e^{im\phi}\int_{-\infty}^{\infty} S_{\ell m}(\cos\theta, \omega) R_{\ell m}(r, \omega) e^{i\omega t}\frac{d\omega}{2\pi}\,. \tag{4.139}$$

Then the angular and radial functions $S_{\ell m}(\cos\theta, \omega)$ and $R_{\ell m}(r, \omega)$ satisfy the ordinary differential equations

$$\begin{aligned}
&\frac{d}{d\cos\theta}\left[(1-\cos^2\theta)\frac{dS_{\ell m}}{d\cos\theta}\right]\\
&+\left[a^2(\omega/c)^2\cos^2\theta + 4a(\omega/c)\cos\theta - 2 - \frac{(m-2\cos\theta)^2}{1-\cos^2\theta} + A_{\ell m}\right]S_{\ell m}\\
&= 0
\end{aligned} \tag{4.140a}$$

$$\begin{aligned}
&\Delta^2\frac{d}{dr}\left(\frac{1}{\Delta}\frac{dR_{\ell m}}{dr}\right)\\
&\quad+\left\{\frac{[(r^2+a^2)(\omega/c) - am]^2 - 4i(r - GM/c^2)[(r^2+a^2)(\omega/c) - am]}{\Delta}\right.\\
&\quad\left.-8ir(\omega/c) + 2am(\omega/c) - a^2(\omega/c)^2 - A_{\ell m}\right\}R_{\ell m} = 0
\end{aligned} \tag{4.140b}$$

(Teukolsky, 1973). If matter sources are present, terms would appear on the right hand side of Eq. (4.140b). Here, $A_{\ell m}(a\omega/c)$ is a separation constant. The solutions to the differential equation for the angular function $S_{\ell m}(\cos\theta, \omega)$ with boundary conditions that $S_{\ell m}$ is finite for $\cos\theta = \pm 1$ are the *spin-weighted spheroidal harmonics* of spin-weight $s = -2$. The solutions to the homogeneous Teukolsky equation represent the quasi-normal modes of oscillation of the Kerr black hole. The frequencies of these quasi-normal modes now additionally depend on the spin parameter a of the black hole and on the angular mode number m. The first few of these quasi-normal mode frequencies are given in Table 4.5.

Table 4.5 The fundamental $n = 1$ quadrupolar $\ell = 2$ quasi-normal mode frequencies $GM\omega_{\ell m}/c^3$ and eigenvalues $A_{\ell m}$ for spinning Kerr black holes of mass M and spin parameter $a = S/(Mc)$. See Leaver (1985) (note that Leaver sets $GM/c^3 = 1/2$ and has the opposite sign convention for ω). The table gives values for positive m only; by symmetry, $\omega_{\ell,-m}(a) = \omega_{\ell,m}(-a)$ and $A_{\ell,-m}(a) = A_{\ell,m}(-a)$.

	$c^2a/(GM)$	$A_{\ell m}$	$GM\omega_{\ell m}/c^3$
$\ell = 2,\ m = +2$	-0.98	$4.71083 + 0.19500i$	$+0.292\,663 + 0.088\,078i$
	-0.90	$4.66623 + 0.18160i$	$+0.297\,244 + 0.088\,281i$
	-0.80	$4.60808 + 0.16428i$	$+0.303\,313 + 0.088\,512i$
	-0.60	$4.48269 + 0.12765i$	$+0.316\,784 + 0.088\,892i$
	-0.40	$4.34266 + 0.08821i$	$+0.332\,458 + 0.089\,131i$
	-0.20	$4.18385 + 0.04574i$	$+0.351\,053 + 0.089\,183i$
	0	4	$\pm 0.373\,672 + 0.088\,962i$
	$+0.20$	$3.78097 + 0.04921i$	$-0.402\,145 + 0.088\,311i$
	$+0.40$	$3.50868 + 0.10185i$	$-0.439\,842 + 0.086\,882i$
	$+0.60$	$3.14539 + 0.15669i$	$-0.494\,045 + 0.083\,765i$
	$+0.80$	$2.58529 + 0.20530i$	$-0.586\,017 + 0.075\,630i$
	$+0.90$	$2.10982 + 0.21112i$	$-0.671\,614 + 0.064\,869i$
	$+0.98$	$1.33362 + 0.14999i$	$-0.825\,429 + 0.038\,630i$
$\ell = 2,\ m = +1$	-0.98	$4.38917 + 0.07868i$	$+0.343\,922 + 0.083\,713i$
	-0.90	$4.36229 + 0.07547i$	$+0.344\,359 + 0.084\,865i$
	-0.80	$4.32786 + 0.07049i$	$+0.345\,356 + 0.086\,003i$
	-0.60	$4.25579 + 0.05767i$	$+0.348\,911 + 0.087\,566i$
	-0.40	$4.17836 + 0.04150i$	$+0.354\,633 + 0.088\,484i$
	-0.20	$4.09389 + 0.02224i$	$+0.362\,738 + 0.088\,935i$
	0	4	$\pm 0.373\,672 + 0.088\,962i$
	$+0.20$	$3.89315 + 0.02520i$	$-0.388\,248 + 0.088\,489i$
	$+0.40$	$3.76757 + 0.05324i$	$-0.407\,979 + 0.087\,257i$
	$+0.60$	$3.61247 + 0.08347i$	$-0.435\,968 + 0.084\,564i$
	$+0.80$	$3.40228 + 0.11217i$	$-0.480\,231 + 0.077\,955i$
	$+0.90$	$3.25345 + 0.11951i$	$-0.516\,291 + 0.069\,804i$
	$+0.98$	$3.07966 + 0.10216i$	$-0.564\,155 + 0.051\,643i$
$\ell = 2,\ m = 0$	0	4	$\pm 0.373\,672 + 0.088\,962i$
	± 0.20	$3.99722 \pm 0.00139i$	$\mp 0.375\,124 + 0.088\,700i$
	± 0.40	$3.98856 \pm 0.00560i$	$\mp 0.379\,682 + 0.087\,827i$
	± 0.60	$3.97297 \pm 0.01262i$	$\mp 0.388\,054 + 0.085\,995i$
	± 0.80	$3.94800 \pm 0.02226i$	$\mp 0.401\,917 + 0.082\,156i$
	± 0.90	$3.93038 \pm 0.02763i$	$\mp 0.412\,004 + 0.078\,483i$
	± 0.98	$3.91269 \pm 0.03152i$	$\mp 0.422\,254 + 0.073\,532i$

Example 4.3 Black hole ringdown radiation

A $M = 10\,M_\odot$ black hole with a spin that is 98% of the extremal Kerr spin (i.e. $cS/(GM^2) = 0.98$) is created during a binary merger. Upon formation at time t_0, the black hole is distorted and begins to radiate. Suppose this ringdown radiation is dominated by the fundamental $\ell = |m| = 2$ mode. At a large distance r, the gravitational waveform will be given by Ψ_4, which is

$$\Psi_4 = \frac{A}{r} e^{-\pi f_{\mathrm{qnm}}(t-t_0)/Q_{\mathrm{qnm}}} \operatorname{Re}\left\{ e^{2i\phi} e^{2\pi i f_{\mathrm{qnm}}(t-t_0)} \times \left[S_{22}\left(\cos\theta, 2\pi f_{\mathrm{qnm}}\right) + S_{22}^*\left(-\cos\theta, 2\pi f_{\mathrm{qnm}}\right)\right]\right\} \quad (t > t_0)\,, \tag{4.141}$$

where A is the initial amplitude, f_{qnm} is the frequency of the quasi-normal mode, and Q_{qnm} is the quality factor of the quasi-normal mode, which is related to its decay time τ_{qnm} by $Q_{\mathrm{qnm}} = \pi f_{\mathrm{qnm}} \tau_{\mathrm{qnm}}$. The quasi-normal mode frequency and quality factor are obtained from the value $GM\omega_{22}/c^3 = -0.825\,429 + 0.038\,630i$ for $c^2 a/(GM) = 0.98$:

$$f_{\mathrm{qnm}} = \frac{|\operatorname{Re}\omega_{22}|}{2\pi} = 2.667\,\mathrm{kHz}\left(\frac{M}{10\,M_\odot}\right)^{-1} \tag{4.142a}$$

$$Q_{\mathrm{qnm}} = \frac{1}{2}\frac{|\operatorname{Re}\omega_{22}|}{\operatorname{Im}\omega_{22}} = 10.7\,. \tag{4.142b}$$

The quasi-normal mode frequency and quality factors for black hole ringdown are approximately equal to the following fits obtained by Berti *et al.* (2006): for the fundamental, $n = 1$, quadrupolar quasi-normal modes $\ell = 2$, they obtain

$$\frac{2\pi GM}{c^3} f_{2m} = \begin{cases} 1.5251 - 1.1568(1-j)^{0.1292} & m = +2 \\ 0.6000 + 0.2339(1-j)^{0.4175} & m = +1 \\ 0.4437 + 0.0739(1-j)^{0.3350} & m = 0 \\ 0.3441 + 0.0293(1-j)^{2.0010} & m = -1 \\ 0.2938 + 0.0782(1-j)^{1.3546} & m = -2 \end{cases} \tag{4.143a}$$

and

$$Q_{2m} = \begin{cases} +0.7000 + 1.4187(1-j)^{-0.4990} & m = +2 \\ -0.3000 + 2.3561(1-j)^{-0.2277} & m = +1 \\ +4.0000 - 1.9550(1-j)^{0.1420} & m = 0 \\ +2.0000 + 0.1078(1-j)^{5.0069} & m = -1 \\ +1.6700 + 0.4192(1-j)^{1.4700} & m = -2 \end{cases}, \tag{4.143b}$$

where $j := c^2 a/(GM)$ is the dimensionless Kerr parameter.

4.3 Numerical Relativity

Non-perturbative analyses of strongly gravitating systems require us to obtain numerical solutions to the Einstein equations. In this section we present a brief overview of numerical relativity, and we refer the reader to Alcubierre (2008) and Baumgarte and Shapiro (2010) for a complete discussion of the topic.

The most natural approach to generating numerical solutions is to cast the Einstein equations as a set of time evolution equations and a set of initial value equations. To do this, spacetime is sliced into spatial slices according to some prescribed time-like vector field (the level surfaces of this vector field prescribe the "time" on each slice), and the geometry of spacetime is similarly decomposed into the geometry of each of these spatial slices, which gives the intrinsic curvature of the slices, and a description of how these slices are embedded in spacetime, which gives the extrinsic curvature of the slices.

4.3.1 The Arnowitt–Deser–Misner (ADM) Formalism

The standard decomposition proceeds as follows: a time coordinate t is introduced which labels spatial slices. The vector $\mathbf{t} := d/dt$ then describes how to associate a point on one time slice $x^\alpha(t)$ with the "same" point on the next time slice $x^\alpha(t + \Delta t)$: since

$$\mathbf{t} := \frac{d}{dt} = \frac{dx^\mu}{dt}\frac{\partial}{\partial x^\mu} = t^\mu \frac{\partial}{\partial x^\mu} \tag{4.144}$$

we see

$$t^\alpha := \frac{dx^\alpha}{dt} = \lim_{\Delta t \to 0} \frac{x^\alpha(t + \Delta t) - x^\alpha(t)}{\Delta t} \,. \tag{4.145}$$

Points on two time slices may be associated in different ways depending on our choice of the spatial coordinates on the slices. Furthermore, the rate of passage of time for the various regions on the spatial slice can be adjusted arbitrarily (thereby changing how we divide spacetime into slices). We can explicitly see these two effects by writing the vector $\mathbf{t}$ in terms of the time-like normal vector to a spatial slice $\mathbf{n}$, $n_\alpha \propto \nabla_\alpha t$ (imagine t as a scalar field on spacetime whose level surfaces are the surfaces of constant coordinate time, and therefore the normal to the surfaces is the gradient with respect to the field t), and another vector, tangent to the spatial slice, called the *shift vector*, $\boldsymbol{\beta}$ (Figure 4.1),

$$\mathbf{t} = \alpha \mathbf{n} + \boldsymbol{\beta} \,, \tag{4.146}$$

where α is known as the *lapse function* (it can change over a spatial surface); since $t^\mu \nabla_\mu t = 1$ and $\beta^\mu \nabla_\mu t = 0$ (because $\boldsymbol{\beta}$ lies entirely on the spatial slice) we see that

$$n_\beta = -\alpha \nabla_\beta t = [-\alpha, 0, 0, 0] \,, \tag{4.147}$$

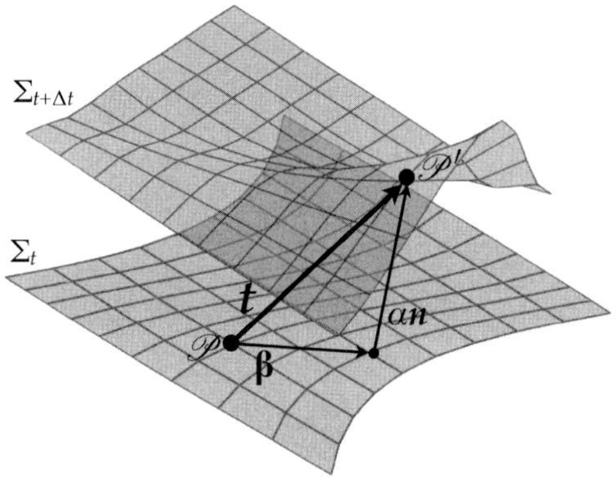

Figure 4.1 The decomposition of the vector **t** into a shift vector **β** and a lapse function α times the time-like unit normal **n**. The vector **t** maps points on one time slice Σ_t to the next time slice $\Sigma_{t+\Delta t}$, for example point $\mathcal{P}$ is mapped to point $\mathcal{P}'$. The shift vector **β** describes how much to move over on the surface Σ_t before going to the next surface and the lapse function α describes the normal distance to the next surface. Specification of the lapse and shift determine the foliation of spacetime.

where the lapse function now plays the role of the normalization constant for time-like normal, that is α is defined so that

$$g_{\mu\nu} n^\mu n^\nu = -1 . \tag{4.148}$$

The normal vector is (by definition) orthogonal to the metric on the spatial slice, $\gamma_{\alpha\beta}$, that is $n^\mu \gamma_{\mu\alpha} = 0$, so we define

$$\gamma_{\alpha\beta} := g_{\alpha\beta} + n_\alpha n_\beta . \tag{4.149}$$

The tensor $\gamma_{\alpha\beta}$ can be viewed as a projection tensor that projects a vector onto the spatial slice, and the spatial components of this tensor, γ_{ij}, make up the intrinsic metric on the spatial slice. We can now compute the components of the spacetime metric:

$$g_{00} = g_{\mu\nu} t^\mu t^\nu = (\gamma_{\mu\nu} - n_\mu n_\nu)(\alpha n^\mu + \beta^\mu)(\alpha n^\nu + \beta^\nu) = \gamma_{ij} \beta^i \beta^j - \alpha^2 \tag{4.150}$$

$$g_{0j} = g_{\mu j} t^\mu = \gamma_{\mu j}(\alpha n^\mu + \beta^\mu) = \beta_j \tag{4.151}$$

$$g_{ij} = \gamma_{ij} \tag{4.152}$$

and therefore

$$ds^2 = -\alpha^2 dt^2 + \gamma_{ij} \left(dx^i + \beta^i dt\right)\left(dx^j + \beta^j dt\right) . \tag{4.153}$$

Note that α and β^i (which we now write as a spatial vector $\boldsymbol{\beta}$) both have units of velocity (m s^{-1}).

The quantity γ_{ij} describes the intrinsic geometry on the spatial slice. The *intrinsic* curvature of this three-dimensional space is described by the Riemann tensor

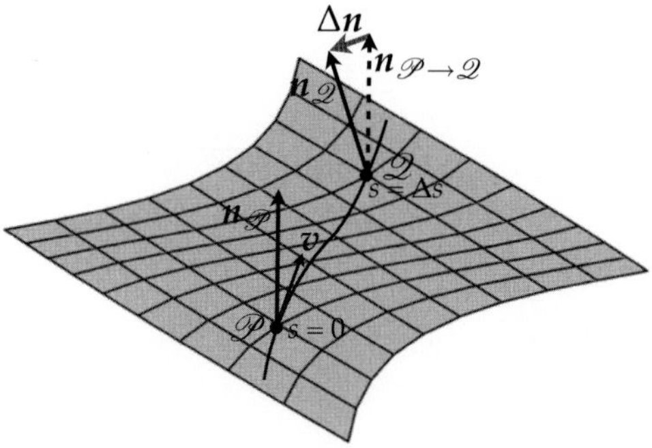

Figure 4.2 The extrinsic curvature of a surface describes how the normal vector to the surface changes over the surface compared to a parallel-transported normal vector. A point $\mathcal{Q}$ (with $s = \Delta s$) is arrived at from point $\mathcal{P}$ (with $s = 0$) along the integral curve of the vector $\mathbf{v} = d/ds$. The difference between the normal vector at point $\mathcal{Q}$, $\mathbf{n}_\mathcal{Q}$ and the vector $\mathbf{n}_{\mathcal{P}\to\mathcal{Q}}$, which is the parallel-transported image of the normal vector at point $\mathcal{P}$, is $\Delta\mathbf{n}$ with components $\Delta n^\alpha = \Delta s v^\mu \nabla_\mu n^\alpha$. The extrinsic curvature $K_{\alpha\beta}$ is defined by $-v^\mu K_{\mu\beta} = v^\mu \nabla_\mu n_\beta$, or, since $\mathbf{v}$ is an arbitrary vector on the spatial surface, $K_{\alpha\beta} = -\gamma^\mu_\alpha \nabla_\mu n_\beta$.

$\mathcal{R}_{ijkl}$ constructed from the spatial metric γ_{ij}. Another measure of the curvature of the space, however, invokes the way that the three-dimensional spatial surface is embedded in the four-dimensional spacetime: the *extrinsic curvature* shows how the unit normal $\mathbf{n}$ to the spatial slice changes over the slice. Consider two points $\mathcal{P}$ and $\mathcal{Q}$, both on the spatial slice, and connected by the integral curve of a vector $\mathbf{v} = d/ds$ that lies on the slice. The difference between the normal vector at point $\mathcal{Q}$, $\mathbf{n}_\mathcal{Q}$, and the vector $\mathbf{n}_{\mathcal{P}\to\mathcal{Q}}$ which is the normal vector at point $\mathcal{P}$ parallel-transported along $\mathbf{v}$ to point $\mathcal{Q}$, is

$$\Delta n^\alpha = \Delta s\, v^\mu \nabla_\mu n^\alpha \,. \tag{4.154}$$

In the limit $\Delta s \to 0$, we define the extrinsic curvature along the integral curve of $\mathbf{v}$ to be $-dn^\alpha/ds = K^\alpha{}_\mu v^\mu$, where $K_{\alpha\beta}$ is the extrinsic curvature tensor; therefore

$$-v^\mu K_{\mu\beta} := v^\mu \nabla_\mu n_\beta \,, \tag{4.155}$$

or, since $\mathbf{v}$ is an arbitrary vector on the spatial slice,

$$K_{\alpha\beta} = -\gamma^\mu_\alpha \nabla_\mu n_\beta \,. \tag{4.156}$$

(See Figure 4.2 for an illustration.) Note that because $n^\nu \nabla_\alpha n_\nu = 0$ (since $\mathbf{n}$ has unit normalization) the extrinsic curvature is purely spatial, that is we can write it as a tensor K_{ij} living entirely on the spatial slice. Nevertheless, the extrinsic curvature depends on the normal vector to the slice and therefore it is a measure of the curvature of the slice *as embedded in the spacetime* (hence the word "extrinsic" in its name). In fact, the extrinsic curvature can be thought of in terms of the time derivative of the intrinsic metric $\gamma_{\alpha\beta}$, that is as a kind of velocity in an initial value problem.

To demonstrate this, first we show that the extrinsic curvature is symmetric: $K_{\alpha\beta} = K_{\beta\alpha}$. This follows from the fact that $\mathbf{n}$ is a normal vector, $n_\mu = -\alpha\nabla_\mu t$, and so

$$\nabla_\mu n_\nu - \nabla_\nu n_\mu = n_\mu \nabla_\nu \ln \alpha - n_\nu \nabla_\mu \ln \alpha \tag{4.157}$$

since $\nabla_\mu \nabla_\nu t = \nabla_\nu \nabla_\mu t$. Note that both terms on the right-hand side are proportional to the unit normal. Now, because $K_{\alpha\beta}$ is spatial, $K_{\alpha\beta} = \gamma^\mu_\alpha \gamma^\nu_\beta K_{\mu\nu} = -\gamma^\mu_\alpha \gamma^\nu_\beta \nabla_\mu n_\nu$, we see that

$$\begin{aligned} K_{\alpha\beta} - K_{\beta\alpha} &= -\gamma^\mu_\alpha \gamma^\nu_\beta (\nabla_\mu n_\nu - \nabla_\nu n_\mu) \\ &= -\gamma^\mu_\alpha \gamma^\nu_\beta (n_\mu \nabla_\nu \ln \alpha - n_\nu \nabla_\mu \ln \alpha) \\ &= 0\,, \end{aligned} \tag{4.158}$$

which shows that $K_{\alpha\beta}$ is symmetric in its indices. Furthermore,

$$\begin{aligned} 2K_{\alpha\beta} &= K_{\alpha\beta} + K_{\beta\alpha} \\ &= -\gamma^\mu_\alpha \nabla_\mu n_\beta - \gamma^\mu_\beta \nabla_\mu n_\alpha \\ &= -\nabla_\alpha n_\beta - \nabla_\beta n_\alpha - n^\mu \nabla_\mu (n_\alpha n_\beta) \\ &= -\nabla_\alpha n_\beta - \nabla_\beta n_\alpha - n^\mu \nabla_\mu \gamma_{\alpha\beta}\,. \end{aligned} \tag{4.159}$$

The last term looks like a time derivative of the spatial metric. In fact, this equation can be re-expressed as

$$\frac{\partial \gamma_{ij}}{\partial t} = -2\alpha K_{ij} + D_i \beta_j + D_j \beta_i \tag{4.160}$$

in our coordinate system, where we have introduced the covariant derivative D_i on the spatial slice, that is, the covariant derivative that is compatible with the spatial-metric: $D_i \gamma_{jk} = 0$. In particular, the covariant derivatives D and ∇ are related by

$$D_\alpha w^\beta = \gamma^\mu_\alpha \gamma^\beta_\nu \nabla_\mu w^\nu \tag{4.161}$$

for arbitrary vector **w**, and this allows us to obtain an expression that relates the intrinsic curvature of the spatial surface $\mathcal{R}_{\alpha\beta\gamma}{}^\delta$, the extrinsic curvature of the spatial slice $K_{\alpha\beta}$, and the curvature of the surrounding spacetime $R_{\alpha\beta\gamma}{}^\delta$ (evaluated on the spatial slice). Starting with the definition of intrinsic curvature (for the spatial slice), we find (after some manipulation) for arbitrary vector **w** *on the spatial slice,*

$$\begin{aligned} \mathcal{R}_{\alpha\beta\sigma}{}^\delta w^\sigma &:= -(D_\alpha D_\beta - D_\beta D_\alpha) w^\delta \\ &= -\gamma^\mu_\alpha \gamma^\nu_\beta \gamma^\delta_\sigma (\nabla_\mu \nabla_\nu - \nabla_\nu \nabla_\mu) w^\sigma + K_\alpha{}^\delta K_{\beta\sigma} w^\sigma - K_\beta{}^\delta K_{\alpha\sigma} w^\sigma \end{aligned} \tag{4.162}$$

so

$$\gamma^\mu_\alpha \gamma^\nu_\beta \gamma^\rho_\gamma \gamma^\delta_\sigma R_{\mu\nu\rho}{}^\sigma = \mathcal{R}_{\alpha\beta\gamma}{}^\delta + K_{\alpha\gamma} K_\beta{}^\delta - K_{\beta\gamma} K_\alpha{}^\delta\,. \tag{4.163}$$

This is known as the *Gauss equation*. Two other important relations are the *Codazzi equation,*

$$\gamma^\mu_\alpha \gamma^\nu_\beta \gamma^\rho_\gamma n_\sigma R_{\mu\nu\rho}{}^\sigma = -D_\alpha K_{\beta\gamma} + D_\beta K_{\alpha\gamma}\,, \tag{4.164}$$

and the *Ricci equation*,

$$\gamma_\alpha^\mu n^\nu \gamma_\beta^\rho n_\sigma R_{\mu\nu\rho}{}^\sigma = \frac{1}{\alpha}\left[\frac{\partial K_{\alpha\beta}}{\partial t} + \beta^\mu \frac{\partial K_{\alpha\beta}}{\partial x^\mu} - K_{\mu\beta}\frac{\partial \beta^\mu}{\partial x^\alpha} - K_{\alpha\mu}\frac{\partial \beta^\mu}{\partial x^\beta}\right] + \frac{1}{\alpha} D_\alpha D_\beta \alpha + K_\alpha{}^\mu K_{\beta\mu} \,. \tag{4.165}$$

Now we decompose Einstein's equations:

$$G_{\mu\nu} n^\mu n^\nu = \frac{8\pi G}{c^4} T_{\mu\nu} n^\mu n^\nu = \frac{8\pi G}{c^2}\rho \tag{4.166}$$

$$G_{\mu\nu} \gamma_\alpha^\mu n^\nu = \frac{8\pi G}{c^4} T_{\mu\nu} \gamma_\alpha^\mu n^\nu = -\frac{8\pi G}{c^3} j_\alpha \tag{4.167}$$

$$G_{\mu\nu} \gamma_\alpha^\mu \gamma_\beta^\nu = \frac{8\pi G}{c^4} T_{\mu\nu} \gamma_\alpha^\mu \gamma_\beta^\nu = \frac{8\pi G}{c^4} S_{\alpha\beta} \,, \tag{4.168}$$

where $\rho c^2 := n^\mu n^\nu T_{\mu\nu}$ is the matter density, $j_\alpha c := -\gamma_\alpha^\mu n^\nu T_{\mu\nu}$ is the matter current, and $S_{\alpha\beta} := \gamma_\alpha^\mu \gamma_\beta^\nu T_{\mu\nu}$ is the matter stress tensor. We obtain the evolution equations

$$\frac{\partial \gamma_{ij}}{\partial t} = -2\alpha K_{ij} + D_i \beta_j + D_j \beta_i \tag{4.169}$$

and

$$\frac{\partial K_{ij}}{\partial t} = \alpha\left\{\mathcal{R}_{ij} - 2K_{ik}K^k{}_j + K K_{ij} - \frac{8\pi G}{c^4}\left[\frac{1}{2}\rho c^2 \gamma_{ij} + S_{ij} - \frac{1}{2}\gamma_{ij} S\right]\right\} - D_i D_j \alpha + \beta^k D_k K_{ij} + K_{kj} D_i \beta^k + K_{ik} D_j \beta^k \tag{4.170}$$

with $K := \gamma^{ij} K_{ij}$ and $S := \gamma^{ij} S_{ij}$, as well as two *constraint equations*: the *Hamiltonian constraint* equation

$$\mathcal{R} + K^2 - K_{ij} K^{ij} = \frac{16\pi G}{c^2}\rho \tag{4.171}$$

and the *momentum constraint* equation

$$\gamma^{jk}(D_j K_{ik} - D_i K_{jk}) = \frac{8\pi G}{c^3} j_i \,. \tag{4.172}$$

Here $\mathcal{R}_{ij} := \mathcal{R}_{ikj}{}^k$ and $\mathcal{R} := \gamma^{ij}\mathcal{R}_{ij}$ are the Ricci tensor and the Ricci scalar of the spatial curvature. The evolution equations form a coupled set of first-order partial differential equations that can be used to evolve the spatial metric γ_{ij} along with the extrinsic curvature K_{ij} once a the lapse function α and the shift vector field $\boldsymbol{\beta}$ are prescribed. The constraint equations must be satisfied for the initial slice and will continue to hold on later slices. These evolution equations along with the constraint equations are known as the *Arnowitt–Deser–Misner equations* or ADM equations.

Example 4.4 Analogy with electromagnetism

Maxwell's equations of electromagnetism are

$$\nabla \cdot \boldsymbol{E} = \rho/\epsilon_0 \tag{4.173a}$$

$$\nabla \times \boldsymbol{B} - \frac{1}{c^2}\frac{\partial \boldsymbol{E}}{\partial t} = \mu_0 \boldsymbol{j} \tag{4.173b}$$

$$\nabla \times \boldsymbol{E} + \frac{\partial \boldsymbol{B}}{\partial t} = 0 \tag{4.173c}$$

$$\nabla \cdot \boldsymbol{B} = 0 \tag{4.173d}$$

where here ρ and $\boldsymbol{j}$ are the electric charge density and current. The last of these suggest that $\boldsymbol{B}$ can be written instead as the curl of a vector potential $\boldsymbol{A}$, $\boldsymbol{B} = \nabla \times \boldsymbol{A}$, so that $\nabla \cdot \boldsymbol{B} = 0$ is identically true. Then, the second homogeneous Maxwell equation, $\partial \boldsymbol{B}/\partial t = -\nabla \times \boldsymbol{E}$ can be written as

$$\frac{\partial \boldsymbol{A}}{\partial t} = -\boldsymbol{E} - \nabla \phi\,, \tag{4.174a}$$

where ϕ is a new, arbitrary, potential, which can be introduced because the curl of a gradient vanishes $\nabla \times \nabla \phi = 0$. This gives us an evolution for the vector potential $\boldsymbol{A}$, but we now need one for the electric field $\boldsymbol{E}$. The second Maxwell equation provides this:

$$\begin{aligned}\frac{1}{c^2}\frac{\partial \boldsymbol{E}}{\partial t} &= \nabla \times \nabla \times \boldsymbol{A} - \mu_0 \boldsymbol{j} \\ &= \nabla(\nabla \cdot \boldsymbol{A}) - \nabla^2 \boldsymbol{A} - \mu_0 \boldsymbol{j}\,.\end{aligned} \tag{4.174b}$$

We now have a coupled set of first order partial differential equations for $\boldsymbol{A}$ and $\boldsymbol{E}$ where the (arbitrary) choice of ϕ specifies a gauge. The first Maxwell equation, $\nabla \cdot \boldsymbol{E} = \rho/\epsilon_0$ is a constraint equation that must be satisfied at an initial time.

Therefore, to evolve Maxwell's equations in this decomposed form, one must first generate initial data by solving $\nabla \cdot \boldsymbol{E} = \rho/\epsilon_0$ (the constraint equation), and then use the two evolution equations to obtain $\boldsymbol{A}$ and $\boldsymbol{E}$ on future time steps. A gauge choice for ϕ must also be prescribed. Unfortunately, the system of equations (4.174), while physically correct are not well-posed for numerical evolution in that the evolution is not stable against any small numerical errors that may arise. If we substitute the time derivative of Eq. (4.174a) into Eq. (4.174b) we find

$$\frac{1}{c^2}\Box \boldsymbol{A} - \nabla(\nabla \cdot \boldsymbol{A}) = -\mu_0 \boldsymbol{j} + \frac{1}{c^2}\nabla\frac{\partial \phi}{\partial t}\,. \tag{4.175}$$

If the second term on the left-hand side were not present, this equation would be a *symmetric hyperbolic equation*, and such equations are known to be well-posed for numerical evolution.

There are various approaches to put the evolution equations into a form that is symmetric and hyperbolic. One approach is to enforce a gauge condition on ϕ

that forces Eq. (4.175) into a symmetric hyperbolic form; common choices are the Lorenz gauge, in which ϕ is chosen to be the solution of $\partial\phi/\partial t = -c^2\,\boldsymbol{\nabla}\cdot\boldsymbol{A}$, or the Coulomb gauge where $\boldsymbol{\nabla}\cdot\boldsymbol{A} = 0$ in which ϕ is chosen to be the solution of the elliptic equation $\nabla^2\phi = -\rho/\epsilon_0$. Note that the Coulomb gauge is a gauge in which $\boldsymbol{\nabla}\cdot\boldsymbol{A}$ is an arbitrary constant – and can be taken to be zero – which can be shown by taking the divergence of Eq. (4.174a). We can combine Eqs. (4.174a) and (4.174b) to obtain a second-order equation for the single variable $\boldsymbol{A}$. In either the Lorenz gauge or the Coulomb gauge a relatively simple result is obtained:

$$\Box\boldsymbol{A} = -\frac{1}{c^2}\frac{\partial^2\boldsymbol{A}}{\partial t^2} + \nabla^2\boldsymbol{A} = \begin{cases} -\mu_0\boldsymbol{j} & \text{Lorenz gauge} \\ -\mu_0\boldsymbol{j} + \frac{1}{c^2}\boldsymbol{\nabla}\frac{\partial\phi}{\partial t} & \text{Coulomb gauge.} \end{cases} \tag{4.176}$$

This shows that with either of these two gauge choices, the equations for $\boldsymbol{A}$ are hyperbolic.

A second approach is to adopt the divergence of the vector potential as a new auxiliary dynamical variable, $\Gamma := \boldsymbol{\nabla}\cdot\boldsymbol{A}$. By taking the divergence of Eq. (4.174a) we obtain the evolution equation for this new variable:

$$\frac{\partial\Gamma}{\partial t} = -\boldsymbol{\nabla}\cdot\boldsymbol{E} - \nabla^2\phi = -\frac{1}{\epsilon_0}\rho - \nabla^2\phi\,. \tag{4.177}$$

Equations (4.174a), (4.174b) (with $\boldsymbol{\nabla}\cdot\boldsymbol{A}$ replaced by Γ) and (4.177) form a system of evolutionary equations. There are now two constraint equations: $\Gamma = \boldsymbol{\nabla}\cdot\boldsymbol{A}$ and $\rho = \epsilon_0\,\boldsymbol{\nabla}\cdot\boldsymbol{E}$.

A third approach is to avoid gauge issues altogether by deriving a hyperbolic equation directly from the covariant form of Maxwell's equations: Take a time derivative of Eq. (4.173b), substitute Eq. (4.173c) to eliminate the magnetic field term, and use Eq. (4.173a) to eliminate the divergence of the electric field (which arises from the use of the vector identity $\boldsymbol{\nabla}\times\boldsymbol{\nabla}\times\boldsymbol{E} = \boldsymbol{\nabla}(\boldsymbol{\nabla}\cdot\boldsymbol{E}) - \nabla^2\boldsymbol{E}$). The result is

$$\Box\boldsymbol{E} = \frac{1}{\epsilon_0}\boldsymbol{\nabla}\rho + \mu_0\frac{\partial\boldsymbol{j}}{\partial t} \tag{4.178}$$

which is a symmetric hyperbolic equation.

Example 4.5 The BSSN formulation

The ADM equations have many of the problematic features described in Example 4.4. As they stand, they do not form a system of equations that can stably be evolved in numerical relativity; and, as with the analogy with electromagnetism, the problem arises owing to the fact that the ADM equations do not form a symmetric hyperbolic system. In parallel with the approaches to obtain a symmetric hyperbolic system of equations explored in Example 4.4, one may choose coordinates, such as harmonic coordinates (i.e. make a gauge choice) in which the system of equation is symmetric and hyperbolic or one may introduce suitable auxiliary variables that result in a symmetric and hyperbolic system.

The *Baumgarte–Shapiro–Shibata–Nakamura formalism* or BSSN formalism (Baumgarte and Shapiro, 1999; Shibata and Nakamura, 1995) is an important example of the latter approach. The first part of this formalism, however, is not related to making the ADM equations into a symmetric hyperbolic system, but instead to separating the radiative and non-radiative degrees of freedom in the evolution by performing a conformal decomposition of the dynamical variables. Specifically, a conformal metric $\hat{\gamma}_{ij}$ that has volume element, $\det\hat{\gamma} = 1$, is defined in terms of spatial metric γ_{ij} and a conformal variable ϕ by

$$\hat{\gamma}_{ij} := \mathrm{e}^{-4\phi}\gamma_{ij} , \tag{4.179}$$

where

$$\phi := \frac{1}{12}\ln\det\gamma . \tag{4.180}$$

The extrinsic curvature K_{ij} is split into its trace, K, and its conformally weighted trace-free part $\hat{A}_{ij}$,

$$\hat{A}_{ij} := \mathrm{e}^{-4\phi}\left(K_{ij} - \frac{1}{3}K\gamma_{ij}\right) . \tag{4.181}$$

In terms of these variables, $\hat{\gamma}_{ij}$, ϕ, $\hat{A}_{ij}$ and K, the evolution equations are

$$\frac{\partial\hat{\gamma}_{ij}}{\partial t} = -2\alpha\hat{A}_{ij} + \beta^k\frac{\partial\hat{\gamma}_{ij}}{\partial x^k} + \hat{\gamma}_{ik}\frac{\partial\beta^k}{\partial x^j} + \hat{\gamma}_{jk}\frac{\partial\beta^k}{\partial x^i} - \frac{2}{3}\hat{\gamma}_{ij}\frac{\partial\beta^k}{\partial x^k} \tag{4.182a}$$

$$\frac{\partial\phi}{\partial t} = -\frac{1}{6}\alpha K + \beta^k\frac{\partial\phi}{\partial x^k} + \frac{1}{6}\frac{\partial\beta^k}{\partial x^k} \tag{4.182b}$$

$$\begin{aligned}\frac{\partial\hat{A}_{ij}}{\partial t} &= \mathrm{e}^{-4\phi}\left\{-D_iD_j\alpha + \frac{1}{3}\gamma_{ij}D_kD^k\alpha\right.\\ &\quad\left.+\alpha\left[\mathcal{R}_{ij} - \frac{8\pi G}{c^4}S_{ij} - \frac{1}{3}\gamma_{ij}\left(\mathcal{R} - \frac{8\pi G}{c^4}S\right)\right]\right\}\\ &\quad+\alpha K\hat{A}_{ij} - 2\alpha\hat{A}_{ik}\hat{A}^k{}_j + \beta^k\frac{\partial\hat{A}_{ij}}{\partial x^k} + \hat{A}_{ik}\frac{\partial\beta^k}{\partial x^j}\\ &\quad+\hat{A}_{jk}\frac{\partial\beta^k}{\partial x^i} - \frac{2}{3}\hat{A}_{ij}\frac{\partial\beta^k}{\partial x^k}\end{aligned} \tag{4.182c}$$

$$\frac{\partial K}{\partial t} = -D_iD^i\alpha + \alpha\left(\hat{A}_{ij}\hat{A}^{ij} + \frac{1}{3}K^2\right) + \frac{4\pi G}{c^4}\alpha(\rho c^2 + S) + \beta^i\frac{\partial K}{\partial x^i} . \tag{4.182d}$$

The indices of $\hat{A}_{ij}$ are raised and lowered with the conformal metric $\hat{\gamma}_{ij}$. In these equations, connection coefficients for the 3-metric γ_{ij} appear in the covariant derivative term $D_i D_j \alpha$ and in the 3-curvature $\mathcal{R}_{ij}$. Explicitly, we have

$$D_i D_j \alpha = \frac{\partial^2 \alpha}{\partial x^i \partial x^j} - \hat{\gamma}^{kl} \hat{\Gamma}_{kij} \frac{\partial \alpha}{\partial x^l} , \tag{4.183}$$

$$\begin{aligned} \mathcal{R}_{ij} = {} & \hat{\mathcal{R}}_{ij} - 2\frac{\partial^2 \phi}{\partial x^i \partial x^j} + 2\hat{\gamma}^{kl}\hat{\Gamma}_{kij}\frac{\partial \phi}{\partial x^l} - 2\hat{\gamma}_{ij}\hat{\gamma}^{kl}\frac{\partial^2 \phi}{\partial x^k \partial x^l} + 2\hat{\gamma}_{ij}\hat{\Gamma}^k\frac{\partial \phi}{\partial x^k} \\ & + 4\frac{\partial \phi}{\partial x^i}\frac{\partial \phi}{\partial x^j} - 4\hat{\gamma}_{ij}\hat{\gamma}^{kl}\frac{\partial \phi}{\partial x^k}\frac{\partial \phi}{\partial x^l} , \end{aligned} \tag{4.184}$$

where

$$\begin{aligned} \hat{\mathcal{R}}_{ij} = {} & -\frac{1}{2}\hat{\gamma}^{kl}\frac{\partial^2 \hat{\gamma}_{ij}}{\partial x^k \partial x^l} + \frac{1}{2}\hat{\gamma}_{ik}\frac{\partial \hat{\Gamma}^k}{\partial x^j} + \frac{1}{2}\hat{\gamma}_{jk}\frac{\partial^2 \hat{\Gamma}^k}{\partial x^i} + \frac{1}{2}\hat{\Gamma}^k\hat{\Gamma}_{ijk} + \frac{1}{2}\hat{\Gamma}^k\hat{\Gamma}_{jik} \\ & + \hat{\gamma}^{kl}\hat{\gamma}^{mn}\left(\hat{\Gamma}_{kmi}\hat{\Gamma}_{jln} + \hat{\Gamma}_{kmj}\hat{\Gamma}_{iln} + \hat{\Gamma}_{kmi}\hat{\Gamma}_{lnj}\right) \end{aligned} \tag{4.185}$$

is the Ricci curvature tensor for the conformal metric,

$$\hat{\Gamma}_{kij} = \frac{1}{2}\left(\frac{\partial \hat{\gamma}_{ik}}{\partial x^j} + \frac{\partial \hat{\gamma}_{jk}}{\partial x^i} - \frac{\partial \hat{\gamma}_{ij}}{\partial x^k}\right) \tag{4.186}$$

are the connection coefficients for the conformal metric, and we have introduced the quantity

$$\hat{\Gamma}^i := -\frac{\partial \hat{\gamma}^{ij}}{\partial x^j} . \tag{4.187}$$

The system of equations given by Eq. (4.182) would be a symmetric hyperbolic system if it weren't for the second and third terms in Eq. (4.185), which contain second-order spatial partial derivatives of $\hat{\gamma}_{ij}$ that are *not* of the form of a Laplacian (such as the first term of Eq. (4.185)). To remedy this, the BSSN formalism promotes the quantity $\hat{\Gamma}^i$ to a new, separately evolved dynamical variable (cf. Example 4.4). The evolution equation for this auxiliary variable is

$$\begin{aligned} \frac{\partial \hat{\Gamma}^i}{\partial t} = {} & -2\hat{A}^{ij}\left(\frac{\partial \alpha}{\partial x^j} - 6\alpha\frac{\partial \phi}{\partial x^j}\right) + 2\alpha\hat{\gamma}^{ij}\left(\hat{\Gamma}_{jkl}\hat{A}^{kl} - \frac{2}{3}\frac{\partial K}{\partial x^j} - \frac{8\pi G}{c^4}j_j\right) \\ & + \beta^j\frac{\partial \hat{\Gamma}^i}{\partial x^j} - \hat{\Gamma}^j\frac{\partial \beta^i}{\partial x^j} + \frac{2}{3}\hat{\Gamma}^i\frac{\partial \beta^j}{\partial x^j} + \frac{1}{3}\hat{\gamma}^{ik}\frac{\partial^2 \beta^j}{\partial x^j \partial x^k} + \hat{\gamma}^{jk}\frac{\partial^2 \beta^i}{\partial x^j \partial x^k} , \end{aligned} \tag{4.188}$$

which is obtained from the definition of $\hat{\Gamma}^i$ in Eq. (4.187), Eq. (4.169), and also incorporates the momentum constraint Eq. (4.172). The BSSN system of equations are Eqs. (4.182) and (4.188).

In a numerical evolution, initial data is provided in the form of a spatial metric γ_{ij} and its time derivative. A prescription is adopted for choosing the lapse and shift functions α and β^i, which essentially chooses the foliation of spacetime, and γ_{ij} and its time derivative are then evolved using evolution equations that are derived from the Einstein equations.

4.3.2 Coordinate Choice

The choice of lapse α and shift $\boldsymbol{\beta}$ describes the slicing of spacetime into spatial surfaces and the mappings between them. Like the choice of the scalar potential ϕ in the electromagnetism analogy of Example 4.4, the choice of lapse and shift is a gauge choice, which, in general relativity, is a choice of coordinates.

An obvious choice,

$$\alpha = c \quad \text{and} \quad \boldsymbol{\beta} = 0 \tag{4.189}$$

is known as *geodesic slicing* because a fixed coordinate point will follow a geodesic. To see this, note that, for constant lapse α, $n_\alpha \propto \nabla_\alpha t$, so $\nabla_\alpha n_\beta = \nabla_\beta n_\alpha$ and therefore $n^\mu \nabla_\mu n_\alpha = n^\mu \nabla_\alpha n_\mu = \frac{1}{2}\nabla_\alpha(n^\mu n_\mu) = 0$ since the vector $\mathbf{n}$ is normalized. Hence $\mathbf{n}$ satisfies the geodesic equation, and, because the shift vector is zero, $\mathbf{t} = c\mathbf{n}$ also satisfies the geodesic equation.

Geodesic slicing is not normally a useful choice for numerical relativity as the grid points tend to fall together (as particles would), forming coordinate singularities. This is seen by considering the evolution of the trace of the extrinsic curvature, $K := \gamma^{ij} K_{ij}$. We find

$$\begin{aligned}
\frac{\partial K}{\partial t} &= \gamma^{ij}\frac{\partial K_{ij}}{\partial t} + K_{ij}\frac{\partial \gamma^{ij}}{\partial t} = \gamma^{ij}\frac{\partial K_{ij}}{\partial t} - K^{ij}\frac{\partial \gamma_{ij}}{\partial t} \\
&= \alpha\left\{\mathcal{R} + K^2 - \frac{4\pi G}{c^4}(3\rho c^2 - S)\right\} - \gamma^{ij} D_i D_j \alpha + \beta^i \frac{\partial K}{\partial x^i} \\
&= \alpha\left\{K_{ij} K^{ij} + \frac{4\pi G}{c^4}(\rho c^2 + S)\right\} - \gamma^{ij} D_i D_j \alpha + \beta^i \frac{\partial K}{\partial x^i}\,,
\end{aligned} \tag{4.190}$$

where we used Eqs. (4.170), (4.169), the Hamiltonian constraint (4.171), and the fact that $(\partial\gamma^{ij}/\partial t) = -\gamma^{ik}\gamma^{il}(\partial\gamma_{kl}/\partial t)$. For a vacuum spacetime and geodesic slicing, this becomes simply

$$\frac{\partial K}{\partial t} = c K_{ij} K^{ij} \quad \text{(geodesic slicing)}. \tag{4.191}$$

Notice that the right-hand side is positive definite so that as soon as the extrinsic curvature achieves a non-zero value, it will increase indefinitely with time. This in turn leads to a shrinking of the coordinate volume element on each time slice: the volume element, $(\det\gamma)^{1/2}$, evolves according to[3)]

$$\begin{aligned}
\frac{\partial}{\partial t}\ln\det\gamma &= \gamma^{ij}\frac{\partial\gamma_{ij}}{\partial t} = -2\alpha K + 2D_i\beta^i \\
&= -2cK \quad \text{(geodesic slicing)}
\end{aligned} \tag{4.192}$$

3) Note that $d(\ln\det\mathbf{A}) = \mathrm{Tr}(d\mathbf{A}\cdot\mathbf{A}^{-1})$ which follows from the identity $\det\mathbf{A} = \exp(\mathrm{Tr}\ln\mathbf{A})$ since $d(\det\mathbf{A}) = d\exp(\mathrm{Tr}\ln\mathbf{A}) = \exp(\mathrm{Tr}\ln\mathbf{A})d(\mathrm{Tr}\ln\mathbf{A}) = \det\mathbf{A}\,\mathrm{Tr}(d\mathbf{A}\cdot\mathbf{A}^{-1})$.

and thus, since K is monotonically increasing, the volume of a coordinate region of the spatial slices is monotonically decreasing.

To prevent the pathology of coordinates falling together that is encountered in geodesic slicing, a common choice of slicing, known as *maximal slicing*, is often made for which the trace of the extrinsic curvature is held fixed with $K = 0$ on all slices. Again, the shift vector is taken to be zero. From Eq. (4.190) we see that the lapse condition that enforces maximal slicing is

$$\gamma^{ij} D_i D_j \alpha = \alpha \left\{ K_{ij} K^{ij} + \frac{4\pi G}{c^4}(\rho c^2 + S) \right\} . \tag{4.193}$$

Maximal slicing has the property that $K = \nabla_\mu n^\mu = 0$, that is, the normal vectors to the spatial slices are divergenceless.

Another common coordinate choice is *harmonic coordinates* in which the coordinates x^α are harmonic functions that satisfy $g^{\mu\nu}\nabla_\mu \nabla_\nu x^\alpha = 0$. As was shown in Example 2.14, harmonic coordinates are ones in which $g^{\mu\nu}\Gamma^\alpha_{\mu\nu} = 0$, which can be expressed as a set of evolution equations for the lapse and the shift. A simpler choice, called *harmonic slicing* sets only the one component $g^{\mu\nu}\Gamma^0_{\mu\nu}$ to zero. For a zero shift vector, this yields an equation for the lapse. Since

$$0 = g^{\mu\nu}\Gamma^\alpha_{\mu\nu} = -\frac{\partial g^{\alpha\mu}}{\partial x^\mu} - \frac{1}{2} g^{\mu\nu} g^{\alpha\rho} \frac{\partial g_{\mu\nu}}{\partial x^\rho} \tag{4.194}$$

(cf. Example 2.14), we find

$$\begin{aligned} 0 &= -\frac{\partial(-\alpha^{-2})}{\partial t} + \frac{1}{2}\frac{1}{\alpha^2} g^{\mu\nu} \frac{\partial g_{\mu\nu}}{\partial t} \\ &= -\frac{2}{\alpha^3}\frac{\partial \alpha}{\partial t} + \frac{1}{2}\frac{1}{\alpha^4}\frac{\partial \alpha^2}{\partial t} + \frac{1}{2}\frac{1}{\alpha^2}\gamma^{ij}\frac{\partial \gamma_{ij}}{\partial t} \\ &= -\frac{1}{\alpha^3}\frac{\partial \alpha}{\partial t} + \frac{1}{2}\frac{1}{\alpha^2}\frac{\partial}{\partial t}\ln\det\gamma = -\frac{1}{\alpha^3}\frac{\partial \alpha}{\partial t} - \frac{1}{\alpha}K \end{aligned} \tag{4.195}$$

so

$$\frac{\partial \alpha}{\partial t} = \frac{1}{2}\alpha\frac{\partial}{\partial t}\ln\det\gamma \quad \text{or} \quad \frac{\partial \alpha}{\partial t} = -\alpha^2 K . \tag{4.196}$$

The first form can be integrated to obtain an equation for the lapse, $\alpha \propto (\det\gamma)^{1/2}$ where the proportionality constant is time-independent.

A modification of the harmonic slicing instead uses the condition

$$\frac{\partial \alpha}{\partial t} = c\frac{\partial}{\partial t}\ln\det\gamma \tag{4.197}$$

for the lapse – this equation drives the lapse toward zero faster as the volume element shrinks, and therefore leads to stronger singularity avoidance. A solution to this equation is

$$\alpha = c(1 + \ln\det\gamma) , \tag{4.198}$$

where the arbitrary constant of integration is simply taken to be unity. This choice of slicing is known as *one-plus-log slicing*.

4.3.3
Initial Data

Initial data for the spatial metric γ_{ij} (six functions of space) and the extrinsic curvature K_{ij} (six other functions of space) must be specified for an evolution; however, these twelve functions cannot be freely specified: the four constraint equations (4.171) and (4.172) must be satisfied. Furthermore, four degrees of gauge freedom are allowed in the choice of coordinates, so there are in fact only four physically independent functional degrees of freedom in the initial data.

We saw earlier in this chapter that, to first-post-Newtonian order, the spatial metric is *conformally flat*, that is $\gamma_{ij} = \Omega^2\delta_{ij}$ where Ω is a conformal factor (a function of spacetime coordinates). A great simplification to the problem of finding initial data for numerical evolutions can be achieved by assuming that the initial spatial slice is conformally flat. A straightforward calculation of the Ricci scalar for the spatial slice results in

$$\mathcal{R} = \Omega^{-2}\left[-4\nabla^2 \ln\Omega - 2(\nabla \ln\Omega)\cdot(\nabla \ln\Omega)\right]. \tag{4.199}$$

For convenience, the conformal factor is normally taken to be $\Omega = \psi^2$ or $\gamma_{ij} = \psi^4\delta_{ij}$ in terms of which the Ricci scalar takes the simple form

$$\mathcal{R} = -8\psi^{-5}\nabla^2\psi \tag{4.200}$$

If we now wanted to solve for time-symmetric initial data in a coordinate condition with zero shift, $K_{ij} = 0$, we only need to solve a Hamiltonian constraint, which becomes the equation

$$\nabla^2\psi = -\psi^5\frac{2\pi G}{c^2}\rho \tag{4.201}$$

for the conformal factor ψ. Outside the matter sources, the conformal factor takes the form of a Laplacian $\nabla^2\psi = 0$ and we can easily construct a solution corresponding to a system of particles

$$\psi(\boldsymbol{x}) = 1 + \frac{1}{2}\sum_A \frac{G m_A}{c^2\|\boldsymbol{x} - \boldsymbol{x}_A\|} \tag{4.202}$$

(cf. Eq. (4.40c) and note that $g_{ij} = \psi^4\delta_{ij}$). Unfortunately, time-symmetric initial data is often undesirable. For example, to model a binary system with two stars or black holes in a circular orbit, we cannot assume time symmetry. Therefore we will need to obtain an expression for the extrinsic curvature and solve both the Hamiltonian and momentum constraints.

If we consider a maximal $K = 0$, conformally flat $\gamma_{ij} = \psi^4\delta_{ij}$ initial slice and define a conformally weighted extrinsic curvature $\hat{K}_{ij}$ by $\hat{K}_{ij} := \psi^2 K_{ij}$, then the

Hamiltonian and momentum constraint equations take the form

$$\nabla^2 \psi = -\frac{1}{8}\psi^{-7}\delta^{ij}\delta^{kl}\hat{K}_{ik}\hat{K}_{jl} - \psi^5 \frac{2\pi G}{c^2}\rho \tag{4.203}$$

$$\frac{\partial \hat{K}_{ij}}{\partial x_j} = \psi^6 \frac{8\pi G}{c^3} j_i \,. \tag{4.204}$$

The vacuum momentum constraint can be solved in vacuum by expressing the conformally weighted trace-free extrinsic curvature in terms of a vector potential $\boldsymbol{A}$:

$$\hat{K}_{ij} = \frac{1}{c}\left(\frac{1}{2}\frac{\partial A_i}{\partial x^j} + \frac{1}{2}\frac{\partial A_j}{\partial x^i} - \frac{1}{3}\delta_{ij}\frac{\partial A_k}{\partial x_k}\right) \tag{4.205}$$

and the momentum constraint becomes

$$\nabla^2 \boldsymbol{A} + \frac{1}{3}\boldsymbol{\nabla}(\boldsymbol{\nabla}\cdot \boldsymbol{A}) = 0 \,, \tag{4.206}$$

which has a solution (corresponding to a single body located at point $\boldsymbol{x}_A$)

$$\boldsymbol{A} = -\frac{7}{2}\frac{G m_A}{c^2\|\boldsymbol{x}-\boldsymbol{x}_A\|}\boldsymbol{v}_A - \frac{1}{2}\frac{G m_A \boldsymbol{v}_A\cdot(\boldsymbol{x}-\boldsymbol{x}_A)}{c^2\|\boldsymbol{x}-\boldsymbol{x}_A\|}(\boldsymbol{x}-\boldsymbol{x}_A). \tag{4.207}$$

The constant vector $\boldsymbol{v}$ represents the velocity of the body. At this point, $\hat{K}_{ij}$ can be constructed and the Hamiltonian constraint can be solved, with suitable boundary conditions near the body's location at $\boldsymbol{x}_A$. This method is known as the *Bowen–York approach* (Bowen and York Jr., 1980). Since the constraint equations are linear equations in ψ and $\boldsymbol{A}$, additional bodies can be added simply by adding solutions of the form given above together.

A useful method of then solving the Hamiltonian constraint equation, known as the *puncture method*, involves writing the conformal factor in terms of a singular (at the location of the bodies present) term plus a smooth term. For example, for binary black hole initial data, one writes

$$\psi = u + \psi_H \quad \text{where} \quad \psi_H = \frac{1}{2}\frac{G m_1}{c^2\|\boldsymbol{x}-\boldsymbol{x}_1\|} + \frac{1}{2}\frac{G m_1}{c^2\|\boldsymbol{x}-\boldsymbol{x}_2\|} \,. \tag{4.208}$$

The term ψ_H is a solution to the homogeneous Laplace equation so it describes two bodies with masses m_1 and m_2 located at positions $\boldsymbol{x}_1$ and $\boldsymbol{x}_2$ if the bodies were not moving (i.e. for time-symmetric initial data). Because the bodies *are* moving, the Hamiltonian constraint is not a homogeneous equation, so a smooth particular solution u must be added. By capturing the singular part exactly in the ψ_H term, one need only find a numerical solution to the Hamiltonian constraint for the smooth part:

$$\nabla^2 \psi = -\frac{1}{8}(u + \psi_H)^{-7}\hat{K}_{ij}\hat{K}^{ij} \,. \tag{4.209}$$

Note that the mass parameter quantities m_1 and m_2 are known as the *bare masses*, and they no longer exactly equal the true masses of the two black holes because they do not contain the relativistic effects of the bodies' motion.

4.3.4 Gravitational-Wave Extraction

At large distances from the dynamical system, at the edge of the numerical grid, the gravitational-wave content can be obtained from the spacetime metric. One approach to measuring the gravitational waveform is to compute the Weyl scalar Ψ_4 which was introduced earlier in Eq. (4.129). At large distances from the source of gravitational waves, the null basis vectors approach the form in Eq. (4.130), and Ψ_4 contains a complete description of the gravitational waveform. Using the Gauss, Codazzi and Ricci equations (4.163), (4.164) and (4.165), the Weyl scalar Ψ_4 can be expressed in terms of the evolution variables.

Rather than evaluate Ψ_4 at each point on the boundary of the numerical simulation it is often spatially averaged over a sphere at large radius and decomposed into spin-weighted spherical harmonic modes,

$$\Psi_{4,\ell m}(t) := \int {}_{-2}Y^*_{\ell m}(\theta,\phi)\Psi_4(t,\theta,\phi)\sin\theta\, d\theta\, d\phi\,. \tag{4.210}$$

These spatial averages contain less numerical noise than the individual points on the grid. Furthermore, the gravitational-wave content tends to be concentrated primarily in the low-ℓ modes. The waves propagating in direction (θ,ϕ) can then be computed from

$$\Psi_4(t,\theta,\phi) = \sum_{\ell=2}^{\infty}\sum_{m=-\ell}^{\ell} {}_{-2}Y_{\ell m}(\theta,\phi)\Psi_{4,\ell m}(t)\,. \tag{4.211}$$

4.3.5 Matter

For simplicity, we will consider only perfect fluid matter in this section. The stress-energy tensor for the matter is therefore

$$T^{\alpha\beta} = (\rho_0 + \rho_0\Pi/c^2 + p/c^2)u^\alpha u^\beta + p g^{\alpha\beta}\,, \tag{4.212}$$

where ρ_0 is the rest mass density, $\rho_0\Pi$ is the internal energy density, and p is the fluid pressure. The four-velocity of the fluid is $u^\alpha = u^0[1, v^i]$ where v^i is the spatial three-velocity of the fluid and $\alpha u^0 = -n_\mu u^\mu = w$ is related to the relativistic Lorentz factor for the fluid moving through the numerical grid. The fluid forms a source for the decomposed Einstein equations:

$$\rho = c^{-2}w^2(\rho_0 + \rho_0\Pi/c^2 + p/c^2) - p \tag{4.213a}$$

$$j_i = c^{-1}\alpha^{-1}w^2(\rho_0 + \rho_0\Pi/c^2 + p/c^2)v_i \tag{4.213b}$$

$$S_{ij} = \alpha^{-2}w^2(\rho_0 + \rho_0\Pi/c^2 + p/c^2)v_i v_j + p\gamma_{ij}\,. \tag{4.213c}$$

The motion of the fluid must also be computed. The equations of motion for the fluid are

$$\nabla_\mu(\rho_0 u^\mu) = 0 \tag{4.214}$$

(conservation of rest mass) and

$$\nabla_\mu T^{\mu\alpha} = 0\,. \tag{4.215}$$

The first of these can immediately be written as (cf. Example 2.6)

$$\frac{\partial}{\partial t}(|\det\gamma|^{1/2} w\rho_0) = -\frac{\partial}{\partial x^j}\left(|\det\gamma|^{1/2} w\rho_0 v^j\right). \tag{4.216a}$$

The second equation (4.215) can be contracted with u_α or projected onto the spatial surface using $\gamma_{\alpha i}$. The resulting equations are, respectively,

$$\begin{aligned}\frac{\partial}{\partial t}\left(|\det\gamma|^{1/2} w\rho_0 \Pi\right) = &-\frac{\partial}{\partial x^j}\left(|\det\gamma|^{1/2} w\rho_0 \Pi v^j\right)\\ &-p\frac{\partial}{\partial t}\left(|\det\gamma|^{1/2} w\right) - p\frac{\partial}{\partial x^j}\left(|\det\gamma|^{1/2} w v^j\right),\end{aligned} \tag{4.216b}$$

which describes the energy transport, and

$$\begin{aligned}\frac{\partial}{\partial t}\left(|\det\gamma|^{1/2} w j_i\right) = &-\frac{\partial}{\partial x^j}\left(|\det\gamma|^{1/2} w j_i v^j\right) - \alpha|\det\gamma|^{1/2}\\ &\times\left[\frac{\partial p}{\partial x^i} + \frac{1}{2}\left(\rho_0 + \rho_0\Pi/c^2 + p/c^2\right)u_\mu u_\nu\frac{\partial g^{\mu\nu}}{\partial x^i}\right]\end{aligned} \tag{4.216c}$$

which are the relativistic Euler equations. In the Newtonian limit, $\rho = \rho_0$, $\boldsymbol{j} = \rho\boldsymbol{v}$, and we find that Eq. (4.216a) becomes the non-relativistic equation of conservation of mass, Eq. (2.107). Furthermore, combining Eq. (4.216a) and Eq. (4.216c) we obtain the non-relativistic Euler equation (2.105) in the Newtonian limit.

4.3.6
Numerical Methods

We have presented some of the basic formalism describing the spacetime split of Einstein's equations. However, most of the complexity of numerical relativity has not been presented. Various schemes can be used to obtain the finite differencing or spectral equations for the discretized versions of the evolution equations. There are different approaches that use different methods of gridding up spacetime, and employ adaptive mesh refinement to focus grid points in regions of space that require high resolution. There are also different approaches to handling the boundary of the grid, for example conformally extending the grid to infinity or matching from a Cauchy evolution to a characteristic evolution at some large distance for the dynamical system. When event horizons form, numerical evolutions must employ some scheme, for example excising black hole interiors, to avoid encountering the inevitable singularity. Simulations involving matter must be able to handle the tremendous complexity of the physical processes that can arise, including shocks, magnetohydrodynamics, viscosity, finite temperature, neutrino transport, and so on.

4.4 Problems

Problem 4.1

Two particles have masses m_1 and m_2 and are located far away from each other on the z-axis. They are released from rest and allowed to fall towards each other. At any time, the location of the particles are $\boldsymbol{x}_1 = (m_2/M)\boldsymbol{r}_{12}$ and $\boldsymbol{x}_2 = -(m_1/M)\boldsymbol{r}_{12}$ where $\boldsymbol{r}_{12}$ is the separation between the two particles, $\boldsymbol{r}_{12} = \boldsymbol{x}_1 - \boldsymbol{x}_2$ and $M = m_1 + m_2$.

a) Use the Einstein–Infield–Hoffman equations (4.43) to compute the *relative* acceleration $\ddot{z}$ of the particles. The result should be

$$\ddot{z} = -(GM/z^2)\left[1 - (4 + 2\eta)(GM/c^2 z) - (3 - 7\eta/2)\dot{z}^2/c^2\right],$$

where $\eta = m_1 m_2/M^2 = \mu/M$, $\boldsymbol{r}_{12} = z\hat{\boldsymbol{e}}_z$, $\boldsymbol{v} = \dot{z}\boldsymbol{e}_z$, and $d\boldsymbol{v}/dt = \ddot{z}\boldsymbol{e}_z$.

b) Compute $z(t)$ for this motion correct to the first post-Newtonian order. Hint: multiply $\ddot{z}$ by $\dot{z}$ and integrate to get an expression for $\dot{z}^2$ (solving first for the leading order behaviour and iterating to get the M/z correction), and then integrate (iteratively) again to obtain $z(t)$. Use the boundary condition that $\dot{z} \to 0$ as $z \to \infty$.

c) Use Eq. (4.50) to compute the value of h^{ij}_{TT} for an observer on the x-axis. (See Wagoner and Will, 1976.)

Problem 4.2

In Example 4.2, a crude estimate of the amplification factor due to cosmic inflation was made. Repeat the analysis there to obtain a more accurate result: initially a mode is described by $u = \mathrm{e}^{i\hat{\omega}\eta}$. When it encounters the effective potential at time η_1 the solution takes the form $u = Aa(\eta) + Ba(\eta)\int^{\eta} a^{-2}(\eta')\,d\eta$, and then, when inflation ends abruptly (and the effective potential vanishes) at time η_2 the solution again becomes oscillating with $u = \alpha \mathrm{e}^{i\hat{\omega}\eta} + \beta \mathrm{e}^{-i\hat{\omega}\eta}$. Use the continuity of u and $du/d\eta$ across the junctions at times η_1 and η_2 to obtain values for α and β in terms of H_{inf}, $\hat{\omega}$, $\eta_1 - \eta_2$, $a(\eta_1)$, $a(\eta_2)$, and $J = \int_{\eta_1}^{\eta_2} a^{-2}(\eta)\,d\eta$. For a long period of inflation, show that $|\alpha|^2 \approx |\beta|^2 \approx 1/4(H_{\mathrm{inf}} a_2/\hat{\omega})^4$ (Grishchuk and Solokhin, 1991).

Problem 4.3

A distorted black hole with Kerr spin parameter a rings down in the fundamental $\ell = |m| = 2$ quasi-normal mode, so that

$$\Psi_4 = \frac{A}{r}\left[S_{22}(\cos\theta, \omega_{22})e^{2i\phi}e^{i\omega_{22}t} + S_{22}(-\cos\theta, \omega_{22})e^{-2i\phi}e^{-i\omega_{22}^* t}\right]$$

and

$$h_+ - i h_\times = -\frac{2c^2}{|\omega_{22}|}\Psi_4$$

for $t > 0$.

a) Show that the energy radiated per unit time per unit solid angle is

$$\left|\frac{dE}{dt d\Omega}\right| = \frac{c^7}{G}\frac{r^2}{4\pi|\omega_{22}|^2}|\Psi_4|^2 .$$

b) Use the normalization convention for the spin-weighted spheroidal harmonics,

$$\int_{-1}^{1} S_{\ell m}(\cos\theta, a)\, d\cos\theta = 1 ,$$

to show that the total energy radiated is related to the amplitude of the radiation by

$$E = \frac{c^7}{G}\frac{A^2}{|\omega_{22}|^2}\frac{1}{2\,\mathrm{Im}\,\omega_{22}} .$$

c) A non-spinning black hole of mass M radiates energy $E = \epsilon M c^2$ in ringdown radiation. Write the gravitational waveform h_+ and $h_\times$ found by an observer at distance r in the direction (θ, ϕ). Your expression should depend on the parameters ϵ, M, r, θ and ϕ alone. What is the largest strain that is produced if $M = 10\,M_\odot$, $\epsilon = 1\%$, and $r = 1\,\mathrm{Mpc}$?

Problem 4.4

Starting with Eq. (4.159), derive Eq. (4.160).

Problem 4.5

Derive the Hamiltonian and the momentum constraint equations.

a) Show that $G_{\mu\nu}n^\mu n^\nu = \frac{1}{2}R_{\mu\rho\nu\sigma}\gamma^{\mu\nu}\gamma^{\rho\sigma}$ and use the Gauss equation, Eq. (4.163), to obtain the Hamiltonian constraint equation (4.171).
b) Use the Codazzi equation (4.164), to obtain the momentum constraint equation (4.172).

References

Alcubierre, M. (2008) *Introduction to 3+1 Numerical Relativity*, Oxford University Press, Oxford.

Baumgarte, T.W. and Shapiro, S.L. (1999) On the numerical integration of Einstein's field equations. *Phys. Rev.*, **D59**, 024 007. doi: 10.1103/PhysRevD.59.024007.

Baumgarte, T.W. and Shapiro, S.L. (2010) *Numerical Relativity: Solving Einstein's Equations on the Computer*, Cambridge University Press, Cambridge.

Berti, E., Cardoso, V. and Will, C.M. (2006) On gravitational-wave spectroscopy of massive black holes with the space interferometer LISA. *Phys. Rev.*, **D73**, 064 030. doi: 10.1103/PhysRevD.73.064030.

Blanchet, L. (2002) Gravitational radiation from post-newtonian sources and inspiralling compact binaries. *Living Rev. Rel.*, **5**(3). http://www.livingreviews.org/lrr-2002-3 (last accessed 2011-01-03).

Bowen, J.M. and York, J.W. Jr. (1980) Time asymmetric initial data for black holes and black hole collisions. *Phys. Rev.*, **D21**, 2047–2056. doi: 10.1103/PhysRevD.21.2047.

Epstein, R. and Wagoner, R.V. (1975) Post-Newtonian generation of gravitational waves. *Astrophys. J.*, **197**, 717–723. doi: 10.1086/153561.

Grishchuk, L.P. and Solokhin, M. (1991) Spectra of relic gravitons and the early history of the Hubble parameter. *Phys. Rev.*, **D43**, 2566–2571. doi: 10.1103/PhysRevD.43.2566.

Kidder, L.E. (2008) Using full information when computing modes of Post-Newtonian waveforms from inspiralling compact binaries in circular orbit. *Phys. Rev.*, **D77**, 044 016. doi: 10.1103/PhysRevD.77.044016.

Leaver, E.W. (1985) An Analytic representation for the quasi-normal modes of Kerr black holes. *Proc. Roy. Soc. Lond.*, **A402**, 285–298.

Misner, C.W., Thorne, K.S. and Wheeler, J.A. (1973) *Gravitation*, Freeman, San Francisco.

Ryan, M.P. (1974) Teukolsky equation and Penrose wave equation. *Phys. Rev.*, **D10**, 1736–1740. doi: 10.1103/PhysRevD.10.1736.

Shibata, M. and Nakamura, T. (1995) Evolution of three-dimensional gravitational waves: harmonic slicing case. *Phys. Rev.*, **D52**, 5428–5444. doi: 10.1103/PhysRevD.52.5428.

Teukolsky, S.A. (1972) Rotating black holes – separable wave equations for gravitational and electromagnetic perturbations. *Phys. Rev. Lett.*, **29**, 1114–1118. doi: 10.1103/PhysRevLett.29.1114.

Teukolsky, S.A. (1973) Perturbations of a rotating black hole. 1. Fundamental equations for gravitational electromagnetic and neutrino field perturbations. *Astrophys. J.*, **185**, 635–647. doi: 10.1086/152444.

Wagoner, R.V. and Will, C.M. (1976) Post-Newtonian gravitational radiation from orbiting point masses. *Astrophys. J.*, **210**, 764–775. doi: 10.1086/154886.

Weinberg, S. (1972) *Gravitation and Cosmology: Principles and Applications of the General Theory of Relativity*, John Wiley & Sons.

Will, C.M. and Wiseman, A.G. (1996) Gravitational radiation from compact binary systems: gravitational waveforms and energy loss to second Post-Newtonian order. *Phys. Rev.*, **D54**, 4813–4848. doi: 10.1103/PhysRevD.54.4813.

5
Sources of Gravitational Radiation

We have seen that gravitational waves are produced by the non-spherically symmetric motion of bodies that create a time-changing quadrupole moment. Roughly put, gravitational waves are generated whenever a dynamical system is seen to have a changing silhouette (a pencil rotating about its axis would not produce gravitational waves, but if it were tumbling then it would). To be a *strong* source of gravitational waves, the masses of the bodies must be large, the motion must be fast, and the gravitational fields must be large. The dynamical processes typically occur over some characteristic dynamical timescale, which sets the frequency band for the gravitational-wave emission. For example, in a binary system, it is the orbital frequency that determines the frequency of the gravitational-wave emission. Other timescales can determine the duration of the signal. For the orbiting binary, the secular timescale is determined by the rate of loss of energy from the orbit, which will ultimately cause the coalescence of the binary system (and the termination of the gravitational radiation). We often classify the gravitational-wave sources by the frequency band in which they produce gravitational waves. For example, systems that produce gravitational waves with frequencies between $\sim 1\,\text{Hz}$ and $\sim 10\,\text{kHz}$ are said to be radiating in the *high frequency band*, which is the band that we are sensitive to with ground-based gravitational-wave detectors. A list of potentially detectable sources in the various frequency bands of gravitational waves are given in Table 5.1 where the frequency bands are chosen to align with the various kinds of gravitational-wave detectors that are operating; for example, Doppler tracking of spacecraft and laser interferometers in space will operate in the *low frequency band* with gravitational-wave frequencies between $\sim 1\,\text{mHz}$ and $\sim 1\,\text{Hz}$, while pulsar timing experiments are sensitive to gravitational waves in the *very low frequency band* between $\sim 1\,\text{nHz}$ and $\sim 1\,\text{mHz}$. These types of gravitational-wave detectors will be discussed in the next chapter.

Another way to classify gravitational-wave sources is by the character of the dynamical processes that are taking place, which imprint themselves on the signal morphology. The typical categories are the following: signals generated by sources that involve periodic motion that holds a more-or-less steady frequency over long timescales (typically this means over a period of time larger than the observational time) are called *continuous-wave signals*. Because their frequency is relatively stable, such signals can be relatively well-modelled. However, another class of continu-

Gravitational-Wave Physics and Astronomy, First Edition. Jolien D. E. Creighton, Warren G. Anderson.

Table 5.1 The expected sources of gravitational waves, organized by frequency band, and detectors that operate in those frequency bands. The QCD energy scale $\sim$ 200 MeV corresponds to $\sim 10^{-8}$ Hz; the electroweak energy scale $\sim$ 200 GeV corresponds to $\sim 10^{-5}$ Hz; the supersymmetry (SUSY) energy scale is $\sim$ 1 TeV.

Band	Typical sources	Detectors
Extremely low frequency $H_0 \sim 10^{-18}$ Hz–10^{-15} Hz	Primordial stochastic background	Gravitational-wave signatures in the Cosmic Microwave Background
Very low frequency 1 nHz–1 mHz	Supermassive black hole binaries ($M \sim 10^9\,M_\odot$); Cosmic string cusps; Stochastic background (supermassive black hole binaries, QCD-scale phase transitions)	Pulsar Timing Arrays
Low frequency 1 mHz–1 Hz	Supermassive black hole binaries ($M \sim 10^3\,M_\odot$–$10^9\,M_\odot$); Extreme mass ratio inspirals; Dwarf/white dwarf binaries; Stochastic background (dwarf binaries, cosmic strings, electroweak phase transitions)	Space-based interferometers (LISA, DECIGO); Spacecraft Doppler ranging
High frequency 1 Hz–10 kHz	Neutron star/black hole binaries ($M \sim 1\,M_\odot$–$10^3\,M_\odot$); Supernovae; Pulsars, X-ray binaries; Stochastic background (cosmic strings, binary mergers, SUSY-scale phase transitions)	Ground-based interferometers (GEO, LIGO, Virgo); Resonant mass detectors

ous signals would be those that are produced by random but on-going processes in the Universe, which typically form a gravitational-wave background similar to the electromagnetic cosmic microwave background radiation. Often such signals are produced by the incoherent superposition of gravitational waves produced by countless individual sources. While these signals also have durations much larger than the observation time, because the radiation is a stochastic process (and cannot be simply modelled), we classify them as a *stochastic background* of gravitational waves. Short duration signals, in which the duration of the signal (at least within a given frequency band of interest) is shorter than the observation time, are known as gravitational-wave *burst signals*. Burst signals can be further divided into those signals from sources that can be modelled well (by theory or numerical simulation), and those that cannot be effectively modelled. This means of classifying gravitational-wave sources is helpful in that it categorizes the sources in the same way as the data analysis methods that we use to search for the sources.

This chapter will provide an overview of the most commonly treated sources of gravitational waves, but we do not attempt to provide a comprehensive survey. Recommended reviews of gravitational-wave sources are Thorne (1987), Cutler and Thorne (2002) and Sathyaprakash and Schutz (2009).

5.1 Sources of Continuous Gravitational Waves

When a source emits gravitational waves continuously over a period of time that is long compared to the observation time and at a nearly constant frequency (or set of frequencies), we say it is a source of continuous gravitational waves. Such sources are rotating systems in which the rotational motion of the system has some particular steady frequency, which then determines the frequencies of the gravitational waves. Rotating neutron stars that have some non-axial distortion (due to a deformation in its structure or from fluid oscillations) or are wobbling about their rotation axis can provide signals in the high frequency band that is probed by ground-based detectors. Binary systems consisting of compact objects such as white dwarfs or black holes produce continuous gravitational waves in the low-frequency or very-low-frequency bands if their orbital decay timescale is longer than the observational timescale (otherwise we classify the radiation from these objects as burst radiation).

Typically the gravitational-wave signal from a continuous-wave source can be modelled with high accuracy as a nearly fixed-frequency sinusoid. However, over the course of a long observation, there will normally be some frequency drift (e.g. due to angular momentum lost to gravitational radiation), so one must also model the spin-down. Additional effects may also affect the gravitational-wave phase, and would need to be incorporated in the model. Nevertheless, in searching for continuous gravitational waves, we typically exploit the fact that we have a well-modelled waveform in constructing optimal searches.

In the low-frequency and the very-low-frequency gravitational-wave bands, continuous gravitational waves are produced by orbiting binary systems, as described in Section 3.5. For a system to be classified as a source of continuous gravitational waves (rather than a source of gravitational-wave bursts) we require the orbital frequency to remain relatively steady over the period of observation, that is, we require $T_{\text{obs}} \ll \tau_{\text{c}}$ where T_{obs} is the observation time and τ_{c} is the time until coalescence that is given by Eq. (3.178a). The gravitational waveform is then given by Eq. (3.172a). If there is no significant spin-down during the observational time, the argument of the sinusoidal functions is $2\varphi = 2\omega t = 2\pi f t$ where $f = \omega/\pi$ is the gravitational-wave frequency and ω is the orbital angular frequency. Small changes in orbital frequency can be modelled by allowing for a $\dot{f}$ correction in the phase evolution, where $\dot{f}$ is given by Eq. (3.189) (assuming that the orbital decay is entirely driven by the loss of energy to gravitational waves). The resulting gravitational-wave signal then has the form

$$h_+ = -h_0 \frac{1}{2}(1 + \cos^2 \iota) \cos\left[\Phi(t) + \Phi_0\right] \tag{5.1a}$$

$$h_\times = h_0 \cos\iota \sin\left[\Phi(t) + \Phi_0\right], \tag{5.1b}$$

where the gravitational-wave phase is Φ_0 at time $t = t_0$ and evolves according to

$$\Phi(t) = 2\pi f(t - t_0) + \pi \dot{f}(t - t_0)^2 . \tag{5.1c}$$

For a binary system with total mass $M := m_1 + m_2$ and reduced mass $\mu := m_1 m_2/M$, the frequency derivative is

$$\dot{f} = \frac{96}{5}\pi^{8/3} \frac{\mu}{M} \left(\frac{GM}{c^3}\right)^{5/3} f^{11/3} , \tag{5.2a}$$

and

$$h_0 = \frac{4G}{c^4}\frac{\mu a^2 \omega^2}{r} = \frac{4G\mu}{c^2 r}\left(\frac{\pi G M f}{c^3}\right)^{2/3} \tag{5.2b}$$

is the characteristic amplitude of the waves. Here, a is the orbital separation.

In an observation time T_{obs} we observe $N_{\text{obs}} = T_{\text{obs}} f$ gravitational-wave cycles. The characteristic number of cycles over which spin-down under gravitational-wave radiation takes place is $N_{\text{spin-down}} = f^2/\dot{f} \sim (GMf/c^3)^{-5/3}$. The requirement that $N_{\text{obs}} \ll N_{\text{spin-down}}$ (that the gravitational-wave frequency is relatively constant over the observation time) therefore determines the characteristic mass of a binary system that will produce continuous gravitational waves in each frequency band. In the very-low frequency band, $f \sim 1\,\mu\text{Hz}$, pulsar timing experiments might have observations over the course of a decade, so $N_{\text{obs}} \sim 100$; to have $N_{\text{spin-down}}$ much larger than this then the mass of the system can be as large as $M \sim 10^9\,M_\odot$. For the low-frequency band, $f \sim 10\,\text{mHz}$, one year of observation would have $N_{\text{obs}} \sim 10^6$, the masses could be as large as $M \sim 1000\,M_\odot$. For the high-frequency band, $f \sim 100\,\text{Hz}$, $N_{\text{obs}} \sim 10^9$ for $T_{\text{obs}} \sim 1\,\text{yr}$, which restricts the mass to much less than a solar mass. Therefore, binary systems of supermassive black holes will be sources of continuous waves in the very-low frequency band, and in the low-frequency band, Galactic binaries such as white dwarf binaries will be an important source of continuous waves. In the high-frequency band, however, binaries are not expected to produce continuous-wave signals (instead, binaries will be burst sources).

In the high-frequency band, the principal sources of continuous waves are rapidly rotating non-axisymmetric neutron stars. Neutron stars are compact remnants of supernovae having masses around a solar mass, radii around 10 km and are supported by neutron degeneracy pressure. Neutron stars are sometimes observed as pulsars where flashes of electromagnetic radiation are seen at regular intervals set at the rotational period of the neutron star. The gravitational radiation from rotating neutron stars will occur at a frequency proportional to the frequency of rotation;

we are interested, therefore, in neutron stars that rotate with millisecond periods as sources in the high-frequency band. Typically, such rapidly rotating neutron stars are either young neutron stars, examples of which include the Crab pulsar and the Vela pulsar, which have not had time to spin-down too much (nascent neutron stars are expected to be rapidly rotating), or old neutron stars which have been spun-up through mass transfer from a companion (which might have since been shed). The neutron stars may be isolated, or they may be in binary systems with companions that are either ordinary stars, white dwarfs, or even other neutron stars. Some systems, such as Scorpius X-1 (or Sco X-1) are low-mass X-ray binaries (LMXBs) in which matter accreting onto a compact object produces X-ray emission.

Gravitational waves can be emitted by rotating neutron stars through various mechanisms. If the neutron star is not axisymmetric (e.g. triaxial) then it will produce gravitational radiation at a frequency equal to twice the rotational frequency (see Section 3.4). The gravitation waveform will have the same form as Eq. (5.1), but now the frequency evolution is given by (cf. Eq. (3.161))

$$\dot{f} = -\frac{32\pi^4}{5}\frac{G I_3 f^5}{c^5}\varepsilon^2 \tag{5.3a}$$

and the characteristic amplitude is

$$h_0 = \frac{4\pi^2 G I_3 f^2}{c^4 r}\varepsilon\,, \tag{5.3b}$$

where I_3 is the moment of inertia about the rotational axis, $\varepsilon = (I_1 - I_2)/I_3$ is the ellipticity (I_1 and I_2 are the other two principal moments of inertia for the triaxial ellipsoid), and r is the distance of the source. Estimates of neutron star structure predict that the largest deformation that can be supported is $\varepsilon \sim 10^{-6}$.

Even if the neutron star is axisymmetric, it can still emit radiation if the axis of symmetry of the star is not the same as the rotational axis. If the neutron star has non-zero eccentricity, $e = (I_3 - I_1)/I_1$, where $I_1 = I_2 \neq I_3$ are the three principal moments of inertia about the x^1, x^2 and x^3 axes respectively, and if the axis of rotation is not aligned with the axis of symmetry (the x^3-axis) with an angle ϑ between them, then the neutron star will wobble: the axis of symmetry will precess about the rotational axis. Gravitational radiation will be produced at both the rotational frequency and twice the rotational frequency. The gravitational waveform is given by (Zimmermann and Szedenits, 1979)

$$h_+ = \frac{2G I_1\omega^2}{c^4 r} e\sin\vartheta\left[(1+\cos^2\iota)\sin\vartheta\cos 2\omega t + \cos\iota\sin\iota\cos\vartheta\cos\omega t\right] \tag{5.4a}$$

$$h_\times = \frac{2G I_1\omega^2}{c^4 r} e\sin\vartheta\,[2\cos\iota\sin\vartheta\sin 2\omega t + \sin\iota\cos\vartheta\sin\omega t]\,. \tag{5.4b}$$

Here, ι is the inclination angle defined as the angle between the direction of the observer and the rotational axis and $\omega = \|J\|/I_1$ where $\|J\|$ is the magnitude of the total angular momentum vector. Unfortunately, internal dissipation tends to decrease the wobble angle ϑ on relatively short timescales, so it is unlikely that gravitational waves from freely precessing neutron stars will be observable.

A third mechanism for the production of gravitational radiation from rotating neutron stars are fluid modes of oscillation that may become unstable. Oscillatory fluid modes which have a pattern that rotates retrograde relative to the star can be dragged into a prograde motion with respect to inertial observers far away from the star by the star's rotation. In such a configuration, the pattern moving in the prograde-sense as seen by the distant observers will generate gravitational waves that carry a positive amount of angular momentum away from the star; however, since the pattern is moving in a retrograde-sense relative to the star, the decrease in angular momentum causes an amplification of the fluid mode. Such modes are secularly unstable under gravitational radiation, and this instability is known as the *Chandrasekhar–Friedman–Schutz instability* or CFS instability. Most modes that exhibit CFS instability are quickly damped out by viscous forces in the star. For example, the fundamental modes, or *f*-modes, are radial modes whose restoring force is gravity; these modes only become unstable in neutron stars that are rotating very near to their maximum angular velocity. On the other hand, *r*-modes, which are toroidal modes that have the Coriolis force as their restoring force, are more promising as a source of gravitational radiation, and might be seen in young neutron stars or in accreting neutron stars in LMXBs where viscosity is low because the neutron star is hot. The *r*-mode amplitude growth is limited due to non-linear coupling between oscillation modes, but they may persist for hundreds of years in newborn neutron stars. The dominant *r*-mode produces gravitational waves with frequency of approximately $f \approx \frac{4}{3} P_{\text{rot}}^{-1}$ where $P_{\text{rot}} = 2\pi/\omega$ is the rotational period of the neutron star.

To detect a continuous-wave signal, we make assumptions about the form of the gravitational waves, for example the frequency, frequency evolution, and, in surveys for unknown sources, the sky position of the source. For searches for continuous-wave signals from rotating neutron stars, therefore, we normally divide the source population into three categories:

Known pulsars (targeted searches) There are a number of millisecond pulsars that have been observed electromagnetically. Periodic flashes of synchrotron electromagnetic radiation (in radio, optical, X-ray, and even in gamma-rays) are observed each time the magnetic axis, offset from the rotation axis, sweeps across the line-of-sight to the pulsar once per rotational period. The canonical example of such a system is the Crab pulsar, which produces electromagnetic pulses of radiation with a frequency of $\simeq 30$ Hz. When we observe electromagnetic pulses, we have direct information about the phase model of the system: we know from these observations the exact frequency and spin-down of the pulsar. If the pulsar has some triaxial ellipticity then we know that the gravitational radiation will have a gravitational-wave phase that is twice the observed rotational phase. Additionally we know the precise sky position of the source. We therefore have a nearly complete model of the gravitational waveform and highly effective gravitational-wave searches can be made that target this exact waveform.

Known/suspected neutron stars (directed searches) In some cases, the location of a neutron star is known (or suspected), but we do not observe that neutron star as a pulsar. In these cases, we do not have a precise phase evolution model for the source. We do, however, know its sky position. Examples of such systems include Cassiopeia A (Cas A), a nearby ($\sim 3\,\mathrm{kpc}$), young ($\sim$ 300 year old) isolated neutron star, and Scorpius X-1 (Sco X-1), a low-mass X-ray binary system. We can therefore perform a *directed* search in which we must search over unknown source parameters (which include the gravitational-wave frequency, f, the rate of change of this frequency, $\dot{f}$, and additional orbital parameters if the neutron star is in a binary system with unknown orbital elements).

In the case of Sco X-1, an estimate of the characteristic gravitational-wave amplitude can be made by assuming that the torque due to accretion is balanced by the angular momentum radiated in gravitational waves: from the observed X-ray flux $F_X \approx 2 \times 10^{-10}\,\mathrm{W\,m^{-2}}$ and an assumed rotational period $P_{\mathrm{rot}} = 4\,\mathrm{ms}$, we find $h_0 \sim 3 \times 10^{-26}$ (see Problem 5.2).

Unknown pulsar surveys In addition to the targeted and directed searches discussed previously, in which we focus on a known neutron star source, we also conduct gravitational-wave searches for *unknown* neutron stars. Searches for unknown pulsars are computationally very challenging (as we will see in Section 7.7.2) because the rotation of the Earth and the revolution about the Sun will induce Doppler modulations of the gravitational-wave frequency that are sky-position-dependent, and, without knowing the location of the source, corrections for these Doppler modulations must be made for an enormous number of patches on the sky.

Nevertheless, there could be a number of potentially detectable but currently unknown neutron stars in the Galaxy. Neutron stars are born in supernovae which occur at a rate of approximately one event per 30 years in our Galaxy, which suggests that there should be a large number of Galactic neutron stars that have not yet been observed electromagnetically – that are not pulsars (or the pulsar beam is never directed toward the Earth) and are not in X-ray binaries. If some of these neutron stars are rotating with millisecond periods and if they have some non-axisymmetric distortion then there is a possibility that they can be observed by their gravitational-wave emission if they are near enough to the Earth. Neutron stars are expected to be born with rapid rotation and to spin-down as angular momentum is lost in either electromagnetic radiation (if the neutron star is a pulsar) or gravitational radiation. If the spin-down is slow then there could be many nearby neutron stars that are still rapidly rotating and producing gravitational waves in the high-frequency band. On the other hand, if the spin-down is large, then the nearest neutron star that is radiating gravitational waves in the high-frequency might be far away, but amplitude of the gravitational waves would be large if the rapid spin-down was caused by gravitational radiation. The balance between the strength of the radiation and the proximity of the nearest source leads to a estimate of the amplitude of gravitational waves that we might hope to detect, which is detailed in an argument put forward by Blandford.

Example 5.1 Blandford's argument

Roger Blandford made the following argument relating the strength of gravitational-wave emission from the nearest neutron star (see Thorne, 1987): consider a population of *gravitars* (neutron stars that are spinning down entirely due to angular momentum loss in gravitational radiation), which we assume all have the same ellipticity and hence the same spin-down rate. If we also assume that the population is approximately distributed as a disk of uniform density, then it turns out that the strength of the gravitational waves from the nearest source depends on the birth rate of gravitars but not on ellipticity.

Blandford's argument proceeds as follows: suppose the ellipticity is small, which means that the spin-down is slow. Then the newborn gravitars will spend a long time in the high-frequency band that ground-based detectors are sensitive to, so at any given time there will be many gravitars radiating in the high-frequency band and the nearest one should be close. On the other hand, if the ellipticity is large then the gravitars spend less time radiating in the high-frequency band, so the number of gravitars radiating in the high-frequency band will be small and the nearest one will likely be far away. These two scenarios predict the same gravitational-wave amplitude because the source is either nearby but radiating weakly or far away and radiating strongly.

Let $\mathcal{R}$ be the birth rate of gravitars in the Galactic disk per unit area of the disk. The nearest gravitar to us will therefore be at a distance $r_{\text{nearest}} \simeq (\mathcal{R}\tau_{\text{GW}})^{-1/2}$ where τ_{GW} is the timescale of spin-down (which is due to gravitational radiation, by assumption), which we define as $\tau_{\text{GW}} = P/\dot{P}$. In Example 3.13, we found that $\dot{P} \sim \varepsilon^2$ and $h \sim r^{-1}\varepsilon$ where ε is the ellipticity of the gravitar, and therefore $h \sim r^{-1}\tau_{\text{GW}}^{-1/2}$. For the nearest pulsar, $r = r_{\text{nearest}}$, the gravitational-wave strain is largest and $h_{\text{largest}} \sim \mathcal{R}^{1/2}$. We find the characteristic amplitude to be

$$h_{0,\text{largest}} = 3 \times 10^{-24} \left(\frac{I_3}{10^{38}\,\text{m}^2\,\text{kg}}\right)^{1/2} \left(\frac{\mathcal{R}}{10^{-4}\,\text{yr}^{-1}\,\text{kpc}^{-2}}\right)^{1/2}, \tag{5.5}$$

where I_3 is the moment of inertia of the neutron star (see Problem 5.3). This sets an optimistic scale for how large the amplitude of gravitational waves from the nearest rotating neutron star could be.

The significant caveats with Blandford's argument are (i) the spin-down must be driven by gravitational radiation (whereas pulsar spin-down is normally thought to be driven by electromagnetic breaking), and (ii) the argument only works for a range of gravitar ellipticities and detector sensitivities such that we can approximate the population of gravitars as a disk (for very small ellipticities, the thickness of the disk of pulsars will be comparable to the range to which we can detect them, so the spatial distribution would be better modelled as uniform in volume).

5.2
Sources of Gravitational-Wave Bursts

Gravitational radiation produced in episodes that are comparatively short (relative to the observation time) are classified as gravitational-wave bursts. Burst radiation is normally produced in violent events such as binary coalescence (the orbital decay and collision of compact binary stars such as white dwarfs, neutron stars, or black holes), in supernovae (explosions of stars due to core-collapse), or during other kinds of highly energetic, short-lived events.

5.2.1
Coalescing Binaries

Of all the possible sources of gravitational waves, the sources in which we have the greatest confidence are binary systems. The gravitational waves produced by white dwarf binaries, neutron star binaries and binaries with black hole primaries are the most promising sources for detection with future detectors. In the low-frequency band, white dwarf binaries will be so prevalent that the jumbled superposition of gravitational waves from the multitude of Galactic systems will produce a kind of gravitational-wave noise in the detector (see Section 5.3.2). In the high-frequency band probed by ground-based detectors, the source we understand best – both in terms of the anticipated waveform and in terms of the expected event rate – is binary neutron stars: binary neutron stars that are close enough to merge in the foreseeable future (i.e. within the age of the Universe) have been found in our Galaxy. In various bands, from the very-low-frequency band to the high-frequency band, binary black holes of a wide range of masses are also likely candidates: we know galaxies harbour supermassive black holes, and we expect that when galaxies collide, their nuclear black holes will eventually be drawn together and collide – these coalescences of supermassive black holes would be an important source for study by pulsar timing arrays and by space-based gravitational-wave detectors. We also know that massive stars end their lives by producing black holes, and it is quite possible that binary systems with two massive stars could produce binary black holes that would eventually merge; these would be a source of gravitational waves detectable by ground-based detectors. Between stellar-mass black holes ($M \lesssim 100\,M_\odot$) and supermassive black holes ($M \gtrsim 10^6\,M_\odot$) there could well be a population of intermediate mass black holes (we don't know how the supermassive black holes form, but they may well be formed by growing intermediate mass black holes through multiple merger episodes).

Some binary systems, such as the white dwarf binaries that will be detected in droves by LISA, will evolve so slowly that they will be a source of continuous waves (as discussed previously in Section 5.1). However, if the binary stars are close enough together, the orbital decay will take place over a timescale that is shorter than the observation time; the binary system is then a source of a gravitational-wave burst. That is, we classify a gravitational wave as a burst when it occurs over a period that is less than (usually much less than) the observation time.

The gravitational waves produced by binary inspiral and merger are, by and large, well modelled by the various techniques that we have described before. When the motion of the companions is not too relativistic, post-Newtonian theory provides an accurate description of the gravitational waves from the *inspiral* phase, as the orbit decays secularly under the energy loss due to gravitational radiation. The late stages of inspiral and merger, which we call the *merger* phase, are explored with numerical relativity. The final object produced by binary neutron star and black hole mergers is a black hole, whose settling to a final stationary state produces *ringdown* radiation, which can be described analytically in terms of the black hole's quasi-normal modes of oscillation. Our relatively complete knowledge of the gravitational waveform produced during binary coalescence is of great assistance in our search for such systems (searches for signals of known form are generally more powerful than for signals of unknown form – see Chapter 7); furthermore, because we have a good prediction of the gravitational waveform, observations of gravitational waves from binary coalescence will provide a strong test of our theory of gravitation.

Nevertheless, our understanding of binary coalescence is incomplete. Post-Newtonian methods predict the waveform only when the component velocities are not near the speed of light; ultimately the post-Newtonian approximation will fail. In addition, for extended bodies such as neutron stars, the tidal interaction between the bodies is expected to be a significant component of the binary evolution at high frequencies (depending on the size of the neutron star). The rapid rotation that black holes may have will also cause changes to the gravitational waveform through spin-orbit and spin-spin (if there are two rapidly rotating black holes) interactions. If the total angular momentum of the system and the orbital angular momentum of the system are not aligned then the precession of the orbital plane will result in a complex waveform. Finally, the inspiral of small bodies into much more massive black holes (known as an *extreme mass ratio inspiral* or *EMRI*) can be extremely complex, and to model these systems requires accurate calculations of the radiation reaction on the orbital dynamics – a challenge that has not yet been overcome. (Numerical relativity has had only limited success in modelling such systems because the large discrepancy in scales between the two bodies creates an enormous computation burden.)

These limitations of our knowledge of the gravitational waveforms presents the opportunity for great discovery: observations of the effects of tidal interactions between neutron stars on the inspiral waveform will tell us about the composition of these stars, ultimately yielding information about the nature of cold matter at extremely high density. The highly complex waveform of an EMRI, in which the small body probes the gravitational field of the black hole over many many orbits, will map out the gravitational field in the vicinity of the massive object it is orbiting and will reveal whether that object truly is a black hole or if it is some other (unexpected) object. The gravitational radiation produced during the merger and ringdown of black holes also gives us a means to observe the strong-field effects of gravitation, and will allow tests of general relativity in the strong-field regime that we have not been able to probe so far.

Binary mergers are also expected to be a highly likely source of gravitational radiation for detection with the next generation of gravitational-wave detectors. Space-based interferometers such as LISA will detect mergers of supermassive black holes as well as extreme mass ratio inspirals. Pulsar timing arrays may also see supermassive binary black hole mergers. Ground-based interferometers that are currently under construction, Advanced LIGO and Advanced Virgo, should also detect binary coalescences routinely. Abadie *et al.* (2010) presents a survey of methods of estimating event rates and gives likely ranges of event rates for the LIGO-Virgo network; we summarize the data given in Tables 2, 4 and 5 of that review. For binary neutron stars inspirals, the rate of events in the Galaxy is likely in the range $R_{\text{G,NSNS}} \sim 1\,\text{Myr}^{-1}$–$1000\,\text{Myr}^{-1}$ with a most likely rate of $R_{\text{G,NSNS}} \sim 100\,\text{Myr}^{-1}$; these correspond to a rate-density in the range $\mathcal{R}_{\text{NSNS}} \sim 0.1\,\text{Myr}^{-1}\,\text{Mpc}^{-3}$–$10\,\text{Myr}^{-1}\,\text{Mpc}^{-3}$ in the local Universe and a most likely rate-density of $\mathcal{R}_{\text{NSNS}} \sim 1\,\text{Myr}^{-1}\,\text{Mpc}^{-3}$. With the anticipated sensitivity of the Advanced LIGO-Virgo network, these correspond to an anticipated detection rate in the range 0.4–400 events per year, with a most likely estimate of 40 events per year. Estimates of the merger rates of binary black hole systems and neutron-star + black-hole binary systems are more uncertain than neutron-star + neutron-star systems. For neutron-star + black-hole binary systems, the estimated range for the Galactic merger rate is $R_{\text{G,NSBH}} \sim 0.05\,\text{Myr}^{-1}$–$100\,\text{Myr}^{-1}$ with a most likely value of $R_{\text{G,NSBH}} \sim 3\,\text{Myr}^{-1}$, and these correspond to rate densities in the range $\mathcal{R}_{\text{NSBH}} \sim 6 \times 10^{-4}\,\text{Myr}^{-1}\,\text{Mpc}^{-3}$–$1\,\text{Myr}^{-1}\,\text{Mpc}^{-3}$ with a most likely value of $\mathcal{R}_{\text{NSBH}} \sim 0.03\,\text{Myr}^{-1}\,\text{Mpc}^{-3}$. For binary black-hole systems, the likely rates are even lower with $R_{\text{G,BHBH}} \sim 0.01\,\text{Myr}^{-1}$–$30\,\text{Myr}^{-1}$ with a most likely value of $R_{\text{G,BHBH}} \sim 0.4\,\text{Myr}^{-1}$, which give rate densities $\mathcal{R}_{\text{BHBH}} \sim 1 \times 10^{-4}\,\text{Myr}^{-1}\,\text{Mpc}^{-3}$–$0.3\,\text{Myr}^{-1}\,\text{Mpc}^{-3}$ with a most likely value of $\mathcal{R}_{\text{BHBH}} \sim 0.005\,\text{Myr}^{-1}\,\text{Mpc}^{-3}$. However, these more massive systems can be detected to a greater distance than binary neutron stars and the anticipated event rates for the Advanced LIGO-Virgo network are comparable: For the neutron-star + black-hole systems, the likely range of detection rates is 0.2 events per year to 300 events per year with a most likely value of 10 events per year, while for the binary black holes the range is 0.4 events per year to 1000 events per year with a most likely value of 20 events per year. Note that while the event rate for binary neutron stars is on comparatively firm ground (we know that binary neutron stars that will eventually merge exist), the data for binary black holes and for neutron-star + black-hole binaries is more speculative. That said, binary inspiral signals are considered to be the most promising source for first detection by the advanced ground-based detectors.

Example 5.2 Rate of binary neutron star coalescences in the Galaxy

There are currently six known binary neutron star systems in the Galaxy that will merge in less then ten billion years. These six systems are detected by pulsar surveys, but such pulsar surveys are only capable of detecting a fraction of the total number of binary neutron star systems. If we wish to make an empirical estimate of the rate of binary mergers in the Galaxy, we must determine (i) the lifetime of each binary system for the systems that we know, and (ii) the fraction of the total Galactic population that our pulsar surveys are sensitive to. In this example, we present a (very) rough estimate of the event rate based on the observed population; a more careful analysis and a review of the various methods of rate estimation can be found in Abadie *et al.* (2010).

The lifetime of a binary system is the sum of its current age and the remaining time until it merges, τ_c, given by Eq. (3.178a), which can be determined from the orbital parameters of the system. The current age can be estimated from the observed spin-down rate of the pulsar, $\dot{P}$, as $P/(2\dot{P})$, under the assumption that the pulsar had been spun up by its companion star before that companion became a neutron star and the pulsar has been spinning down since then.

The fraction of the total population that the pulsar surveys could have detected is found by modelling a distribution of Galactic binary pulsars and determining the number that would be detected by the current surveys based on their sensitivities. There are considerable uncertainties in estimates of this fraction, particularly due to uncertainties in the luminosity distribution of pulsars.

It turns out that the estimated rate of Galactic mergers is essentially determined by the single system J0737–3039, a double pulsar system that has a current age of ~ 200 Myr and will merge in 85 Myr. It is estimated that the current surveys would have only detected one in $\sim 10^4$ such systems, so an estimated Galactic merger rate is

$$R_G \sim \frac{10^4}{200\,\text{Myr} + 85\,\text{Myr}} \approx 40\,\text{Myr}^{-1}\,. \tag{5.6}$$

There are very large uncertainties in this calculation, and the estimated rate could easily be an order of magnitude too large or too small.

There are several methods to convert a Galactic merger rate to a rate density in the local Universe. One is to compare the star formation rate in the Galaxy ($\sim 3\,M_\odot\,\text{yr}^{-1}$) to that of the local Universe ($\sim 0.03\,M_\odot\,\text{yr}^{-1}\,\text{Mpc}^{-3}$); if star formation rate determines the population of binary neutron stars then we have

$$\mathcal{R} \approx 0.01\,\text{Mpc}^{-3}\,R_G \tag{5.7}$$

or $\mathcal{R} \sim 4 \times 10^{-7}\,\text{yr}^{-1}\,\text{Mpc}^{-3}$. To get a sense of what this means in terms of detection rate, consider that Initial LIGO detectors were sensitive to binary neutron star mergers over a volume $V \approx 10^5$ Mpc while Advanced LIGO detectors will detect mergers over a volume $V \approx 3 \times 10^7\,\text{Mpc}^{-3}$, so the event rate would be $\sim 1/(200\,\text{yr})$ with Initial LIGO detectors and $\sim 1/\text{month}$ with Advanced LIGO. Again, there are very large uncertainties in these numbers.

Methods of determining the binary coalescence event rate involve either empirical estimation (for binary neutron stars) using the observed binary neutron star systems (Kim *et al.*, 2003), or estimates based on population synthesis models (Postnov and Yungelson, 2006). The uncertainty in the empirical estimates of binary neutron star coalescence rates arises primarily from accounting for the fraction of low-luminosity pulsars that cannot be detected by pulsar surveys. The range of results obtained through population synthesis analyses reflects details of various scenarios of stellar evolution in binary systems. Therefore, measuring the event rates and component mass distributions from gravitational-wave observations will be highly informative about the population of binary neutron stars, binary black holes, and neutron-star black-hole binaries, and this population information will, in turn, constrain the models of binary evolution.

Until now we have focused discussion on binary formation in galactic fields; however, binaries can also form in star clusters such as globular clusters. In a cluster, stars are close enough together that their interactions become dynamically important. Dynamical friction causes a mass segregation in which the more massive stars fall toward the centre of the cluster. Three body interactions in which a binary star encounters a third star can lead to partner exchange and can also "harden" the binary (drive the companions closer together). This may result in efficient production of tight binary systems that will eventually merge and produce detectable signals. A potentially interesting scenario would involve the creation of an intermediate mass black hole with mass $\sim 100\,M_{\odot}$. If a smaller object (perhaps a neutron star with $M \sim 1.4\,M_{\odot}$ or a small black hole with mass $M \sim 10\,M_{\odot}$) coalesced with such an intermediate-mass black hole, the result would be an *intermediate mass ratio inspiral* or *IMRI*, which would be similar to the extreme mass ratio inspirals described earlier but could be detectable in the high-frequency band by the Advanced LIGO-Virgo network.

As mentioned earlier, binary coalescences are normally divided into phases during which the dynamics of the system have distinct characteristics. At early times, the orbiting bodies are slowly driven together through a secular process resulting from the loss of energy to gravitational radiation. At the end of this inspiral, strong-field gravity comes to dominate the evolution and the bodies merge on a dynamical timescale. If one of the bodies is a neutron star, it is possibly pulled apart by tidal forces during this merger phase. The final remnant will be a black hole but, in the case of binary neutron merger, a single hypermassive neutron star that eventually collapses into a black hole might be produced. In either case, the post-merger object will initially be distorted and the oscillations of this object will produce periodic, damped gravitational radiation at a spectrum of frequencies set by the natural oscillation frequencies. Here we discuss these phases is more detail.

5.2.1.1 **Binary Inspiral**

The binary inspiral is driven by the secular process of energy loss to gravitational radiation. We have studied this process for Newtonian orbital dynamics with energy loss in radiation given by the quadrupole formula in Section 3.5 and in more detail for post-Newtonian equations of motion and radiation in Section 4.1.2; Appendix B presents the waveform for the current state of post-Newtonian calculations. These results are obtained for non-spinning point particles in circular orbits for relatively slow motion (compared to the speed of light), and only account for the dissipative part of radiation reaction through an energy balance argument. Here we will briefly touch on some of these limitations.

Binary orbits are normally assumed to be circular by the time the orbital frequency enters the high-frequency gravitational-wave band because gravitational radiation effectively circularizes (see Problem 3.9) the orbit during millions of years of inspiral prior to the final few minutes (the time spent in the high-frequency band). However, if the binary was formed by capture (a possible scenario for forming tight binaries in star clusters) then there could be some vestigial amount of eccentricity. To model eccentric orbits, one needs to include two additional parameters: the eccentricity e and the longitude of the periapsis ϖ.

If one or both of the binary companions have significant spin then the orbital dynamics can be affected by spin-orbit and spin-spin interactions. Neutron stars are not likely to have large enough spins to have an appreciable effect on the orbit, but black holes *can* have large spins. In the case that the spin (or spins) are aligned or anti-aligned with the orbital angular momentum, then the spin-orbit effect produces a $O((v/c)^3)$ correction to the leading order phase evolution – that is the spin-orbit term is a (post)$^{3/2}$ effect – and the spin-spin effect produces a $O((v/c)^4)$ (second post-Newtonian) correction to the phase evolution. These effects are known and can be incorporated into the post-Newtonian phase evolution for the gravitational waveform. To a reasonably good approximation, the phase evolution caused by spin is similar to that expected for a non-spinning binary with slightly different component masses. Waveforms constructed for non-spinning binary systems are remarkably similar to those of spinning binary systems if we allow for a bias in the masses, so we should be able to detect spinning systems with searches designed to detect signals from non-spinning systems.[1)] Estimation of the masses of the binary components is compromised by this degeneracy with spin, however.

The preceding argument applies only to spins that are aligned or anti-aligned with the orbital angular momentum – in such a case, the total angular moment and the orbital angular momentum are co-aligned and the orbital plane is steady. When there is a mis-alignment between the orbital angular momentum axis and the total angular momentum axis, the orbital plane will precess. Such precession can produce significant modulation of the waveform, which can make the waveforms quite complex. To model the general case of inspiral of two spinning bodies requires a dramatic increase in the number of parameters required: for non-spinning systems, only the two masses m_1 and m_2 are needed, while for spinning systems one additionally requires the spin vectors $\boldsymbol{S}_1$ and $\boldsymbol{S}_2$ and the orientation of the orbital plane at some reference time $\hat{\boldsymbol{L}}$ (eight additional parameters).[2)] The additional pa-

1) To successfully model spinning systems with waveforms for non-spinning systems, however, it is sometimes necessary to take the symmetric mass ratio $\eta = m_1 m_2/(m_1 + m_2)^2$ to unphysical values $\eta > 1/4$.

2) In fact, for the non-spinning case, the two parameters given by the propagation direction of the gravitational wave, $\hat{\boldsymbol{n}}$, or equivalently the inclination and azimuthal angles ι and ϕ, would be required but these parameters can be largely ignored for single detector searches that focus only on the leading order quadrupole amplitude evolution.

rameters needed to model binaries with generic spin vectors makes the search for such systems considerably more challenging.

Fortunately, neutron stars are not expected to have significant spin, and furthermore their masses are expected to be confined to a fairly narrow range between one and three solar masses. However, neutron stars are extended bodies, and the tidal interaction between the neutron star and its companion can affect the inspiral waveform. The leading order tidal correction terms affect both the binding energy and the gravitational-wave flux at the fifth post-Newtonian order, $O((v/c)^{10})$. However, the corrections to these two functions are $\sim (c^2 R/GM)^5 (v/c)^{10}$ where R is the size of the neutron star and M is the mass of the binary system, and if the ratio $c^2 R/GM$ is large then the tidal correction can become important during the late inspiral (typically at gravitational-wave frequencies above 400 Hz for realistic neutron star equations of state). This is not expected to have an adverse effect on the detectability of these systems but it raises the intriguing possibility of measuring the size (in fact, the tidal deformability) of neutron stars from the high-frequency gravitational-wave signatures, which would then constrain the nuclear equation of state.

Ultimately the post-Newtonian approach to constructing the binary evolution will fail as motion of the bodies becomes relativistic with $(v/c) \sim 1$. At some point it is necessary to use other techniques, such as numerical relativity, to model the late inspiral. Fortunately, recent advances in numerical relativity have made this a tractable problem for binary systems with components of comparable mass. It is now becoming possible to construct entire waveforms that smoothly interpolate between post-Newtonian models of the early inspiral and numerical simulations of the late inspiral and the merger.

Extreme mass ratio binary systems, in which one companion (a black hole) has a mass that much greater than the other (either a much smaller black hole or a neutron star), present a new challenge: such systems can evolve to the point where the motion becomes relativistic, which means that post-Newtonian methods cannot be used, but the orbit evolves slowly (since the luminosity is proportional to the symmetric mass ratio squared). In such cases, the orbit of the smaller mass can be obtained to high accuracy by computing the geodesics of the black hole background spacetime produced by the larger mass object, and the gravitational waveform can be obtained by solving the Teukolsky equation (see Section 4.2.2) with a matter source given by the orbiting small object. Although this procedure allows one to compute the energy and angular momentum radiated, the detailed balance in which the orbit of the small body is adjusted to account for the decrease in the energy and angular momentum becomes problematic: one issue is that the generic orbits about spinning black holes are characterized by more conserved quantities than simply their energy and angular momentum. The second issue is that the detailed balance only accounts for the dissipative term in the gravitational self-force on the small body – there is also conservative term which affects the orbit of the body but does not result in radiation. To correctly model extreme mass ratio inspiral systems requires the correct handling of the gravitational self-force.

5.2.1.2 Binary Merger

At the end of the inspiral phase, the binary companions will rapidly plunge together and collide on a dynamical timescale. In the test particle limit, the dynamical instability occurs when the particle comes to the innermost stable circular orbit, but for binary systems with companions of comparable masses there is less of a dramatic shift from the slow secular evolution to a dynamical one. Still, the modelling of the merger of binary systems requires numerical simulation.

Numerical relativity is now capable of stably and accurately evolving systems of black holes for many orbits leading up to a final plunge, merger and subsequent ringdown. Such simulations have been performed for a variety of black hole spins (aligned, anti-aligned, and with arbitrary orientation), and for various mass ratios. Some interesting effects have been observed: For spins and orbital angular momentum that are all aligned, the frequency evolution proceeds more slowly toward the end of the inspiral compared to the non-spinning case. This occurs because the system must shed a larger amount of angular momentum before the merger can take place. The opposite occurs when the two spins are anti-aligned with the orbital angular momentum: the frequency increases more rapidly during the late stages of inspiral. Finally, for certain configurations of component spins (specifically, when the spins are anti-aligned with each other and lie in the orbital plane), the gravitational radiation produced during the merger of the two holes is beamed in a particular direction, which imparts a momentum on the final black hole. In these *super kick* scenarios, the black hole remnant can have a large final velocity of $(v/c) \approx 0.008 \simeq 2500\,\mathrm{km\,s^{-1}}$.[3]

Mergers involving at least one companion that is a neutron star may lead to the tidal disruption of the neutron star that can distribute matter in a disk about the final remnant. For neutron-star + black-hole systems, in order for the neutron star to be torn apart, the black hole must be either relatively small or rapidly rotating, otherwise the neutron star is simply sucked into the black hole without leaving a disk of matter behind. Numerical simulations of binary neutron star and neutron-star black-hole coalescences are being performed, but there are more complex physical processes that need to be taken into consideration when matter is involved. Among these are the equation of state of the neutron star (of which there is considerable uncertainty), finite temperature effects on the equation of state, shocks, the effects of magnetic fields that may be present in the system, the effects of neutrino cooling, and so on. Despite the difficulty of these simulations, there is considerable effort devoted to the study of neutron star disruption during binary coalescence because this is thought to be the progenitor for most of the short-hard gamma ray bursts. Detecting a gravitational wave inspiral in association with an observed short-hard gamma ray burst would confirm this hypothesis.

3) Non-spinning black holes can also receive a kick from gravitational recoil, which is along the orbital plane, if the components have unequal masses. This effect is largest when the mass ratio is $m_1/m_2 \approx 0.36$, in which case the kick speed is $(v/c) \approx 5.8 \times 10^{-4} \simeq 175\,\mathrm{km\,s^{-1}}$.

5.2.2
Gravitational Collapse

We now consider the process of gravitational-wave emission during the collapse of a star, or the core of a star, to form a neutron star or a black hole. There are several different scenarios here, spanning a range of masses of the progenitor star. White dwarfs – compact stars supported by electron degeneracy pressure – are the remnants of relatively light ($\lesssim 8\,M_\odot$) stars. If the white dwarfs are in binary systems where there is mass transfer from the companion onto the white dwarf, *accretion-induced collapse* (AIC) occurs if the accretion of matter pushes the mass of the white dwarf beyond its stability limit, the *Chandrasekhar mass limit*; if, as the temperature and pressure increase, a nuclear burning detonates the star, a Type Ia supernova is produced.[4] Stars with masses above $\sim 8\,M_\odot$ end their lives with core collapse when nuclear burning eventually fails to support them. Type II supernovae (as well as Type Ib and Ic supernovae) have their origin in such scenarios, and the remnant of these supernovae is a neutron star or a black hole. The long-duration gamma ray bursts are also believed to be formed by core collapse of massive stars when black holes are created. Very massive ($\sim 50\,M_\odot$) stars are expected to collapse directly to black holes without producing supernovae, and even smaller mass stars may ultimately collapse to black holes if the mantle of the star is not ejected. Further discussion of gravitational-wave emission from core collapse is given in reviews such as Fryer *et al.* (2002), Fryer and New (2003) and Ott (2009) and the references therein.

In accretion-induced collapse, a white dwarf accretes material from a companion star until its mass exceeds the Chandrasekhar mass (see Example 5.3; at this point, the star can no longer support its weight and will collapse. During the collapse, the pressure and temperature will rise. If the temperature rises enough to ignite nuclear burning then the star may explode; such explosions are seen as Type I supernovae. However, neutrino emission may limit the amount of heating, and if the neutrino cooling is sufficient to prevent an explosion, a neutron star is formed.

In stellar core collapse, the onset of the collapse occurs as the nuclear burning in the core becomes insufficient to support the star. The core implodes and forms a proto-neutron star. Infalling matter encounters this now-rigid proto-neutron star and bounces off of it. The resulting shock wave does *not* explode the star, but instead stalls as it encounters the star's mantle. In at least some cases, the shock is rejuvenated (there are several possible mechanisms that can do this but it is not known which one – or ones – actually occur) and this rejuvenated shock is responsible for a Type II supernova (or perhaps a Type Ib or Ic supernova if the star's envelope has been lost). In other cases, there is a failure to explode and the infalling matter will eventually form a black hole. For massive stars, black hole formation may occur directly (these are known as collapsars). Black hole formation during core-collapse episodes are normally considered to be the progenitor of long-duration gamma-ray bursts.

4) Type Ia supernovae may also be produced in the collision of two white dwarf stars.

Example 5.3 Chandrasekhar mass

White dwarfs are supported by non-thermal pressure arising from electron degeneracy; in fact, for high-density white dwarfs, the electrons are relativistic, and the non-thermal equation of state that describes the matter is that of a cold, relativistic degenerate gas:

$$p \sim \hbar c n_{\mathrm{e}}^{4/3} , \tag{5.8}$$

where n_{e} is the number density of electrons. Since $n_{\mathrm{e}} \simeq \rho/(2\,\mathrm{amu})$, that is about one electron per two atomic mass units (amu), the equation of state for a relativistic degenerate gas is $p \sim \rho^{4/3}$. The star must be supported, so it must be in hydrostatic equilibrium. This means that the central regions of the star must support the weight of the star, $\sim G M^2/R^2$, where M and R are the star's mass and radius, and so the gas must provide a central pressure $p \sim G M^2/R^4 \sim G M^{2/3} \rho^{4/3}$. Notice that both the degeneracy pressure and the required hydrostatic pressure scale as $\rho^{4/3}$: as mass is added to a star, the required hydrostatic pressure cannot be generated by compressing the star (i.e. by increasing the density). The critical mass that can be supported, then, is

$$M_{\mathrm{Ch}} \sim \left(\frac{\hbar c}{G}\right)^{3/2} \frac{1}{(2\,\mathrm{amu})^2} . \tag{5.9}$$

To obtain the important factor of order unity one must solve the actual equations of hydrostatic equilibrium; when this is done the resulting mass limit is

$$M_{\mathrm{Ch}} = 1.4\, M_{\odot} , \tag{5.10}$$

which is known as the *Chandrasekhar mass limit.*

Numerical models of stellar collapse are extremely challenging because of the complexity of the physical processes that are important for such systems. The simulations must capture general relativistic effects as well as the effects of neutrino transport, microphysics and magnetic fields. The matter must be described by realistic equations of state (which are not known). The initial stellar structure must also be realistic. Ideally, simulations should be conducted in $3 + 1$ dimensions. Steady progress over the years has resulted in substantial gains in our understanding of what kinds of processes might occur during stellar collapse, but there is still considerable uncertainty about important aspects such as the exact mechanism (and perhaps there are several mechanisms) that causes the shock that generates supernovae. Likewise, the gravitational-wave signature from core collapse is expected to be quite varied, depending on the particular scenario, but observations of the gravitational wave from core collapse can potentially give a rare observation of the dynamics that are occurring in the core of the star, which is entirely obscured from electromagnetic observations.

Although the detailed gravitational-wave signature – and indeed, the amplitude of gravitational waves – produced by stellar collapse is not well modelled, we do know that the gravitational waves will form a reasonably short burst of radiation. We can also explore various kinds of scenarios of gravitational-wave production and estimate the characteristics of the radiation produced by these scenarios.

Consider an axisymmetric collapse in which an initially rotating stellar core (supported by electron degeneracy pressure) with angular momentum J, collapses to form a neutron star. As the collapse progresses, the angular momentum is conserved, so shape of the stellar core changes with the matter near the rotation axis falling inward faster than the matter near the equator. Collapse continues until the core reaches nuclear density, at which point neutron degeneracy pressure begins to support the star against collapse. During the collapse the proto-neutron star will be an axisymmetric ellipsoid with eccentricity $e \propto J/(GM^3R)^{1/2}$, and, since J and M are fixed, the eccentricity increases as the size decreases. The collapse takes place on the free-fall timescale,

$$\tau_{\text{ff}} = \sqrt{\frac{3\pi}{32}\frac{1}{G\rho}}\,, \tag{5.11}$$

where ρ is the density of the matter; near the end of collapse, $\rho \sim 10^{18}\,\text{kg}\,\text{m}^{-3}$ (nuclear density), so the dynamical timescale is $\tau_{\text{dyn}} \sim 0.1\,\text{ms}$. The proto-neutron star is aspherical, with principle moments of inertia (assuming it is uniform density) of $I_{11} = I_{22} = \frac{2}{5}MR^2(1-\frac{1}{2}e^2)$ and $I_{33} = \frac{2}{5}MR^2$ where $R \sim 15\,\text{km}$ is the radius of the proto-neutron star and $M \sim 1\,M_\odot$. An observer along the axis of the collapse will not see any gravitational radiation, but other observers will see linearly polarized gravitational waves, with the strongest emission toward observers located on the equator of the system:

$$h_+ \sim \sin^2\iota\,\frac{G}{c^4}\frac{\ddot{I}_{22}-\ddot{I}_{33}}{r} \sim \sin^2\iota\,\frac{GM}{c^2r}\left(\frac{eR}{c\tau_{\text{dyn}}}\right)^2, \tag{5.12}$$

where ι is the angle between the rotational axis of the system and the direction to the observer. For $e \sim 0.1$, we find $rh_+ \sim 1\,\text{m}$ for an observer on the equator, or

$$h_+ \sim 10^{-20}\left(\frac{r}{10\,\text{kpc}}\right)^{-1}, \tag{5.13}$$

where $r \sim 10\,\text{kpc}$ would be the typical distances to supernovae in our Galaxy. The waveform would be a rapid rise to this peak value, followed by a second sharp spike as the infalling matter rebounds off of the newly formed, rigid, proto-neutron star. The impulse on the proto-neutron star will also set it into oscillations, and these oscillations will produce gravitational radiation at twice the oscillation frequency.

Gravitational radiation from non-axisymmetric motion can also be produced if the stellar core is rapidly rotating. In such a situation, the collapsing stellar core may centrifugally hang up and bar-mode instabilities can form, or they may form after the proto-neutron star is formed if this is spun-up by accumulating matter with high angular momentum. Instabilities are expected to form when $\beta = E_{\text{rot}}/|E_{\text{grav}}|$, the ratio of rotational kinetic energy of the star, E_{rot}, to its gravitational potential energy E_{grav}, is above some critical value (typically a bar mode grows due to viscosity or gravitational radiation on a secular time-scale if $\beta \gtrsim 0.14$, and on a dynamical time-scale if $\beta \gtrsim 0.27$). If this happens, then the gravitational radiation produced will be that of a rotating bar:

$$h_+ = -\frac{1}{3}\frac{1+\cos^2\iota}{2}\frac{GM}{c^2 r}\left(\frac{L\omega}{c}\right)^2 \cos 2\omega t \tag{5.14a}$$

$$h_\times = \frac{1}{3}\cos\iota\frac{GM}{c^2 r}\left(\frac{L\omega}{c}\right)^2 \sin 2\omega t\ , \tag{5.14b}$$

where the bar of length L and mass M is rotating with an angular velocity ω and ι is the angle between the rotation axis and the direction to the observer. The radiation directed along the axis of rotation is circularly polarized, while the radiation directed along equatorial directions is linearly polarized. If mass in the rotating bar is $M \sim 0.1\,M_\odot$ and the bar has $L\omega \sim 0.1c$ then the amplitude of the gravitational waves will be $rh \sim 1\,\text{m}$. The rotating bar may last for many cycles, which would make the signal more detectable (roughly by the square root of the number of cycles).

If a black hole is formed during stellar core collapse (either prompt collapse of the core to form a black hole, or delayed black hole formation in which a proto-neutron star is formed but eventually collapses to a black hole, possibly due to continued infall of matter onto the neutron star) then the quasi-normal modes of the black hole will be excited, and the associated ringdown radiation may be observable. To estimate the amplitude of such ringdown radiation, suppose that the collapse is axisymmetric and that the black hole that is formed is a non-spinning black hole of mass M. The fundamental $\ell = 2$ quasi-normal mode of oscillation of the black hole has a frequency $f_{\text{qnm}} \simeq 12\,\text{kHz}\,(M/M_\odot)^{-1}$ and quality factor $Q_{\text{qnm}} \simeq 2$ (cf. Table 4.4) and so the ringdown waveform has the form

$$h_+ \propto h_0 \sin^2\iota e^{-\pi f_{\text{qnm}} t/Q_{\text{qnm}}} \cos(2\pi f_{\text{qnm}} t + \varphi_0) \quad (t > 0)\ , \tag{5.15}$$

where h_0 is the initial amplitude we seek. By integrating the time derivative of this waveform over the duration of the ringdown we can obtain a relationship between the energy radiated ΔE and the amplitude of the waveform via Eq. (3.112):

$$rh_0 = \sqrt{\frac{8}{Q_{\text{qnm}}}\frac{G\Delta E}{c^3 f_{\text{qnm}}}}\ . \tag{5.16}$$

If a clump of matter of mass μ falls onto the hole, most of its mass-energy is simply incorporated into the hole, but a fraction of the mass energy, proportional to μ/M, is radiated. For a head-on collision of an object of mass μ into an object of mass $M \gg \mu$, $\Delta E \sim 0.01(\mu/M)\mu c^2$. Then

$$rh_0 \sim 10\,\text{m}\left(\frac{\mu}{0.01\,M_\odot}\right) \tag{5.17}$$

is the initial amplitude of the gravitational wave produced by a blob of matter of mass μ falling onto a black hole. In order for such ringdown radiation to be detectable, however, the black hole mass would need to be quite large in order for the radiation to lie in a band that would be detectable by current detectors (recall that f_{qnm} is inversely proportional to the black hole mass).

Other possible sources of gravitational radiation from stellar core collapse can possibly arise from acoustic instabilities and turbulence in the convective regions of the star, anisotropic neutrino emission and magnetohydrodynamic effects. The various different effects would be expected to exhibit different gravitational-wave signatures, so gravitational-wave observations could assist our understanding of the physical processes involved during core collapse episodes. Unfortunately, supernovae only occur in our Galaxy at a rate of approximately once every thirty years, and the expected gravitational-wave amplitudes suggest that we would only be able to detect gravitational radiation from core collapses that occur in our Galaxy, or perhaps very nearby galaxies, with advanced detectors, so these sources will be rare.

5.2.3
Bursts from Cosmic String Cusps

Beyond the expected sources of gravitational-wave burst radiation, there is the potential that there are unknown violent phenomena that produce gravitational-wave bursts. Gravitational-wave observations may uncover aspects of the Universe that are speculative or even entirely unsuspected – there is a great potential for discovery with gravitational-wave observations.

An illustrative example of a *possible* source of gravitational-wave bursts is the radiation from cosmic string cusps. Cosmic strings, and cosmic string cusps, will be described in more detail later in Section 5.3.1.5, and a full review is found in Vilenkin and Shellard (2000). They are string-like one-dimensional topological defects that may have been formed in the early Universe during phase transitions and possibly also in certain superstring-inspired cosmologies. The principal quantity characterizing the cosmic strings is their mass-per-unit-length, $\mu = m/\ell$, which is normally expressed in the dimensionless constant $G\mu/c^2$. The value of $G\mu/c^2$ depends on the energy scale at which the phase transition that produced the cosmic strings occurred; for breaking of a grand unified theory, the cosmic strings would have $G\mu/c^2 \sim 10^{-6}$ (though observational bounds place the value of $G\mu/c^2$ somewhat lower than this).

Cosmic strings can occasionally form points where the string doubles back on itself and forms a *cusp*. Because of the high tension in the strings, the tips of these cusps acquire a very large Lorentz boost and produce copious amounts of radiation, which is highly beamed along the motion of the cusp. If we happened to lie in the beam of radiation, we would experience a burst of gravitational waves of the form

$$h(t - t_0) \sim \frac{G\mu\ell^{2/3}}{c^2 r} c^{1/3} |t - t_0|^{1/3} , \tag{5.18}$$

where t_0 is the moment of peak cusp radiation and ℓ is the length scale of the string curvature in the region where the cusp is formed. At frequencies above f_{low}, the change in the metric perturbation is $\Delta h \sim 6(G\mu\ell^{2/3}/c^2 r)c^{1/3} f_{\text{low}}^{-1/3}$, and the frequency spectrum is $\tilde{h}(f) \sim (G\mu\ell^{2/3}/c^2 r)c^{1/3}|f|^{-4/3}$. These equations have assumed that the line of sight of the observer is relatively close to the direction of the cusp, and that the cusp is not at cosmological distances. However both the length scale of string curvature ℓ and the distance r are likely to be on cosmological scales, and cosmological corrections to both f and r are needed to obtain an appropriate expression for the strain.

For a cusp produced at a cosmological redshift of z, the string length scale is $\ell \sim \alpha c/H(z)$ where $H(z)$ is the value of the Hubble parameter (see Section 5.3.1.2) at that redshift, so that c/H is the size of the Universe at that time, and α is the fraction of the size of the Universe of the cosmic string loop. Various predictions place the value of α to be $\alpha \sim 10^{-3}$–10^{-1} (known as "large loops") or find that loop sizes are determined by gravitational radiation back-reaction and have $\alpha \sim 50G\mu/c^2$. Equation (5.18) is expressed only for relatively small distances (not cosmological distances) from the cosmic string cusp; a source at a cosmological redshift z produces a gravitational waveform in the frequency domain of

$$\tilde{h}(f) \sim C(z)(G\mu/c^2)\alpha^{2/3}H_0^{1/3}|f|^{-4/3}\,, \tag{5.19}$$

where H_0 is the present day Hubble constant and

$$C(z) \sim \begin{cases} z^{-1} & z \ll 1 \quad \text{(nearby events)} \\ z^{-4/3} & 1 \ll z \ll z_{\text{eq}} \quad \text{(matter era events)} \\ z_{\text{eq}}^{1/3} z^{-5/3} & z \gg z_{\text{eq}} \quad \text{(radiation era events)}, \end{cases} \tag{5.20}$$

where $z_{\text{eq}} \sim 3200$ is the redshift at which the energy density of matter and radiation were equal in the Universe. For example, if we can detect a burst with $\tilde{h}(f) \sim 10^{-23}$ s at $f = 100$ Hz, then we could potentially detect cosmic string cusp radiation to a redshift of $z \sim 10^4$ with $G\mu/c^2 = 10^{-6}$ and $\alpha = 50G\mu/c^2$, that is from cusp gravitational waves produced during the radiation era (Siemens *et al.*, 2006).

Although it is unclear how likely a source of gravitational waves cosmic strings will turn out to be, they are not wholly unexpected, and the observational discovery of cosmic strings would have a dramatic impact on our understanding of fundamental physics and the processes that occurred during the early Universe.

5.2.4
Other Burst Sources

There are many violent events observed by astronomers that may have associated gravitational-wave radiation, including flares from soft gamma-ray repeaters (believed to be magnetars, or neutron stars with high magnetic fields), pulsar glitches (in which a sudden change in a pulsar's rotational period is observed), gravitational bremsstrahlung radiation from stars in hyperbolic orbits, and so on. There are potentially any number of violent events in the Universe that we *do not* witness in electromagnetic observations, but could well be sources of gravitational waves. Gravitational-wave searches are intentionally designed to be capable of detecting unanticipated sources.

5.3
Sources of a Stochastic Gravitational-Wave Background

Up until now in this chapter, we have discussed sources which produce gravitational waves that are bounded either in frequency (continuous gravitational waves) or in time (gravitational-wave bursts). When sufficient numbers of such sources overlap in both time and frequency, it can become impossible to distinguish which gravitational waves come from which sources – we call this *source confusion*. In the case of confused sources, it makes little sense to think of gravitational waves from individual sources in the sky. Rather, we cast our discussion in terms of a *gravitational-wave background* distributed over the sky. Excellent reviews of gravitational-wave backgrounds are presented in Allen (1997) and Maggiore (2000).

Let us now consider the case where the confusion is so strong that many sources contribute to the gravitational-wave signal (detector strain) at any given time or at any given frequency. This is the case for perhaps the most interesting background distributions. Then, whatever the nature of the statistical distribution governing the signal (strain) produced by each gravitational-wave source in a detector (providing that distribution has a finite variance), the *central limit theorem* guarantees that the sum of the signals at any given time or frequency is a random variable drawn from a *Gaussian distribution*. For this reason, we call it a *stochastic background* of gravitational waves.

This is both good and bad for gravitational-wave astronomers. It is good because we need not know the details of the distribution of gravitational waves from the background sources nor the distribution of the background sources themselves – those details are washed out and every set of sources produces a gravitational-wave stream which is detected as a Gaussian-distributed signal. It is bad because noise has exactly the same property – a large number of independent overlapping noise sources always contribute to the noise of the detector, and those noise sources therefore also present a Gaussian-distributed noise profile.

One could therefore be forgiven for wondering if it is even possible to determine whether a Gaussian stream is signal or noise. If there is no way to distinguish between these possibilities then it would be more appropriate to consider such gravitational waves an additional noise component and to consider the sources undetectable. Certainly, if the noise can be modelled well enough from additional measurements, we might be able to detect the additional contributions of a stochastic background of gravitational waves. But this is a daunting prospect – all major noise sources would need to be well understood as would their transfer functions into the detector. The sensitivity to a stochastic background one could expect from real instruments would be poor at best. Fortunately, there is a much more sensitive procedure that can often be used. It turns out that we can distinguish stochastic signals from noise if there are two or more detectors with uncorrelated noise sources that are observing in the same frequency band at the same time. The attentive reader will already have at least an inkling of how from the wording of the previous sentence. The full details are in Section 7.9.

In the case where there are not sufficiently many confused sources at any given time and frequency to create a Gaussian signal, the detection problem changes, and in fact becomes somewhat more difficult. On one hand, the signal is no longer Gaussian, so one might hope that it was easily distinguished from the noise. Unfortunately, real detectors also suffer from non-Gaussian noise sources, so searching for non-Gaussian stochastic processes does not help tremendously. On the other hand, the failure of the central limit theorem in this case means that the signal from the confused sources now depends somewhat on the details of the source distribution in the sky and the signal that each source produces. Since at least one of these and usually both are almost never known in advance, we have limited ability to predict the signal. Nonetheless, techniques have been developed to look for classes of such signals which, although more difficult to implement, give us similar detection sensitivities. And, the greater dependence of the signal on the source means that if such a signal is detected, more information about the source may be available.

Let us now turn our attention to the cosmological and astrophysical sources of stochastic backgrounds of gravitational waves.

5.3.1 Cosmological Backgrounds

The discovery of the *cosmic microwave background* (CMB), a bath of electromagnetic radiation left over from the hot dense early Universe, has an interesting history. It may have been indirectly measured as early as 1941 by Canadian astronomer Andrew McKellar, who measured the bolometric temperature of the Universe via interstellar absorption lines to be 2.3 K. It was first recognized as a theoretical possibility five years later by Robert Dicke. George Gamow, Ralph Alpher and Robert Herman estimated its temperature in a series of papers from 1948 to 1956. In 1957, Tigran Shmaonov measured a radio-emission background with an effective temperature of 4 ± 3 K, independent of either time or direction, which is now considered to be the first direct measurement of the CMB. Seven years later, in 1964, A.G. Doroskevich and Igor Novikov finally published a paper stating that the CMB should be detectable. Around the same time Dicke came to the same conclusion, and this led his Princeton colleagues David Wilkinson and Peter Roll to begin construction of a radiometer to make this measurement in 1964. But before construction was completed, Bell Labs physicists Arno Penzias and Robert Woodrow Wilson discovered an antenna temperature of 3.5 K in their newly-built radiometer that could not be explained. Upon consulting with Dicke and collaborators, they realized they had unintentionally scooped Wilkinson and Roll. They went on to win the Nobel prize for their discovery of the predicted CMB.

You might wonder "if the hot early Universe produced a bath of electromagnetic waves that has survived until today, might it not have also produced a bath of gravitational waves that we might also detect?" The answer is yes, it most probably did. The quantum fluctuations in the density of the early Universe that we believe are responsible for the (very small) anisotropy of the CMB, and for the eventual formation of galaxies, are predicted to be accompanied by fluctuations in the spacetime geometry. We will discuss here two different mechanisms for the generation of a *cosmic gravitational-wave background* (CGWB) which is analogous to the CMB.

One of the most attractive features of the CMB is that it gives us a relatively early glimpse of the Universe. The photons of the CMB were "trapped" by Thompson scattering in a proton-electron plasma in the early Universe until it cooled down to about 3000 K, at which point the protons and electrons formed neutral hydrogen atoms and the photons were free to propagate basically undisturbed. This freeing of the photons is called *electromagnetic decoupling* and it occurred about $t_d \sim 10^{13}$ s (a few hundred thousand years) after the Big Bang.

By comparison to the electromagnetic fields of the CMB, gravitation is an extremely weakly coupled interaction. Gravitational waves would in principle be expected to begin decoupling at the *Planck time*, $t_{\text{Planck}} = \sqrt{\hbar G/c^5} \sim 10^{-43}$ s after the Big Bang. Unfortunately, waves from the Planck era will likely not be visible today because of the postulated epoch of *inflation* that our Universe probably underwent. Inflation was proposed to explain why we can't find particles that many particle theories predict were created in the early Universe, such as magnetic monopoles. The same applies to gravitational waves created before inflation – none will be left after the inflationary period (we'll discuss how this works in a bit more detail later in this section). We have not yet found a "smoking gun" that proves inflation happened, but recent results from CMB experiments like COBE and WMAP are providing a good deal of supporting evidence for it. It is quite likely therefore that gravitational waves from before inflation are not available to us.

If inflation happened, it was likely just after the symmetry that united the strong force and the electroweak force was broken. The precise time at which this symmetry breaking would have occurred is unknown – as yet, we do not have a *Grand Unified Theory* (GUT) that describes the symmetry, but on general grounds it is believed that the symmetry breaking happened at about 10^{-35} s after the Big Bang. This would represent the earliest time from which a significant amount of gravitational radiation could be detected. So, can we probe GUT scale physics with gravitational-wave astronomy? The answer is "in principle, probably we can, but not with current or planned instruments".

Why not, you ask? The answer is that the length scale for gravitational waves from cosmic sources will be set by the size of the Universe (i.e. the *Hubble length*) at the time they were produced. To calculate the size of the Universe at any previous time t_1, we simply need to know how the Hubble length scales with time and that the current value, ℓ_0, is about 10^{26} m. The scaling law for the Hubble parameter depends on the cosmological model, but let us assume a standard post-inflationary cosmology (known to cosmologists as the spatially flat *Friedmann–Lemaître–Robertson–Walker cosmology* or FLRW cosmology). Then, as we shall show a bit later in this section, the Hubble length scales linearly with time, so $\lambda_1 = (t_1/t_0)\ell_0$, where $\lambda_1 = \ell_1$ is the characteristic wavelength of the CGWB at time t_1 and t_0 is the present time.

Although this gives us the characteristic wavelength of gravitational waves created at time t_1, it is not the wavelength we need our detectors to be sensitive to now. This is because gravitational waves undergo a cosmological redshift as the Universe grows from t_1 to present. The redshift increases the wavelength of the gravitational waves. How λ_1 increases in time is again dependent on the details of the cosmological model. Up until the time of matter-radiation equality, t_{eq}, the Universe was *radiation dominated,* and the redshift of the wavelength scaled as $\sqrt{t}$. So at the time of matter-radiation equality, the characteristic wavelength of cosmological gravitational waves created at some time before matter-radiation equality would be

$$\lambda_1(t_{eq}) = \sqrt{\frac{t_{eq}}{t_1}}\left(\frac{t_1}{t_0}\right)\ell_0 = \frac{\sqrt{t_{eq}t_1}}{t_0}\ell_0 \,. \tag{5.21}$$

We do not observe at the time of matter-radiation equality, however, we observe at the present. Since the time of matter-radiation equality, the Universe has been *matter dominated,* and the redshift scales wavelengths as $t^{2/3}$. Thus, the characteristic wavelength at present (t_0) of gravitational waves created at time t_1 would be

$$\lambda_1(t_0) = \left(\frac{t_0}{t_{eq}}\right)^{2/3}\frac{\sqrt{t_{eq}t_1}}{t_0}\ell_0 = \frac{t_1^{1/2}}{t_0^{1/3}t_{eq}^{1/6}}\ell_0 \,. \tag{5.22}$$

Substituting in the GUT timescale for t_1 we find that to look for the CGWB from the GUT scale we should look for wavelengths of about $\lambda_{GUT}(t_0) \approx 300\,\text{cm}$, or at frequencies of $f_{GUT}(t_0) = c/\lambda_{GUT}(t_0) = 10^8\,\text{Hz}$. This frequency is well above the sensitive band of any current or planned gravitational-wave detectors.

So, what is the timescale back to which we can look with actually existing or planned detectors? To answer that question, we simply substitute $\lambda_1(t_0) = c/f_1(t_0)$ into Eq. (5.22), solve it for t_1 and then plug in the frequency for which the instrument in question is most sensitive. The result is

$$t_1 = t_0^{2/3}t_{eq}^{1/3}\frac{c^2}{f_1^2(t_0)\ell_0^2} \,. \tag{5.23}$$

For ground-based interferometers such as LIGO and Virgo, the peak sensitivity is at $f_1(t_0) \sim 100\,\text{Hz}$, which corresponds to an emission time of about 10^{-23} s for the gravitational waves. For LISA, the peak sensitivity is in the 0.01 Hz range, making it most sensitive to a CGWB originating at $t = 10^{-15}$ s. The Big Bang Explorer, still in the concept phase, would be a space-based interferometer designed specifically to explore the CGWB. Its most sensitive frequency would be about 1 Hz, and would see back to about 10^{-19} s. Thus, the CGWB should allow us to probe much earlier times than the CMB, times when such exciting events as the electroweak phase transition or even GUT symmetry breaking occurred.

Before addressing specific source models for the CGWB, there is one more general point that we wish to illustrate. Earlier in this section, we said that the signal would be Gaussian if there were sufficiently many independent sources emitting gravitational waves overlapping in time and frequency. Does a CGWB satisfy this criterion? The answer is a resounding "yes!"

It is difficult to discuss CGWBs in more depth, however, without first setting the scene by discussing modern cosmology. Indeed, we have already started using results from the FLRW cosmological model above without a clear explanation of where they come from. Because cosmology is a huge topic and could fill this volume itself, our treatment will be necessarily brief, and will focus on concepts that will be important to understanding the sources of gravitational-wave backgrounds of cosmological origin. Those already familiar with modern cosmology but not the cosmological production of gravitational waves can skip ahead to Section 5.3.1.3.

5.3.1.1 **FLRW Cosmologies**

Cosmology is the study of the origin and large scale structure of the Universe. It begins with our observations of other places and times in the Universe. When you first look up at the sky on a cloudless night from a city street, you see stars and planets and a lot of empty space in between. The Universe appears to be almost empty and where it is not, there are fairly dense blobs of matter. However, in this case, appearances deceive us, because we are so incredibly small compared to the Universe as a whole (10^{-26} times smaller) and can see so little of it with our eyes. It's a little as though you were one millionth the size of a water molecule, were sitting on one and were able to observe several thousand of the closest water molecules. From that perspective, a glass of water appears mostly empty and pretty lumpy. But from the perspective of the whole glass, it is a uniform fluid. So too with the universe.

We therefore model the matter content of the Universe with a perfect fluid stress-energy tensor. But what about the spacetime geometry? Well, one clue is that the cosmic fluid appears to be isotropic as we look out to cosmological distances in any direction. Actually, since all the light that reaches us comes to us along our past light cone, as we look farther out we look further back in time as well. So we can say that the Universe appears to be isotropic out to the farthest distances and the earliest times we can see. Unless we are in a privileged place, the Universe must look isotropic at all distances (and past times) from every place. Isotropy at every point implies homogeneity of the fluid as well. But if matter has a symmetry at every time, then the spacetime too must have that symmetry at every time (since matter and geometry respond to each other in general relativity). So the cosmological geometry must be isotropic and homogeneous at every time.

As stated in Section 4.2.1, homogeneity and isotropy immediately imply that the line element for the cosmological geometry can be written as:

$$ds^2 = -c^2 dt^2 + a(t)\left[f(r)dr^2 + r^2 d^2\Omega\right], \tag{5.24}$$

where t is the proper time of a set of observers with respect to which the cosmic fluid is locally stationary and appears isotropic. The expression $d^2\Omega$ is a common way of writing a line element on a two-dimensional sphere. The spatial part of the metric

$$d\varsigma^2 = f(r)dr^2 + r^2 d\Omega^2 \tag{5.25}$$

is written in terms of polar coordinates on a surface of homogeneity (and hence isotropy). We have written the spatial metric as a general spherically-symmetric line element, which is certainly allowed since a homogeneous space must be spherically symmetric about some point (in fact, it is symmetric about every point, although this metric form only requires it to be so about one point).

Note that if there were off-diagonal elements $dtdx^i$, then the time coordinate would not be normal to the surfaces of homogeneity. If this were true, the projection of dt onto the surfaces would define a preferred direction, which breaks isotropy, and therefore there are no $dtdx^i$ terms. Also, if the coefficient of the dt^2 term depended on the spatial coordinates, then time would flow at different rates at different places in the Universe, and homogeneity would soon be broken. Finally, the coefficient of the dt^2 term cannot depend on t because if it did, t would not be the proper time for a locally stationary observer.

There are just two functions to be determined, then. First, we deal with $f(r)$, which is the only remaining part of the metric which can determine the intrinsic geometry of the homogeneous and isotropic spatial slices. That geometry must have constant curvature, since if it did not, one could differentiate points based on their different curvature (which is not very homogeneous). There are exactly three types of constant curvature three-dimensional spaces (there are actually more if you allow topologically non-trivial spaces, but such spaces are generally not isotropic): the

3-sphere, with positive constant curvature; flat three space, with zero curvature; and the Lobachevsky–Bolyai 3-saddle, with constant negative curvature. The metric for each of these cases is given by Eq. (5.25) with

$$f(r) = \left(1 - kr^2\right)^{-1}, \tag{5.26}$$

where k is the constant value of the curvature. It is possible to rescale r so that k only takes the values -1, 0 or 1 (rather than any negative value, 0 and any positive value). This is often done in other treatments, but we will not bother. Spacetimes with the line element given by Eqs. (5.24) and (5.26) are called *Robinson–Walker* spacetimes.

The final function to determine, $a(t)$, is called the *scale factor*. It determines how the constant-curvature spatial surfaces evolve from moment to moment. To find it, we must use the Einstein field equations (2.118) and the conservation equations for the fluid. Substituting our metric and the perfect fluid stress-energy tensor into these equations give two equations of evolution:

$$\left(\frac{\dot{a}}{a}\right)^2 = \frac{8\pi G}{3}\rho - \frac{kc^2}{a^2}\,, \tag{5.27a}$$

$$\dot{\rho} = -3\frac{\dot{a}}{a}\left(\rho + \frac{p}{c^2}\right), \tag{5.27b}$$

where ρ is the mass density of the fluid and p is the pressure. These equations are called the *Friedmann equations.*

Notice that in order to solve for the scale factor of the universe, we need to involve the matter content of the Universe as well – this is the dynamical evolution of the cosmology. Since Eq. (5.27) are two equations in three unknowns (a, ρ and p), we apparently need another equation. For a perfect fluid, we can use the equation of state, which relates ρ and p.

There are three simple cases of interest for cosmology. The first corresponds to the present era, when the dynamics of the Universe are driven by matter that has no significant non-gravitational interactions over cosmological distances. In this case, which we call *matter-dominated,* the pressure of the fluid vanishes so

$$p = 0\,. \tag{5.28}$$

For $k = 0$ (flat universe), the universe expands as $a(t) = t^{2/3}$ and the density decreases as $\rho \propto t^{-2}$.

The second case of interest corresponds to an earlier epoch, in which the Universe was much hotter and denser. Under these conditions, the contents of the Universe are strongly interacting, and the associated radiation drives the gravitational dynamics of the Universe. In this *radiation-dominated* universe, the pressure of the fluid does not vanish, but rather

$$p = \frac{1}{3}\rho c^2\,. \tag{5.29}$$

For this equation of state with $k = 0$, the Universe expands as $a(t) = t^{1/2}$ and the density again decreases as $\rho \propto t^{-2}$.

Finally, we consider the case where

$$p = -\rho c^2\,. \tag{5.30}$$

Recalling from Eq. (2.99) that the stress-energy tensor for a perfect fluid is

$$T_{\alpha\beta} = \left(\rho + \frac{p}{c^2}\right) u_\alpha u_\beta + p g_{\alpha\beta} \tag{5.31}$$

we see that the stress-energy tensor for this equation of state is proportional to the FLRW metric itself. Solving the Friedmann equations (5.27) with $k = 0$ for this matter tells us two interesting things. The first is that the density of the Universe (and hence the entire stress-energy tensor) for this matter is constant at all times. Indeed, such matter is completely equivalent to Einstein's famous *cosmological constant.* Perhaps even more interesting is the evolution of the scale factor $a(t) \propto \exp(\pm\kappa t)$, where $\kappa = \sqrt{8\pi G\rho/3}$. In the case where $a(t) \propto \exp\kappa t$, this exponential growth is characteristic of the process known as *inflation,* about which we will have more to say shortly.

5.3.1.2 Cosmological Quantities

There are a few quantities that you will almost surely encounter in any discussion of cosmology. One is the *Hubble parameter,*

$$H := \frac{\dot{a}}{a} , \tag{5.32}$$

which is the rate at which the Universe expands at a given time. The value of the Hubble parameter right now is denoted by H_0 and is called the *Hubble constant.* H_0 cannot be measured directly, but must be inferred from other measurements. Astronomers performing these measurements often make slightly different assumptions about the correct model of the Universe, and it is therefore not surprising that different measurements lead to different and sometimes inconsistent estimates of the Hubble constant. The simplest model that explains all measurements made to date (2009) is the *concordance model,* which requires a Hubble constant of $H_0 \approx (71 \pm \text{a few})\,\text{km}\,\text{s}^{-1}\,\text{Mpc}^{-1}$. When an explicit value is needed, we will use $70\,\text{km}\,\text{s}^{-1}\,\text{Mpc}^{-1}$, such as in the estimation of the *Hubble time*

$$t_{\text{H}} := \frac{1}{H_0} \approx 4.4 \times 10^{17}\,\text{s} , \tag{5.33a}$$

and the *Hubble length*

$$\ell_{\text{H}} := \frac{c}{H_0} \approx 1.3 \times 10^{26}\,\text{m} , \tag{5.33b}$$

which set typical cosmological scales.

A quantity related to the Hubble parameter is the *critical density,* ρ_{crit}. It arises from the consideration of Eq. (5.27a), which can be rewritten as

$$\frac{c^2 k}{\dot{a}^2} = \frac{8\pi G}{3H^2}\rho - 1 . \tag{5.34}$$

From it, we see that the sign of k, and hence the geometry of the spatial sections, depends on whether the density of the cosmological fluid is greater than, equal to, or less than

$$\rho_{\text{crit}} := \frac{3H^2}{8\pi G} , \tag{5.35}$$

or equivalently, whether the *density parameter,*

$$\Omega := \frac{\rho}{\rho_{\text{crit}}} , \tag{5.36}$$

is greater than, equal to, or less than 1. This is of special interest because the geometry of the spatial surfaces of a Robertson–Walker spacetime determines what the fate of that spacetime will be: if $k < 0$ the Universe will expand forever, if $k > 0$ the Universe will reach a maximum radius and then recollapse and if $k = 0$ the Universe will asymptotically approach a static flat space in infinite time. From this point forward, we will restrict our attention to the case $k = 0$, since this is the scenario that is best supported by present evidence.

For convenience, the density parameter is often divided into components depending on the type of matter that is being considered, because the density of each type of matter scales differently as the scale factor changes. For instance, matter with $p = -\rho c^2$ (cosmological constant) produces a density ρ_Λ that does not change with scale factor, whereas the energy density that $p = 0$ matter (dust) produces scales as $\rho_{\rm M} \propto a^{-3}$ and radiation, for which $p = \frac{1}{3}\rho c^2$, has a density that scales as $\rho_{\rm R} \propto a^{-4}$. If we create density parameters for each type of matter, $\Omega_\Lambda = \rho_\Lambda(a = 1)/\rho_{\rm crit}$, $\Omega_{\rm M} = \rho_{\rm M}(a = 1)/\rho_{\rm crit}$ and $\Omega_{\rm R} = \rho_{\rm R}(a = 1)/\rho_{\rm crit}$, then we can write the first Friedmann equation (5.27a) as

$$\frac{H^2}{H_0^2} = \Omega_\Lambda + \Omega_{\rm M} a^{-3} + \Omega_{\rm R} a^{-4} , \tag{5.37}$$

which shows explicitly how the Hubble parameter evolves with each type of matter.

Finally, for $\Omega_{\rm R}$, it is often useful to consider not only the total density of radiation, but also its spectrum as a function of frequency. Although one could express this as a linear function of frequency, when dealing with cosmic radiation one is often considering frequencies varying by many orders of magnitude. A logarithmic spectrum is therefore usually used. In particular, we define

$$\Omega_{\rm R}(f) := \frac{1}{\rho_{\rm crit}} \frac{d\rho_{\rm R}}{d \ln f} . \tag{5.38}$$

5.3.1.3 **Inflation**

As well as observational considerations, there is a theoretical reason for believing that $k = 0$ in our universe. A number of problems have arisen in the study of Big Bang cosmologies which must be resolved to explain cosmological observations. These problems suggest that FLRW might not be the complete story for the evolution of our cosmos.

For instance, sufficiently close in time to the Big Bang, the energy density of the Universe would be so high that symmetry breaking that separated the strong nuclear force from the electroweak force would have not yet occurred. In the early Universe, these forces are expected to be replaced by a simpler gauge model described by a *Grand Unified Theory* (GUT). A myriad of GUT candidates exist, and some general features of GUTs are observed. One is that GUTs generically lead to the copious production of magnetic monopoles. Estimates indicate that these heavy and stable relic particles would dominate the matter content of the Universe today if FLRW expansion tells the whole story between the GUT era and today. However, magnetic monopoles do not appear to exist at all today – what happened to them?

Another problem is that on a cosmic scale the Universe appears extremely homogeneous. The cosmic microwave background (CMB) radiation, produced when atoms first formed and photons decoupled from matter at about 10^{13} s after the Big Bang, gives us our earliest view of the Universe. That view indicates that the Universe was homogeneous at that time (called the *surface of last scattering*) to one

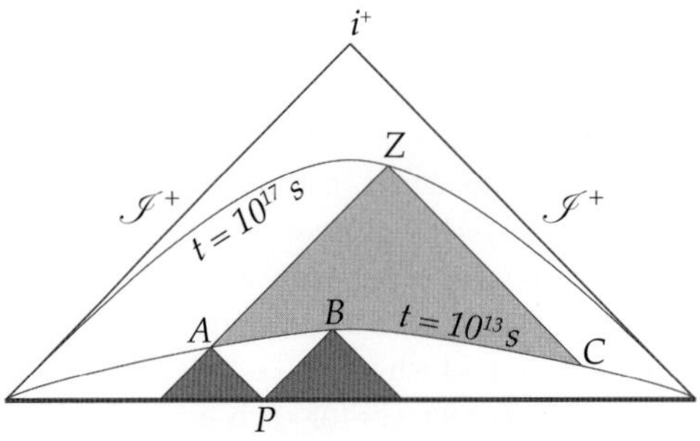

Figure 5.1 Conformal diagram of flat FLRW spacetime. The thick line at the bottom is the Big Bang at time $t = 0$. The edges marked $\mathcal{I}^+$ are called future null infinity, and are the infinite future of anything travelling at the speed of light. The point at the top labeled i^+ is called future time-like infinity, and is the infinite future of everything travelling slower than the speed of light. A and B are points at time $t = 10^{13}$ s whose past light cones are shaded dark grey. These light cones share only one common point, P (at time $t = 0$). Z represents a point in the present era where an observer on Earth sits. The intersection of its past light cone (light grey) and the $t = 10^{13}$ s surface, which extends from A to C, indicates the portion of the Universe that is visible to the observer.

part in 10^5. That is remarkably homogeneous. Further, since gravitation causes inhomogeneities to grow with time, it indicates that the Universe was much more homogeneous at earlier times – more homogeneous, in fact, than would be allowed by quantum fluctuations.

Inhomogeneities are naturally smoothed in fluids by the approach to thermal equilibrium, so this would seem to be the solution to the problem. However, in a pure FLRW cosmology, the entire surface of last scattering is not in causal contact, so it is not possible that thermal equilibrium could be responsible for the degree of homogeneity there. The situation is illustrated in Figure 5.1. If FLRW correctly describes the expansion of the Universe at all times up to the surface of last scattering, there are of the order of ten thousand causally separate regions on the sky. This is called the *horizon problem*.

Finally, there is the question of how the Universe can appear to be so flat ($k = 0$). From Eq. (5.34) we see that a spatially flat universe implies that the density of the Universe is the critical density, or that $\Omega = 1$. We can measure Ω by measuring the CMB radiation, because the angular scale of fluctuations of the CMBR is determined by the spatial curvature of the Universe. Current observations indicate that $\Omega = 1.00 \pm 0.01$. It is curious, if not troubling, that the density of the Universe is so close to the critical density – were it not for what that implies at earlier times.

Invoking Eq. (5.34) again we can see that

$$k \propto a^2 \rho \left(1 - \Omega^{-1}\right) \sim a^2 \rho (\Omega - 1) . \tag{5.39}$$

We note, however, from Eq. (5.27b) that for pure FLRW expansion $a^2\rho$ is not constant in time – ρ decreases as a^3 for dust or a^4 for photons, not a^2. From the time of the Big Bang to the present, $a^2\rho$ would have decreased by approximately sixty orders of magnitude in an FLRW universe. But recall that k is a constant. So, if $a^2\rho$ has decreased by sixty orders of magnitude, $|\Omega - 1|$ must have increased by sixty orders of magnitude. Since we know that $|\Omega - 1| \leq 10^{-2}$ now, this means that just after the Big Bang $|\Omega - 1| \leq 10^{-62}$. How can one explain why the density of the Universe started so incredibly close to the critical density? This is called the *flatness problem.*

The monopole problem, horizon problem and flatness problem all point to the implausibility of a pure FLRW evolution for the Universe. And all three can be explained away by *cosmic inflation* (often shortened to just "inflation"). While the exact mechanism for inflation is unknown, the basic feature is that the scale factor a increases exponentially by at least twenty-six orders of magnitude over a short time in the early Universe. If this occurred, and the evidence mounts that it did, then the entire observable volume of the Universe today began as a tiny causally connected region.

This solves the monopole problem (provided the GUT symmetry was broken before inflation) because even though the density of magnetic monopoles was high, the region which inflated into the observable Universe was so small that a negligible number of monopoles would be contained. It solves the horizon problem because the entire observable Universe was in causal contact before last scattering, and thus thermal process could lead to the observed level of isotropy of the CMB. And finally, it solves the flatness problem because, whatever the value $\Omega - 1$ before inflation, an exponential increase in a would be accompanied by an exponential decrease in $\Omega - 1$, thus leading the Universe after inflation to have a density very close to the critical density.

5.3.1.4 Inflationary Background

While, generally speaking, inflation homogenizes whatever early structure there was in the Universe, it also leads to perturbations in whatever fields exist within the inflating spacetime by essentially promoting quantum fluctuations to classical scales. In matter fields, these perturbations are thought to be the source of the anisotropy of the CMB, which eventually led to galaxy formation. More interesting from the perspective of this book, however, is the perturbation of the spacetime itself, which would be a source of gravitational waves.

A toy model for illustrating how quantum fluctuations would be amplified by inflation is that of a pendulum. The string of the pendulum, which is massless, can be lengthened arbitrarily and at an arbitrary rate. Classically, the frequency of the pendulum when the string is of length ℓ is $\omega = \sqrt{g/\ell}$, where g is, of course, the acceleration due to gravity. Quantum mechanically, the pendulum is a harmonic oscillator. Recall that the energy eigenstates of the harmonic oscillator are

$$\psi_n(x) = \sqrt{\frac{\alpha}{2^n n! \sqrt{\pi}}} e^{-\alpha^2 x^2/2} H_n(\alpha x) , \tag{5.40}$$

where $H_n(x)$ are the Hermite polynomials and

$$\alpha = \sqrt{\frac{m\omega}{\hbar}} . \tag{5.41}$$

The corresponding energy eigenvalues are $E_n = \hbar\omega(n + 1/2)$.

Now, consider a quantum pendulum of initial length ℓ_{in} which is in the ground state

$$\psi_0^{\text{in}}(x) = \frac{\sqrt{\alpha_{\text{in}}}}{\pi^{1/4}} e^{-\alpha_{\text{in}}^2 x^2/2} , \tag{5.42}$$

where $\alpha_{\text{in}} = \sqrt{m\omega_{\text{in}}/\hbar}$ and $\omega_{\text{in}} = \sqrt{g/\ell_{\text{in}}}$. Due to the uncertainty principle, even in this lowest energy state, the pendulum will have some energy due to quantum fluctuations, the zero-point energy $E_{\text{in}} = \hbar\omega_{\text{in}}/2$.

If we lower the pendulum to a new length, ℓ_{out}, the new classical frequency becomes $\omega_{\text{out}} = \sqrt{g/\ell_{\text{out}}}$. The new state will depend on how quickly we lower it. If the pendulum is lowered sufficiently slowly, over a time $\Delta t \gg \omega_{\text{out}}^{-1}$, then the quantum state of the pendulum will be able to continuously deform from the ground state it originally occupied, Eq. (5.42), to the new ground state ψ_0^{out}. We call such a process *adiabatic*, and as one would expect for an adiabatic process, the number of energy quanta (in this case zero) is conserved.

However, if the pendulum is lowered quickly, so that $\Delta t \ll \omega_{\text{in}}^{-1}$, then the quantum state will be unable to change before the transition to the new pendulum length is made. In this case, the quantum state of the pendulum remains the same, but that state is no longer the ground state of the pendulum, or even (in general) an energy eigenstate at all. Like any state, however, it must be expressible as a superposition of energy eigenstates

$$\psi_0^{\text{in}}(x) = \sum_{n=0}^{\infty} c_n \psi_n^{\text{out}}(x) , \tag{5.43}$$

where,

$$\psi_n^{\text{out}}(x) = \sqrt{\frac{\alpha_{\text{out}}}{2^n n! \sqrt{\pi}}} e^{-\alpha_{\text{out}}^2 x^2/2} H_n(\alpha_{\text{out}} x) , \tag{5.44}$$

with $\alpha_{\text{out}} = \sqrt{m\omega_{\text{out}}/\hbar}$ and $\omega_{\text{out}} = \sqrt{g/\ell_{\text{out}}}$. As usual, the coefficients c_n are given by

$$c_n = \int_{-\infty}^{\infty} \psi_0^{\text{in}\dagger}(x)\psi_n^{\text{out}}(x)dx , \tag{5.45}$$

$$= \begin{cases} \frac{1}{(n/2)!}\sqrt{\frac{n!}{2^{n-1}}} \frac{(\omega_{\text{out}}\,\omega_{\text{in}})^{1/4}}{(\omega_{\text{out}}+\omega_{\text{in}})^{1/2}} \left(\frac{\omega_{\text{out}}-\omega_{\text{in}}}{\omega_{\text{out}}+\omega_{\text{in}}}\right)^{n/2} & n \text{ even,} \\ 0 & n \text{ odd.} \end{cases} \tag{5.46}$$

Now, an interesting question to ask is "what is the expectation value of the energy of the pendulum after it is quickly lowered?" Recall that the expectation value of the energy of a particle in state ψ_0^{in} is

$$\langle E \rangle = \int_{-\infty}^{\infty} \psi_0^{\text{in}\dagger}(x)\hat{H}\psi_0^{\text{in}}(x)dx , \tag{5.47}$$

where $\hat{H}$ is the Hamiltonian (or energy) operator. Substituting Eq. (5.43) into Eq. (5.49) and recalling the orthonormality of the eigenstates given by Eq. (5.44), we have

$$\langle E \rangle = \sum_{n=0}^{\infty} c_n^2 E_n^{\text{out}} , \tag{5.48}$$

where $E_n^{\text{out}} = \hbar\omega_{\text{out}}(n + 1/2)$ are the energy eigenvalues of the pendulum after it is lowered. Fortunately, the sum in Eq. (5.48) can be done analytically, and the end result is

$$\langle E \rangle = \hbar\omega_{\text{out}} \left[\frac{(\omega_{\text{in}} - \omega_{\text{out}})^2}{4\omega_{\text{in}}\omega_{\text{out}}} + \frac{1}{2} \right] . \tag{5.49}$$

From this, we can deduce that $(\omega_{\text{in}} - \omega_{\text{out}})^2/(4\omega_{\text{in}}\omega_{\text{out}})$ quanta of energy were produced in the sudden lengthening of the pendulum. In the limit where the pendulum is lengthened very much, so that $\omega_{\text{in}} \gg \omega_{\text{out}}$, the number of quanta created goes as $\omega_{\text{in}}/(4\omega_{\text{out}})$.

In quantum field theory, a field acts as an infinite collection of harmonic oscillators. In the lowest energy state, or vacuum, the quantum field point has only the zero point energy. However, if spacetime expands, the field must expand with it. This is the equivalent process to lowering the pendulum. If the field expands quickly, quanta of the field will be created. In the case of the gravitational field, the quanta are gravitons. As the gravitons are created, their wavelength expands in the inflating universe to cosmic length scales and they become classical gravitational waves. Thus, it seems almost certain that if inflation occurred, a cosmological background of gravitational waves will exist.

The question of how many gravitons are created by inflation, or to be more specific, what the energy density in gravitational waves $d\rho_{\text{GW}}$ between frequencies ω and $\omega + d\omega$ is, depends on the details of how inflation occurred. Unfortunately, although there is mounting evidence for inflation, nothing is known about the mechanism behind it or the energy scale (or equivalent time scale) at which it occurred. In fact, it is not even certain that inflation did occur. However, we can get some feeling for what to expect from a simple model.

The four quantities that go into calculating $d\rho_{\text{GW}}$ are the number of polarizations of a graviton (2 polarizations), the energy per graviton mode of frequency ω ($E_\omega = \hbar\omega$), the number density $n_\omega\, d\omega$ of modes in the frequency between ω and $\omega + d\omega$, and the number of gravitons created at frequency ω, N_ω. The number of modes in the frequency interval between ω and $\omega + d\omega$ is the same for any kind of quantum. Recalling the derivation of the Planck black-body spectrum, the number of quanta per unit volume in the frequency interval between ω and $\omega + d\omega$ is

$$n_\omega\, d\omega = \frac{\omega^2}{2\pi^2 c^3}\, d\omega . \tag{5.50}$$

Finally, we need to know the number of gravitons created by the rapid expansion of inflation at frequency ω. This is the part of the calculation that depends on the inflationary model one chooses. We will consider a universe in which there was an initial inflationary period followed by radiation-dominated FLRW. For such a model, one may calculate the number of quanta of a given angular frequency ω that are created. The expression one obtains is

$$N_\omega = \frac{H_0^2 H_{\text{inf}}^2}{4\omega^4} \,, \tag{5.51}$$

where H_0 is the Hubble parameter today and $H_{\text{inf}} = \kappa$ is the (constant) Hubble parameter during the inflationary epoch.

Putting this all together, we get that at the end of the radiation-dominated era

$$d\rho_{\text{GW}} = \frac{2 N_\omega E_\omega}{c^2} n_\omega \, d\omega = \frac{\hbar H_0^2 H_{\text{inf}}^2}{4\pi^2 c^5} \frac{d\omega}{\omega} \,, \tag{5.52}$$

or equivalently

$$\Omega_{\text{GW}} = \frac{\hbar}{4\pi^2 c^5} \frac{H_0^2 H_{\text{inf}}^2}{\rho_{\text{crit}}} \,. \tag{5.53}$$

Note that this expression is proportional to Planck's constant, $\hbar$. This is because the gravitational waves are being created in a quantum process.

Also notice that, for this cosmological model at least, Ω_{GW} is independent of the frequency at which we measure. Clearly, this cannot be the full story, or the energy density integrated over all frequencies of gravitational waves would be infinite. As was stated earlier, the form of Ω_{GW} depends on the cosmological model we use. In reality, the transition from the inflationary epoch to the radiation-dominated epoch is not instantaneous, as was assumed to obtain Eq. (5.52), but rather takes some time $\Delta\tau$. This introduces a frequency scale, $f_{\text{max}} = (\Delta\tau)^{-1} a(t_{\text{inf}})$ above which Ω_{GW} drops off precipitously.

Finally, we can ask what is the constant value of Ω_{GW} one could expect to obtain for frequencies below f_{max}. This depends on when inflation ended. Suppose that it was at the GUT energy scale of $E_{\text{GUT}} = 10^{16}$ GeV. Then, at the time inflation ended, the natural frequency of the gravitational waves produced in inflation would be $f = E_{\text{GUT}} \approx 10^{39}$ Hz. This scale must be set by the size of the Hubble parameter. Taking the Hubble parameter at the end of inflation, therefore, to be $H_{\text{inf}} = 10^{39}$ Hz, we can calculate Ω_{GW} using Eq. (5.53) to be $\Omega_{\text{GW}} \approx 10^{-9}$.

Recall, however, that we have assumed that the Universe is radiation dominated for all times after inflation in our calculation. This is not correct – when the expansion parameter $a(t)$ was of the order of 3200 times smaller than it is now, the Universe entered the matter-dominated phase which it has been in ever since. If we denote the time of transition from radiation dominated to matter dominated by t_{eq}, then it can be shown that the Ω_{GW} is diluted by a further factor or $a(t_{\text{eq}})/a(t_0)$, or in other words, at any frequencies at which we are likely to measure gravitational waves directly, the gravitational waves from inflation will give a constant background of $\Omega_{\text{GW}} \approx 10^{-12}$. The matter-dominated era also creates one other change to $\Omega_{\text{GW}}(f)$. It turns out that for frequencies which are most important

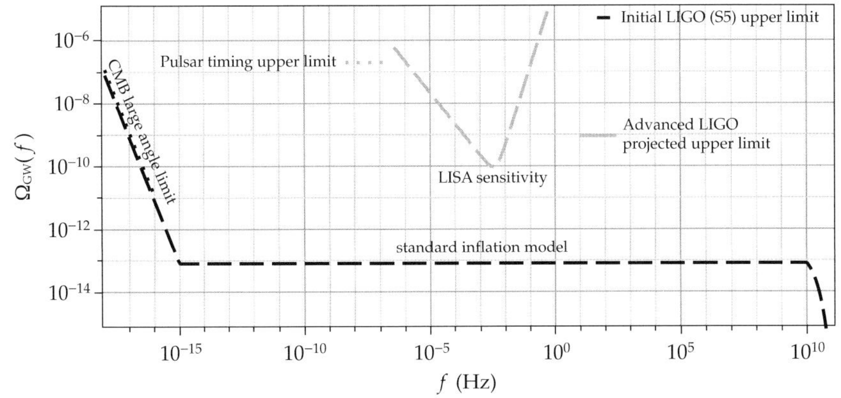

Figure 5.2 The gravitational-wave background spectrum as predicted by inflation (black dashed line). This represents the strongest such background that is consistent with CMB large-angle measurements (black dotted line). Note that the predicted value is well below the levels measured by Initial LIGO in the fifth science run (S5) (black line) or by pulsar timing (grey dotted line). Furthermore, it is also well below the predicted sensitivity for Advanced LIGO (grey solid line) and LISA (grey dashed line). The gravitational-wave background from standard inflation is, in fact, unlikely to be directly measured by any existing or future gravitational-wave observatory.

during the matter-dominated phase (those whose wavelengths are greater than the Hubble length for $t_{eq} < t < t_0$), the spectrum is not flat, as we found, but rather $\Omega_{GW} \propto 1/f^2$. Note, however, that these frequencies, lower than approximately 10^{-16} Hz, are inaccessible to any gravitational-wave observatory currently existing or planned.

We end this section by noting that, while the gravitational-wave background from inflation is not limited by any direct observations, current or planned, there are indirect constraints from large-angle observations of the cosmic microwave background. This is because gravitational waves are metric perturbations, and thus change the spacetime geometry over the course of their wavelength. If the wavelengths are of cosmological scale, they should cause anisotropic redshift of light coming from widely separated parts of the sky. Large-angle observations of CMB do not exhibit such an effect, and effectively limit the strength of ultra-low-frequency gravitational waves. When these constraints are considered, the level of the inflationary gravitational-wave background in the regimes accessible to our observatories falls to almost $\Omega_{GW} < 10^{-14}$. The current limit on an inflationary gravitational-wave background, along with experimental sensitivity of some current and planned gravitational-wave observatories, can be seen in Figure 5.2.

5.3.1.5 **Stochastic Backgrounds from Other Cosmological Sources**

While all the evidence seems to point to inflation as a real mechanism, there are other potential sources of cosmological gravitational-wave backgrounds that are somewhat more speculative. The first that we will discuss are *strings*. In the last decades of the twentieth century, the strings of interest for gravitational-wave pro-

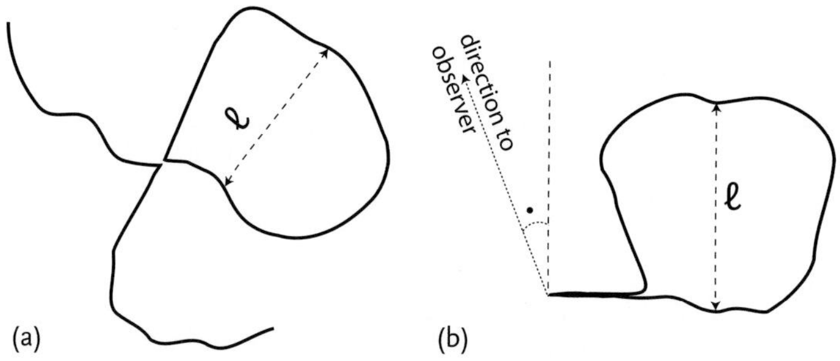

Figure 5.3 In panel (a) there is an illustration of a cosmic string which self intersects and produces a new loop of scale ℓ. In panel (b), a cusp has formed in a loop of scale ℓ. Due to the extreme tension that cosmic strings are under, such a feature will move relativistically and create gravitational waves. The maximum frequency of the observed gravitational waves depends on the angle of observation, θ.

duction were thought to be only *cosmic strings*. Cosmic strings are one-dimensional topological defects that may have resulted from a phase transition due to symmetry breaking in the early Universe. They are analogous to vortices in a superfluid in which rotational symmetry is broken.

Cosmic strings are remarkable in a number of fascinating ways. They are a relatively generic prediction of quantum field theory models of the early Universe. Every viable supersymmetric GUT, and most other GUT theories, predict the creation of cosmic strings. A cosmic string is characterized by a single parameter, its mass per unit length, μ, which is predicted to be an incredible 10^{21} kg m^{-1} for GUT scale strings. The string tension is equal to μc^2, so cosmic strings are dynamically highly relativistic. In most models, cosmic strings are always closed. If they exist, they are predicted to be of the order of tens of cosmic string loops of circumference much greater than the Hubble distance per Hubble volume, in addition to a number of loops that are smaller than the cosmological horizon size. When strings meet, they re-connect and can form loops; there are also cusps, where the string doubles-back on itself and forms a point that instantaneously moves at the speed of light (Figure 5.3).

When a cosmic string self-intersects to form a loop, that loop separates from the rest of the string and starts an independent life. Thus, there will be a myriad of string loops of different sizes as large loops separate into smaller loops, which separate into smaller loops, eventually reaching stable trajectories that are not self-intersecting having a size that depends on the size of the smallest significant wiggles on the parent loop. When they reach that stable trajectory, the loops dissipate under gravitational radiation. When a cusp forms in a cosmic string, it creates a burst of gravitational waves. Gravitational waves are also produced by kinks (twists) in the string and by loop vibrations. Through these mechanisms, the string loop eventually loses all its energy and shrinks to zero size.

In particular, the gravitational wave emitted by a cosmic string cusp has a particular form in the frequency domain,

$$\tilde{h}(f) = \begin{cases} A f^{-4/3} & f_\mathrm{l} < f < f_\mathrm{h} \\ 0 & \text{otherwise,} \end{cases} \tag{5.54}$$

where A is some undetermined amplitude. The lower frequency cutoff f_l is cosmological, so in effect it will depend on the lower frequency band cutoff of the detector for all practical purposes. The upper frequency cutoff f_h depends on the angle between the direction of observation and the direction of the cusp (see Figure 5.3) – it goes roughly like

$$f_\mathrm{h} \propto \frac{c}{\theta^3 \ell}\,, \tag{5.55}$$

where ℓ is the loop size and θ is the angle shown in Figure 5.3. The amplitude, A, also scales with the size of the loop, in this case

$$A \propto \frac{G\mu}{c r}\left(\frac{\ell}{c}\right)^{2/3}, \tag{5.56}$$

where r is the distance from the cosmic string cusp to the observer. Note that the amplitude is not dimensionless – it has units of $\mathrm{s}^{-1/3}$.

It is clear, then, how cosmic strings can create a cosmological gravitational-wave background – string loops shatter into smaller loops through self intersection. Each small string loop that results can vibrate and can form cusps and kinks, all of which act as sources of gravitational waves. Many independent sources radiating at once from all over the sky are exactly the sort of mechanism that gives rise to a stochastic background of gravitational waves.

It is clear from Eqs. (5.55) and (5.56) that ℓ is crucial for understanding the background of gravitational waves emitted by cosmic strings cusps. In fact, the size of the loops is a critical factor in the gravitational radiation produced by all mechanisms. There are two proposals in the literature for the cosmic string length scale. The first is that ℓ is of the order of the Hubble length, $\ell \propto ct$. The second is that the scale is set by gravitational back reaction, $\ell \propto G\mu t/c$, which is much smaller. Each of these proposals has arguments in their favor in the literature, and no consensus has been reached by researchers. Nonetheless, the remarkable scale invariance of cosmic string networks allows one to calculate that the spectrum of the gravitational-wave background from a network of cosmic strings will be approximately constant in the band between approximately 10^{-4} Hz and 10^9 Hz at the level of

$$\Omega_\mathrm{GW} = \frac{16\pi}{9} N \Gamma \left(\frac{G\mu}{c^2}\right)^2 \left(\frac{\beta^{3/2}-1}{\alpha}\right) \frac{a(t_\mathrm{eq})}{a(t_0)} \tag{5.57}$$

(Allen, 1997; Vilenkin and Shellard, 2000). In this expression, N is the number of strings whose circumference is longer than the Hubble length (of order 50),

Γ is the radiation rate from a typical loop (also of order 50), $G\mu/c^2 \approx 10^{-6}$ for GUT strings, and $a(t_{\text{eq}})/a(t_0)$ is the dilution factor for matter-dominated expansion, which is given by the ratio of scale factors relating the size of the Universe when it first became matter-dominated to the size today, roughly 3000. The parameter α characterizes the size of the loop as a fraction of the size of the Universe, $\ell = \alpha c t$, and has values in the range $\alpha \sim 10^{-3}$–10^{-1} for long loop predictions, or $\alpha \sim \Gamma G\mu/c^2$ for small loop predictions, while $\beta = 1 + \alpha c^2/(\Gamma G\mu)$ sets the lifetime of a loop with $t_{\text{death}} = \beta t_{\text{birth}}$. If the loops are small (and $\alpha = \epsilon \Gamma G\mu/c^2$ with $\epsilon \ll 1$) then a simple expression

$$\Omega_{\text{GW}} = \frac{8\pi}{3} N \frac{G\mu}{c^2} \frac{a(t_{\text{eq}})}{a(t_0)} \tag{5.58}$$

describes the gravitational-wave background.

We end this discussion of gravitational-wave backgrounds from cosmic strings with some comments on recent developments in our understanding of string cosmology. It has turned out that precision measurements of the angular spectrum of the CMB have ruled out cosmic strings with string tensions of $G\mu/c^2 > 10^{-7}$. This means that for practical purposes, strings produced during GUT scale phase transitions are no longer viable, and the cosmic background of gravitational waves from cosmic strings at other energy scales (if they exist at all) are probably unmeasurable.

However, since the first few years of the twenty-first century, there has been a renewed interest in gravitational-wave backgrounds from strings because of the discovery that gravitational waves *are* detectable for $G\mu/c^2 < 10^{-7}$ (Damour and Vilenkin, 2005), and also because superstring theorists have realized that string-theory-inspired cosmologies can contain cosmic superstrings. In the last few years, it has become apparent that cosmic superstrings have many of the same features as cosmic strings from phase transitions, and thus everything we have discussed in this section appears to continue to hold apart from two significant changes. The first is that, because cosmic superstrings are fundamentally quantum objects, they do not always form loops when they self intersect. In fact, they form loops with reconnection probability $10^{-3} < p < 1$. This leads to a second effect – it turns out that the mass density of cosmic superstrings is inversely proportional to the reconnection probability, $\rho_{\text{css}} \propto p^{-1}$. Thus, the size of the background can be quite large if p is low.

String theorists have just recently started exploring the menagerie of cosmic superstring states that can exist, so there may be more surprises in store. Also, ground-based interferometers are already ruling out parts of the cosmic superstring parameter space. These are exciting developments indeed. Who knows, it could be that gravitational waves will provide the first experimentally verifiable prediction of superstring theory!

The last cosmological possibility of gravitational-wave backgrounds that we will discuss is first-order phase transitions in the early Universe. A first-order phase transition is characterized by a latent energy, that is, the phase transition absorbs energy so that, at the transition point, adding or removing energy does not change the temperature but rather induces the phase change itself. Such phase changes do not happen uniformly in the system, but rather nucleate in particular places and then expand, creating bubbles of one phase which are expanding in the other phase. A typical example is the transition of water into vapour – it does not occur everywhere simultaneously, but rather expanding bubbles of vapour are created within the water.

In quantum field theory in the early Universe, phase transitions are associated with spontaneous symmetry-breaking as the Universe expands and cools, for instance when the electroweak symmetry broke resulting in the separation of the electromagnetic and weak nuclear interactions. Before the symmetry breaking, the quantum field was in the lowest energy (vacuum) state for the full electroweak Hamiltonian. However, after the transition, this is no longer the vacuum state – there is a lower energy state, the broken vacuum, which the field will attain. If this phase transition is first order, then bubbles of true vacuum will nucleate and expand in the false (electroweak) vacuum as the Universe continues to expand and the energy density is reduced. The rate of expansion of these bubbles increases exponentially because the energy liberated by the expanding bubble (that is, the difference in the energy of the true and false vacua integrated over the volume of the bubble) is converted into kinetic energy of the bubble walls in expansion. In short order, the bubble walls are expanding outwards at relativistic speeds. Within a Hubble time the walls from nearby bubbles start colliding and the bubbles start merging.

Bubble mergers can cause gravitational waves through three different mechanisms. The first is that the collisions break the spherical symmetry of the bubble walls, generating anisotropic stresses that act as a source for gravitational waves. The second is that the collisions inject energy into the matter fields inside the bubbles. In the early Universe, the matter will be in the form of a primordial plasma, and the energy will lead to magnetohydrodynamic turbulence which will also be a source for gravitational waves. Finally, the bubble wall will cause charge separation which is amplified by the magnetohydrodynamic turbulence, causing additional anisotropic stresses which act as a further source of gravitational waves. Which of these mechanisms is most important depends on details of the physics, but all can contribute significantly to the gravitational-wave background.

Unlike the scale-invariant gravitational-wave spectrum of inflation, the gravitational-wave spectrum from colliding bubbles in a first-order phase transition is peaked about a characteristic frequency. There are two natural scales that could set the characteristic frequency: the duration of the phase transition, which is typically denoted by β^{-1} in the literature, or the typical size of the colliding bubbles, $R_\star$. In the case of the former, the characteristic frequency is

$$f_\star \simeq 10^{-2}\,\mathrm{mHz}\left(\frac{\beta}{H_\star}\right)\left(\frac{k T_\star}{100\,\mathrm{GeV}}\right), \tag{5.59}$$

where $H_\star$ is the Hubble parameter during the phase transition, k is Boltzmann's constant and $T_\star$ is the characteristic temperature of the phase transition. If the characteristic frequency is set by the size of the colliding bubbles, then one simply replaces β with $c/R_\star$.

The spectral shape depends on the mechanisms that are contributing, but to a first approximation tends to be rather simple. For instance, the spectrum from the bubble collisions is characterized by an $\Omega_{\text{GW}}(f) \propto f^3$ rise of Ω_{GW} with frequency below $f_\star$, which in this case is set by β. The magnetohydrodynamic turbulence spectrum initially also rises like $\Omega_{\text{GW}}(f) \propto f^3$ at low frequency, but for frequencies above $H_\star$ the rise slows to $\Omega_{\text{GW}} \propto f^2$ until $f_\star$, which in this case is set by $R_\star$. Above $f_\star$, the spectrum from bubble collisions falls off as $\Omega_{\text{GW}}(f) \propto f^{-1}$ while the magnetohydrodynamic turbulence spectrum falls off somewhat faster ($f^{-5/3}$ or $f^{-3/2}$, depending on the details of the model).

The strength of the gravitational-wave background from a first-order phase transition can also be estimated on general grounds. The result, which we simply quote here, is

$$\Omega_{\text{GW}}(f_\star) \sim \Omega_{\text{rad}} \left(\frac{H_\star}{\beta}\right)^2 \left(\Omega_{\text{s}}^\star\right)^2 , \tag{5.60}$$

where Ω_{rad} is the radiation energy density at the current epoch and $\Omega_{\text{s}}^\star$ is the portion of the energy density that is converted into gravitational waves. Again, if the scaling is set by $R_\star$ rather than β^{-1}, one should substitute β with $c/R_\star$.

What kind of numerical values can we expect? Let us consider the electroweak phase transition and see. For this symmetry breaking, the energy scale is $kT_\star \approx 100$ GeV, and $\beta/H_\star \approx 100$. This gives us a frequency scale of order mHz, which is the most sensitive band for the planned LISA mission. In order to estimate the strength of a background from a first-order electroweak symmetry breaking, we also need Ω_{s} and Ω_{rad}. The amount of energy in the form of radiation today is measured to be $\Omega_{\text{rad}} \approx 10^{-5}$. Ω_{s} will depend on how strongly first order the phase transition is. From Figure 5.2 we see that LISA will be sensitive to backgrounds down to about $\Omega_{\text{GW}} = 10^{-10}$. This means that we need $\Omega_{\text{s}} > 0.01$ for a detection. Realistically, something like a few percent of the energy density of the Universe would need to be made available as gravitational waves through this mechanism for LISA to have a realistic chance of seeing the background. This would correspond to a phase transition that is strongly first order.

Unfortunately, in the standard model of particle physics, the electroweak symmetry breaking is known not to be strongly first order, and the peak gravitational-wave spectral contribution has been calculated to only be $\Omega_{\text{GW}}(f_\star) \approx 10^{-22}$. There are, however, a plethora of alternative models in the literature which do have first-order electroweak phase transitions. These models are being explored not because they produce gravitational waves, but because they produce baryons. This is currently one of the most popular ideas for dealing with the deeply puzzling baryon asymmetry problem in the Universe, the name given to the fact that the Universe is almost entirely composed of matter, even though most processes that produced matter in the early Universe are believed to have produced matter and antimatter in equal amounts.

Finally, besides electroweak phase transitions, there are others which might be first order, for instance the QCD phase transition which is estimated to occur at $kT_\star \approx 100$ MeV, which would make it a candidate for detection with pulsar timing arrays. To peak in the sensitive band for ground based detectors, a phase transition would have to occur when the Universe had a characteristic temperature in the PeV range.

Is it possible that such a phase transition occurred? The only honest answer we can give is that we don't know. Also, there are other mechanisms that create gravitational waves that we have not discussed in this book. Pre-Big Bang models are quantum gravity models in which the Big Bang is actually a quantum bounce which is preceded by an inflationary pre-Big Bang epoch. The same mechanism as described for inflationary backgrounds leads to gravitational waves in pre-Big Bang models. Another class of phenomena which produce cosmological backgrounds are post-inflationary preheating – the period after inflation which must occur if the Universe is to create the matter that we see today. Which of these mechanisms create significant gravitational-wave backgrounds? Again, we simply don't know. We have many unanswered questions about the earliest history of the Universe, before radiation and matter decoupled. If we are lucky, gravitational waves may be one of tools we can use to answer those questions.

5.3.2
Astrophysical Backgrounds

We have so far concentrated on cosmological backgrounds of gravitational waves. While these backgrounds are enticing for the glimpses they could give us of the early Universe, and hence the questions they might help us answer, there are many unknowns. Furthermore, in many cases there are either severe constraints on these backgrounds from indirect observations, such as the CMB experiments or Big Bang nucleosynthesis, or calculations of the background indicate that they are unlikely to be strong enough to be observed with foreseeable technologies.

On the other hand, there are stochastic backgrounds of gravitational waves that we expect to be easily seen with, for instance, the LISA observatory – backgrounds from astrophysical sources. As stated earlier, these will be produced whenever there are multiple sources radiating in a given frequency bin for a detector at a given time – that is, whenever there are an unresolvable number of astrophysical gravitational-wave emitters.

In this section, we will explore two astrophysical backgrounds that are expected to be important (for more details, see Postnov, 1997). The first is the gravitational-wave background from rotating neutron stars. Because neutron stars are not expected to be perfectly axisymmetric, they will emit gravitational waves as they spin. The degree to which they deviate from axisymmetry will dictate their time-changing quadrupole moment, and thus the amplitude of the gravitational waves they emit. It is unknown at this time how non-axisymmetric the average neutron star might be. There are conditions under which the quadrupole moment of a neutron star might grow precipitously – when they exhibit a Chandrasekhar–Friedman–Schutz style instability. However, it appears at this time as though these conditions might be extremely rare if they actually exist at all, and in some cases non-gravitational-wave interactions (e.g. viscosity) appear to dampen the quadrupole moment before it can become significant. Nonetheless, there are likely of the order of 10^8 neutron stars in our Galaxy alone, so even if each is radiating only a small amount of energy in gravitational waves, it could be significant. The rotational periods of isolated spinning neutron stars, as determined by pulsar observations, span a range from a few milliseconds to a few seconds, making a background from these sources most likely visible in the ground-based interferometer band.

The second source of astrophysical gravitational-wave background we will consider is white dwarf binaries. Unlike binaries containing only more compact objects, white dwarf binaries can have complicated dynamics, with mass transfer between stars which can lead to unstable dynamics such as common-envelope evolution. However, these phases are typically short-lasting and are not expected to contribute significantly to a gravitational-wave background. For our purposes here, therefore, the inspiral phase, in which the stars orbit without significant mass transfer, and a stable mass-transfer phase, during which the less massive star (which will have filled its Roche lobe) transfers mass slowly to the more massive companion, will dominate the background. The number of white dwarf binaries in these phases in our Galaxy is estimated to be of the order of 10^8 detached binaries (i.e. binaries in the inspiral phase) and 10^6–10^7 binaries with stable mass transfer. Because binaries have large changing mass quadrupoles, these binaries will emit strongly, and form a significant background. With periods of less than a few minutes, such binaries are expected to form a formidable background for space-based detectors such as LISA.

The calculation of spectra associated with populations of sources such as neutron stars or white dwarf binaries might seem to require detailed knowledge of the populations and their evolutions. However, our task can be simplified with the following three assumptions: (i) we assume that the dominant mechanism for the loss of angular momentum for a given source is the emission of gravitational waves; (ii) we assume that the rate of source formation is constant; and (iii) we assume that the sources are distributed in the same way as luminous stars within the galaxy.

Denoting the formation rate by R, then, the number N of sources per unit logarithmic frequency f is simply given by

$$\frac{dN}{d\ln f} = R\frac{f}{\dot{f}}\,, \tag{5.61}$$

and the total energy emitted in gravitational waves per unit time per unit logarithmic frequency is

$$\dot{E}(f) = L_{\mathrm{GW}}(f)R\frac{f}{\dot{f}}\,, \tag{5.62}$$

where $L_{\mathrm{GW}}(f)$ is the average gravitational-wave luminosity of a source at frequency f. In terms of the standard measure of the spectral strength of a stochastic background, Ω_{GW}, this yields

$$\Omega_{GW}(f)\rho_{crit}c^2 = \frac{f}{\dot{f}}\frac{L_{GW}(f)R}{4\pi c\langle r\rangle^2}, \tag{5.63}$$

where ρ_{crit} is again the critical density to close the Universe, c is the speed of light, and $\langle r\rangle$ is the mean distance to a source. Since we assume the distribution of sources follows the distribution of luminous stars in the Galaxy, we take the mean photometric distance of a spheroidal distribution of the form $dN \propto \exp(-\varpi/\varpi_0)\exp(-z^2/z_0^2)$ where ϖ is the radial distance from the axis of the Galaxy, ϖ_0 is the radius of the Galaxy, which we will take to be $\varpi_0 = 5$ kpc, z is the distance from the Galactic plane, and z_0 is the thickness of the Galaxy, which we will take to be $z_0 = 4.2$ kpc, giving an average source distance of $\langle r\rangle = 7.9$ kpc.

Now, consider the case where the amount of energy that can be radiated in gravitational waves is a power law in frequency, $E \propto f^{\alpha}$. This is the case for both spinning neutron stars ($\alpha = 2$) and detached white dwarf binaries ($\alpha = 2/3$). Then, using Eq. (5.62), we have that $f/\dot{f} = \alpha E/\dot{E}$. Substituting this into Eqs. (5.62) and (5.63) one gets

$$\Omega_{GW}(f)\rho_{crit}c^2 = \frac{\alpha E}{4\pi c\langle r\rangle^2}\frac{\dot{f}}{f}. \tag{5.64}$$

Finally, the total radiated energy is related to the average energy radiated per source E_{GW} by $E(f) = E_{GW}R f/\dot{f}$, so that the astrophysical stochastic spectrum can be expressed as

$$\Omega_{GW}\rho_{crit}c^2 = \frac{\alpha R E_{GW}}{4\pi c\langle r\rangle^2}. \tag{5.65}$$

Returning to our two source populations, all we need to do is estimate the rate R, the spectral power α and the average energy radiated per source, E_{GW} to determine the spectra. For neutron stars, a reasonable rate of formation in the galaxy is $R = (30\,\text{yr})^{-1}$, $\alpha = 2$ (as indicated in the previous paragraph), and the energy radiated per source depends on the moment of inertia, I, about the axis of rotation. The rate of energy emission for a given neutron star is

$$\dot{E}_{GW} = \pi I f\dot{f}. \tag{5.66}$$

However, there will be equal numbers of neutron stars radiating at essentially every frequency < 1 kHz within the band of a ground-based detector, and the average energy emitted will be constant across all frequencies. Performing an average over the population, one eventually arrives at

$$\Omega_{NS} \approx 2\times 10^{-7}\left(\frac{R}{(30\,\text{yr})^{-1}}\right)^{1/2}\left(\frac{I}{10^{38}\,\text{kg m}^2}\right)\left(\frac{f}{100\,\text{Hz}}\right)^2\left(\frac{r}{10\,\text{kpc}}\right)^{-2}. \tag{5.67}$$

For white dwarf binaries, the power law exponent is $\alpha = 2/3$, as we indicated above, and a realistic rate for formation of these sources within the galaxy is $R = (100\,\text{yr})^{-1}$. In this case the energy radiated by a source at frequency f is just the orbital energy of the binary, $E_{\text{GW}} \approx M(GMf)^{2/3}$. Substituting these values into Eq. (5.65) yields

$$\Omega_{\text{WDB}} \approx 4 \times 10^{-8} \left(\frac{R}{(100\,\text{yr})^{-1}}\right)\left(\frac{M}{M_\odot}\right)^{5/3}\left(\frac{f}{10^{-3}\,\text{Hz}}\right)^{2/3}\left(\frac{r}{10\,\text{kpc}}\right)^{-2} . \tag{5.68}$$

We end this section with a note of caution. One of our assumptions, that the energy loss of a source is accounted for entirely by the emission of gravitational waves, is not always valid. Of the sources discussed here, this is a particularly suspect assumption for neutron stars, where it is quite likely that electromagnetic breaking dominates. In fact, at least for the Crab pulsar, upper limits placed by the LIGO instruments already preclude gravitational waves as the dominant energy loss mechanism. The primary difference to expressions above is that if there are other energy loss mechanisms, Ω_{GW} must be reduced by the ratio of energy lost in gravitational waves to total energy lost. Since this may be frequency dependent, it may change the shape of the spectrum of the astrophysical background as well as its overall value.

5.4 Problems

Problem 5.1

Derive the expression for the gravitational waveform produced by a freely precessing axisymmetric neutron star where the axis of symmetry makes a non-zero angle ϑ (the wobble angle) to the axis of rotation given by Eq. (5.4). See Zimmermann and Szedenits (1979).

Problem 5.2

Consider a low-mass X-ray binary (LMXB) system in which a neutron star is accreting matter at a rate $\dot{M}$. As matter falls onto the neutron star its potential energy is converted into X-ray emission with a luminosity $L_X \approx GM\dot{M}/R$ where M and R are the mass and radius of the neutron star. The X-ray flux observed at the Earth, a distance r away, is F_X. The accretion of matter produces a torque $\dot{M}\sqrt{GMR}$ on the neutron star, which spins the neutron star up to some angular velocity ω at which point it is balanced by the loss of angular momentum to gravitational radiation. Show that the characteristic amplitude of the gravitational waves produced,

$h_0 = 4G I_3 \omega^2 \varepsilon / (c^4 r)$, is given by

$$h_0^2 = \frac{5G}{c^3} P_{\mathrm{rot}} \sqrt{\frac{R^3}{GM}} F_{\mathrm{X}} \,, \tag{5.69}$$

where $P_{\mathrm{rot}} = 2\pi/\omega$ is the rotational period of the neutron star. Compute the value of h_0 for a LMXB with an X-ray flux of $F_{\mathrm{X}} = 2 \times 10^{-10}\,\mathrm{W\,m^{-2}}$ if the neutron star has mass $M = 1.4\,M_{\odot}$, radius $R = 10\,\mathrm{km}$, and rotational period $P_{\mathrm{rot}} = 4\,\mathrm{ms}$.

Problem 5.3

A gravitar is a slightly non-axisymmetric neutron star that spins-down primarily due to gravitational radiation. Suppose that the population of gravitars is a disk of uniform surface density along the Galactic plane (out to some large distance) and that the birth-rate of pulsars per unit area of the disk is $\mathcal{R}$. Gravitars are born at some high frequency and spin-down until the frequency of the gravitational waves is too low to be detected by a ground-based detector. Let $\tau_{\mathrm{GW}} = P/\dot{P}$ be the "lifetime" of a gravitar, that is, the time that a gravitar is radiating gravitational waves in the high-frequency band that we are sensitive to. Here, $\dot{P}$ is the rate of change of period of rotation of the gravitar, which is given by Eq. (3.162). Let r_{nearest} be defined so that there is one gravitar, still radiating in our sensitive band, in an area $\pi r_{\mathrm{nearest}}^2$. For this nearest gravitar, show that the characteristic gravitational-wave amplitude, given by $h_{0,\mathrm{largest}} = 4G I_3 \omega^2 \varepsilon / (c^4 r_{\mathrm{nearest}})$ with I_3 the moment of inertia of the gravitar, $\omega = 2\pi/P$ being its rotational angular velocity, and ε being its ellipticity, is

$$h_{0,\mathrm{largest}} = \sqrt{\frac{5\pi}{2} \frac{G}{c^3} I_3 \mathcal{R}} \,. \tag{5.70}$$

If gravitars are created in the Galactic disk (with radius $\simeq 10\,\mathrm{kpc}$) at a rate of $\simeq 1/(30\,\mathrm{yr})$, and have a moment of inertia of $I_3 \simeq 10^{38}\,\mathrm{m^2\,kg}$, compute the value of $h_{0,\mathrm{largest}}$.

Problem 5.4

Suppose a uniform rotating axisymmetric spheroid of mass $M = 1.4\,M_{\odot}$ and initial radius $R = 1\,R_{\oplus}$ undergoes free-fall collapse until it forms a neutron star with $R = 15\,\mathrm{km}$. During the collapse, the eccentricity of the spheroid increases with $e \propto R^{-1/2}$ until $e = 0.1$ at the end of collapse. The principal moments of inertia are $I_{11} = I_{22} = \frac{2}{5} M R^2 (1 - \frac{1}{2} e^2)$ and $I_{33} = \frac{2}{5} M R^2$ where x^3 is the axis of rotation. Compute the gravitational waveform seen by an observer at a distance $r = 10\,\mathrm{kpc}$ on the axis of the collapsing spheroid.

Problem 5.5

A bar of length L and mass M lies in the x–y plane, with the middle of the bar at the origin, and rotates with angular velocity ω about the z-axis. Show that the gravitational waveform is given by Eq. (5.14). Find the gravitational-wave luminosity produced by this rotating bar.

References

Abadie, J. *et al.* (2010) Predictions for the rates of compact binary coalescences observable by ground-based gravitational-wave detectors. *Class. Quant. Grav.*, **27**, 173 001. doi: 10.1088/0264-9381/27/17/173001.

Allen, B. (1997) The stochastic gravity-wave background: sources and detection, in *Relativistic Gravitation and Gravitational Radiation* (eds J.A. Marck and J.P. Lasota), Cambridge University Press, Cambridge.

Cutler, C. and Thorne, K.S. (2002) An overview of gravitational-wave sources, in *General Relativity and Gravitation. Proceedings of the 16th International Conference* (eds N.T. Bishop and S.D. Maharaj), World Scientific, Singapore.

Damour, T. and Vilenkin, A. (2005) Gravitational radiation from cosmic (super)strings: bursts, stochastic background, and observational windows. *Phys. Rev.*, **D71**, 063 510. doi: 10.1103/PhysRevD.71.063510.

Fryer, C.L., Holz, D.E. and Hughes, S.A. (2002) Gravitational wave emission from core-collapse of massive stars. *Astrophys. J.*, **565**, 430–446. doi: 10.1086/324034.

Fryer, C.L. and New, K.C. (2003) Gravitational waves from gravitational collapse. *Living Rev. Rel.*, **6** (2). http://www.livingreviews.org/lrr-2003-2 (last accessed 2011-01-03).

Kim, C., Kalogera, V. and Lorimer, D.R. (2003) The probability distribution of binary pulsar coalescence rates. I. Double neutron star systems in the galactic field. *Astrophys. J.*, **584**, 985–995. doi: 10.1086/345740.

Maggiore, M. (2000) Gravitational wave experiments and early universe cosmology. *Phys. Rept.*, **331**, 283–367. doi: 10.1016/S0370-1573(99)00102-7.

Ott, C.D. (2009) The gravitational wave signature of core-collapse supernovae. *Class. Quant. Grav.*, **26**, 063 001. doi: 10.1088/0264-9381/26/6/063001.

Postnov, K.A. (1997) Astrophysical sources of stochastic gravitational radiation in the universe, http://arxiv.org/abs/astro-ph/9706053v1.

Postnov, K.A. and Yungelson, L.R. (2006) The evolution of compact binary star systems. *Living Rev. Rel.*, **9**(6). http://www.livingreviews.org/lrr-2006-6 (last accessed 2011-01-03).

Sathyaprakash, B.S. and Schutz, B.F. (2009) Physics, astrophysics and cosmology with gravitational waves. *Living Rev. Rel.*, **12**(2). http://www.livingreviews.org/lrr-2009-2 (last accessed 2011-01-03).

Siemens, X. *et al.* (2006) Gravitational wave bursts from cosmic (super)strings: quantitative analysis and constraints. *Phys. Rev.*, **D73**, 105 001. doi: 10.1103/PhysRevD.73.105001.

Thorne, K.S. (1987) Gravitational radiation, in *300 Years of Gravitation* (eds S. Hawking and W. Israel), Cambridge University Press, Cambridge, Chap. 9.

Vilenkin, A. and Shellard, E.P.S. (2000) *Cosmic Strings and Other Topological Defects*, Cambridge University Press, Cambridge.

Zimmermann, M. and Szedenits, E. (1979) Gravitational waves from rotating and precessing rigid bodies: simple models and applications to pulsars. *Phys. Rev.*, **D20**, 351–355. doi: 10.1103/PhysRevD.20.351.

6
Gravitational-Wave Detectors

Among the variety of gravitational-wave detectors, the first detectors to be built were cylindrical resonant mass gravitational-wave detectors, or bar detectors. Bar detectors are set into oscillation by gravitational waves that are near the frequency of the natural frequency of the bar, and it is these oscillations that we measure. These detectors are sensitive to gravitational waves with relatively high frequency (around 1 kHz), and their sensitivity is relatively narrow-band (although recent advances have expanded the sensitive band to around a 100 Hz bandwidth).

While some bar detectors are still operating, and there is recent development of spherical resonant mass detectors, most attention is now focused on gravitational detectors that use laser interferometry to detect differential changes in the positions of test masses at the ends of the interferometer arms. Kilometre-scale ground-based interferometric gravitational-wave detectors have been operating for several years, and are the most sensitive detectors operating in the frequency band from $\sim$10 Hz to $\sim$1 kHz. We will focus mostly on these detectors in this chapter. Other gravitational-wave experiments include pulsar timing observations, which observe gravitational waves in the nHz–mHz frequency band; efforts to detect B-polarization modes in the cosmic microwave background radiation, which result from gravitational waves at 10^{-18} Hz; and planned space-based interferometric detectors such as LISA, which will observe in the 1–100 mHz frequency band. Further reading on gravitational-wave detectors includes the classic text Saulson (1994) and a detailed treatment in Maggiore (2007, Chaps. 8 and 9).

Table 6.1 Resonant mass detectors of the International Gravitational Event Collaboration (IGEC) (Allen *et al.*, 2000).

Detector	Location	Type
ALLEGRO	Baton Rouge, LA, USA	Aluminum Bar
AURIGA	Padova, Italy	Aluminum Bar
EXPLORER	Geneva, Switzerland	Aluminum Bar
NAUTILUS	Rome, Italy	Aluminum Bar
NIOBE	Perth, Australia	Niobium Bar

Gravitational-Wave Physics and Astronomy, First Edition. Jolien D. E. Creighton, Warren G. Anderson.

Table 6.2 First generation kilometre-scale interferometric detectors.

Detector	Location	Type
Laser Interferometric Gravitational-wave Observatory (LIGO)	Hanford, WA, USA & Livingston, LA, USA	Power-recycled Fabry–Pérot Michelson Interferometer (4 km and 2 km arms)
Virgo	Pisa, Italy	Power-recycled Fabry–Pérot Michelson Interferometer (3 km arms)
GEO 600	Hannover, Germany	Dual-recycled Michelson Interferometer (600 m arms)
TAMA 300	Tokyo, Japan	Dual-recycled Michelson Interferometer (300 m arms)

Table 6.3 Anticipated second and third generation kilometre-scale interferometric detectors.

Detector	Location	Type
Advanced LIGO (aLIGO)	Hanford, WA, USA & Livingston, LA, USA	Dual-recycled Fabry–Pérot Michelson Interferometer (4 km arms)
Advanced Virgo	Pisa, Italy	Dual-recycled Fabry–Pérot Michelson Interferometer (3 km arms)
Large Cryogenic Gravitational Telescope (LCGT)	Kamioka, Japan	Cryogenic, Underground Dual-recycled Fabry–Pérot Michelson Interferometer (3 km arms)

The major resonant mass detector experiments, first generation interferometric detector experiments and planned advanced interferometric detector experiments are listed in Tables 6.1–6.3. Figure 6.1 shows the sensitivity curves for some contemporary and planned gravitational-wave detectors, spanning twelve orders of magnitude in the gravitational-wave spectrum.

6.1
Ground-Based Laser Interferometer Detectors

The major efforts to detect gravitational waves currently use the technique of *laser interferometry* to sense the small changes in the lengths of optical paths that would be produced by gravitational waves. An interferometer transmits light down two non-aligned arms to a distant object (a mirror) which returns the light to the source;

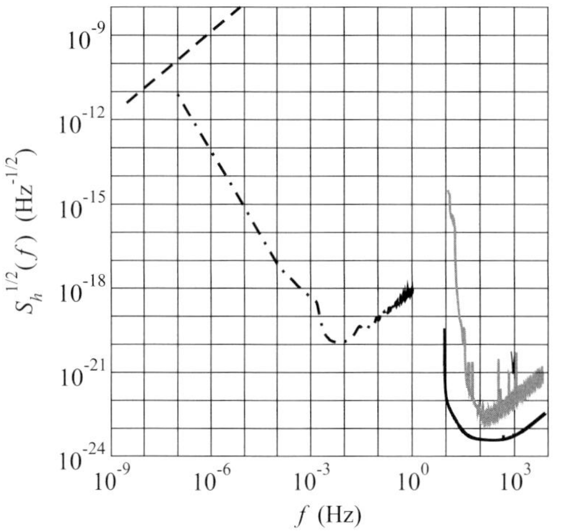

Figure 6.1 Representative sensitivity curves for various contemporary or planned gravitational-wave detectors: a representative noise amplitude spectrum for a resonant mass detector (thin black line); a representative spectrum for the Initial LIGO detector during its fifth science run (grey line) (4 km Hanford interferometer on 18 March 2007[1]); a spectrum for Advanced LIGO operating at high power and tuned for broadband operation (black solid line) (LIGO Document T0900288-v3[2]); expected spectrum for the LISA detector (dot-dashed line) (Source: Shane L. Larson *Online Sensitivity Curve Generator*,[3] based on Larson *et al.* (2000)); sensitivity for pulsar timing with a *single* pulsar with ten years of monthly observations (dashed line).

the returned light beams from the two arms are then interfered to form a measure of the relative lengths of the arms. A change in the interference of the light thus forms a readout of the change in the relative lengths of the arms. Ground-based laser interferometers that measure the lengths of arms several kilometres long are the state-of-the-art, and there are designs for space-based interferometers with arms that would be millions of kilometres long.

To understand the technique of laser interferometry, consider first the relatively simple *Michelson interferometer* (see Figure 6.2). In such an instrument, a beam of light produced by a laser is split into two orthogonal paths by a beam-splitter. The two light beams then traverse the two orthogonal paths, or arms, and are reflected back toward the beam-splitter by mirrors positioned at the ends of the arms. When the light beams are reunited at the beam-splitter, they interfere. If the arms are exactly the same length (up to an integer number of wavelengths), the light returns entirely toward the laser (the *symmetric output* direction): there is destructive interference of the light that would travel in the direction away from the laser (the *antisymmetric output* direction). If, however, the arms differ in length by an amount that is not an integer number of wavelengths, then the destructive interference of the light travelling toward the antisymmetric output direction will be incomplete;

1) http://www.ligo.caltech.edu/~jzweizig/distribution/LSC_Data
2) https://dcc.ligo.org/cgi-bin/DocDB/ShowDocument?docid=2974
3) http://www.srl.caltech.edu/~shane/sensitivity

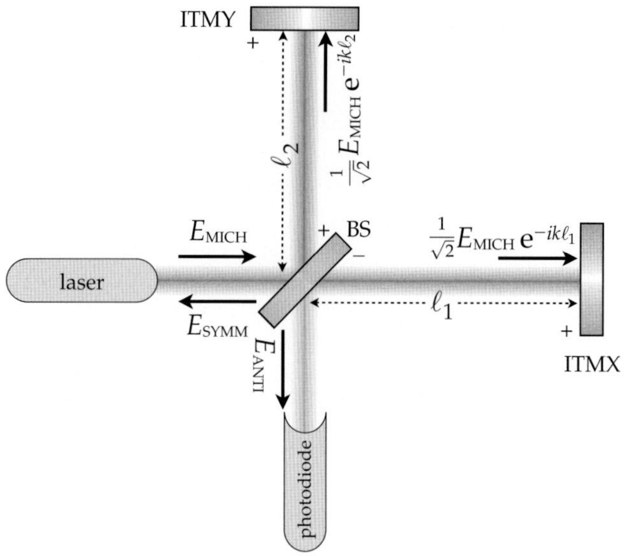

Figure 6.2 Optical topology of a Michelson interferometer.

a photodiode placed in the path of the antisymmetric output direction will detect some light. By arranging the arm lengths so that there is no light transmitted to the antisymmetric output initially, one can attempt to detect a gravitational wave by its differential effect on the lengths of the two arms. If the wave stretches one arm while squeezing the other, then a trickle of light will be observed by the photodiode.

It is instructive to estimate the sensitivity of such an interferometer. Suppose that the lengths of the two arms, ℓ_1 and ℓ_2, are on the kilometre scale. The interference of the light traversing the two arms depends on the difference in the lengths of the arms, $\ell_1 - \ell_2$, and the change in the interference of the light is therefore proportional to $\Delta\ell := \Delta\ell_1 - \Delta\ell_2$. A gravitational wave will induce a strain $h := \Delta\ell/\ell$ on the detector where h is of the order of the gravitational-wave metric perturbation. A measurable change in the interference pattern would be of the order of the wavelength of the laser light, $\Delta\ell \sim \lambda_{\text{laser}}$. If an infrared laser with wavelength 1 μm were used then metric perturbations of the order

$$h := \frac{\Delta\ell}{\ell} \sim \frac{\lambda_{\text{laser}}}{\ell} \sim \frac{10^{-6}\,\text{m}}{10^{3}\,\text{m}} = 10^{-9} \tag{6.1}$$

would be detectable. This is nowhere near sufficient sensitivity to detect anticipated sources of gravitational waves, which would produce anticipated metric perturbations of the order $h \sim 10^{-20}$ or smaller.

There are several significant improvements that can be made:

1. The optical path can be significantly increased. Although physical constraints may limit the size of ground-based interferometric detectors to $\ell \sim 1\,\text{km}$, the

effective length of the arm can be increased by forcing the light to make many traversals of the arm before allowing it to interfere with the light from the other arm. By creating an optical cavity within each arm and allowing the light in the arms to bounce several times before exiting the cavity, the effective optical path ℓ_{eff} can be made significantly larger than the arm length ℓ. There is a limitation, however. When the effective arm length becomes comparable to the wavelength λ_{GW} of the gravitational wave, $\ell_{\text{eff}} \sim \lambda_{\text{GW}} = c/f_{\text{GW}}$ (where f_{GW} is the frequency of the gravitational wave), the metric perturbation when the light emerges from the cavity will be significantly different from when it entered the cavity and this will begin to degrade the sensitivity of the detector. The largest useful effective length is $\ell_{\text{eff}} \sim \lambda_{\text{GW}}$ and for ground-based detectors we are interested in gravitational waves of $\lambda_{\text{GW}} \sim 1000\,\text{km}$ ($f_{\text{GW}} \sim 300\,\text{Hz}$). Allowing multiple bounces therefore allows us to measure gravitational waves with amplitude

$$h \sim \frac{\Delta\ell}{\ell_{\text{eff}}} \sim \frac{\lambda_{\text{laser}}}{\lambda_{\text{GW}}} \sim \frac{10^{-6}\,\text{m}}{10^{6}\,\text{m}} = 10^{-12}\,. \tag{6.2}$$

This is still orders of magnitude away from where we need to be.

2. Improved measurement of the optical fringe. To make further progress, we need to improve our ability to measure the change in the optical path lengths. Previously we assumed that we could measure the change in the optical path if it was on the order of the wavelength of the laser light. However, with a sensitive photodiode, better results can be achieved.
The primary limitation to our ability to measure the interference of the two beams comes from *shot noise*. Suppose that we have some light falling on the photodiode. What is the smallest change in the amount of light that is detectable? Imagine that the photodiode collects light over a time τ and counts the number of photons N_{photons} incident on it in this time. The rate at which photons arrive follows a Poisson process, so the natural fluctuation in the number of photons observed will be $\sim N_{\text{photons}}^{1/2}$. In order to detect a change in light beyond the natural fluctuations, we require a change in the optical path on the order of

$$\Delta\ell \sim \frac{N_{\text{photons}}^{1/2}}{N_{\text{photons}}}\lambda_{\text{laser}}\,. \tag{6.3}$$

We can collect photons for a time of the order of the period of the gravitational wave (before we start washing out the signal), $\tau \sim 1/f_{\text{GW}}$, so the total number of photons will depend on the frequency of the gravitational wave we are trying to detect and the power of the laser, P_{laser}. Since each photon has energy $hc/\lambda_{\text{laser}}$,

$$N_{\text{photons}} = \frac{P_{\text{laser}}}{hc/\lambda_{\text{laser}}}\tau \sim \frac{P_{\text{laser}}}{hc/\lambda_{\text{laser}}}\frac{1}{f_{\text{GW}}}\,. \tag{6.4}$$

For $P_{\text{laser}} = 1\,\text{W}$, $\lambda_{\text{laser}} = 1\,\mu\text{m}$, and $f_{\text{GW}} \sim 300\,\text{Hz}$, $N_{\text{photons}} \sim 10^{16}$ photons, and we find

$$h \sim \frac{\Delta\ell}{\ell_{\text{eff}}} \sim \frac{N_{\text{photons}}^{-1/2}\lambda_{\text{laser}}}{\lambda_{\text{GW}}} \sim \frac{10^{-8} \times 10^{-6}\,\text{m}}{10^{6}\,\text{m}} = 10^{-20}\,. \tag{6.5}$$

The interferometer is now close to the required sensitivity.

3. Higher laser power. Further improvements to the sensitivity of the interferometer require further reduction of the shot noise, which is achieved by increasing the laser power. There are practical limits to how powerful a laser can be made (and have the required stability), but the effective power can be increased by recovering the light that would otherwise exit the interferometer through the symmetric output direction and reflecting this back into the interferometer. The net result of such *power recycling* is a build-up of power circulating in the interferometer, and thus an increase in the power of the light being interfered on the beam splitter. Practical limitations mean that the light power can only be increased by a couple of orders of magnitude, which yields approximately one order of magnitude improvement in sensitivity. As the light power increases, eventually the mirrors begin to be buffeted by the quantum fluctuations of the light intensity and this *radiation pressure* forms a source of noise that competes with the shot noise; heavier mirrors can be used to reduce the effect of radiation pressure noise. This will be discussed in more detail later.
4. Advanced interferometry. Additional improvements can be achieved by reflecting the light leaving the interferometer through the antisymmetric output path back into the interferometer, which, when done correctly, can improve the sensitivity to signals in particular frequency bands. This is known as *signal recycling*. Quantum squeezing of the light can also be used to reduce the effect of the shot noise and improve sensitivity. These techniques are being developed for use in advanced interferometers.

Now that we have performed an order of magnitude estimate of the anticipated sensitivity, we turn to a more detailed examination of an interferometric detector. Although there have been several first generation ground-based interferometers in operation – the Laser Interferometric Gravitational-wave Observatory (LIGO), GEO 600, TAMA 300 and Virgo – and second and third generation interferometers being planned or under construction – advanced LIGO (aLIGO), advanced Virgo, Large Cryogenic Gravitational Telescope (LCGT) and the Einstein Telescope (ET) – we will, for definiteness, focus on the LIGO interferometers. We will consider first the configuration of the initial LIGO interferometer, which is a power-recycled Michelson interferometer with Fabry–Pérot cavities as the arms, and then discuss the advances planed for aLIGO. We begin with an overview of the concepts in optics that are necessary for our discussion.

6.1.1
Notes on Optics

A monochromatic, linearly polarized light beam travelling along the z-axis has an electric field

$$E(t,z) = e_{\text{pol}}\sqrt{2c\epsilon_0}\,\text{Re}\left\{E_{(+z)}(t)e^{i(\omega t - kz)} + E_{(-z)}(t)e^{i(\omega t + kz)}\right\}, \tag{6.6}$$

where e_{pol} is the polarization vector, the angular frequency of the light is given by ω, and $k = \omega/c$. The two complex amplitudes $E_{(+z)}(t)$ and $E_{(-z)}(t)$ describe the field travelling in the +z-direction and the $-z$-direction, respectively, and it is these amplitudes that we use to describe the radiation field inside the interferometer. The normalization constant has been chosen so that, for a wave travelling in one direction, say, the +z-direction (so that $E_{(-z)}(t) = 0$), the intensity (power per unit area) of the radiation is $I = |E_{(+z)}|^2$. We make the simplifying assumption that the transverse profile of the beam is fixed (Gaussian beams are used) so we will refer to I as the power or the intensity of the beam interchangeably. (Technically the power is IA where A is some effective area of the beam spot.) Since the amplitudes contain the essential information about the radiation field, in this section we show how various optical components can be described by their effect on the amplitudes.

Delay lines The complex amplitudes pick up a phase shift of $\phi = kz = \omega z/c$ after traversing a distance z so that

$$E(z) = E(0)e^{-ikz}. \tag{6.7}$$

Mirrors A mirror is characterized by its *amplitude reflectivity*, r, and *transmissivity*, t. Figure 6.3 shows the light beams that are incident on, reflected by and transmitted through a simple mirror. The reflectivity is defined by

$$r := E_{\text{R}}/E_{\text{I}} \tag{6.8}$$

and the transmissivity is defined by

$$t := E_{\text{T}}/E_{\text{I}}. \tag{6.9}$$

These are the mirror's reflectivity and transmissivity for a beam incident *from the left*; if the beam strikes the mirror on the right side, the reflectivity r' and transmis-

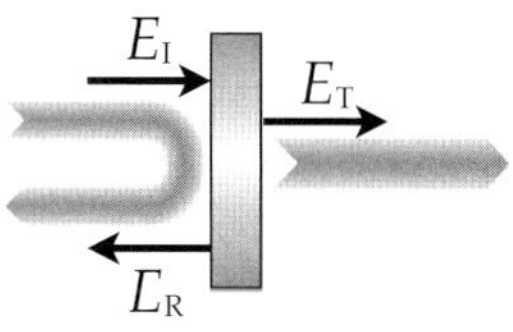

Figure 6.3 The incident, E_{I}, reflected, E_{R}, and transmitted, E_{T} field amplitudes for a simple mirror.

sivity t' are different. If the mirror is lossless, conservation of energy implies

$$\begin{aligned}|r| &= |r'| \,, \\ |t| &= |t'| \,, \\ 1 &= |r|^2 + |t|^2 = |r'|^2 + |t'|^2 \,, \\ 0 &= r^* t' + t^* r' \,. \end{aligned} \tag{6.10}$$

For simple mirrors we will be dealing with real-valued reflection and transmission coefficients so the last of these relations means that we can identify a "+" side of the mirror and a "−" side of the mirror so that $t' = t$ and

$$\begin{aligned} r &= +|r| \quad \text{on "+" side,} \\ r' &= -|r| \quad \text{on "−" side.} \end{aligned} \tag{6.11}$$

In the diagrams that follow, the sign convention will be shown by marking the "+" and "−" sides of each mirror.

Example 6.1 Stokes relations

Consider the reflection and refraction of light at the boundary between two different media. When a beam of light is incident on the boundary from above, the transmissivity and reflectivity are given by t and r; when a beam of light is incident on the boundary from below, they are t' and r'. Figure 6.4a shows a beam E_I incident on the boundary producing a reflected beam $E_\mathrm{R} = r E_\mathrm{I}$ and a transmitted beam $E_\mathrm{T} = t E_\mathrm{I}$. If there are no losses then the process must be reversible: Figure 6.4b shows the (now reversed) beams $r E_\mathrm{I}$ and $t E_\mathrm{I}$ incident on the boundary forming the single outgoing beam E_I. Figure 6.4c then shows the reflected and transmitted beams for both of these incident beams. By comparing Figure 6.4b and c (which both depict the same scenario), it is clear that

$$\begin{aligned} r(r E_\mathrm{I}) + t'(t E_\mathrm{I}) &= E_\mathrm{I} \,, \\ r'(t E_\mathrm{I}) + t(r E_\mathrm{I}) &= 0 \,, \end{aligned} \tag{6.12}$$

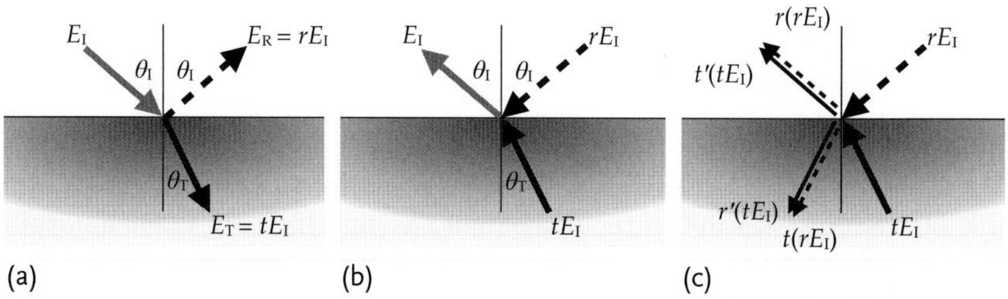

Figure 6.4 (a) A field incident on a medium, E_I (grey arrow), produces a reflected field, $E_\mathrm{R} = rE_\mathrm{I}$ (dashed arrow), and a transmitted field, $E_\mathrm{T} = tE_\mathrm{I}$ (black arrow). (b) Time-reversed version of the previous situation: now the two fields rE_I (dashed arrow) and tE_I (black arrow) are incident and combine to form the single field E_I. (c) Same as (b), but with the reflected and transmitted fields for both of the two incident fields shown.

and therefore

$$r^2 + tt' = 1\,,$$
$$r' = -r\,. \tag{6.13}$$

These are known as *Stokes relations.*

Example 6.2 Dielectric mirror

Consider a simple mirror made of a dielectric slab of material of thickness ℓ and index of refraction $n > 1$. Light travelling within the dielectric material has a wavelength $\lambda' = \lambda/n$ and $k' = 2\pi/\lambda' = nk$. We put the origin of our coordinates so that the left side of the mirror is at $z = 0$ and the right hand side of the mirror is at $z = \ell$, and we define r and t to be the reflectivity and transmissivity of the boundary for light incident on the boundary from outside the dielectric material and r' and t' are the reflectivity and transmissivity of the boundary for light incident on the boundary from inside the dielectric material. These are related by the Stokes relations, Eq. (6.13). Suppose that light is incident on the mirror from the left with field E_I. Part of the light is reflected by the mirror, E_R, and part of the light is transmitted through the mirror to the right side, E_T. Within the mirror there is a circulating field E_M. See Figure 6.5. The circulating field equals the transmitted part of the incident light plus the reflected part of the returning circulating field,

$$E_\mathrm{M}(0) = t\,E_\mathrm{I}(0) + (r')^2\,E_\mathrm{M}(0)\mathrm{e}^{-2ik'\ell}\,. \tag{6.14}$$

Notice that the returning circulating field has also reflected off of the right boundary at $z = \ell$, so there have been *two* reflections and hence the factor of $(r')^2$. In addition it has accumulated an overall phase $-2k'\ell$. This equation can be solved for the circulating field

$$E_\mathrm{M}(0) = \frac{t}{1 - r^2\mathrm{e}^{-2ik'\ell}}\,E_\mathrm{I}(0)\,. \tag{6.15}$$

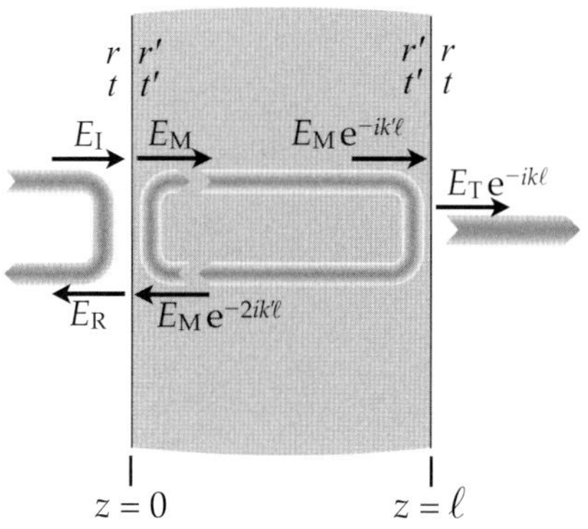

Figure 6.5 The incident, E_I, reflected, E_R, transmitted, E_T, and circulating E_M field amplitudes for a dielectric mirror.

We also have

$$E_{\mathrm{R}}(0) = r E_{\mathrm{I}}(0) + r' t' E_{\mathrm{M}}(0) \mathrm{e}^{-2ik'\ell} \tag{6.16}$$

for the reflected field and

$$E_{\mathrm{T}}(\ell) = t' E_{\mathrm{M}}(\ell) = t' E_{\mathrm{M}}(0) \mathrm{e}^{-ik'\ell} \tag{6.17}$$

for the transmitted field. We want to refer all external fields to the plane $z = 0$ so we write $E_{\mathrm{T}}(\ell) = E_{\mathrm{T}}(0)\mathrm{e}^{ik\ell}$ (even though the transmitted field actually starts at $z = \ell$). We now obtain the effective transmissivity t_{M} and reflectivity r_{M} of the mirror as the ratio of the transmitted field to the incident field and the ratio of the reflected field to the incident field, respectively:

$$t_{\mathrm{M}} = \frac{E_{\mathrm{T}}(0)}{E_{\mathrm{I}}(0)} = t' t \frac{\mathrm{e}^{-i(k'-k)\ell}}{1 - (r')^2 \mathrm{e}^{-2ik'\ell}} , \tag{6.18}$$

$$r_{\mathrm{M}} = \frac{E_{\mathrm{R}}(0)}{E_{\mathrm{I}}(0)} = r + t' t \frac{\mathrm{e}^{-2ik\ell}}{1 - (r')^2 \mathrm{e}^{-2ik'\ell}} = r \frac{1 - \mathrm{e}^{-2ik\ell}}{1 - r^2 \mathrm{e}^{-2ik'\ell}} . \tag{6.19}$$

Now imagine reversing the situation so that the light is incident on the mirror from the other side. By symmetry the transmissivity and reflectivity of the mirror are still t_{M} and r_{M} *if we refer the fields to the incident surface*, now at $z = \ell$. Translating the fields to the original reference plane, $z = 0$, then gives us the transmissivity t'_{M} and reflectivity r'_{M} of the mirror for fields incident from the right:

$$t_{\mathrm{M}} = \frac{E'_{\mathrm{T}}(\ell)}{E'_{\mathrm{I}}(\ell)} = \frac{E'_{\mathrm{T}}(0)\mathrm{e}^{-ik\ell}}{E'_{\mathrm{I}}(0)\mathrm{e}^{-ik\ell}} = \frac{E'_{\mathrm{T}}(0)}{E'_{\mathrm{I}}(0)} = t'_{\mathrm{M}} \tag{6.20a}$$

$$r_{\mathrm{M}} = \frac{E'_{\mathrm{R}}(\ell)}{E'_{\mathrm{I}}(\ell)} = \frac{E'_{\mathrm{R}}(0)\mathrm{e}^{ik\ell}}{E'_{\mathrm{I}}(0)\mathrm{e}^{-ik\ell}} = \frac{E'_{\mathrm{R}}(0)}{E'_{\mathrm{I}}(0)} \mathrm{e}^{2ik\ell} = r'_{\mathrm{M}} \mathrm{e}^{2ik\ell} . \tag{6.20b}$$

Notice the difference here: the transmitted and incident light are travelling in the same direction so the phase factor accumulated in the translation from $z = \ell$ to $z = 0$ cancels in the ratio for t_{M}, but the reflected light is travelling in the opposite direction to the incident light so the phase accumulations add in forming the ratio for r_{M}.

Given the expressions for t_{M}, t'_{M}, r_{M} and r'_{M} it is straightforward to show

$$r^*_{\mathrm{M}} r_{\mathrm{M}} + t^*_{\mathrm{M}} t'_{\mathrm{M}} = 1 , \tag{6.21a}$$

$$r^*_{\mathrm{M}} t'_{\mathrm{M}} + t^*_{\mathrm{M}} r'_{\mathrm{M}} = 0 . \tag{6.21b}$$

We have then obtained the relations of Eq. (6.10).

Compound mirrors Combinations of mirrors and delay lines can be collectively treated as a single *compound mirror* with complex-valued and frequency-dependent reflectivities and transmissivities. For example, a delay line of length ℓ is trivially represented as a compound mirror with $r = 0$ and $t = e^{-ik\ell}$. A more interesting example is the Fabry–Pérot cavity discussed in the next section.

6.1.2
Fabry–Pérot Cavity

A Fabry–Pérot interferometer is comprised of two mirrors along an optical axis which form a cavity, as shown in Figure 6.6. The input mirror (for a gravitational-wave detector, the mirrors are test masses so the input mirror is referred to as the *input test mass* or ITM) is partly transmissive and allows some of the light to enter the cavity. The end mirror (or *end test mass* ETM) is highly reflective. If the separation between the two mirrors is correctly tuned, there is a build-up of light power within the cavity. In such a configuration, the incident light upon the cavity and the reflected light, which is a combination of light reflected from the ITM and light transmitted through the ITM from within the Fabry–Pérot cavity, are related by a phase shift that is highly sensitive to small changes in the length of the cavity.

Let r_{ITM} and t_{ITM} be the reflectivity and the transmissivity of the ITM and, similarly, let r_{ETM} and t_{ETM} be the reflectivity and the transmissivity of the ETM. For simplicity we will make the following assumptions: (i) the optical components are lossless so that $|r|^2 + |t|^2 = 1$, and (ii) the ETM is perfectly reflective[4] so that $r_{\text{ETM}} = 1$ and $t_{\text{ETM}} = 0$. (Later we will have to revisit the assumption that the mirrors are lossless as we will see that the sensitivity of the interferometer is determined by the degree of absorption of light in the optical components.) The field within the Fabry–Pérot cavity $E_{\text{FP}\triangleright}$ is a superposition of the incident light E_{I} that is transmitted through the ITM and the circulating light within the cavity, $E_{\text{FP}\triangleleft}$, that is reflected by the ITM. (We use the symbol $\triangleright$ to indicate the outgoing light at the ITM and the symbol $\triangleleft$ to indicate the returning light at the ITM.) In travelling from the ITM to the ETM, the circulating light picks up a phase factor of e^{-ikL}, where L is the length of the cavity, and a similar phase factor on the return journey from ETM back to ITM. Therefore

$$E_{\text{FP}\triangleright} = t_{\text{ITM}} E_{\text{I}} - r_{\text{ITM}} E_{\text{FP}\triangleleft} = t_{\text{ITM}} E_{\text{I}} - r_{\text{ITM}} e^{-2ikL} E_{\text{FP}\triangleright} \,. \tag{6.22}$$

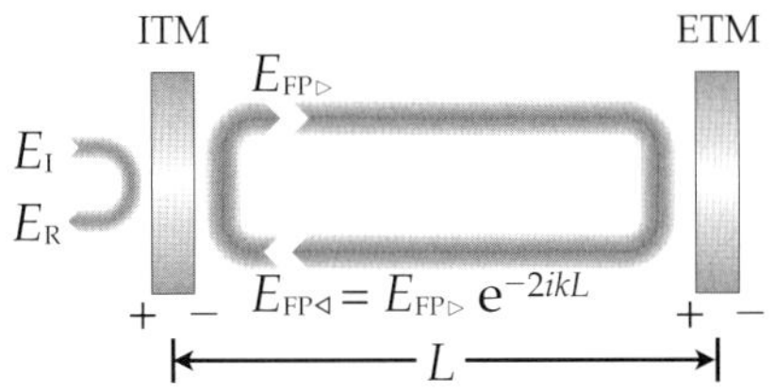

Figure 6.6 Optical configuration of a Fabry–Pérot cavity.

4) For LIGO this is not entirely true – some light is transmitted through the cavity – but the transmissivity of the ETM is so small that we choose to neglect it.

Notice that the second term of the right-hand side involves the returning circulating field that has been reflected off both the ETM and the ITM and has picked up a phase shift of $-2kL$. Notice also the negative sign of this term: it arises due to the sign convention chosen for the ITM mirror (the "+" side is chosen to be on the left side of the ITM as shown in in Figure 6.6). This equation can be solved for the circulating field, the field within the Fabry–Pérot cavity:

$$E_{\mathrm{FP}\triangleright} = \frac{t_{\mathrm{ITM}}}{1 + r_{\mathrm{ITM}} e^{-2ikL}} E_{\mathrm{I}} \,. \tag{6.23}$$

The field within the cavity is maximized by tuning the cavity – that is, by adjusting the length L between the two mirrors – so that the circulating light is in resonance. This is achieved for a length L_{res} that satisfies

$$e^{-2ikL_{\mathrm{res}}} = -1 \,, \tag{6.24}$$

which creates a circulating field

$$E_{\mathrm{FP}\triangleright,\mathrm{res}} = \frac{t_{\mathrm{ITM}}}{1 - r_{\mathrm{ITM}}} E_{\mathrm{I}}. \tag{6.25}$$

If r_{ITM} is close to unity then the power build-up within the cavity can be large.

The relationship between the field E_{I} upon the Fabry–Pérot cavity and the field E_{R} reflected by the Fabry–Pérot cavity can now be computed. The reflected field is the superposition of the portion of the incident field that is directly reflected by the ITM and the portion of the circulating field within the cavity that is transmitted through the ITM:

$$\begin{aligned} E_{\mathrm{R}} &= r_{\mathrm{ITM}} E_{\mathrm{I}} + t_{\mathrm{ITM}} e^{-2ikL} E_{\mathrm{FP}\triangleright} \\ &= \left[r_{\mathrm{ITM}} + \frac{t_{\mathrm{ITM}}^2 e^{-2ikL}}{1 + r_{\mathrm{ITM}} e^{-2ikL}} \right] E_{\mathrm{I}} \\ &= \frac{r_{\mathrm{ITM}} + e^{-2ikL}}{1 + r_{\mathrm{ITM}} e^{-2ikL}} E_{\mathrm{I}} \,. \end{aligned} \tag{6.26}$$

The effective reflectivity of the Fabry–Pérot cavity, viewed as a compound mirror, is defined by the ratio $E_{\mathrm{R}}/E_{\mathrm{I}}$, which depends on the length, L, of the cavity:

$$r_{\mathrm{FP}}(L) = \frac{r_{\mathrm{ITM}} + e^{-2ikL}}{1 + r_{\mathrm{ITM}} e^{-2ikL}} \,. \tag{6.27}$$

When the cavity is at resonance, $L = L_{\mathrm{res}}$ where $\exp(-2ikL_{\mathrm{res}}) = -1$,

$$r_{\mathrm{FP,\,res}} := r_{\mathrm{FP}}(L_{\mathrm{res}}) = \frac{r_{\mathrm{ITM}} - 1}{1 - r_{\mathrm{ITM}}} = -1 \,. \tag{6.28}$$

The cavity is perfectly reflecting and, at resonance, the reflected light experiences a phase shift of π.

The phase response of the Fabry–Pérot cavity is a measure of how much additional phase shift the reflected light experiences relative to the incident light when the length of the Fabry–Pérot cavity is changed by a very small amount ΔL from resonance, $L = L_{\text{res}} + \Delta L$. The effective reflectivity near resonance is

$$\begin{aligned} r_{\text{FP}}(L_{\text{res}} + \Delta L) &\simeq r_{\text{FP}}(L_{\text{res}}) + \left.\frac{d r_{\text{FP}}}{d L}\right|_{L=L_{\text{res}}} \Delta L \\ &= -1 + (-2ik\Delta L)\left[\frac{e^{-2ikL}(1 - r_{\text{ITM}}^2)}{(1 + r_{\text{ITM}} e^{-2ikL})^2}\right]_{L=L_{\text{res}}} \\ &= -1 + \frac{1 + r_{\text{ITM}}}{1 - r_{\text{ITM}}} 2ik\Delta L \\ &\simeq -\exp\left[-2ik\left(\frac{1 + r_{\text{ITM}}}{1 - r_{\text{ITM}}}\right)\Delta L\right]. \end{aligned} \tag{6.29}$$

From this it can be seen that a small change in the cavity length ΔL near resonance will cause a large change in the phase of the reflected light: if there was only an ETM and no ITM – a configuration that would be a delay line of length $2L$ – then the phase shift due to a change in length ΔL would be $-2ik\Delta L$; however, the Fabry–Pérot cavity amplifies this phase shift by a factor

$$G_{\text{arm}} = \frac{1 + r_{\text{ITM}}}{1 - r_{\text{ITM}}} = \frac{t_{\text{ITM}}^2}{(1 - r_{\text{ITM}})^2} . \tag{6.30}$$

This quantity G_{arm} is called the *arm cavity gain*. For Initial LIGO, $t_{\text{ITM}}^2 = 2.8\%$ which yields $G_{\text{arm}} = 140$. By having Fabry–Pérot cavities for arms, LIGO is 140 times more sensitive than it would be if the light simply travelled the length of the arm a single time.

Example 6.3 Anti-resonant Fabry–Pérot cavity

Light is said to be *anti-resonant* in the Fabry–Pérot cavity when condition $\exp(-2ikL_{\text{antires}}) = +1$. In this case,

$$E_{\text{FP}\triangleright,\text{antires}} = \frac{t_{\text{ITM}}}{1 + r_{\text{ITM}}} E_{\text{I}} . \tag{6.31}$$

When r_{ITM} is close to unity and t_{ITM} is small, the circulating field is much smaller than the incident field. The effective reflectivity of the anti-resonant cavity is

$$r_{\text{FP,antires}} = 1 \tag{6.32}$$

so, unlike the case in resonance, the reflected field does not have a phase shift. For small deviations from anti-resonance, the reflectivity is

$$r_{\text{FP}}(L_{\text{antires}} + \Delta L) \simeq \exp\left[-2ik\left(\frac{1 - r_{\text{ITM}}}{1 + r_{\text{ITM}}}\right)\Delta L\right] \tag{6.33}$$

so instead of the phase shift being magnified by a factor of G_{arm}, it is suppressed by the factor $1/G_{\text{arm}}$. This makes the Fabry–Pérot cavity essentially insensitive to changes in its length.

As the wavelength of the light incident upon the Fabry–Pérot cavity is changed, the light stored in the cavity goes from resonance to anti-resonance to resonance. One full cycle is known as the *free spectral range* or FSR, which is the change in wavelength between resonances, $4\pi L/\lambda_{\mathrm{FSR}} = 2\pi$. The FSR frequency is

$$f_{\mathrm{FSR}} := \frac{c}{2L} \,. \tag{6.34}$$

For the four-kilometre-long LIGO arms, $f_{\mathrm{FSR}} = 37\,\mathrm{kHz}$. The *finesse* of a Fabry–Pérot cavity describes the sharpness of the interference fringes that it produces. As the wavelength of the incident light changes, resonance of the light is quickly lost and power contained in the Fabry–Pérot cavity drops: relative to the free spectral range, there is significant power build-up within the Fabry–Pérot cavity only for a narrow range of frequencies of incident light. The power in the Fabry–Pérot cavity is half of the maximum resonant power at the frequencies $f_{1/2} = f_{\mathrm{res}} \pm \frac{1}{2}\Delta f_{\mathrm{FWHM}}$, where Δf_{FWHM} is the full-width at half-maximum (FWHM) of the narrow spectral range in which there is a power build-up in the Fabry–Pérot cavity. The ratio of the free spectral range to the FWHM is the finesse:

$$\mathcal{F} := \frac{f_{\mathrm{FSR}}}{f_{\mathrm{FWHM}}} = \frac{\pi}{2\arcsin\left(\frac{1-r_{\mathrm{ITM}}}{2\sqrt{r_{\mathrm{ITM}}}}\right)} \approx \frac{\pi\sqrt{r_{\mathrm{ITM}}}}{1-r_{\mathrm{ITM}}} \,. \tag{6.35}$$

For Initial LIGO, the arm cavity finesse is $\mathcal{F} = 220$. Each photon that enters the cavity will remain in the cavity for an average number of round-trips

$$\langle N_{\text{round-trips}} \rangle = \frac{r_{\mathrm{ITM}}^2}{1-r_{\mathrm{ITM}}^2} \approx \frac{1}{4} G_{\mathrm{arm}} \tag{6.36}$$

and the *light storage time* for a Fabry–Pérot cavity, τ_{s}, is the typical amount of time that any photon remains within the cavity,

$$\tau_{\mathrm{s}} := \frac{2L}{c} \langle N_{\text{round-trips}} \rangle \approx \frac{1}{2}\frac{L}{c} G_{\mathrm{arm}} \approx \frac{L}{c}\frac{\mathcal{F}}{\pi} \tag{6.37}$$

(see Problem 6.1).

In principle, it would be possible to use a single large Fabry–Pérot cavity as a gravitational-wave detector by monitoring the phase of the reflected light. A gravitational-wave incident on the cavity would affect its length slightly and this would produce a phase shift of the reflected light. Unfortunately, any amplitude and frequency noise in the laser light that is incident on the cavity would also produce phase shifts in the reflected light, and it is very difficult to produce a light source that has the degree of stability that would be required to achieve the sensitivity that is needed for gravitational wave astronomy. For this reason, LIGO uses two Fabry–Pérot cavities that form the ends of a Michelson interferometer. The Michelson interferometer, we will see, has two output ports, and the amplitude and frequency of the light is only present in the light exiting through one of these ports (the symmetric port). Thus, the use of a Michelson interferometer is essential to mitigating the problem of laser noise.

6.1.3 Michelson Interferometer

A Michelson interferometer operates by splitting an incoming beam of light via a *beam splitter* (BS) – a partly transparent mirror – into two beams travelling in orthogonal directions. These beams travel along their orthogonal arms, reflect off end mirrors and return to the beam splitter. Two beams then emerge: one beam exits the interferometer along the same axis as the incoming beam (called the X-axis), this is called the beam leaving along the *symmetric output port* of the interferometer, and the other beam exits the interferometer along the orthogonal axis (called the Y-axis), which is called the beam leaving along the *antisymmetric output port.*

We denote the light entering the Michelson interferometer as E_{MICH}, the light leaving the interferometer through the symmetric port as E_{SYMM} and the light leaving the interferometer through the antisymmetric output port as E_{ANTI} (see Figures 6.2 and 6.7). Let ℓ_1 be the length of the interferometer along the X-axis – that is, the length of the optical path from the point at which the light enters the interferometer to the end mirror on the X-axis – and let ℓ_2 be the length of the interferometer along the Y-axis – that is, the length of the optical path from the point at which the light enters the interferometer to the mirror on the Y-axis. If we refer the incoming light to the point at which it strikes the beam splitter then ℓ_1 is the distance between the beam splitter and the mirror on the X-axis and ℓ_2 is the distance between the beam splitter and the mirror on the Y-axis, and this is a convenient simplification (though the results do not require this). We also assume that the beam splitter is half-reflective and half-transmissive so that $r_{\text{BS}} = t_{\text{BS}} = 1/\sqrt{2}$. Then, the light transmitted/reflected into the X-arm and the Y-arm are,

$$E_1 = E_2 = \frac{1}{\sqrt{2}} E_{\text{MICH}} \,. \tag{6.38}$$

After traversing the two arms, these two beams return to the beam splitter and recombine to form the fields

$$\begin{aligned} E_{\text{SYMM}} &= r_{\text{BS}} r_2 E_2 \mathrm{e}^{-2ik\ell_2} + t_{\text{BS}} r_1 E_1 \mathrm{e}^{-2ik\ell_1} \\ &= \frac{1}{2} E_{\text{MICH}} \left[r_2 \mathrm{e}^{-2ik\ell_2} + r_1 \mathrm{e}^{-2ik\ell_1} \right] \end{aligned} \tag{6.39a}$$

and

$$\begin{aligned} E_{\text{ANTI}} &= t_{\text{BS}} r_2 E_2 \mathrm{e}^{-2ik\ell_2} - r_{\text{BS}} r_1 E_1 \mathrm{e}^{-2ik\ell_1} \\ &= \frac{1}{2} E_{\text{MICH}} \left[r_2 \mathrm{e}^{-2ik\ell_2} - r_1 \mathrm{e}^{-2ik\ell_1} \right], \end{aligned} \tag{6.39b}$$

where r_1 is the reflectivity of the mirror at the end of the X-arm and r_2 is the reflectivity of the mirror at the end of the Y-arm. These expressions are often written in a form where ℓ_1 and ℓ_2 are re-expressed in terms of the mean arm length

$\bar{\ell} := \frac{1}{2}(\ell_1 + \ell_2)$ and the difference of the arm lengths $\delta := \ell_1 - \ell_2$,

$$\ell_1 = \bar{\ell} + \frac{1}{2}\delta\,, \tag{6.40a}$$

$$\ell_2 = \bar{\ell} - \frac{1}{2}\delta\,. \tag{6.40b}$$

We then have effective complex reflectivity (for reflection back toward the input light – the symmetric output direction) and transmissivity (for transmission to the antisymmetric output direction) for the Michelson interferometer treated as a complex mirror:

$$r_{\text{MI}} := \frac{E_{\text{SYMM}}}{E_{\text{MICH}}} = \frac{1}{2}e^{-2ik\bar{\ell}}\left[r_2 e^{ik\delta} + r_1 e^{-ik\delta}\right] \tag{6.41a}$$

and

$$t_{\text{MI}} := \frac{E_{\text{ANTI}}}{E_{\text{MICH}}} = \frac{1}{2}e^{-2ik\bar{\ell}}\left[r_2 e^{ik\delta} - r_1 e^{-ik\delta}\right]. \tag{6.41b}$$

Example 6.4 Michelson interferometer gravitational-wave detector

A simple Michelson interferometer gravitational-wave detector can be constructed by choosing perfectly reflective end mirrors, $r_1 = r_2 = 1$. In this case, the effective reflectivity and transmissivity of the interferometer are

$$r_{\text{MI}} = e^{-2ik\bar{\ell}}\cos(k\delta) \quad \text{and} \tag{6.42}$$

$$t_{\text{MI}} = ie^{-2ik\bar{\ell}}\sin(k\delta)\,. \tag{6.43}$$

If the arms are initially tuned to be the same length (in the absence of a gravitational wave) then $|r_{\text{MI}}| = 1$ and $t_{\text{MI}} = 0$ and there is no light transmitted to the antisymmetric port. A gravitational wave that changes the lengths of the arms in a differential sense will then affect the effective transmissivity and will allow some light to leave through the antisymmetric port, which will be a signal of a gravitational wave (or some other effect that changes the lengths of the arms).

Such a detector can be made more sensitive by using Fabry–Pérot cavities with long arm lengths in place of the perfectly reflecting end mirrors.

In LIGO, the end mirrors of the Michelson interferometer are actually the 4-kilometre-long Fabry–Pérot cavities, so r_1 and r_2 are the effective reflectivities of these two Fabry–Pérot compound mirrors, $r_1 = r_{\text{FP}}(L_1)$ and $r_2 = r_{\text{FP}}(L_2)$ where L_1 and L_2 are the lengths of the Fabry–Pérot cavity arms along the X- and Y-axes, respectively. Although a gravitational wave will affect the lengths of the arms in the Michelson interferometer, ℓ_1 and ℓ_2, as well as the lengths of the arms in the Fabry–Pérot cavity, L_1 and L_2, the Fabry–Pérot Michelson interferometer is far more sensitive to the change in the lengths of the arms in the Fabry–Pérot cavities (for two reasons: because these arms are about three orders of magnitude longer than the arms of the Michelson, and because the cavity-gain magnifies the sensitivity to changes in the Fabry–Pérot cavity lengths by another two orders of magnitude). Therefore, we will make the approximation that ℓ_1 and ℓ_2 are constant, even in the presence of a gravitational wave.

Although ℓ_1 and ℓ_2 are taken to be constant, they are intentionally made unequal in LIGO; this imbalance is known as the *Schnupp asymmetry* and it is important for locking the interferometer and for the radio-frequency readout scheme. For Initial LIGO, the difference in the Michelson arm lengths is $\delta = 0.278\,\text{m}$. The Michelson arm lengths are chosen so that, for the carrier light, $\exp(ik\bar{\ell}) = 1$ and $\exp(ik\delta) = 1$, though we will keep the expressions general for now.

For the Fabry–Pérot Michelson interferometer (see Figure 6.7), the effective reflection and transmission coefficients are

$$r_{\text{FPMI}}(L_1, L_2) := \frac{E_{\text{SYMM}}}{E_{\text{MICH}}} = \frac{1}{2}\mathrm{e}^{-2ik\bar{\ell}}\left[r_{\text{FP}}(L_2)\mathrm{e}^{ik\delta} + r_{\text{FP}}(L_1)\mathrm{e}^{-ik\delta}\right] \tag{6.44a}$$

and

$$t_{\text{FPMI}}(L_1, L_2) := \frac{E_{\text{ANTI}}}{E_{\text{MICH}}} = \frac{1}{2}\mathrm{e}^{-2ik\bar{\ell}}\left[r_{\text{FP}}(L_2)\mathrm{e}^{ik\delta} - r_{\text{FP}}(L_1)\mathrm{e}^{-ik\delta}\right]. \tag{6.44b}$$

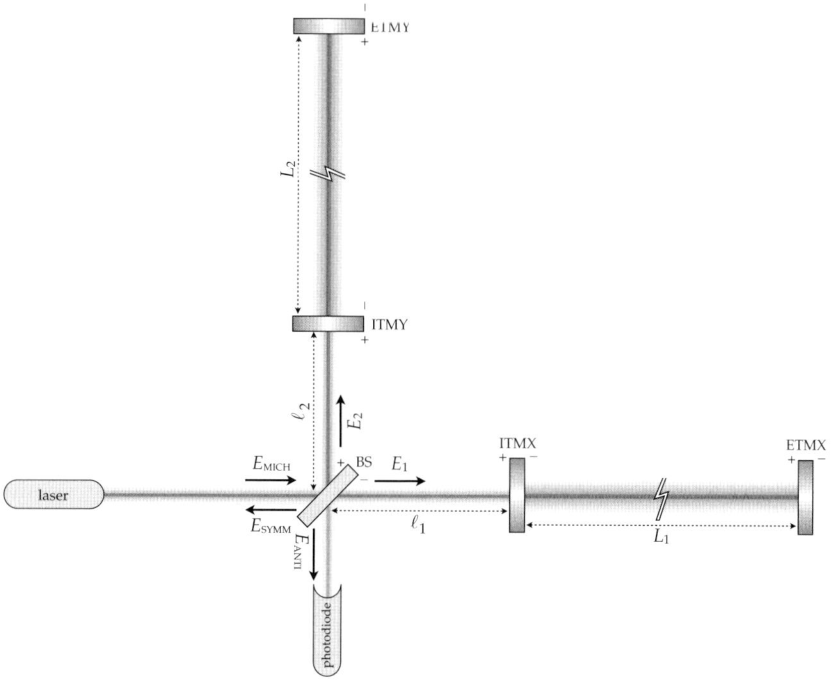

Figure 6.7 Optical topology of a Fabry–Pérot Michelson interferometer.

When the light is resonant in the Fabry–Pérot arms, $r_{FP} = -1$ and

$$r_{FPMI,\,res} := r_{FPMI}(L_{1,res}, L_{2,res}) = -e^{-2ik\bar{\ell}}\cos(k\delta) \tag{6.45a}$$

and

$$t_{FPMI,\,res} := t_{FPMI}(L_{1,res}, L_{2,res}) = -ie^{-2ik\bar{\ell}}\sin(k\delta)\,. \tag{6.45b}$$

When the cavities are displaced slightly from resonance,

$$\begin{aligned} &r_{FPMI}(L_{1,res} + \Delta L_1, L_{2,res} + \Delta L_2) \\ &= r_{FPMI,\,res} + \frac{1}{2}e^{-2ik\bar{\ell}}G_{arm}\left(2ik\Delta L_2 e^{ik\delta} + 2ik\Delta L_1 e^{-ik\delta}\right) \end{aligned} \tag{6.46a}$$

and

$$\begin{aligned} &t_{FPMI}(L_{1,res} + \Delta L_1, L_{2,res} + \Delta L_2) \\ &= t_{FPMI,\,res} + \frac{1}{2}e^{-2ik\bar{\ell}}G_{arm}\left(2ik\Delta L_2 e^{ik\delta} - 2ik\Delta L_1 e^{-ik\delta}\right). \end{aligned} \tag{6.46b}$$

In a perfectly symmetric Michelson interferometer, amplitude and frequency noise in the input laser light would not be present in the light in the antisymmetric readout; this noise is transmitted entirely to the symmetric readout (see Problem 6.2). However, in ground-based interferometric detectors, slight differences in the optical properties of the two arms will allow some noise to creep into the antisymmetric output channel, so a highly stable laser is required. Fortunately the frequency noise in the laser can be stabilized in a feedback control system by measuring the frequency noise in the symmetric channel.

6.1.4
Power Recycling

The purpose of power recycling is to increase the light power in the interferometer, which, as we described before, will lead to an improved sensitivity. A power recycling mirror, PRM, is added between the laser and the beam splitter, which creates an effective cavity between the recycling mirror and the compound mirror representing the Fabry–Pérot Michelson interferometer. The laser light incident on and reflected by the power recycling cavity are E_{INC} and E_{REFL}, respectively, while within the recycling cavity the field E_{MICH} is incident on the interferometer, E_{SYMM} emerges from the interferometer (and is returned to the recycling mirror), and E_{ANTI} is the field that is effectively transmitted through the power-recycled Fabry–Pérot Michelson interferometer (see Figure 6.8). The following relations hold amongst these fields:

$$E_{MICH} = t_{PRM}E_{INC} - r_{PRM}E_{SYMM}\,, \tag{6.47a}$$

$$E_{REFL} = r_{PRM}E_{INC} + t_{PRM}E_{SYMM}\,, \tag{6.47b}$$

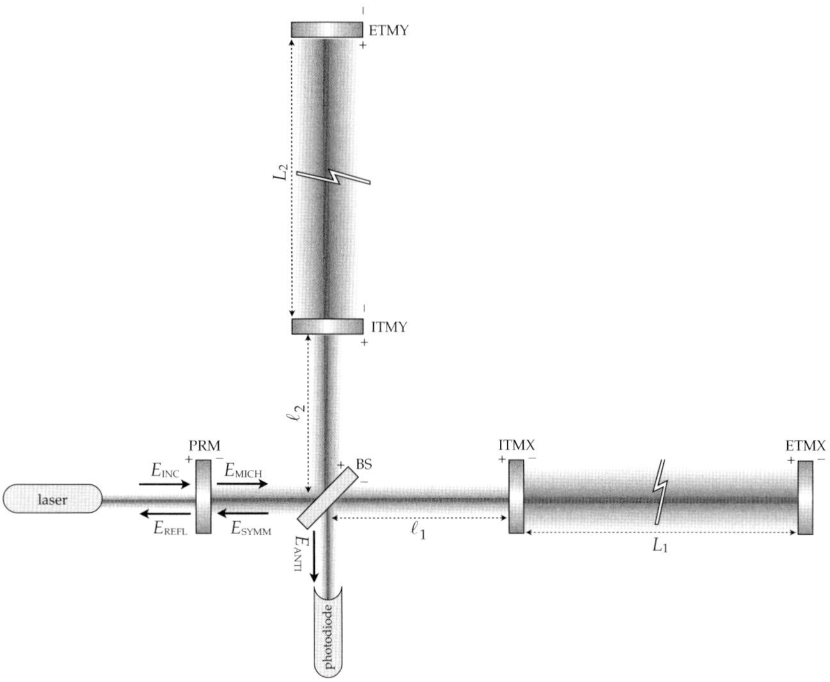

Figure 6.8 Optical topology of a power-recycled Fabry–Pérot Michelson interferometer.

$$E_{\rm SYMM} = r_{\rm FPMI}\, E_{\rm MICH}\,, \tag{6.47c}$$

$$E_{\rm ANTI} = t_{\rm FPMI}\, E_{\rm MICH}\,. \tag{6.47d}$$

From these it can be shown that

$$\frac{E_{\rm MICH}}{E_{\rm INC}} = \frac{t_{\rm PRM}}{1 + r_{\rm PRM}\, r_{\rm FPMI}}\,, \tag{6.48a}$$

$$r_{\rm PRFPMI} := \frac{E_{\rm REFL}}{E_{\rm INC}} = \frac{r_{\rm PRM} + r_{\rm FPMI}}{1 + r_{\rm PRM}\, r_{\rm FPMI}}\,, \tag{6.48b}$$

and

$$t_{\rm PRFPMI} := \frac{E_{\rm ANTI}}{E_{\rm INC}} = \frac{t_{\rm PRM}\, t_{\rm FPMI}}{1 + r_{\rm PRM}\, r_{\rm FPMI}}\,. \tag{6.48c}$$

To achieve a power build-up within the recycling cavity, we wish to maximize the ratio $E_{\rm MICH}/E_{\rm INC}$. When the light is resonant in the Fabry–Pérot cavities, we have $r_{\rm FPMI,\,res} = -\exp(-2ik\bar{\ell})\cos(k\delta)$. We choose $\bar{\ell}$ and δ so that, for the light at the carrier frequency, $\exp(-2ik\bar{\ell}) = 1$ and $\cos(k\delta) = 1$; therefore we have $r_{\rm FPMI,\,res} = -1$. This results in a power build up of

$$\left|\frac{E_{\rm MICH}}{E_{\rm INC}}\right|^2 = \left(\frac{t_{\rm PRM}}{1 - r_{\rm PRM}}\right)^2\,. \tag{6.49}$$

It would seem that the power recycling gain can be made arbitrarily great simply by letting $r_{\text{PRM}} \to 1$. In fact, this cannot be done in practice due to light loss (primarily) from the Fabry–Pérot cavity. The absorptivities in the optical components are ~few ppm, but, after hundreds of bounces within the Fabry–Pérot cavity, ~ 1% of the light may be lost. The result is that $r_{\text{FPMI, res}}$ is close to -1, but not quite. Suppose the effective absorptivity of the Fabry–Pérot Michelson interferometer is A_{FPMI}. Then

$$r_{\text{FPMI, res}}^2 + A_{\text{FPMI}} = 1 \tag{6.50}$$

(there is no transmission through the cavity). The recycling power gain is

$$\begin{aligned} G_{\text{prc}} &:= \left| \frac{E_{\text{MICH}}}{E_{\text{INC}}} \right|^2 = \left(\frac{t_{\text{PRM}}}{1 + r_{\text{PRM}} r_{\text{FPMI}}} \right)^2 \\ &= \frac{1 - r_{\text{PRM}}^2}{(1 + r_{\text{PRM}} r_{\text{FPMI}})^2} \,. \end{aligned} \tag{6.51}$$

To maximize the gain, the reflectivity of the recycling mirror must be chosen so that $r_{\text{PRM}} = -r_{\text{FPMI, res}}$; doing so yields a power recycling gain of

$$G_{\text{prc}} = \frac{1}{A_{\text{FPMI}}} \,. \tag{6.52}$$

The overall sensitivity of the interferometer is therefore limited by the losses in the optical components. For Initial LIGO, $G_{\text{prc}} \simeq 50$.

An interesting consequence of the choice $r_{\text{PRM}} = -r_{\text{FPMI, res}}$ is that, when the light is resonant in the arms, we have $r_{\text{PRFPMI, res}} = t_{\text{PRFPMI, res}} = 0$. That is, every photon entering the interferometer eventually ends up being absorbed by some optical component; there is no light transmitted to either the antisymmetric (readout) port, or to the reflected port.

6.1.5
Readout

As we have just seen, the antisymmetric port of the power-recycled Fabry–Pérot Michelson interferometer is dark when the light is resonant in the interferometer. This poses a problem because the readout device is a photodiode that measures the intensity of the light at the antisymmetric port. The photocurrent that is measured is

$$I_{\text{AS}} \propto |E_{\text{ANTI}}|^2 \,, \tag{6.53}$$

where

$$E_{\text{ANTI}} = E_{\text{INC}} G_{\text{prc}}^{1/2} t_{\text{FPMI}}(L_1, L_2) \,. \tag{6.54}$$

For our choice of $\bar{\ell}$ and δ such that $\exp(-2ik\bar{\ell}) = 1$ and $\cos(k\delta) = 1$ we have

$$t_{\text{FPMI}}(L_{1,\text{res}}, L_{2,\text{res}}) = 0 \tag{6.55}$$

and

$$t_{\mathrm{FPMI}}(L_{1,\mathrm{res}} + \Delta L_1, L_{2,\mathrm{res}} + \Delta L_2) = -i\, G_{\mathrm{arm}} k(\Delta L_1 - \Delta L_2) \tag{6.56}$$

so, therefore,

$$E_{\mathrm{ANTI}} = -i\, E_{\mathrm{INC}} G_{\mathrm{prc}}^{1/2} G_{\mathrm{arm}} k \Delta L\,, \tag{6.57}$$

where

$$\Delta L := \Delta L_1 - \Delta L_2 \tag{6.58}$$

is the differential length displacement of the arms from their resonant length.

The photocurrent is thus quadratic in the differential length displacement:

$$I_{\mathrm{AS}} = I_0 G_{\mathrm{prc}} G_{\mathrm{arm}}^2 (k\Delta L)^2\,, \tag{6.59}$$

where I_0 is the power of the laser light ($|E_{\mathrm{INC}}|^2$). This poses a problem in the detection of a gravitational wave: the wave produces only a slight change in the length of the arm, and the change in the photocurrent is *quadratic* in that small quantity. In fact, the real problem is that imperfections in the alignment of the optical components will always result is a small amount of stray light leaking into the antisymmetric port; this *contrast defect* means that one would have to measure a change in the photocurrent that is greater than the light power leaking out of the antisymmetric port.

In Initial LIGO, the laser light input into the interferometer is modulated at radio frequencies (RF) in such a way that the RF sidebands on the carrier light are anti-resonant in the Fabry–Pérot cavities and are transmitted (via the Schnupp asymmetry) to the antisymmetric port. Then, when there is a differential change in the arm length, the carrier light that leaves through the antisymmetric port will beat against the RF sidebands and can be demodulated to form a measurable signal.

In Enhanced LIGO, a simpler readout scheme is used (though the RF modulation is still used in locking the interferometer): the differential arm length ΔL is deliberately kept near some small offset ΔL_0, which allows a little light to escape the interferometer through the antisymmetric port. Then, when there is a small shift x from the offset ΔL_0, $\Delta L = \Delta L_0 + x$,

$$\frac{I_{\mathrm{AS}}}{I_0} = G_{\mathrm{prc}} G_{\mathrm{arm}}^2 \left[(k\Delta L_0)^2 + 2(k\Delta L_0)(kx) + O(x^2)\right] + C_{\mathrm{d}}\,, \tag{6.60}$$

where C_{d} is the contrast defect.

To determine whether a signal that produces a differential change in the lengths of the arms x is detectable, it is important to consider the noise contributions inherent in the readout I_{AS}. The most fundamental source of noise (though not the dominant source of noise for low frequency gravitational-wave signals) is shot noise:

$$S_I = 2\hbar ck\, I_{\mathrm{AS}} = 2\hbar ck\, I_0 G_{\mathrm{prc}} G_{\mathrm{arm}}^2 (k\Delta L_0)^2\,, \tag{6.61}$$

where we have assumed that the contrast defect is small compared to the fringe offset, $k\Delta L_0$, and we also assume that the small differential length change x produced by a gravitational wave, if present, would also have an insignificant effect on the level of the shot noise. On the other hand, the signal we are trying to measure, the gravitational-wave-induced differential strain on the detector, $h = x/L$, generates a readout I_{AS}

$$I_{\text{AS}} = Ch + \text{constant}\,, \tag{6.62}$$

where

$$C = 2 I_0 G_{\text{prc}} G_{\text{arm}}^2 (k\Delta L_0)(kL) \tag{6.63}$$

is known as the sensing function. In strain-equivalent units, the noise power spectrum is

$$S_h = \frac{S_I}{|C|^2} = \frac{\hbar c k}{2 I_0 G_{\text{prc}} G_{\text{arm}}^2 (kL)^2}\,. \tag{6.64}$$

The shot noise, in strain-equivalent units, decreases with increasing laser power, recycling cavity gain, arm cavity gain, and arm length. Another expression for the shot noise in terms of the power incident on the beam-splitter, $I_{\text{BS}} = I_0 G_{\text{prc}}$, and the light storage time, τ_{s} (which is related to the arm gain by $G_{\text{arm}} \approx 2c\tau_{\text{s}}/L$), is

$$S_h^{1/2} = \sqrt{\frac{\pi\hbar\lambda_{\text{laser}}}{I_{\text{BS}}c}\frac{1}{4\pi\tau_{\text{s}}}}\,. \tag{6.65}$$

The square-root of the power spectral density, $S_h^{1/2}(f)$, is known as the *amplitude strain sensitivity*.

Recall that under optimal conditions, the power recycling cavity gain is determined by the light losses in the instrument, A_{FPMI}, via $G_{\text{prc}} = 1/A_{\text{FPMI}}$. These losses are the accumulated result of the loss in each optical component. Roughly speaking, for each traversal of the Fabry–Pérot cavity, a photon encounters two optical components (the end mirrors) and, on average, a photon will make G_{arm} traversals. It can be shown that $A_{\text{FPMI}} = 2G_{\text{arm}}A$ where A is the absorptivity of each optical component. The amplitude strain sensitivity can therefore be expressed as

$$S_h^{1/2} = \frac{\lambda_{\text{laser}}}{L}\sqrt{\frac{\hbar c/\lambda_{\text{laser}}}{2\pi I_0}\frac{A}{G_{\text{arm}}}}\,. \tag{6.66}$$

For Initial LIGO, $\lambda_{\text{laser}} \simeq 1\,\mu\text{m}$, $I_0 \simeq 5\,\text{W}$, $A \sim$ few $\times\, 10\,\text{ppm}$, $G_{\text{arm}} \simeq 140$, and $L = 4000\,\text{m}$. This yields a value of $S_h^{1/2} \sim 10^{-23}\,\text{Hz}^{-1/2}$.

The main free parameter that enters into this sensitivity estimate – once the characteristics of the laser (I_0 and λ_{laser}), the quality of the optics (A), and the size of the facility (L) are chosen – is the arm cavity gain G_{arm}, or, equivalently, the reflectivity of the ITM mirrors, r_{ITM}. It would seem preferable to make this quantity as large as possible. However, as we will see in the next section, as G_{arm} increases, the bandwidth of the interferometer – those frequencies of gravitational-wave signals to which the detector is sensitive – decreases (for fixed absorptivity A). At low frequencies, the dominant noise sources are *not* shot noise, so tuning the instrument to have high sensitivity to low frequency gravitational waves is a false optimization. To quantify this, we now need to compute the frequency response of the interferometer.

Example 6.5 Radio-frequency readout

Initial LIGO used a *radio-frequency readout* or *RF readout* scheme. The light that is incident on the interferometer is modulated at radio frequencies ω_{RF} so that

$$E_{\text{INC}} = E_0 e^{i\Gamma \cos \omega_{\text{RF}} t} \tag{6.67}$$

where Γ is the (small) modulation depth and E_0 is the electric field amplitude of the laser light. Recalling that $\exp[\frac{1}{2}x(t-t^{-1})] = \sum_{n=-\infty}^{\infty} J_n(x)t^n$ is the generating function for Bessel functions, we see that

$$\begin{aligned} E_{\text{INC}} &= E_0 \sum_{n=-\infty}^{\infty} J_n(\Gamma)\left(i e^{i\omega_{\text{RF}} t}\right)^n \\ &= E_0 \left\{J_0(\Gamma) + i J_1(\Gamma) e^{i\omega_{\text{RF}} t} - i J_{-1}(\Gamma) e^{-i\omega_{\text{RF}} t}\right\} + O(\Gamma^2) \\ &= E_0 \left\{J_0(\Gamma) + i J_1(\Gamma) e^{i\omega_{\text{RF}} t} + i J_1(\Gamma) e^{-i\omega_{\text{RF}} t}\right\} + O(\Gamma^2) \,. \end{aligned} \tag{6.68}$$

The first term represents the light at the carrier frequency and the next two terms represent two sidebands on this carrier light: the upper sideband has frequency $\omega_1 = \omega_0 + \omega_{\text{RF}}$ and the lower sideband has frequency $\omega_{-1} = \omega_0 - \omega_{\text{RF}}$. Here, ω_0 is the frequency of the carrier light. Note that $J_1(\Gamma) \simeq \frac{1}{2}\Gamma$ for $\Gamma < 1$.

In the RF readout scheme, the lengths of the Fabry–Pérot arms are chosen so that the carrier light is resonant in the cavities and the modulation frequency is chosen so that the sideband light is anti-resonant. This renders the upper and lower sideband light essentially insensitive to any small change in the length of the Fabry–Pérot cavity, and hence these sidebands will not be affected by a gravitational-wave signal. We can therefore ignore the Fabry–Pérot cavities when considering the light in the sidebands.

The lengths of the Michelson interferometer arms are designed so that there is no light at the carrier frequency transmitted to the antisymmetric port. However, as a result of the Schnupp asymmetry, some sideband light *is* transmitted. The reflectivity and transmissivity of the Michelson interferometer for the sideband light is

$$r_{\text{FPMI},\,k=k_0 \pm k_{\text{RF}}} = e^{\mp 2 i k_{\text{RF}} \bar{\ell}} \cos(k_{\text{RF}}\delta) \tag{6.69a}$$

and

$$t_{\text{FPMI},\,k=k_0 \pm k_{\text{RF}}} = +i e^{\mp 2 i k_{\text{RF}} \bar{\ell}} \sin(k_{\text{RF}}\delta) \,, \tag{6.69b}$$

where $k_0 = \omega_0/c$ and $k_{RF} = \omega_{RF}/c$. The addition of a recycling mirror to the Michelson interferometer yields some build-up of the sideband light within the interferometer, but this is limited by the light that is now transmitted to the antisymmetric port. The average length of the Michelson arms $\bar{\ell}$ is chosen so that $\exp(-2ik_{RF}\bar{\ell}) = -1$; the effective reflection and transmission coefficients for the sideband light are then

$$r_{PRFPMI,\,k=k_0\pm k_{RF}} = \frac{r_{PRM} - \cos(k_{RF}\delta)}{1 - r_{PRM}\cos(k_{RF}\delta)} \tag{6.70a}$$

and

$$\pm t_{sb} := t_{PRFPMI,\,k=k_0\pm k_{RF}} = \frac{\mp i t_{PRM}\sin(k_{RF}\delta)}{1 - r_{PRM}\cos(k_{RF}\delta)}\,, \tag{6.70b}$$

where in the second equation we define t_{sb}: the sideband transmission to the antisymmetric port.

The electric field at the antisymmetric port now has contributions from both the carrier field (which is zero when $\Delta L = 0$) as well as the upper- and lower-radio-frequency sidebands:

$$\begin{aligned} E_{ANTI} &= E_{ANTI,\,k=k_0} + E_{ANTI,\,k=k_0+k_{RF}} + E_{ANTI,\,k=k_0-k_{RF}} \\ &= E_0\left\{-iJ_0(\Gamma)G_{prc}^{1/2}G_{arm}k_0\Delta L + iJ_1(\Gamma)e^{i\omega_{RF}t}t_{sb} - iJ_1(\Gamma)e^{-i\omega_{RF}t}t_{sb}\right\}\,. \end{aligned} \tag{6.71}$$

The intensity on the photodiode located at the antisymmetric port now contains (i) a zero-frequency (DC) component that is effectively insensitive to gravitational waves (it is quadratic in the small change ΔL); (ii) cross-terms between the carrier light and the sideband light that are linear in ΔL and exhibit beating at the radio frequency; and (iii) sideband-sideband beating at twice the radio frequency, which is also insensitive to gravitational waves:

$$\begin{aligned} I_{ANTI} &= |E_{ANTI,\,k=k_0} + E_{ANTI,\,k=k_0+k_{RF}} + E_{ANTI,\,k=k_0-k_{RF}}|^2 \\ &= |E_{ANTI,\,k=k_0}|^2 + |E_{ANTI,\,k=k_0+k_{RF}}|^2 + |E_{ANTI,\,k=k_0-k_{RF}}|^2 \\ &\quad + \left[E^*_{ANTI,\,k=k_0}(E_{ANTI,\,k=k_0+k_{RF}} + E_{ANTI,\,k=k_0-k_{RF}}) + \text{cc}\right] \\ &\quad + E^*_{ANTI,\,k=k_0+k_{RF}}E_{ANTI,\,k=k_0-k_{RF}} + E_{ANTI,\,k=k_0+k_{RF}}E^*_{ANTI,\,k=k_0-k_{RF}} \\ &= (\text{DC terms}) \\ &\quad + 4|E_0|^2 J_0(\Gamma)J_1(\Gamma)|t_{sb}|G_{prc}^{1/2}G_{arm}k_0\Delta L\sin(\omega_{RF}t) \\ &\quad + (\text{sideband-sideband cross-terms})\,. \end{aligned} \tag{6.72}$$

A change in the differential length of the interferometer arms, ΔL, induces beating between the carrier light and the sideband light at the radio frequency. The photodiode output is demodulated by mixing the signal with the same RF oscillator that put the sidebands on the incident light. This produces two antisymmetric

(AS) readout signals, an in-phase signal and a quadrature-phase signal:

$$\mathrm{AS_I} := \frac{1}{T}\int_0^T I_{\mathrm{ANTI}}\cos(\omega_{\mathrm{RF}}t)\,dt\,, \tag{6.73}$$

$$\mathrm{AS_Q} := \frac{1}{T}\int_0^T I_{\mathrm{ANTI}}\sin(\omega_{\mathrm{RF}}t)\,dt\,, \tag{6.74}$$

where T is an integration time of many cycles of the radio frequency beating. Notice that a gravitational wave will induce no signal in the $\mathrm{AS_I}$ output, but it will induce a signal in the $\mathrm{AS_Q}$ output:

$$\mathrm{AS_Q} = 2I_0 J_0(\Gamma)J_1(\Gamma)|t_{\mathrm{sb}}|G_{\mathrm{prc}}^{1/2}G_{\mathrm{arm}}k_0\Delta L\,. \tag{6.75}$$

From this equation we can identify the sensing function for the RF-readout scheme:

$$\mathrm{AS_Q} = Ch \tag{6.76}$$

with

$$C = 2I_0 J_0(\Gamma)J_1(\Gamma)|t_{\mathrm{sb}}|G_{\mathrm{prc}}^{1/2}G_{\mathrm{arm}}(k_0 L)\,. \tag{6.77}$$

The shot noise in the photodiode is proportional to $|E_{\mathrm{ANTI}}|^2$, which is non-stationary due to the demodulation procedure. It is

$$\begin{aligned} S_{\mathrm{AS}} &= 2\times\frac{3}{2}\times 2\hbar c k_0|E_{\mathrm{ANTI}}|^2 \\ &= 2\times\frac{3}{2}\times 2\hbar c k_0 I_0 J_1^2(\Gamma)|t_{\mathrm{sb}}|^2\,, \end{aligned} \tag{6.78}$$

where the factor of 3/2 corrects for the non-stationarity of the shot noise, and the folding of the frequencies in the down-conversion results in the factor of 2. The resulting strain-equivalent noise power spectrum is

$$S_h = \frac{S_{\mathrm{AS}}}{|C|^2} = \frac{3\hbar c k_0}{2I_0 J_0^2(\Gamma)G_{\mathrm{prc}}G_{\mathrm{arm}}^2(k_0 L)^2}\,. \tag{6.79}$$

This is the same value as we found for the DC readout scheme apart from the factor of three (and the factor of $J_0^{-2}(\Gamma)\approx 1$).

6.1.6
Frequency Response of the Initial LIGO Detector

So far we have considered only the response of the LIGO interferometer to very low-frequency gravitational waves. Here we now consider the frequency response of the

LIGO detector. The response of LIGO is essentially determined by the response of the Fabry–Pérot cavity to a small change in length so it will be sufficient to consider the frequency response of the Fabry–Pérot cavity.

Suppose that an externally-induced motion of the end mirror produces a time-dependent phase shift $\phi_{\text{ext}}(t)$ in the returning cavity light at the input mirror.[5] The field in the Fabry–Pérot interferometer, $E_{\text{FP}\triangleleft}$, satisfies the relation

$$E_{\text{FP}\triangleleft}(t) = -r_{\text{ITM}} e^{-2ikL - i\phi_{\text{ext}}(t)} E_{\text{FP}\triangleleft}(t - 2L/c) + t_{\text{ITM}} e^{-2ikL - i\phi_{\text{ext}}(t)} E_{\text{I}} \,, \quad (6.80)$$

where we are now evaluating the returning (incident on the ITM) field in the Fabry–Pérot cavity (recall that before we evaluated the $E_{\text{FP}\triangleright}$ as the outbound field at the ITM). The slow time-dependence of the field in the interferometer is generated by the small externally-induced phase shift, and this generates sidebands on the carrier frequency of the light. The light at the carrier frequency is constant in time so we can write

$$E_{\text{FP}\triangleleft}(t) = E_{\text{FP}\triangleleft,0} + E'_{\text{FP}\triangleleft}(t) \,, \quad (6.81)$$

where $E'_{\text{FP}\triangleleft}(t)$ contains all the time dependence and is first-order in the small external phase shift $\phi_{\text{ext}}(t)$. Thus we can separate the time-dependent and time-independent parts of field in the Fabry–Pérot cavity:

$$E_{\text{FP}\triangleleft,0} = -r_{\text{ITM}} e^{-2ikL} E_{\text{FP}\triangleleft,0} + t_{\text{ITM}} e^{-2ikL} E_{\text{I}} \quad (6.82)$$

and

$$\begin{aligned} E'_{\text{FP}\triangleleft}(t) &\simeq -r_{\text{ITM}} e^{-2ikL} E'_{\text{FP}\triangleleft}(t - 2L/c) \\ &\quad + i r_{\text{ITM}} e^{-2ikL} E_{\text{FP}\triangleleft,0} \phi_{\text{ext}}(t) - i t_{\text{ITM}} e^{-2ikL} E_{\text{I}} \phi_{\text{ext}}(t) \\ &= -r_{\text{ITM}} e^{-2ikL} E'_{\text{FP}\triangleleft}(t - 2L/c) - i E_{\text{FP}\triangleleft,0} \phi_{\text{ext}}(t) \,. \end{aligned} \quad (6.83)$$

The first of these equations recovers our previous equation for the field in the Fabry–Pérot cavity at the carrier frequency,

$$E_{\text{FP}\triangleleft,0} = -\frac{t_{\text{ITM}}}{1 - r_{\text{ITM}}} E_{\text{I}} \,, \quad (6.84)$$

where we have set $\exp(-2ikL) = -1$, which is the condition for the carrier to be resonant in the cavity. Now, the Fourier transform of the expression for E'_{FP} results in

$$\tilde{E}'_{\text{FP}\triangleleft}(f) \simeq +r_{\text{ITM}} e^{-4\pi i f L/c} \tilde{E}'_{\text{FP}\triangleleft}(f) - i E_{\text{FP}\triangleleft,0} \tilde{\phi}_{\text{ext}}(f) \,, \quad (6.85)$$

which can be solved to obtain

$$\tilde{E}'_{\text{FP}\triangleleft}(f) \simeq i \frac{t_{\text{ITM}}}{1 - r_{\text{ITM}}} \frac{1}{1 - r_{\text{ITM}} e^{-4\pi i f L/c}} \tilde{\phi}_{\text{ext}}(f) E_{\text{I}} \,. \quad (6.86)$$

5) For low frequency changes in the cavity length, so that the length of the cavity is constant over timescales longer than the light storage time τ_s, then $\phi_{\text{ext}}(t) = 2k\Delta L(t)$.

The reflected field, $E_R(t)$, is also time dependent and can be expressed as a sum of the reflected field at the carrier frequency, $E_{R,0}$, and the sideband caused by the external motion, $E'_R(t)$,

$$E_R(t) = E_{R,0} + E'_R(t)\,. \tag{6.87}$$

At the carrier frequency we have

$$E_{R,0} = r_{ITM}\,E_I + t_{ITM}\,E_{FP\triangleleft,0} = -E_I\,, \tag{6.88}$$

and at the sideband frequencies we have

$$\tilde{E}'_R(f) = t_{ITM}\,\tilde{E}'_{FP\triangleleft}(f) = i\frac{t^2_{ITM}}{1-r_{ITM}}\frac{1}{1-r_{ITM}e^{-4\pi i f L/c}}\tilde{\phi}_{ext}(f)\,E_I\,. \tag{6.89}$$

Therefore, the effective reflectivity of the Fabry–Pérot cavity has a frequency-dependent component induced by the external motion:

$$\begin{aligned} r_{FP}(f) &= \frac{\tilde{E}_R(f)}{E_I} \\ &= -1 + i\frac{t^2_{ITM}}{1-r_{ITM}}\frac{1}{1-r_{ITM}e^{-4\pi i f L/c}}\tilde{\phi}_{ext}(f) \\ &\simeq -\exp\left\{-i\,G_{arm}\frac{1-r_{ITM}}{1-r_{ITM}e^{-4\pi i f L/c}}\tilde{\phi}_{ext}(f)\right\}, \end{aligned} \tag{6.90}$$

where G_{arm} is the Fabry–Pérot gain. It is convenient to express the frequency-dependent factor multiplying $\tilde{\phi}_{ext}(f)$ as a normalized sensing transfer function $\hat{C}_{FP}(f)$ so that

$$r_{FP}(f) = -\exp\left\{-i\,G_{arm}\,\hat{C}_{FP}(f)\tilde{\phi}_{ext}(f)\right\}, \tag{6.91}$$

where the normalized sensing transfer function is

$$\begin{aligned} \hat{C}_{FP}(f) &= \frac{1-r_{ITM}}{1-r_{ITM}e^{-4\pi i f L/c}} \\ &= e^{2\pi i f L/c}\frac{\sinh(2\pi f_{pole}\,L/c)}{\sinh\left[(2\pi f_{pole}\,L/c)(1+i\,f/f_{pole})\right]} \\ &\simeq \frac{1+i\,f/f_{zero}}{1+i\,f/f_{pole}} \qquad f \ll f_{FSR} \end{aligned} \tag{6.92}$$

and

$$f_{zero} := \frac{f_{FSR}}{\pi} = \frac{c}{2\pi L}\,, \tag{6.93}$$

$$f_{pole} := \frac{|\ln r_{ITM}|}{4\pi L/c} \simeq \frac{1-r_{ITM}}{1+r_{ITM}}\frac{c}{2\pi L} = \frac{f_{zero}}{G_{arm}} \simeq \frac{1}{4\pi\tau_s} \tag{6.94}$$

are the Fabry–Pérot arm cavity zero- and pole-frequencies when the Fabry–Pérot cavity is approximated as a zero-pole filter. In Initial LIGO, $f_{pole} = 85\,\text{Hz}$ and for the 4 km LIGO interferometers, $f_{\text{zero}} = 12\,\text{kHz}$. Notice that for a small, zero-frequency offset of the mirror, $\phi_{\text{ext}} = 2k\Delta L$, we recover our earlier result $r_{\text{FP}} = -\exp(-2ik\,G_{\text{arm}}\Delta L)$ in the low-frequency limit.

The power-recycled Fabry–Pérot Michelson interferometer is modelled just as it was previously, for the zero-frequency response, but now we simply replace $2k\Delta L$ with $\hat{C}_{\text{FP}}(f)\tilde{\phi}_{\text{ext}}(f)$ and express all quantities that depend on these as functions of frequency. The field at the antisymmetric port is now

$$\tilde{E}_{\text{ANTI}} = -\frac{1}{2} i\, E_{\text{INC}}\, G_{\text{prc}}^{1/2}\, G_{\text{arm}}\, \hat{C}_{\text{FP}}(f)\tilde{\phi}_{\text{ext}}(f)\,, \tag{6.95}$$

where now

$$\phi_{\text{ext}} := \phi_{\text{ext},1} - \phi_{\text{ext},2} \tag{6.96}$$

is the externally-induced *differential* phase shift between the two arms. As before, the readout system is based on the photocurrent in the antisymmetric port, $I_{\text{AS}} = |E_{\text{ANTI}}|^2$. In the long-wavelength limit, when the wavelength of the gravitational wave is much larger than the arm length of the Fabry–Pérot cavities, or, equivalently, when the frequency of the gravitational wave is smaller than f_{FSR}, then $\tilde{\phi}_{\text{ext}}(f) \simeq 2kL\tilde{h}(f)$ and

$$I_{\text{AS}}(f) = C(f)\tilde{h}(f) + \text{constant} \tag{6.97}$$

where

$$\begin{aligned} C(f) &= 2I_0\, G_{\text{prc}}\, G_{\text{arm}}^2 (k\Delta L_0)(kL)\hat{C}_{\text{FP}}(f) \\ &\simeq 2I_0\, G_{\text{prc}}\, G_{\text{arm}}^2 (k\Delta L_0)(kL)\frac{1}{1 + i f/f_{\text{pole}}} \quad (f \ll f_{\text{FSR}})\,. \end{aligned} \tag{6.98}$$

The shot noise on the photodiode remains $S_I = 2\hbar c k\, I_0\, G_{\text{prc}}\, G_{\text{arm}}(k\Delta L_0)^2$, which is frequency-independent, and so the noise power spectrum in strain-equivalent units is

$$\begin{aligned} S_h(f) = \frac{S_I}{|C(f)|^2} &\simeq \frac{\hbar c k}{2I_0\, G_{\text{prc}}\, G_{\text{arm}}^2 (kL)^2}\left[1 + (f/f_{\text{pole}})^2\right] \\ &\simeq \frac{\pi\hbar\lambda_{\text{laser}}}{I_{\text{BS}}\, c}\,\frac{1 + (4\pi f \tau_s)^2}{(4\pi\tau_s)^2} \quad (f \ll f_{\text{FSR}})\,. \end{aligned} \tag{6.99}$$

For an optimally coupled recycling cavity, $G_{\text{prc}} = 1/(2G_{\text{arm}}A)$, the amplitude strain sensitivity is

$$S_h^{1/2}(f) \simeq \frac{\lambda_{\text{laser}}}{L}\sqrt{\frac{\hbar c/\lambda_{\text{laser}}}{2\pi I_0}\,\frac{A}{G_{\text{arm}}}}\,\sqrt{1 + (f/f_{\text{pole}})^2} \quad (f \ll f_{\text{FSR}})\,. \tag{6.100}$$

At low frequencies $f \ll f_{\text{pole}}$, this approaches the zero-frequency limit obtained earlier (of course); for larger frequencies the amplitude strain sensitivity increases, that is, the detector is less sensitive to gravitational waves, because of the Fabry–

Pérot cavity transfer function. The arm gain G_{arm} and the cavity pole frequency f_{pole} are not independently tunable quantities as they both are determined by the reflectivity of the ITM. In fact, $f_{\text{pole}} \simeq c/(2\pi L G_{\text{arm}})$. This means that we can express the amplitude strain sensitivity as

$$S_h^{1/2}(f) \simeq \sqrt{A\frac{\hbar}{I_0}\frac{\lambda_{\text{laser}}}{L}}\sqrt{f_{\text{pole}}[1+(f/f_{\text{pole}})^2]} \quad (f \ll f_{\text{FSR}})\,. \tag{6.101}$$

This shows that the sensitivity of the instrument is limited by (i) the absorptivity A of the optics, (ii) the laser power I_0, and (iii) the ratio λ_{laser}/L of the wavelength of the laser light to the arm length. The sensitivity also depends on the choice of the arm cavity pole frequency f_{pole}: for fixed A and assuming an optimally coupled power recycling cavity, small values of this frequency will give high sensitivity at frequencies $f \ll f_{\text{pole}}$ but with a relatively narrow band of sensitivity, while large values of f_{pole} give less sensitivity at zero frequency but more broad band sensitivity. This is shown in Figure 6.9.

The choice of the value of f_{pole} is made by considering other sources of instrumental noise. At high frequencies, the dominant source of noise is the shot noise that we have considered so far; however, at low frequencies, there are several other sources of noise (principally seismic and thermal noise in Initial LIGO) that dominate. Therefore, choosing a small value of f_{pole} in order to maximize sensitivity at low frequencies is a false optimization – the sensitivity at low frequencies will still be dominated by seismic noise, and the instrument will have such a narrow band that it will not be sensitive at higher frequencies either. Therefore, it is best to choose a value for f_{pole} that is just around the point when the shot noise becomes comparable to the other sources of low-frequency noise. The dominant sources of noise will be considered in the next two sections.

At all frequencies, including high frequencies (beyond the long-wavelength limit), the sensing function is

$$I_{\text{AS}}(f) = C(f)\frac{\tilde{\varphi}_{\text{ext}}(f)}{2kL} + \text{constant}\,, \tag{6.102}$$

with

$$C(f) = 2I_0 G_{\text{prc}} G_{\text{arm}}^2 (k\Delta L_0)(kL)\hat{C}_{\text{FP}}(f) \tag{6.103}$$

$$= 2I_0 G_{\text{prc}} G_{\text{arm}}^2 (k\Delta L_0)(kL) e^{2\pi i f L/c} \frac{\sinh(2\pi f_{\text{pole}} L/c)}{\sinh\left[(2\pi f_{\text{pole}} L/c)\,(1 + i f/f_{\text{pole}})\right]}\,. \tag{6.104}$$

For gravitational-wave frequencies comparable to f_{FSR}, the relationship between $\tilde{\varphi}_{\text{ext}}(f)$ and $\tilde{h}(f)$ is not simple as it depends on a frequency-dependent function of the direction of the gravitational wave relative to the detector arms as will be shown in Section 6.1.10.

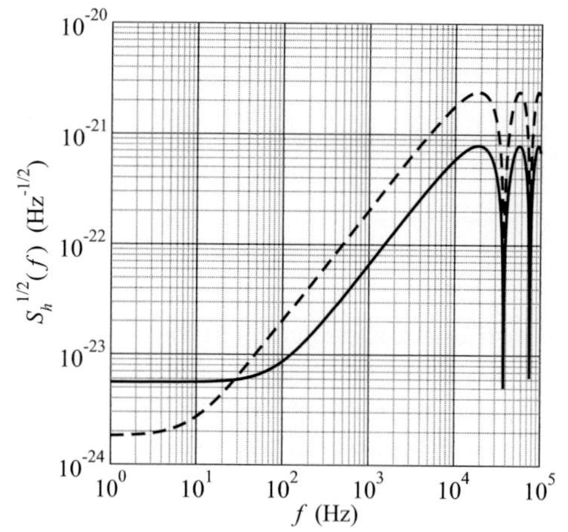

Figure 6.9 Amplitude spectral density of the shot noise in strain-equivalent units for Initial LIGO with a laser producing $I_0 = 5$ W of light at wavelength $\lambda = 1$ μm and with a recycling gain of $G_{\text{prc}} = 1/(2AG_{\text{arm}})$ where the absorptivity of the optics is taken to be $A = 70$ ppm. The solid curve is for ITM power transmission of $t^2_{\text{ITM}} = 2.8\%$ while the dashed curve is for $t^2_{\text{ITM}} = 0.3\%$.

6.1.7 Sensor Noise

We have seen that by changing the arm cavity gain (and hence the cavity pole frequency) there is a trade-off between the sensitivity at low frequencies and the bandwidth of the detector (provided that the power recycling cavity is optimally coupled). However, we have so far neglected an important source of noise associated with the interferometer readout which results from the back-reaction of the readout system on the positions of the test masses. If we were to decrease the shot noise by increasing the light power in the Fabry–Pérot cavities (which reduces the pole frequency), we would ultimately encounter a new source of noise at low frequencies resulting from the high power in the cavity buffeting the mirrors. The shot noise in the readout is therefore one of two types of sensor noise present in an interferometer: the other is known as *radiation pressure* noise and it describes this buffeting of the mirrors due to the fluctuations in the radiation field. The radiation pressure force on a mirror is

$$F_{\text{rad}} = \frac{2I}{c}, \tag{6.105}$$

where I is the light power incident on the mirror. We consider the mirror to be free, so that its position x is related to the induced force F_{rad} by $d^2x/dt^2 = F_{\text{rad}}/M$, or, in the frequency domain, $-4\pi^2 f^2 \tilde{x} = \tilde{F}_{\text{rad}}/M$, where M is the mass of the mirror. Then we find that the power spectral density of the fluctuations in the mirror position, $S_x(f)$, is related to the power spectral density of the fluctuations in the light power, $S_I(f)$, by

$$S_x(f) = \frac{1}{16\pi^4 f^4 M^2} \frac{4}{c^2} S_I(f) . \tag{6.106}$$

For a Fabry–Pérot cavity, the electric field incident on the cavity has two components: a constant radiation field $E_{\text{I},0} = \sqrt{I}$ associated with the incident carrier light of power I, along with (quantum) fluctuations at the sideband frequencies $\tilde{E}'_{\text{I}}(f)$, which have a power spectrum $S_{E_{\text{I}}} = \frac{1}{2}\hbar c k$. The field in the Fabry–Pérot cavity also has a carrier field $E_{\text{FP},0} = G_{\text{arm}}^{1/2} E_{\text{I},0}$ and a sideband field $\tilde{E}'_{\text{FP}}(f) = G_{\text{arm}}^{1/2} \hat{C}(f) \tilde{E}'_{\text{I}}(f)$; therefore the carrier power in the cavity is $I_{\text{FP},0} = I\, G_{\text{arm}}$ while the fluctuating sideband light power has a power spectrum $S_{I_{\text{FP}}}(f) = 2\hbar c k\, I\, G_{\text{arm}}^2 |\hat{C}(f)|^2$ (see Problem 6.4). The buffeting of one of the mirrors of the cavity by these light fluctuations causes random motions with power spectrum

$$S_x = \frac{1}{4\pi^4 f^4 M^2 c^2} S_{I_{\text{FP}}}(f) = \frac{\hbar c k\, I_0 G_{\text{prc}} G_{\text{arm}}^2 |\hat{C}(f)|^2}{4\pi^4 f^4 M^2 c^2} , \tag{6.107}$$

where we note the light power incident on a Fabry–Pérot cavity is half of the total light power within the recycling cavity, $I = \frac{1}{2} I_0 G_{\text{prc}}$.

Within the Fabry–Pérot cavity, a fluctuation in the light power affects both the ETM and ITM coherently, resulting in a change in the length of the cavity that is twice that of the displacement of a single mirror. The power spectrum for the length of a cavity is therefore *four* times the displacement spectrum for a single mirror. The two cavities experience different radiation pressure forces so the fluctuations in the lengths of the two cavities are uncorrelated. This means that the power spectrum for $\Delta L = \Delta L_1 - \Delta L_2$ is *twice* the power spectrum for the length fluctuations of a single cavity. In terms of strain-equivalent units, where $h := \Delta L/L$, we therefore have

$$S_{h,\text{rad}}(f) = \frac{8}{L^2} S_x(f) . \tag{6.108}$$

The resulting radiation pressure noise is

$$S_{h,\text{rad}}(f) = \frac{1}{L^2} \frac{2\hbar k\, I_0 G_{\text{prc}} G_{\text{arm}}^2 |\hat{C}_{\text{FP}}(f)|^2}{\pi^4 f^4 M^2 c} . \tag{6.109}$$

The readout shot noise was found to be

$$S_{h,\text{shot}}(f) = \frac{1}{L^2} \frac{\hbar c}{2k\, I_0 G_{\text{prc}} G_{\text{arm}}^2 |\hat{C}_{\text{FP}}(f)|^2} , \tag{6.110}$$

so between these two sources of noise the total *sensor noise* is

$$S_{h,\text{sensor}}(f) = \frac{\hbar}{\pi^2 f^2 M L^2} \cdot \left(\frac{\pi^2 f^2 M c}{2k\, I_0 G_{\text{prc}} G_{\text{arm}}^2 |\hat{C}_{\text{FP}}(f)|^2} + \frac{2k\, I_0 G_{\text{prc}} G_{\text{arm}}^2 |\hat{C}_{\text{FP}}(f)|^2}{\pi^2 f^2 M c} \right) . \tag{6.111}$$

At low frequencies, the radiation pressure noise will dominate the sensor noise while at higher frequencies it is the shot noise that will dominate. The shot noise can be reduced by increasing the laser power I_0, but this results in an increased radiation pressure noise. Therefore, at any fixed frequency at which we choose to optimize the sensitivity, $f = f_{\text{opt}}$, the optimal choice for the laser power that yields the smallest sensor noise is

$$I_{0,\text{opt}} = \frac{\pi^2 M c}{2k} \frac{1}{G_{\text{prc}} G_{\text{arm}}} \frac{f_{\text{opt}}^2}{|\hat{C}_{\text{FP}}(f_{\text{opt}})|^2} \tag{6.112}$$

and, at this frequency, the *standard quantum limit* (SQL) strain sensitivity is obtained

$$S_{h,\text{SQL}} := \frac{2\hbar}{\pi^2 f_{\text{opt}}^2 M L^2} . \tag{6.113}$$

Notice that this quantity is the minimum of the noise power spectrum for an interferometer tuned to the specified fixed frequency, but the frequency dependence in $S_{h,\text{SQL}}$ is not the spectral density of the sensor noise of such an interferometer.

An important point to notice about $S_{h,\text{SQL}}$ is that it depends on the physical parameters of the interferometer (its mass and length) and on Planck's constant, but not on any detail of the readout system (such as the laser power, the recycling cavity gain, the Fabry–Pérot cavity gain, or any other details of the interferometry) beyond the timescale on which the measurement is being made $\sim 1/f_{\text{opt}}$. This indicates that the standard quantum limit (as its name suggests) has its origin in quantum measurement theory rather than the particular method that interferometers use to make the measurement. However, having said this, it turns out it *is* possible to construct a readout scheme that can beat the standard quantum limit – the standard quantum limit is not a fundamental limit imposed by quantum mechanics, though it can be understood in terms of the Heisenberg uncertainty principle.

Example 6.6 Standard quantum limit

We can see how the standard quantum limit comes about from the Heisenberg uncertainty principle. In an interferometric gravitational-wave detector, one can imagine attempting to detect a gravitational wave of frequency f by repeatedly measuring the difference in arm lengths at time intervals $\tau \simeq 1/f$. Suppose that q_1 and q_2 represent the positions of the mirrors that make up one arm's Fabry–Pérot cavity and q_3 and q_4 represent the positions of the mirrors that make up the other arm's Fabry–Pérot cavity. Then, the observable that is being monitored and its conjugate momentum are

$$q = (q_1 - q_2) - (q_3 - q_4) , \tag{6.114}$$

$$p = \frac{1}{4}[(p_1 - p_2) - (p_3 - p_4)] = \frac{M}{4}\dot{q} , \tag{6.115}$$

where M is the mass of each mirror. The generalized coordinate and momentum satisfy the commutation relation $[p, q] = i\hbar$ and therefore the Heisenberg uncertainty principle $\Delta q \Delta p \geq \frac{1}{2}\hbar$. If the initial value of the observable is q then at some time τ later it is

$$q' = q + \dot{q}\tau = q + \frac{4\tau}{M} p \tag{6.116}$$

so

$$[q, q'] = i\hbar \frac{4\tau}{M} \tag{6.117}$$

and the Heisenberg uncertainty relationship is

$$\Delta q \Delta q' \geq \frac{2\hbar\tau}{M} . \tag{6.118}$$

The strain sensitivity is then

$$S_{h,\mathrm{SQL}} \sim \frac{(\Delta q)^2 \tau}{L^2} \sim \frac{(\Delta q')(\Delta q)\tau}{L^2} \sim \frac{2\hbar\tau^2}{ML^2} . \tag{6.119}$$

With $\tau \sim 1/f$ we see that this has the same form as our previous expression of the standard quantum limit strain sensitivity (with factors of π differences arising from the crude approximation of the power spectral density of the quantum fluctuations).

Note that the reason that the standard quantum limit arises is because attempts to measure the *position* of the mirrors accurately give rise to an uncertainty in the momentum of the mirrors which, in time, is converted into uncertainties in the mirror positions. If we could measure the *momentum* of the mirrors rather than their position, there would not be any fundamental limit to how small the uncertainty in subsequent measurements could be made. Therefore, while the standard quantum limit is quantum mechanical in origin, it does not impose a fundamental uncertainty in the ability to measure the effects of a gravitational wave on an interferometer.

For Initial LIGO, the standard quantum limit is entirely irrelevant: when optimized for $f_{\mathrm{fixed}} = 100$ Hz and adopting the values $G_{\mathrm{arm}} = 140$, $G_{\mathrm{prc}} = 50$ and $M = 11$ kg the laser power would have to be $I_0 \sim 4$ kW (which would correspond to a power of $\sim$10 MW stored in the arm cavities), far greater than the $I_0 \sim 5$ W that is actually used in the Initial LIGO interferometers. If the standard quantum limit could be achieved, however, the strain sensitivity would be $S_{h,\mathrm{SQL}}^{1/2} \simeq 3.5 \times 10^{-24}\,\mathrm{Hz}^{1/2}$ at 100 Hz. Advanced LIGO will exceed this limit over some frequency range by (i) increasing the mirror masses to 30 kg, and (ii) using a more sophisticated readout scheme (which includes a signal recycling cavity) that can help circumvent the standard quantum limit at some frequencies.

6.1.8 Environmental Sources of Noise

Although the photon shot noise and the radiation pressure noise are the fundamental sources of noise affecting the interferometric gravitational-wave detector, they are not always the dominant source of noise. At low frequencies, below around 40 Hz for Initial LIGO, seismic motion shifts the positions of the interferometer optics and produces strains that are of terrestrial origin, which can mask gravitational-wave signals. At intermediate frequencies, thermal vibrations in the optics and their suspensions form the dominant noise sources. We consider here the major sources of noise caused by the environment.

6.1.8.1 Seismic Noise

Ground motion is created intermittently by earthquakes and human activity but also continuously by wind and waves. These motions affect an interferometer readout by shaking the optical components. A point on the ground will jitter about due to seismic noise with a displacement power spectrum

$$S_X(f) \sim 10^{-18}\,\mathrm{m}^2\,\mathrm{Hz}^{-1} \begin{cases} 1 & 1\,\mathrm{Hz} < f \leq 10\,\mathrm{Hz} \\ (10\,\mathrm{Hz}/f)^4 & f > 10\,\mathrm{Hz} \end{cases}, \tag{6.120}$$

where X is the position of the ground point. The exact magnitude depends on the level of seismic activity at a given time, but this is a characteristic scale in the vicinity of the existing facilities. Since a displacement of the end mirror by x would result in a strain $h = x/L$ on the instrument, were the optics resting on the ground, the strain-equivalent amplitude spectrum from seismic noise (assuming that the mirror is fixed to the ground, so $x = X$) is

$$S^{1/2}_{h,\,\mathrm{seis}}(f) \sim 10^{-12}\,\mathrm{Hz}^{-1/2} \left(\frac{10\,\mathrm{Hz}}{f}\right)^2 \quad (f > 10\,\mathrm{Hz}) . \tag{6.121}$$

Although this noise source decreases with increasing frequency, at 100 Hz it would be $S_h^{1/2} \sim 10^{-14}\,\mathrm{Hz}^{-1/2}$, which is much, much larger than the shot noise at that frequency.

To mitigate the effects of seismic ground motion, the various optical components are isolated from seismic activity by both active- and passive-isolation systems. The simplest passive system is simply to suspend each optic as a pendulum. Suppose that X describes the position of the pivot point of the pendulum while x describes the position of the suspended optical component. If ℓ is the length of the pendulum then the position of the optical component is described by the equation of motion

$$\ell \frac{d^2 x}{dt^2} = -g(x - X) , \tag{6.122}$$

where g is the standard free-fall. Taking the Fourier transform of this equation, we find

$$-4\pi^2 \ell \tilde{x} = -g(\tilde{x} - \tilde{X}) \tag{6.123}$$

or

$$\tilde{x}(f) = A(f)\tilde{X}(f) , \quad (6.124)$$

where $A(f)$ is the transfer function of the pendulum, called the *actuation function,*

$$A(f) = \frac{1}{1 - (f/f_{\text{pend}})^2} , \quad (6.125)$$

and

$$f_{\text{pend}} := \frac{1}{2\pi}\sqrt{\frac{g}{\ell}} \quad (6.126)$$

is the pendulum frequency. Note that for $f \ll f_{\text{pend}}$, $\tilde{x} \simeq \tilde{X}$ so the pendulum has no effect at reducing the ground motion noise, but for $f \gg f_{\text{pend}}$, $\tilde{x} \approx -(f_{\text{pend}}/f)^2\tilde{X}$, and the ground motion is suppressed with increasing frequency. The relevant amplitude spectrum is now the amplitude spectrum of the position x of the mirror:

$$S_x^{1/2}(f) = |A(f)|\,S_X^{1/2}(f) . \quad (6.127)$$

If the pendulum frequency is chosen to be below 10 Hz, the strain-equivalent amplitude spectrum is

$$S_{h,\,\text{seis}}^{1/2}(f) \sim 10^{-12}\,\text{Hz}^{-1/2}\left(\frac{f_{\text{pend}}}{f}\right)^2\left(\frac{10\,\text{Hz}}{f}\right)^2 \quad (f > 10\,\text{Hz}) . \quad (6.128)$$

For Initial LIGO, $f_{\text{pend}} = 0.76$ Hz and so, at 100 Hz, the amplitude seismic noise level would be $S_h^{1/2} \sim 10^{-19}\,\text{Hz}^{-1/2}$.

Unfortunately, this is still insufficient suppression. In Initial LIGO, the pendulum support point is further isolated from the ground by four alternating mass-spring layers, each of which gives another attenuation factor of $\propto f^{-2}$. The attenuation at 40 Hz from this isolation stack is a factor of $\sim 10^5$. Therefore, at 40 Hz, the strain-amplitude noise spectrum is $S_h^{1/2} \sim 10^{-22}$, which is now comparable to the shot noise, and this spectrum decreases sharply as f^{-12} at higher frequencies:

$$S_{h,\,\text{seis}}^{1/2} \sim 10^{-12}\,\text{Hz}^{-1/2}\left(\frac{f_{\text{pend}}}{f}\right)^2\left(\frac{10\,\text{Hz}}{f}\right)^{10} \quad (f > 10\,\text{Hz}) . \quad (6.129)$$

6.1.8.2 Thermal Noise

At frequencies above ~40 Hz, another source of noise comes to dominate. The random Brownian motion of the molecules on the surface of the mirror and in the wires that suspend the optics produce a thermal noise source.

The *fluctuation–dissipation theorem* (Callen and Welton, 1951; Callen and Greene, 1952) states that the power spectral density of fluctuations of a system in equilibrium at temperature T is related to the dissipative terms when the system is driven out of equilibrium by

$$S_x(f) = \frac{4k_B T}{(2\pi f)^2}|\,\text{Re}[Y(f)]| , \quad (6.130a)$$

where $Y(f)$ is the *admittance* of the system,

$$Y(f) := 2\pi i f \frac{\tilde{x}(f)}{\tilde{F}_{\text{ext}}(f)} , \tag{6.130b}$$

which describes the response of the system $\tilde{x}(f)$ to an externally applied force $\tilde{F}_{\text{ext}}(f)$. Note that if there is dissipation then $Y(f)$ will not be purely imaginary, that is there will be a phase-lag between the external force and the displacement response.

Example 6.7 Derivation of the fluctuation–dissipation theorem

Suppose that the velocity v of an object responds to an applied force F_{ext} by

$$v(t) = \int_{-\infty}^{t} Y(t-t') F_{\text{ext}}(t') dt' , \tag{6.131}$$

where this equation can be taken as the definition of the admittance $Y(t-t')$ (in which case Eq. (6.130b) follows). If the object is in contact with a heat bath then the force has a stochastic component, but the average value of the velocity of the object follows a similar equation:

$$\langle v(t) \rangle = \int_{-\infty}^{t} Y(t-t') \langle F_{\text{ext}}(t') \rangle dt' . \tag{6.132}$$

Suppose now that a constant-mean force, $\langle F_{\text{ext}}(t') \rangle = F$ is applied for all time $t' < 0$ but then turned off at $t' = 0$; we find that

$$\langle v(t) \rangle = -F \int_{0}^{t} Y(t') dt' \quad (t > 0) , \tag{6.133}$$

where we assume that for $t < 0$ the object was in equilibrium and $\langle v(t) \rangle = 0$.

We can also compute the expected position (and hence velocity) of the particle at time $t > 0$ using an alternate approach: let $p(x, t|x', t')$ represent the probability of finding the particle at position x at time t given that it was at position x' at time t'. Then, if $p(x', 0)$ is the probability distribution for the position of the particle at time $t' = 0$,

$$\langle x(t) \rangle = \int\int x p(x, t|x', 0) p(x', 0) dx dx' . \tag{6.134}$$

The probability distribution $p(x, 0)$ is the *Boltzmann distribution*,

$$\begin{aligned} p(x,0) &= \frac{\exp\{-E(x)/(k_{\text{B}} T)\}}{\int \exp\{-E(x)/(k_{\text{B}} T)\} dx} \\ &\approx \frac{\exp\{-E_0(x)/(k_{\text{B}} T)\}}{\int \exp\{-E_0(x)/(k_{\text{B}} T)\} dx} \left(1 + \frac{xF}{k_{\text{B}} T}\right) , \end{aligned} \tag{6.135}$$

where $E(x) = E_0(x) + xF$ is the energy of the system when the force F is present, $E_0(x)$ is the energy of the system when the force is not present, and we assume that $xF/(k_B T) \ll 1$. We therefore write

$$p(x,0) \approx p_0(x)\left(1 - \frac{xF}{k_B T}\right), \qquad (6.136)$$

where $p_0(x)$ is the probability distribution for the position of the object in the absence of an external force. Therefore, the expectation value of the position of the object at time $t > 0$ is

$$\begin{aligned} \langle x(t)\rangle &\approx \int\int x\,p(x,t|x',0)\,p_0(x')\,dx\,dx' \\ &\quad - \frac{F}{k_B T}\int\int x x' p(x,t|x',0)\,p_0(x')\,dx\,dx' \\ &= -\frac{F}{k_B T} R_x(t)\,, \end{aligned} \qquad (6.137)$$

where $R_x(t)$ is the autocorrelation function,

$$R_x(t) = \langle x(t)x(0)\rangle = \int\int x x' p(x,t|x',0)\,p_0(x')\,dx\,dx' \qquad (6.138)$$

and we have assumed that, in the absence of an externally applied force, $\langle x\rangle = 0$.

We can now combine Eq. (6.133) for the expectation value of the velocity with the time derivative of Eq. (6.137):

$$\Theta(t)F\int_0^t Y(t')\,dt' = \Theta(t)\frac{F}{k_B T}\frac{d}{dt}R(t)\,, \qquad (6.139)$$

where we have introduced the Heaviside step function $\Theta(t)$ to enforce the fact that the equation holds only for $t > 0$. However, since the autocorrelation function is an even function of time, $R_x(t) = R_x(-t)$, so its derivative is an odd function of time, we see that

$$\begin{aligned} \frac{d}{dt}R(t) &= \Theta(t)\left.\frac{dR}{dt}\right|_t - \Theta(-t)\left.\frac{dR}{dt}\right|_{-t} \\ &= k_B T\,\Theta(t)\int_0^t Y(t')\,dt' - k_B T\,\Theta(-t)\int_0^{-t} Y(t')\,dt' \\ &= k_B T\int_{-t}^t Y(t')\,dt'\,. \end{aligned} \qquad (6.140)$$

Now we take the Fourier transform of this equation and, using the reality of the function $Y(t)$ and the definition of the one-sided power spectral density, $S_x(f) = \frac{1}{2}\tilde{R}_x(f)$, we find

$$2\pi i f \times \frac{1}{2}S_x(f) = \frac{k_B T}{2\pi i f}\left[\tilde{Y}(f) + \tilde{Y}^*(f)\right] \qquad (6.141)$$

and therefore

$$S_x(f) = \frac{k_B T}{\pi^2 f^2} | \operatorname{Re}[\tilde{Y}(f)]| \,. \tag{6.142}$$

For example, a simple harmonic oscillator can be described by the equation of motion

$$M \frac{d^2 x}{dt^2} + K x = F_{\text{ext}} \,, \tag{6.143}$$

where M is the mass of the oscillator and K is the spring constant. For an elastic spring, the spring constant is purely real; an anelastic spring, however, has a complex spring constant. For an anelastic spring, Hooke's law is

$$F_{\text{spring}} = -K(1 + i\phi)x \,, \tag{6.144}$$

where ϕ is known as a *loss angle*: it is the phase angle that the sinusoidal response x lags a sinusoidal force F_{spring} impressed upon the spring. In general, the loss angle is a function of frequency; however, for most materials, the loss angle tends to be relatively independent of frequency. The loss angle results from dissipative effects in the material. For high quality materials, the loss angle can be quite small, $\phi \sim 10^{-6}$ or even smaller.

The equation of motion for an anelastic spring is written

$$M \frac{d^2 x}{dt^2} + K(1 + i\phi)x = F_{\text{ext}} \,. \tag{6.145}$$

Taking the Fourier transform of this yields

$$\left\{-4\pi^2 M f^2 + K\left[1 + i\phi(f)\right]\right\} \tilde{x}(f) = \tilde{F}_{\text{ext}}(f) \tag{6.146}$$

and hence the admittance of the system is

$$Y(f) = 2\pi i f \frac{\tilde{x}(f)}{\tilde{F}_{\text{ext}}(f)} = \frac{2\pi i f}{4\pi^2 M} \frac{1}{\left(f_{\text{res}}^2 - f^2\right) + i f_{\text{res}}^2 \phi(f)} \,, \tag{6.147}$$

where $f_{\text{res}} = (2\pi)^{-1}(K/M)^{1/2}$ is the resonant frequency of the oscillator. The admittance function is sharply peaked at the resonant frequency, and this peak has a width of $Q := 1/\phi(f_{\text{res}})$. Oscillations at the resonant frequency die down slowly over a timescale given by $\tau_{\text{decay}} = Q/(\pi f_{\text{res}})$.

The real part of the admittance is

$$\operatorname{Re}[Y(f)] = \frac{1}{2\pi M} \frac{f f_{\text{res}}^2 \phi(f)}{(f_{\text{res}}^2 - f^2)^2 + f_{\text{res}}^4 \phi^2(f)} \tag{6.148}$$

and the fluctuation dissipation theorem shows that the power spectral density of the position fluctuations is

$$S_x(f) = \frac{k_B T}{2\pi^3 M f} \frac{f_{res}^2 \phi(f)}{(f_{res}^2 - f^2)^2 + f_{res}^4 \phi^2(f)}$$
$$\approx \frac{k_B T}{2\pi^3 M} f_{res}^{-3} \begin{cases} \phi(f)(f_{res}/f) & f \ll f_{res} \\ Q/\{1 + 4Q^2[(f/f_{res}) - 1]^2\} & f \simeq f_{res} \\ \phi(f)(f_{res}/f)^5 & f \gg f_{res} \,. \end{cases} \tag{6.149}$$

From the middle case we see that the spectral density has a sharp peak (for large values of Q) at the frequency $f = f_{res}$ and that the quality factor is related to the full width at half maximum (FWHM), Δf_{FWHM} via $Q = f_{res}/\Delta f_{FWHM}$. The spectrum of total energy (kinetic and potential) of the fluctuations is $4\pi^2 f^2 M S_x(f)$ and the total amount of thermal energy in the vibrational mode is

$$\langle E \rangle = 4\pi^2 M \int_0^\infty f^2 S_x(f) df \simeq \frac{1}{2} k_B T \tag{6.150}$$

as expected. For high-Q materials, most of the thermal energy is concentrated at frequencies near the resonance frequency; if the latter frequency is kept away from the band of sensitivity then the effects of thermal noise can be mitigated.

In LIGO, the two most important sources of thermal noise are in the pendulum suspension system for the mirrors and in the internal vibrational modes of the mirrors. The losses in the pendulum largely result from the bending of the suspension wire where it connects to the support and to the mirror. The mirror position depends on both pendulum motion and the elastic motion of the suspension wire:

$$M \frac{d^2 x}{dt^2} + M g \frac{x}{\ell} + K_{wire}(1 + i\phi_{wire})x = F_{ext} \,, \tag{6.151}$$

where ℓ is the length of the pendulum (note that we have assumed that the pendulum motion is essentially lossless). If the pendulum frequency, $f_{pend} = (2\pi)^{-1}(g/\ell)^{1/2}$, is much greater than the elastic frequency of the wire $f_{wire} = (2\pi)^{-1}(K/M)^{1/2}$ then we find that the admittance function is

$$Y(f) \approx \frac{2\pi i f}{4\pi^2 M} \frac{1}{(f_{pend}^2 - f^2) + i f_{pend}^2[(f_{wire}/f_{pend})^2 \phi_{wire}(f)]} \tag{6.152}$$

which is the same as the admittance function for the simple harmonic oscillator we had previously where the resonance frequency is the pendulum frequency and the effective loss angle is the loss angle of the wire multiplied by the dilution factor $(f_{wire}/f_{pend})^2$,

$$\phi_{pend}(f) := \left(\frac{f_{wire}}{f_{pend}}\right)^2 \phi_{wire}(f) \,. \tag{6.153}$$

In Initial LIGO, the loss angle of the suspension wires is $\phi_{\text{wire}} \sim 10^{-4}$, but $f_{\text{pend}} \sim 10 f_{\text{wire}}$ so the effective loss angle is rather small, $\phi_{\text{pend}} \sim 10^{-6}$. The pendulum's natural frequency is made quite low (0.76 Hz in Initial LIGO) so that it is outside the sensitive band of LIGO. Since there are four mirrors that dominate the noise in the interferometer, the input and end test masses on the two arms, and since the thermal noise in these mirrors is independent, the strain-equivalent thermal noise spectrum is

$$S_{h,\text{pend}}(f) = \frac{4}{L^2} S_x(f) \approx \frac{2 k_B T \phi_{\text{pend}} f_{\text{pend}}^2}{\pi^3 M L^2} \frac{1}{f^5} \quad (f \gg f_{\text{pend}}) . \tag{6.154}$$

At $T = 300\,\text{K}$, $\phi_{\text{pend}} = 10^{-6}$, $L = 4\,\text{km}$, $M = 11\,\text{kg}$, $f_{\text{pend}} = 0.76\,\text{Hz}$, we find

$$S_{h,\text{pend}}^{1/2}(f) \approx 10^{-23}\,\text{Hz}^{-1/2} \left(\frac{100\,\text{Hz}}{f}\right)^{5/2} . \tag{6.155}$$

In addition to this continuous noise spectrum, there will be suspension thermal noise at the frequencies of the violin modes of the suspension wires. A full picture of the suspension thermal noise is presented by Gonzalez (2000).

Internal vibrational modes of the mirror substrate contribute an additional source of thermal noise. The resonant frequencies f_{int} of such oscillations are much higher than the sensitive band of LIGO, around 10 kHz. In this regime the strain-equivalent thermal noise spectrum

$$S_{h,\text{int}}(f) = \frac{2 k_B T \phi}{\pi^3 M L^2 f_{\text{int}}^2} \frac{1}{f} \quad (f \ll f_{\text{int}}) . \tag{6.156}$$

Although the mirrors are made of high-quality material with very small losses, the coatings that are placed on the optical components are more dissipative and so the mirrors are more lossy. With the values $T = 300\,\text{K}$, $L = 4\,\text{km}$, $M = 11\,\text{kg}$, $f_{\text{int}} = 10\,\text{kHz}$, and $\phi_{\text{int}} = 10^{-6}$, we find

$$S_{h,\text{int}}^{1/2}(f) \approx 10^{-23}\text{Hz}^{-1/2} \left(\frac{100\,\text{Hz}}{f}\right)^{1/2} . \tag{6.157}$$

As can be seen in Figure 6.10, thermal noise is the dominant source of noise at intermediate frequencies up to $f \sim 100\,\text{Hz}$, which is in the sensitive band of LIGO. Although the pole frequency for the Fabry–Pérot cavity could be reduced, thereby reducing the shot noise at low frequencies, there would be no overall benefit as the thermal noise would continue to dominate at low frequencies; indeed, the situation would be worse as the shot noise would be higher at high frequencies. Initial LIGO was designed so that shot noise and thermal noise would be approximately equal at around $f \sim 100\,\text{Hz}$, with the shot noise dominating at higher frequencies and the thermal noise dominating at lower frequencies.

6.1.8.3 **Gravity Gradient Noise**

Fluctuations in the density of the ground due to seismic activity or in the density of the atmosphere will affect the test masses of an interferometer simply by their

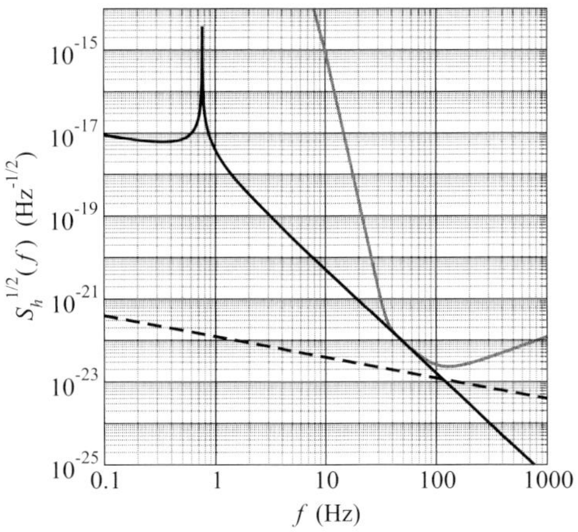

Figure 6.10 Amplitude spectral density of the thermal noise in strain-equivalent units for the suspension (pendulum) thermal noise (solid black curve) and the internal thermal noise (dashed curve). Noise associated with violin modes of the suspension are not shown here. The Initial LIGO sensitivity curve (grey) is also shown.

Newtonian force. This is known as *gravity gradient noise*. As with the direct effect of ground motion shaking the mirrors, gravity gradient noise is a low-frequency noise source. However, it cannot be mitigated simply by adding layers of seismic isolation, since it is due to the actual gravitational attraction between the test masses and the density perturbations in the vicinity of the detector. While we will see that gravity gradient noise is not an important noise source for first generation detectors, it becomes the dominant low-frequency noise in second and third generation interferometers as the seismic isolation is improved.

This description of the gravity gradient noise follows Saulson (1984) (see also Hughes and Thorne, 1998). Consider a perturbation $\delta\rho(\boldsymbol{x})$ in the mass (airmass or ground mass) surrounding a mirror. The mirror feels a gravitational force that causes an acceleration

$$\frac{d^2\boldsymbol{x}}{dt^2} = G\int \frac{\delta\rho(\boldsymbol{x}')}{\|\boldsymbol{x}-\boldsymbol{x}\|^3}\left(\boldsymbol{x}-\boldsymbol{x}'\right)d^3\boldsymbol{x} \tag{6.158}$$

(cf. Eq. (1.3)). Here we are ignoring the suspension so we treat the mirror as a freely falling body, at least in the one degree of motion of interest. In the frequency-domain, we have, for motion in the x-axis,

$$\tilde{x}(f) = -\frac{G}{4\pi^2 f^2}\int \frac{\widetilde{\delta\rho}(f)}{r^2}\sin\theta\cos\phi\, dV\ , \tag{6.159}$$

where we have written $x = r\sin\theta\cos\phi$. To relate the power spectrum of the induced motion of the mirror $S_x(f)$ with the power spectrum of density perturbations $S_\rho(f)$, we assume that the perturbations are coherent over length scales $\lesssim \lambda/2$, where $\lambda = v_s/f$ and v_s is the speed of sound of the material (air or ground), and they become independent over larger length scales. That is, regions of volume

$(\lambda/2)^3$ tend to move as a single mass, but a different region of such volume will move independently. Furthermore, the nearest region that affects the test mass is a distance $\lambda/4$ away. We divide up the volume integral into sums of these regions, integrate over the half-plane (below ground for ground motion or above ground for atmospheric motion), and find

$$\begin{aligned} S_x(f) &\approx \left[\frac{G}{(2\pi f)^2}\right]^2 S_\rho(f) \left(\frac{\lambda}{2}\right)^3 \int\limits_{r>\lambda/4} \frac{1}{r^4} \sin^2\theta \cos^2\phi \, dV \\ &\approx \left[\frac{G}{(2\pi f)^2}\right]^2 S_\rho(f) \left(\frac{\lambda}{2}\right)^3 \frac{2\pi}{3} \frac{4}{\lambda} \\ &= \frac{4\pi^3}{3} \left[\frac{G v_s}{(2\pi f)^3}\right]^2 S_\rho(f) \,. \end{aligned} \tag{6.160}$$

The interferometer is primarily sensitive to the positions of the two input test masses and the two end test masses, and we assume that the gravity gradient noise on the positions of these mirrors is independent. The strain-equivalent noise due to gravity gradients is therefore $4/L^2$ times the single-mirror displacement noise spectrum given in Eq. (6.160):

$$S_h(f) \approx \frac{16\pi^3}{3} \frac{1}{L^2} \left[\frac{G v_s}{(2\pi f)^3}\right]^2 S_\rho(f) \,. \tag{6.161}$$

Atmospheric changes are normally measured as pressure fluctuations. Density perturbations are related to pressure perturbations by $\delta\rho/\rho = \gamma^{-1}\delta p/p$ where $\gamma \approx 1.4$ is the adiabatic index for air. The atmospheric gravity gradient noise is therefore

$$S_h(f) \approx \frac{16\pi^3}{3} \frac{1}{L^2} \left[\frac{G \rho v_s}{(2\pi f)^3 \gamma p}\right]^2 S_p(f) \,. \tag{6.162}$$

Typical pressure fluctuations have $S_p^{1/2} \sim 10^{-3}\,\mathrm{Pa\,Hz^{-1/2}}$; with $p = 10^5\,\mathrm{Pa}$, $\rho = 1.3\,\mathrm{kg\,m^{-3}}$, $v_s = 340\,\mathrm{m\,s^{-1}}$, and $L = 4\,\mathrm{km}$, we find

$$S_h^{1/2}(f) \approx 2 \times 10^{-24}\,\mathrm{Hz^{-1/2}} \left(\frac{f}{10\,\mathrm{Hz}}\right)^{-6} \left(\frac{S_p^{1/2}(f)}{10^{-3}\,\mathrm{Pa\,Hz^{-1/2}}}\right)^2 . \tag{6.163}$$

At higher frequencies, the atmospheric gravity gradient noise will be even smaller than projected by this formula because the mirrors are enclosed in buildings that keep the air relatively stationary. At low frequencies, however, temperature fluctuations *carried past* the instrument can be a larger source of atmospheric gravity gradient noise than the pressure fluctuations. This happens for frequencies $f < v_{air}/r_{min}$ where v_{air} is the speed of the airflow and r_{min} distance from the test mass to the airflow (Creighton, 2008).

Seismic waves travelling through the ground are characterized in terms of their displacement spectrum $S_X(f)$. For one component of horizontal motion,

$$\frac{\delta\rho}{\rho} \sim \frac{\delta X}{\lambda/2} \tag{6.164}$$

(recall that $\lambda/2$ is the size of the parcel of ground that is moving coherently), so the power spectrum of density perturbations from ground motion is $S_\rho(f) \sim (2\rho/\lambda)^2 S_X(f) = (2\rho f/v_s)^2 S_X(f)$ and this yields a strain-equivalent noise power spectral density

$$S_h(f) \approx \frac{16\pi}{3}\frac{1}{L^2}\left[\frac{G\rho}{(2\pi f)^2}\right]^2 S_X(f)\,. \tag{6.165}$$

Assuming a ground density of $\rho \simeq 1800\,\mathrm{kg\,m^{-3}}$ and a ground displacement spectrum given by Eq. (6.120), we have

$$S_h^{1/2}(f) \approx 3\times 10^{-23}\,\mathrm{Hz}^{-1/2}\begin{cases}(f/10\,\mathrm{Hz})^{-2} & f < 10\,\mathrm{Hz}\\ (f/10\,\mathrm{Hz})^{-4} & f > 10\,\mathrm{Hz}\end{cases} \tag{6.166}$$

for the LIGO $L = 4\,\mathrm{km}$ interferometers. Compared to the seismic motion acting on the initial LIGO interferometer's mirrors through the seismic isolation stack, the gravity gradient effect is insignificant. With Advanced LIGO, however, more aggressive seismic isolation will reduce the amount of seismic noise to the point where the gravity gradient noise may become a limitation. For third generation interferometers, it is likely that gravity gradient noise will need to be mitigated, possibly by locating the mirrors underground in geologically quiet locations, or by measuring the nearby density fluctuations and correcting the mirror positions to compensate for the local gravitational influences.

6.1.9 Control System

The LIGO interferometer must control many different degrees of freedom of the motion of the mirrors in order to keep itself and its various cavities (the Fabry–Pérot cavities, the Michelson interferometer, and the recycling cavity) in resonance. Left uncontrolled, the different sources of environmental noise would immediately affect the lengths of these cavities, the light in the interferometers would drop out of resonance, and the interferometer would "lose lock". In order to maintain lock, many readout channels (beyond the antisymmetric port readout, which is the channel used for gravitational-wave detection) are continuously monitored and used to control the various degrees of freedom.

The primary feedback loop controls the differential arm movement and is known as the *differential arm feedback loop*. It is important to keep the differential arm length steady so that the antisymmetric port remains dark (otherwise light spills out of the recycling cavity). Of course, the antisymmetric port is where the gravitational signal is read out so the effect of the feedback on a true gravitational-wave signal must be determined in order to recover an estimate of the external strain on the detector. In the absence of a feedback loop, the sensing function $C(f)$ describes how the interferometer filters an externally applied strain $\tilde{s}(f)$ to produce a readout signal $\tilde{e}(f)$, which is known as an *error signal* (here we work in the frequency domain as this makes it simple to express the linear filter that the loop

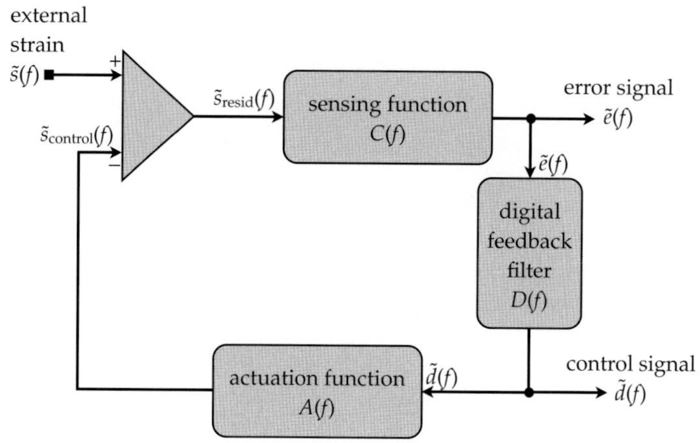

Figure 6.11 The differential arm feedback loop.

applies to the incident signal). The differential error signal is known as `DARM_ERR` in LIGO, and the purpose of the feedback loop is to keep this signal small. Normally the primary noise sources that must be controlled are at low frequency (seismic noise); the error signal is low-passed filtered by a digital filter $D(f)$, which produces the differential arm *control signal* $\tilde{d}(f)$. In LIGO, this signal is also recorded as `DARM_CTRL`. The control signal is used to move the ETM mirrors to counter the effect of the environmentally induced motions (and low-frequency gravitational waves!) on the interferometer. Another transfer function, the actuation function, $A(f)$, relates the actuation force to the motion of the mirror (recall that the mirrors are suspended from pendula so there is a frequency-dependent relationship between the applied force and the induced mirror motion). Figure 6.11 shows the elements of this feedback loop.

We now describe the feedback loop mathematically. The residual strain $\tilde{s}_{resid}(f)$ that the interferometer actually experiences is the difference between the true incident external strain $\tilde{s}(f)$ and the control strain $\tilde{s}_{control}$. The error signal $\tilde{e}(f)$ results from this residual strain via the sensing function $C(f)$ which is essentially the Fabry–Pérot cavity transfer function. The digital feedback filter $D(f)$ is applied to the error signal to produce the differential arm control signal $\tilde{d}(f)$, which is then related to the control strain via the actuation function $A(f)$:

$$\tilde{s}_{resid}(f) = \tilde{s}(f) - \tilde{s}_{control}(f) , \tag{6.167a}$$

$$\tilde{s}_{control}(f) = A(f)\tilde{d}(f) , \tag{6.167b}$$

$$\tilde{d}(f) = D(f)\tilde{e}(f)\,, \tag{6.167c}$$

$$\tilde{e}(f) = C(f)\tilde{s}_{\text{resid}}(f)\,. \tag{6.167d}$$

Therefore,

$$\begin{aligned}\tilde{e}(f) &= C(f)[\tilde{s}(f) - \tilde{s}_{\text{control}}(f)] \\ &= C(f)\tilde{s}(f) - C(f)A(f)\tilde{d}(f) \\ &= C(f)\tilde{s}(f) - C(f)A(f)D(f)\tilde{e}(f)\,.\end{aligned} \tag{6.168}$$

The external strain is thus reconstructed from the readout error signal as

$$\tilde{s}(f) = \frac{1 + C(f)A(f)D(f)}{C(f)}\tilde{e}(f) = R(f)\tilde{e}(f)\,, \tag{6.169}$$

where $R(f)$ is the *response function*,

$$R(f) := \frac{1 + G(f)}{C(f)} \tag{6.170}$$

and

$$G(f) := C(f)A(f)D(f) \tag{6.171}$$

is known as the *open loop gain*. The external strain $\tilde{s}$ contains the strain induced by the various noise sources $\tilde{n}$ as well as by any gravitational waves $\tilde{h}$; in the next chapter we will discuss how to identify a signal in the noise. We now show how a metric perturbation creates the strain $h(t)$ on the interferometer.

6.1.10 Gravitational-Wave Response of an Interferometric Detector

For low-frequency gravitational waves, where the wavelength of the wave is much larger than the arm length of the interferometer arms, we can imagine that light traversing the arm moves in an essentially constant gravitational potential; that is, the metric perturbation when the light is emitted at one end of the arm is essentially the same as it is when it is reflected by the end mirror and is essentially the same as it is when it returns to the emitter. In this case, the slowly varying phase of the returning light depends on slowly varying strain $h(t)$ caused by the gravitational wave: $\phi(t) = 2kL + \phi_{\text{ext}}(t)$ with $\phi_{\text{ext}} = 2kLh(t)$. The strain is

$$h(t) = \frac{1}{2}\hat{p}^i\hat{p}^j h_{ij}(t)\,, \tag{6.172}$$

where $\hat{\boldsymbol{p}}$ is the unit vector from the emitter to the end-mirror (cf. Eq. (3.44)).

For high-frequency gravitational waves, the response of an interferometer is somewhat more complicated. Consider a photon emitted at the emitter at spacetime event 0 and being received at the end-mirror at spacetime event 1. A plane-parallel gravitational wave travelling in the direction $\hat{\boldsymbol{n}}$ produces the metric perturbation $h_{ij}(t, \boldsymbol{x}) = h_{ij}(t - \hat{\boldsymbol{n}} \cdot \boldsymbol{x}/c, \boldsymbol{0})$. Recall from Eq. (3.61) that

$$\mathfrak{z} := \frac{\nu_0 - \nu_1}{\nu_1} = \frac{1}{2}\frac{\hat{p}^i \hat{p}^j}{1 - \hat{\boldsymbol{p}} \cdot \hat{\boldsymbol{n}}}\left(h^1_{ij} - h^0_{ij}\right), \tag{6.173}$$

where h^0_{ij} is the metric perturbation at spacetime event 0, when the photon is emitted at the emitter, and h^1_{ij} is the metric perturbation at spacetime event 1, when the photon is received at the receiver (note that the direction from the receiver to the emitter is $-\hat{\boldsymbol{p}}$ in this case). This is the *up-link*. The photon is reflected by the end mirror and returns to the emitter. The redshift of the *down-link* is

$$\frac{\nu_1 - \nu_2}{\nu_2} = \frac{1}{2}\frac{\hat{p}^i \hat{p}^j}{1 + \hat{\boldsymbol{p}} \cdot \hat{\boldsymbol{n}}}\left(h^2_{ij} - h^1_{ij}\right), \tag{6.174}$$

where h^2_{ij} is the metric perturbation at spacetime event 2 when the photon returns to the emitter. The overall redshift of the transit (the up-link and the down-link) is (note that all the frequencies are very close)

$$\frac{\nu_0 - \nu_2}{\nu_0} = \frac{1}{2}\hat{p}^i \hat{p}^j \left(\frac{h^2_{ij} - h^1_{ij}}{1 + \hat{\boldsymbol{p}} \cdot \hat{\boldsymbol{n}}} + \frac{h^1_{ij} - h^0_{ij}}{1 - \hat{\boldsymbol{p}} \cdot \hat{\boldsymbol{n}}}\right). \tag{6.175}$$

Now note that

$$h^0_{ij} = h_{ij}(t), \tag{6.176}$$

$$h^1_{ij} = h_{ij}\left[t + (L/c)(1 - \hat{\boldsymbol{p}} \cdot \hat{\boldsymbol{n}})\right], \tag{6.177}$$

$$h^2_{ij} = h_{ij}(t + 2L/c), \tag{6.178}$$

or, in the frequency domain,

$$\tilde{h}^0_{ij} = \tilde{h}_{ij}(f), \tag{6.179}$$

$$\tilde{h}^1_{ij} = \mathrm{e}^{2\pi i f L(1 - \hat{\boldsymbol{p}} \cdot \hat{\boldsymbol{n}})/c}\, \tilde{h}_{ij}(f), \tag{6.180}$$

$$\tilde{h}^2_{ij} = \mathrm{e}^{4\pi i f L/c}\, \tilde{h}_{ij}(f). \tag{6.181}$$

We see that

$$\frac{\widetilde{(\nu_0 - \nu_2)}(f)}{\nu_0} = \frac{1}{2}\hat{p}^i \hat{p}^j \tilde{h}_{ij}(f) \cdot \left(\frac{\mathrm{e}^{4\pi i f L/c} - \mathrm{e}^{2\pi i f L(1 - \hat{\boldsymbol{p}} \cdot \hat{\boldsymbol{n}})/c}}{1 + \hat{\boldsymbol{p}} \cdot \hat{\boldsymbol{n}}} + \frac{\mathrm{e}^{2\pi i f L(1 - \hat{\boldsymbol{p}} \cdot \hat{\boldsymbol{n}})/c} - 1}{1 - \hat{\boldsymbol{p}} \cdot \hat{\boldsymbol{n}}}\right). \tag{6.182}$$

This frequency shift of the light results from a phase shift induced by the gravitational wave

$$\varphi_{\text{ext}} = 2\pi \int \left[\nu_0 - \nu_2(t)\right] dt \tag{6.183}$$

or

$$\tilde{\varphi}_{\text{ext}}(f) = \frac{\widetilde{(\nu_0 - \nu_2)}(f)}{i f} . \tag{6.184}$$

This yields

$$\tilde{\varphi}_{\text{ext}} = k L \hat{p}^i \hat{p}^j \tilde{h}_{ij}(f) D(\hat{\boldsymbol{p}} \cdot \hat{\boldsymbol{n}}, f L/c) \tag{6.185}$$

with

$$\begin{aligned} D(\hat{\boldsymbol{p}} \cdot \hat{\boldsymbol{n}}, f L/c) := \frac{1}{2} e^{2\pi i f L/c} \Big\{ & e^{i\pi f L(1-\hat{\boldsymbol{p}}\cdot\hat{\boldsymbol{n}})/c} \operatorname{sinc}\left[\pi f L(1 + \hat{\boldsymbol{p}} \cdot \hat{\boldsymbol{n}})/c\right] \\ & + e^{-i\pi f L(1+\hat{\boldsymbol{p}}\cdot\hat{\boldsymbol{n}})/c} \operatorname{sinc}\left[\pi f L(1 - \hat{\boldsymbol{p}} \cdot \hat{\boldsymbol{n}})/c\right] \Big\} , \end{aligned} \tag{6.186}$$

where $\operatorname{sinc}(x) := \sin(x)/x$.

A Michelson-like interferometer, such as the LIGO and Virgo interferometers, has two orthogonal arms. The readout at the antisymmetric port senses the difference in the phase shifts encountered in each arm

$$\begin{aligned} \tilde{\varphi}_{\text{ext}}(f) &:= \tilde{\varphi}_{\text{ext},1}(f) - \tilde{\varphi}_{\text{ext},2}(f) \\ &= k L \tilde{h}_{ij}(f) \left[\hat{p}^i \hat{p}^j D(f L/c, \hat{\boldsymbol{p}} \cdot \hat{\boldsymbol{n}}) - \hat{q}^i \hat{q}^j D(f L/c, \hat{\boldsymbol{q}} \cdot \hat{\boldsymbol{n}})\right] \\ &= 2k L \tilde{h}(f) , \end{aligned} \tag{6.187}$$

where $\tilde{h}(f)$ is the effective induced strain,

$$\tilde{h}(f) := G_+(\hat{\boldsymbol{n}}, \psi, f) \tilde{h}_+(f) + G_\times(\hat{\boldsymbol{n}}, \psi, f) \tilde{h}_\times(f) , \tag{6.188}$$

where $\hat{\boldsymbol{p}}$ and $\hat{\boldsymbol{q}}$ are the unit vectors pointing along each of the two arms, ψ is a *polarization angle*,[6] and

$$\begin{aligned} G_+(\hat{\boldsymbol{n}}, \psi, f) := \frac{1}{2} & \left[\hat{p}^i \hat{p}^j D(\hat{\boldsymbol{p}} \cdot \hat{\boldsymbol{n}}, f L_1/c) - \hat{q}^i \hat{q}^j D(\hat{\boldsymbol{q}} \cdot \hat{\boldsymbol{n}}, f L_2/c)\right] \\ & \cdot e^+_{ij}(\hat{\boldsymbol{n}}, \psi) , \end{aligned} \tag{6.189a}$$

$$G_\times(\hat{\boldsymbol{n}}, \psi, f) := \frac{1}{2} \left[\hat{p}^i \hat{p}^j D(\hat{\boldsymbol{p}} \cdot \hat{\boldsymbol{n}}, f L_1/c) - \hat{q}^i \hat{q}^j D(\hat{\boldsymbol{q}} \cdot \hat{\boldsymbol{n}}, f L_2/c)\right] e^\times_{ij}(\hat{\boldsymbol{n}}, \psi) \tag{6.189b}$$

6) The polarization angle ψ is the angle, measured counter-clockwise about the vector $\hat{\boldsymbol{n}} = \boldsymbol{e}_3$ in the transverse plane of the gravitational wave, between the *line of nodes* and the $\boldsymbol{e}_1$-direction (the x-axis) in the wave frame. Here, the line of nodes is the axis $\hat{\boldsymbol{N}} \times \hat{\boldsymbol{n}}$ where $\hat{\boldsymbol{N}}$ is the unit vector pointing toward the North Celestial Pole (the direction of Earth's rotation axis). The polarization angle thus defines the plus- and cross-polarization states of the gravitational wave.

are the detector *beam pattern* responses to the plus- and cross-polarization states of the gravitational wave. In the long-wavelength limit these beam pattern functions become frequency-independent:

$$F_+(\hat{\boldsymbol{n}}, \psi) := G_+(\hat{\boldsymbol{n}}, \psi, 0) = \frac{1}{2}\left(\hat{p}^i\hat{p}^j - \hat{q}^i\hat{q}^j\right) e^+_{ij}(\hat{\boldsymbol{n}}, \psi)\,, \tag{6.190a}$$

$$F_\times(\hat{\boldsymbol{n}}, \psi) := G_\times(\hat{\boldsymbol{n}}, \psi, 0) = \frac{1}{2}\left(\hat{p}^i\hat{p}^j - \hat{q}^i\hat{q}^j\right) e^\times_{ij}(\hat{\boldsymbol{n}}, \psi) \tag{6.190b}$$

for $f \ll f_{\text{FSR}}$.

6.1.11
Second Generation Ground-Based Interferometers (and Beyond)

Second generation detectors, such as Advanced LIGO, will achieve improved sensitivity over the initial detectors in the following major ways: (i) improved seismic isolation, including active seismic isolation, will reduce the low-frequency noise in the detectors; (ii) similarly, higher-quality optical components will reduce the thermal noise at low frequencies; (iii) mirrors with more mass (which reduces radiation pressure noise) and lower absorptivity will allow the interferometer to be run at higher power, which will reduce the shot noise at high frequencies; and (iv) the addition of a signal recycling mirror will change the nature of the shot noise and will allow the low-frequency sensitivity and the bandwidth to be independently tuned.

In Advanced LIGO, nearly 1 MW of light power will be stored in the Fabry–Pérot cavities, dramatically reducing the shot noise. To keep the radiation pressure noise comparable to the thermal noise at low frequencies, the test mass mirrors will be 40 kg, rather than the 11 kg mirrors of Initial LIGO. The mirrors will be suspended by fused silica ribbons which will have a higher quality than the wires used in Initial LIGO and so will reduce the thermal noise within the LIGO band. Active seismic isolation will push the seismic wall to frequencies below ~10 Hz. Overall, the sensitivity will be limited by radiation pressure noise at low frequencies and shot noise at high frequencies.

At intermediate frequencies, Advanced LIGO will be operating near the standard quantum limit, and the dominant noise source is expected to be quantum noise. A complete treatment of the readout noise at the intermediate frequencies would require a quantum mechanical treatment of the radiation field and the readout apparatus (Buonanno and Chen, 2001). In this section we will instead present a heuristic classical treatment. We will ignore radiation pressure noise and consider only the shot noise.

Recall that the frequency spectrum of the shot noise was determined by the Fabry–Pérot transfer function, the sensing function $C(f)$, by $S_{h,\text{shot}} \propto |C(f)|^{-2}$. The sensing function itself has essentially one tunable parameter, the pole frequency f_{pole}, which is set by the choice of the transmissivity of the input test mass mirror. The pole frequency determines not only the bandwidth of the interferometer but also the arm gain G_{arm} in such a way that by increasing the bandwidth (increasing the pole frequency), the arm gain (and therefore the low-frequency sensitivity) is de-

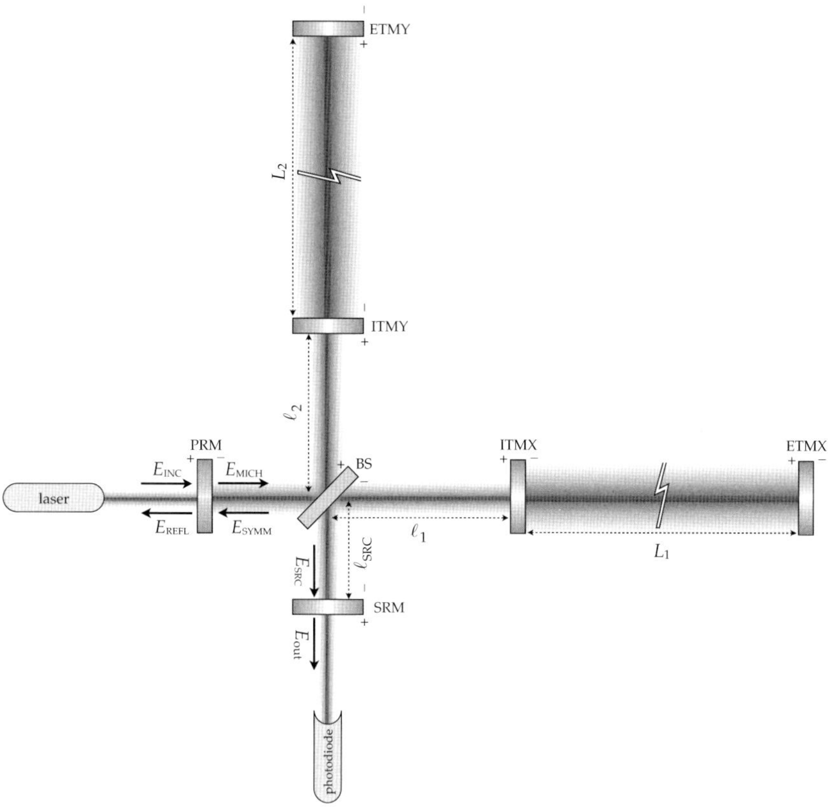

Figure 6.12 Optical topology of a dual-recycled Fabry–Pérot Michelson interferometer.

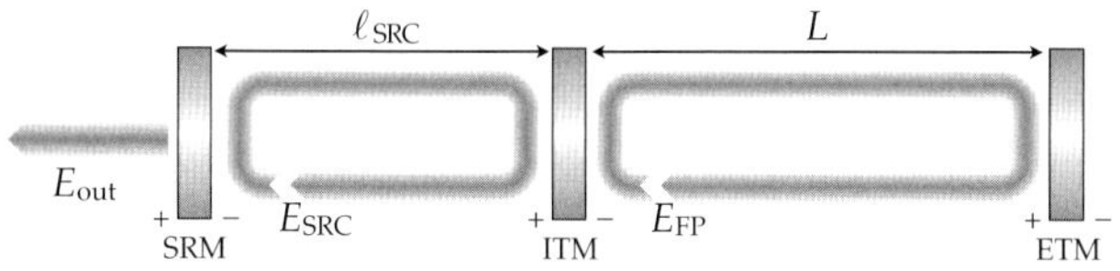

Figure 6.13 Optical topology of coupled Fabry–Pérot and signal recycling cavities.

creased. To decouple these two parameters, Advanced LIGO will introduce an additional mirror at the antisymmetric port, the *signal recycling mirror* (see Figure 6.12), which effectively allows the audio-frequency sidebands of the carrier light to be stored in the interferometer for a longer (or shorter) time than the carrier light.

For simplicity of illustration we approximate the LIGO interferometer as two coupled cavities, the Fabry–Pérot cavity, where the gravitational wave will produce sidebands on the carrier light, and a signal recycling cavity, which can be tuned to either return the sidebands into the Fabry–Pérot cavity (to build up the signal), or to resonantly extract them. The optical configuration we consider is shown in Figure 6.13.

Note that the signal recycling cavity is actually placed at the antisymmetric port of the Michelson interferometer, which is (nearly) dark at the carrier frequency. Therefore, we need to consider only the time-dependent audio sideband frequencies that are generated by the externally-induced motion of the mirror. We have

$$E'_{\text{SRC}}(t) = t_{\text{ITM}} e^{-ik\ell_{\text{SRC}}} E'_{\text{FP}}(t - \ell_{\text{SRC}}/c) - r_{\text{SRM}} r_{\text{ITM}} e^{-2ik\ell_{\text{SRC}}} E'_{\text{SRC}}(t - 2\ell_{\text{SRC}}/c) \tag{6.191}$$

and

$$E'_{\text{FP}}(t) \simeq -r_{\text{SRM}} t_{\text{ITM}} e^{-ik\ell_{\text{SRC}}} e^{-2ikL} E'_{\text{SRC}}(t - \ell_{\text{SRC}}/c - 2L/c) - r_{\text{ITM}} e^{-2ikL} E'_{\text{FP}}(t - 2L/c) - i e^{-2ikL} E_{\text{FP},0} \phi_{\text{ext}}(t) , \tag{6.192}$$

where we have expressed the second equation to first order in the externally-induced phase shift; recall that the primes indicate that we are considering the time-dependent modulation of the fields (at audio frequencies). The length of the recycling cavity is ℓ_{SRC}. The Fourier transform of these equations gives a linear system which we can express as

$$\mathbf{M} \begin{bmatrix} \tilde{E}'_{\text{SRC}}(f) \\ \tilde{E}'_{\text{FP}}(f) \end{bmatrix} = \begin{bmatrix} 0 \\ i\tilde{\phi}_{\text{ext}}(f) E_{\text{FP},0} \end{bmatrix} , \tag{6.193}$$

where

$$\mathbf{M} := \begin{bmatrix} 1 + r_{\text{SRM}} r_{\text{ITM}} e^{-2i(2\pi f \ell_{\text{SRC}}/c + \phi_{\text{SRC}})} & -t_{\text{ITM}} e^{-i(2\pi f \ell_{\text{SRC}}/c + \phi_{\text{SRC}})} \\ -r_{\text{SRM}} t_{\text{ITM}} e^{-i(2\pi f \ell_{\text{SRC}}/c + \phi_{\text{SRC}})} e^{-4\pi i f L/c} & 1 - r_{\text{ITM}} e^{-4\pi i f L/c} \end{bmatrix} \tag{6.194}$$

and we have imposed the condition $\exp(-2ikL) = -1$, which is required for resonance of the carrier light in the Fabry–Pérot cavity. The signal recycling cavity tuning phase ϕ_{SRC} is defined as $\phi_{\text{SRC}} = k\ell_{\text{SRC}} = 2\pi\ell_{\text{SRC}}/\lambda_{\text{laser}}$.

The fields in the cavities can now be computed by inverting the matrix $\mathbf{M}$; we obtain

$$\tilde{E}'_{\text{SRC}}(f) = i E_{\text{FP},0} \frac{-M_{12}}{\det \mathbf{M}} \tilde{\phi}_{\text{ext}}(f) , \tag{6.195}$$

$$\tilde{E}'_{\text{FP}}(f) = i E_{\text{FP},0} \frac{M_{11}}{\det \mathbf{M}} \tilde{\phi}_{\text{ext}}(f) , \tag{6.196}$$

where

$$\det \mathbf{M} = 1 - r_{\text{ITM}} e^{-4\pi i f L/c} + (r_{\text{ITM}} - e^{-4\pi i f L/c}) r_{\text{SRM}} e^{-2i(2\pi f \ell_{\text{SRC}}/c + \phi_{\text{SRC}})} . \tag{6.197}$$

In particular, we have

$$\tilde{E}'_{\text{SRC}}(f) = i E_{\text{FP},0} \frac{t_{\text{ITM}} e^{-i(2\pi f \ell_{\text{SRC}}/c + \phi_{\text{SRC}})}}{\det \mathbf{M}} \tilde{\phi}_{\text{ext}}(f) . \tag{6.198}$$

The carrier light in the Fabry–Pérot cavity is

$$E_{\text{FP},0} = G_{\text{prc}}^{1/2} \frac{t_{\text{ITM}}}{1 - r_{\text{ITM}}} E_{\text{INC}} , \tag{6.199}$$

where we have included the gain from the power recycling cavity. The field that leaves the signal recycling cavity is therefore

$$\begin{aligned}\tilde{E}'_{\text{out}}(f) &= t_{\text{SRM}} \tilde{E}'_{\text{SRC}}(f) \\ &= i E_{\text{INC}} G_{\text{prc}}^{1/2} \frac{t_{\text{ITM}}^2}{1 - r_{\text{ITM}}} \frac{t_{\text{SRM}} e^{-i(2\pi f \ell_{\text{SRC}}/c + \phi_{\text{SRC}})}}{\det \mathbf{M}} \tilde{\phi}_{\text{ext}}(f) \\ &= i E_{\text{INC}} G_{\text{prc}}^{1/2} G_{\text{arm}} \frac{(1 - r_{\text{ITM}})\, t_{\text{SRM}} e^{-i(2\pi f \ell_{\text{SRC}}/c + \phi_{\text{SRC}})}}{\det \mathbf{M}} \tilde{\phi}_{\text{ext}}(f) .\end{aligned} \tag{6.200}$$

To connect this formula with our result for the non-signal-recycled interferometer, we express the output field as

$$\begin{aligned}\tilde{E}'_{\text{out}}(f) &= i E_{\text{INC}} G_{\text{prc}}^{1/2} G_{\text{arm}} \frac{1 - r_{\text{ITM}}}{1 - r_{\text{ITM}} e^{-4\pi i f L/c}} \hat{C}_{\text{SR}}(f) \tilde{\phi}_{\text{ext}}(f) \\ &= i E_{\text{INC}} G_{\text{prc}}^{1/2} G_{\text{arm}} \hat{C}_{\text{FP}}(f) \hat{C}_{\text{SR}}(f) \tilde{\phi}_{\text{ext}}(f) ,\end{aligned} \tag{6.201}$$

where $\hat{C}_{\text{SR}}(f)$ is the effective transfer function from signal recycling

$$\hat{C}_{\text{SR}}(f) = \frac{t_{\text{SRM}} e^{-i(2\pi f \ell_{\text{SRC}}/c + \phi_{\text{SRC}})}}{1 - r_{\text{SRM}} \left(\frac{r_{\text{ITM}} - e^{-4\pi i f L/c}}{1 - r_{\text{ITM}} e^{-4\pi i f L/c}} \right) e^{-2i(2\pi i f \ell_{\text{SRC}}/c + \phi_{\text{SRC}})}} . \tag{6.202}$$

The tuning phase ϕ_{SRC} is a critical parameter in the transfer function arising from the signal recycling cavity. The signal recycling cavity effectively makes the ITM a compound mirror with a frequency-dependent reflectivity as seen by the light in the Fabry–Pérot cavity. At low frequencies, when $\phi_{\text{SRC}} = 0$, the effective transmissivity of this compound mirror is high and the signal is "sucked out" of the interferometer. This is known as *resonant sideband extraction* (RSE). Because the signal does not build up so much in the Fabry–Pérot arms, the sensitivity of the interferometer is lower at low frequencies for RSE; however, the instrument has a larger frequency bandwidth (the effective cavity pole of the Fabry–Pérot cavity is shifted to higher frequencies) so the RSE interferometer has greater sensitivity at high frequencies. The other limit, when $\phi_{\text{SRC}} = 90°$, is known as *signal recycling* (SR) proper. At low frequencies, the signal is reflected back into the Fabry–Pérot cavity, allowing it to build up. This increases low-frequency sensitivity at the expense of bandwidth. So far, the effect of the signal recycling cavity appears to be simply that it effectively allows for dynamic tuning (by changing the length of the signal recycling cavity) of the transmissivity of the ITM. However, unlike simply changing the transmissivity of the ITM, this is a frequency-dependent tuning, which allows for some unique features. Specifically, for ϕ_{SRC} between 0 and 90°,

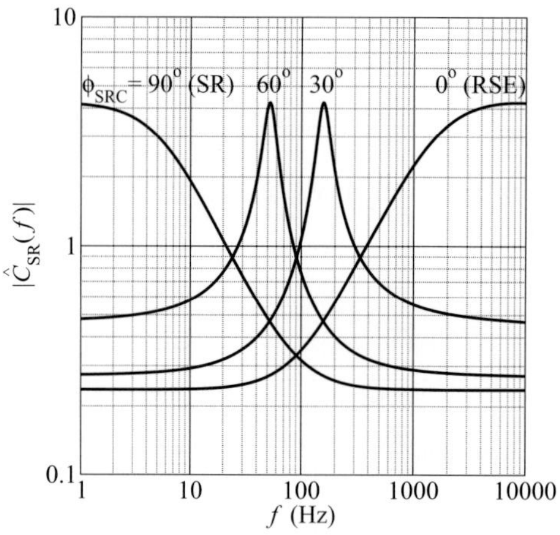

Figure 6.14 The magnitude of the factor $\hat{C}_{SR}(f)$ in the instrumental response arising from the introduction of a signal recycling cavity as a function of frequency. Several transfer functions are shown for different values of the signal recycling cavity tuning phase ϕ_{SRC} from 0° (RSE) to 90° (SR). The parameters used here are $L = 4$ km, $\ell_{SRC} = 50.6$ m, $t^2_{ITM} = 3\%$, $t_{SRM} = 20\%$, which are baseline numbers for the Advanced LIGO detector.

signals of a particular frequency can be held in the Fabry–Pérot cavity longer, effectively tuning the interferometer to be sensitive to those frequencies. These effects are visible in Figure 6.14, which shows the transfer function $|\hat{C}_{SR}(f)|$, which is the additional frequency-dependent factor in the instrumental response that is produced by the introduction of a signal recycling cavity, for various different tunings. As can be seen, RSE ($\phi_{SRC} = 0$) results in greater sensitivity at high frequencies but worse sensitivity at low frequencies; SR ($\phi_{SRC} = 90°$) results in higher sensitivity at low frequencies at the expense of high frequencies; and for values of the tuning parameter ϕ_{SRC} between these extremes, the sensitivity can be improved at a narrow band near some frequency of interest.

If we assume that the shot noise and the radiation pressure noise in Advanced LIGO are independent, then the total sensor noise is

$$S_{h,\text{sensor}}(f) = \frac{\hbar}{\pi^2 f^2 M L^2} \cdot \left(\frac{\pi^2 f^2 M c}{2k I_0 G_{\text{prc}} G^2_{\text{arm}} |\hat{C}_{\text{SRFP}}(f)|^2} + \frac{2k I_0 G_{\text{prc}} G^2_{\text{arm}} |\hat{C}_{\text{SRFP}}(f)|^2}{\pi^2 f^2 M c} \right), \tag{6.203}$$

where

$$\hat{C}_{\text{SRFP}}(f) := \hat{C}_{\text{SR}}(f) \hat{C}_{\text{FP}}(f) . \tag{6.204}$$

Figure 6.15 shows the sensor noise for various different Advanced LIGO configurations. In reality, these sensitivity curves are not accurate: Advanced LIGO will be a quantum measuring device, and a fully quantum mechanical model of the electro-

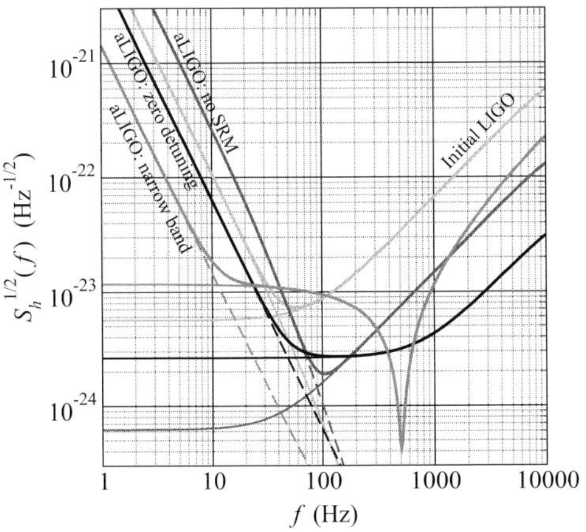

Figure 6.15 Sensor noise curves for various configurations. Thin solid lines show shot noise; dashed lines show radiation pressure noise; thick solid lines are the total sensor noise. Light grey lines show the Initial LIGO interferometer (only the shot and radiation pressure components for an ideal DC-readout scheme are shown). Dark grey lines show the Advanced LIGO (aLIGO) configuration with $t_{\mathrm{ITM}}^2 = 1.4\%$, $t_{\mathrm{PRM}}^2 = 3\%$, $G_{\mathrm{arm}} = 280$, $G_{\mathrm{prc}} = 40$, $L = 4\,\mathrm{km}$, and $I_0 = 125\,\mathrm{W}$, but with no signal recycling cavity. Black lines show the Broadband Advanced LIGO configuration, with a signal recycling cavity that has $t_{\mathrm{SRM}}^2 = 20\%$, $\ell_{\mathrm{SRC}} = 56\,\mathrm{m}$, and $\phi_{\mathrm{SRC}} = 0$. Medium grey lines indicate the narrow-band Advanced LIGO configuration with a signal recycling cavity that has $t_{\mathrm{SRM}}^2 = 1.1\%$, $\ell_{\mathrm{SRC}} = 56\,\mathrm{m}$, and $\phi_{\mathrm{SRC}} = 4.7^\circ$.

magnetic field and readout device is necessary to correctly compute the quantum correlations that exist between the shot noise and the radiation pressure noise. (These are absent in the case of pure signal recycling, $\phi_{\mathrm{SRC}} = 90^\circ$, and pure resonant sideband extraction, $\phi = 0^\circ$; our semi-classical treatment is actually valid in these limits!) When the full quantum mechanical noise calculation is done, the sensitivity curves are significantly different; the quantum noise for three Advanced LIGO configurations, along with the two dominant sources of environmental noise – suspension thermal noise and coating thermal noise – are shown in Figure 6.16.

Third-generation gravitational-wave detectors, such as the European Einstein Telescope (Punturo *et al.*, 2010) (ET) and the Japanese Large Cryogenic Gravitational Telescope (LCGT) are now under development. These detectors will improve sensitivity in all frequency bands by addressing the limiting noise sources at each frequency. Environmental noise from ground motion can be reduced by constructing facilities underground where there is less seismic activity. With active seismic isolation, the dominant low-frequency noise source will be gravity gradient noise. It may be possible to reduce this by monitoring seismic motion in the vicinity of each mirror so that the gravity gradients affecting the mirror can be actively cancelled by moving the mirrors in opposition to the deduced gravity gradient. Thermal noise in the mirror can be reduced be cryogenic cooling of the mirrors and suspension. High-frequency shot noise can be suppressed by increased light power, while ra-

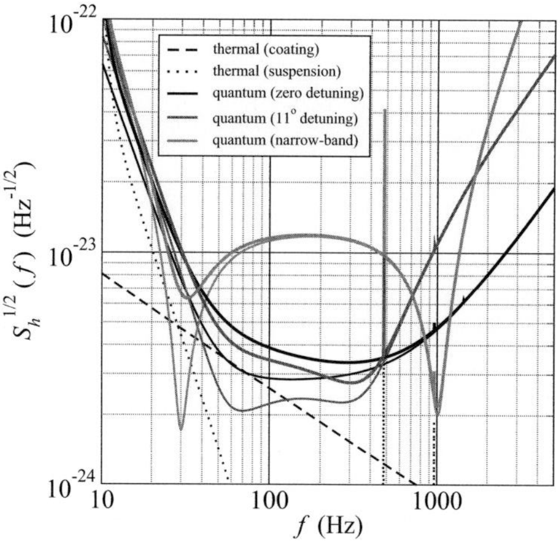

Figure 6.16 Advanced LIGO noise budget for various configurations. Thin lines show the dominant noise sources including coating thermal noise (dashed), suspension thermal noise (dotted), and quantum noise for different optical configurations (black, dark grey and light grey). Thick lines show the overall sensitivity curve for each configuration. Black lines show the Broadband Advanced LIGO configuration, with a signal recycling cavity that has $t^2_{\mathrm{SRM}} = 20\%$, $\ell_{\mathrm{SRC}} = 56$ m, and $\phi_{\mathrm{SRC}} = 0$. Dark grey lines show the Advanced LIGO configuration with a signal recycling cavity that has $t^2_{\mathrm{SRM}} = 20\%$, $\ell_{\mathrm{SRC}} = 56$ m, and $\phi_{\mathrm{SRC}} = 11°$. Light grey lines show the narrow-band Advanced LIGO configuration with a signal recycling cavity that has $t^2_{\mathrm{SRM}} = 1.1\%$, $\ell_{\mathrm{SRC}} = 56$ m, and $\phi_{\mathrm{SRC}} = 4.7°$. Other Advanced LIGO parameters are $t^2_{\mathrm{ITM}} = 1.4\%$, $t^2_{\mathrm{PRM}} = 3\%$, $G_{\mathrm{arm}} = 280$, $G_{\mathrm{prc}} = 40$, $L = 4$ km, and $I_0 = 125$ W. (LIGO Document T0900288-v3[7].)

diation pressure noise can be reduced by increasing the masses of the mirrors. In addition, there are other mechanisms for reducing the readout noise.

Quantum readout noise can be further suppressed, and there are a number of methods to do so. One method involves changing the vacuum state of the quantum fields entering the interferometer through the antisymmetric port. In a quantum mechanical treatment of the radiation fields in the interferometer, both shot noise and radiation pressure noise are found to arise from the vacuum fluctuations "entering" the interferometer from the (open) dark port. The natural vacuum state fluctuations have an uncertainty relationship in which conjugate variables associated with the amplitude fluctuations of the vacuum (which are responsible for radiation pressure) and the phase fluctuations of the vacuum (which affect the phase measurements of the interferometer and are therefore responsible for the shot noise) are equally distributed. If the vacuum is *squeezed* so that the phase fluctuations are decreased at the cost (due to the uncertainty principle) of increased amplitude fluctuations, then the shot noise can be decreased at the cost of increased radiation pressure noise. This would lead to an overall improvement in sensitivity at high frequencies where the quantum noise is dominated by shot noise, though

7) https://dcc.ligo.org/cgi-bin/DocDB/ShowDocument?docid=2974 (last accessed 2011-01-03)

the low-frequency radiation pressure noise would be increased. Conversely, one could squeeze the vacuum such that the amplitude fluctuations are reduced and the phase fluctuations are increased, and this would improve sensitivity at low frequencies while worsening sensitivity at high frequencies.

In fact, it is possible to perform frequency-dependent squeezing so that at low frequencies it is the amplitude fluctuations that are decreased while at high frequencies it is phase fluctuations that are decreased. In this way the limiting quantum noise can be decreased over the entire sensitivity band.

Squeezed vacuum states are difficult to maintain stably in practice, and current experiments have only been able to realize a modest gain in quantum noise compensation. Other possibilities are also being explored such as a xylophone strategy, which involves combining two interferometers, one cryogenically cooled interferometer with low laser power that has high sensitivity at low frequencies, and a second interferometer with high laser power that has high sensitivity at high frequencies, so that the resulting synthetic detector has high sensitivity at both high and low frequencies.

6.2 Space-Based Detectors

6.2.1 Spacecraft Tracking

Doppler tracking of spacecraft provides long baseline $L \gtrsim 10^{10}$ m gravitational-wave detectors, which are sensitive to gravitational waves at $f \sim$ mHz frequencies. Beginning with tracking data for the Voyager 1 spacecraft obtained in 1979 and 1980, there have been several searches for gravitational waves using spacecraft Doppler tracking techniques; Armstrong (2006) gives a detailed review of these techniques and the experiments that have been performed to date.

In spacecraft Doppler tracking experiments, a microwave radio link is used to monitor the redshift $\mathfrak{z} = (\nu_{\rm em} - \nu_{\rm rec})/\nu_{\rm rec}$ between a signal emitted from the Earth and the return signal that is transponded by a spacecraft. We have seen that a gravitational wave affects both the up-link and the down-link of a tracking system, and it will induce a redshift given by (see Eq. (6.175))

$$\mathfrak{z}(t) = \frac{1}{2}\hat{p}^i\hat{p}^j \cdot \left[\frac{h_{ij}(t)}{1-\hat{\boldsymbol{p}}\cdot\hat{\boldsymbol{n}}} - 2\frac{(\hat{\boldsymbol{p}}\cdot\hat{\boldsymbol{n}})h_{ij}\left[t+(L/c)(1-\hat{\boldsymbol{p}}\cdot\hat{\boldsymbol{n}})\right]}{1-(\hat{\boldsymbol{p}}\cdot\hat{\boldsymbol{n}})^2} + \frac{h_{ij}(t+2L/c)}{1+\hat{\boldsymbol{p}}\cdot\hat{\boldsymbol{n}}}\right], \tag{6.205}$$

where $\hat{\boldsymbol{p}}$ is the unit vector directed from the Earth to the spacecraft. The redshift measurements are affected by several noise sources: frequency noise in the radio signal; antenna noise in the Earth-based transmitter, spacecraft-based transponder and Earth-based receiver; noise caused by the atmosphere and by interplanetary plasma; and noise associated with buffeting of the spacecraft.

6.2.2 LISA

The Laser Interferometer Space Antenna (LISA) is a planned NASA/ESA mission to measure gravitational waves through spacecraft tracking using laser interferometry. LISA will consist of a triangular constellation of three spacecraft in heliocentric orbits that trail the Earth's orbit by 20°. The inter-spacecraft distance will be 5×10^9 m, and laser interferometry between pairs of detectors is used to track the inter-spacecraft separations. Contained within each spacecraft will be two proof masses that move along spacetime geodesics (the surrounding spacecraft follow the motion of the proof masses and shield them from external buffeting, e.g. from the solar wind and radiation). LISA will be sensitive to gravitational waves between 0.03 mHz and 0.1 Hz.

The inter-spacecraft distance is tracked using laser ranging. A master laser on one spacecraft directs a beam of light toward the remote spacecraft; a slave laser on the remote is phase-locked to the light received from the master laser, and produces a return beam directed toward the original spacecraft. This returning beam is then compared to the master laser's phase to determine the round-trip redshift.

This procedure, which is exactly Doppler tracking of spacecraft, is limited by frequency noise in the laser, and the remedy is exactly the same as in ground-based beam detectors: the interference of the signals from different arms can be used to cancel the laser frequency noise. With ground based detectors, the interference from the two arms is performed optically at a beam splitter, but with LISA it is performed using *time delay interferometry* (TDI) (Armstrong *et al.*, 1999). We now describe this method.

We begin by considering the six one-way Doppler shifts: let $\mathfrak{z}_{2\to 1}(t)$ be the one-way Doppler shift that is measured on spacecraft 1 from the laser beam emitted by spacecraft 2, $\mathfrak{z}_{3\to 1}(t)$ be the one-way Doppler shift that is measured on spacecraft 1 from the laser beam emitted by spacecraft 3, and so on. Because we are interested in combining (interfering) all six one-way Doppler shift measurements, we adopt a coordinate system with origin $\mathcal{O}$ at the centre of the triangular constellation so that all three spacecraft are the same distance ℓ from the origin and the direction to the three spacecraft are $\hat{\boldsymbol{r}}_1$, $\hat{\boldsymbol{r}}_2$ and $\hat{\boldsymbol{r}}_3$. The spacecraft 1 and 2 pair is separated by length L_{12} and the separation unit vector is $\hat{\boldsymbol{p}}_{12} = (\hat{\boldsymbol{r}}_1 - \hat{\boldsymbol{r}}_2)/\|\hat{\boldsymbol{r}}_1 - \hat{\boldsymbol{r}}_2\|$; we make similar definitions for the pair of spacecraft 1 and 3, and the pair of spacecraft 2 and 3.[8] The gravitational-wave contribution to the Doppler shift for the light emitted at

8) In much of the LISA literature, the variables are described in terms of links and receivers. The links are the sides of the triangle of the constellation, and they are labelled by the opposite vertex. So, for example, the link $1 \to 2$ is labelled 3 (the opposite vertex). That is, L_{12} is written L_3, and the Doppler data streams, instead of being labelled in terms of the sending and receiving spacecraft, are labelled by the link and the receiving spacecraft, for example the Doppler data stream $\mathfrak{z}_{2\to 1}$ would be written as y_{31}.

spacecraft 2 and received at spacecraft 1 is given by

$$\mathfrak{z}_{2\to 1,\mathrm{GW}}(t) = \frac{1}{2}\hat{p}^i_{12}\hat{p}^j_{12}\frac{h_{ij}\left[t-(\ell/c)\hat{\boldsymbol{r}}_1\cdot\hat{\boldsymbol{n}}\right]-h_{ij}\left[t-(\ell/c)\hat{\boldsymbol{r}}_2\cdot\hat{\boldsymbol{n}}-L_{12}/c\right]}{1-\hat{\boldsymbol{p}}_{12}\cdot\hat{\boldsymbol{n}}} \,. \tag{6.206}$$

Note that $h_{ij}(t)$ is the metric perturbation at time t at the origin $\mathcal{O}$. In addition to the gravitational-wave signal, we consider the frequency noise: the frequency of the laser light on spacecraft 2 is $\nu_2 = \nu_0[1+C_2(t)]$ while the frequency of the laser light on spacecraft 1 is $\nu_1 = \nu_0[1+C_1(t)]$, where the functions $C_2(t)$ and $C_1(t)$ encode the frequency noise. The contribution to the measured Doppler shift from laser frequency noise depends on the frequency noise at the emitter, spacecraft 2, and also the frequency noise at the receiver, spacecraft 1 (since the laser at the receiver is maintaining the standard for measuring the Doppler shift):

$$\mathfrak{z}_{2\to 1,\mathrm{laser}}(t) = C_2(t-L_{12}/c) - C_1(t) \,. \tag{6.207}$$

We wish to construct combinations of the one-way Doppler data streams in which the laser noise terms cancel. For example, if we construct $\mathfrak{z}_{3\to 1}(t) - \mathfrak{z}_{2\to 1}(t)$ then the frequency noise in the laser in spacecraft 1 will cancel, though there will remain a residual noise component $C_3(t-L_{13}/c) - C_2(t-L_{12}/c)$ from the frequency noise in the lasers in spacecrafts 2 and 3. However, the combination

$$\begin{aligned}\alpha(t) := {}& \mathfrak{z}_{3\to 1}(t) - \mathfrak{z}_{2\to 1}(t)\\ &+ \mathfrak{z}_{2\to 3}(t-L_{13}/c) - \mathfrak{z}_{3\to 2}(t-L_{12}/c)\\ &+ \mathfrak{z}_{1\to 2}(t-L_{13}/c-L_{23}/c) - \mathfrak{z}_{1\to 3}(t-L_{12}/c-L_{23}/c)\end{aligned} \tag{6.208a}$$

cancels the frequency noise from *all* the lasers, which can be verified by directly substituting laser noise contributions of the form Eq. (6.207) into Eq. (6.208a). This is obviously not a unique combination: by permutation of the labels of the spacecraft, one easily obtains two more combinations:

$$\begin{aligned}\beta(t) := {}& \mathfrak{z}_{1\to 2}(t) - \mathfrak{z}_{3\to 2}(t)\\ &+ \mathfrak{z}_{3\to 1}(t-L_{21}/c) - \mathfrak{z}_{1\to 3}(t-L_{23}/c)\\ &+ \mathfrak{z}_{2\to 3}(t-L_{21}/c-L_{31}/c) - \mathfrak{z}_{2\to 1}(t-L_{23}/c-L_{31}/c) \,,\end{aligned} \tag{6.208b}$$

$$\begin{aligned}\gamma(t) := {}& \mathfrak{z}_{2\to 3}(t) - \mathfrak{z}_{1\to 3}(t)\\ &+ \mathfrak{z}_{1\to 2}(t-L_{32}/c) - \mathfrak{z}_{2\to 1}(t-L_{31}/c)\\ &+ \mathfrak{z}_{3\to 1}(t-L_{32}/c-L_{12}/c) - \mathfrak{z}_{3\to 2}(t-L_{31}/c-L_{12}/c) \,.\end{aligned} \tag{6.208c}$$

Because the Doppler data must be combined with appropriate time delays, this procedure is known as time delay interferometry.

The above TDI variables use all three links between the spacecraft. A Michelson-like TDI variable, which also cancels the laser frequency noise, can be constructed by using only the links between spacecrafts 1 and 2 and spacecrafts 1 and 3, but avoiding the link between spacecrafts 2 and 3:

$$\begin{aligned}X(t) := {} & \mathfrak{z}_{3\to1}(t) - \mathfrak{z}_{2\to1}(t) \\ & + \mathfrak{z}_{1\to3}(t - L_{13}/c) - \mathfrak{z}_{1\to2}(t - L_{12}/c) \\ & - \mathfrak{z}_{3\to1}(t - 2L_{12}/c) + \mathfrak{z}_{2\to1}(t - 2L_{13}/c) \\ & - \mathfrak{z}_{1\to3}\,(t - 2L_{12}/c - L_{13}/c) + \mathfrak{z}_{1\to2}\,(t - 2L_{13}/c - L_{12}/c)\,.\end{aligned} \tag{6.209a}$$

By permuting the spacecraft labels two other Michelson-like TDI variables are obtained:

$$\begin{aligned}Y(t) := {} & \mathfrak{z}_{1\to2}(t) - \mathfrak{z}_{3\to2}(t) \\ & + \mathfrak{z}_{2\to1}(t - L_{21}/c) - \mathfrak{z}_{2\to3}(t - L_{23}/c) \\ & - \mathfrak{z}_{1\to2}(t - 2L_{23}/c) + \mathfrak{z}_{3\to2}(t - 2L_{21}/c) \\ & - \mathfrak{z}_{2\to1}\,(t - 2L_{23}/c - L_{21}/c) + \mathfrak{z}_{2\to3}\,(t - 2L_{21}/c - L_{23}/c)\end{aligned} \tag{6.209b}$$

and

$$\begin{aligned}Z(t) := {} & \mathfrak{z}_{2\to3}(t) - \mathfrak{z}_{1\to3}(t) \\ & + \mathfrak{z}_{3\to2}(t - L_{32}/c) - \mathfrak{z}_{3\to1}(t - L_{31}/c) \\ & - \mathfrak{z}_{2\to3}(t - 2L_{31}/c) + \mathfrak{z}_{1\to3}(t - 2L_{32}/c) \\ & - \mathfrak{z}_{3\to2}\,(t - 2L_{31}/c - L_{32}/c) + \mathfrak{z}_{3\to1}\,(t - 2L_{32}/c - L_{31}/c)\,.\end{aligned} \tag{6.209c}$$

We can interpret the synthesized interferometers $\alpha(t)$, $\beta(t)$ and $\gamma(t)$ as the difference between two beams travelling cyclically in opposite directions, which is known as a *Saganac interferometer*. For example, we can rearrange the terms in Eq. (6.208a) to obtain the following expression for $\alpha(t)$:

$$\begin{aligned}\alpha(t) = {} & \left[\mathfrak{z}_{3\to1}(t) + \mathfrak{z}_{2\to3}(t - L_{13}/c) + \mathfrak{z}_{1\to2}\,(t - L_{13}/c - L_{23}/c)\right] \\ & - \left[\mathfrak{z}_{2\to1}(t) + \mathfrak{z}_{3\to2}(t - L_{12}/c) + \mathfrak{z}_{1\to3}\,(t - L_{12}/c - L_{23}/c)\right].\end{aligned} \tag{6.210}$$

In effect, spacecraft 1 produces two beams, one which travels in the cycle $1 \to 2 \to 3 \to 1$, and the other that travels in the opposite cycle $1 \to 3 \to 2 \to 1$, and these two beams are interfered. However, if the constellation of spacecraft is rotating, as it will be in LISA, then the beams travelling in opposing cycles will experience a different delay – this is known as the *Saganac effect*. Because of this the TDI variables $\alpha(t)$, $\beta(t)$ and $\gamma(t)$ given above are not sufficient for cancelling the laser frequency noise. The rotation of the LISA constellation, along with changing distances between the spacecraft, also causes the Michelson-like TDI variables to fail to cancel the laser frequency noise. New *second generation* combinations, which cancel laser frequency noise even when the spacecraft are moving, can be constructed, but they are more complicated than the above *first generation* TDI variables.

After constructing TDI readout variables that cancel the laser frequency noise, two main sources of noise will dominate. At high frequencies, laser shot noise is the most important noise source. Each spacecraft emits a $I_0 = 1\,\mathrm{W}$, $\lambda = 1.064\,\mu\mathrm{m}$ laser beam directed at each of the other spacecraft, and receives similar beams directed from the other two spacecraft; however, the incoming beams are very weak, $I \sim 100\,\mathrm{pW}$, so the shot noise limit is $S_h \sim \hbar c k/[2I(kL)^2] \sim (10^{-21}\,\mathrm{Hz}^{-1/2})^2$, where $k = 2\pi/\lambda$; since this noise is encountered in each link that is combined to form a TDI variable, typical shot noise levels will be a factor of $\sim$10 higher. At low frequencies, acceleration noise in the motion of the test masses is the dominant *instrumental* noise source. However, at frequencies below $\sim$3 mHz the dominant noise source is actually due to a stochastic background of gravitational waves from white dwarf binary systems in the Galaxy. Because galactic binaries with orbital periods of several minutes or greater are so numerous, the unresolvable background of signals from these binaries provides a confusion noise that dominates all other noise sources at low frequencies.

The light travel time of 16.7 s between the LISA spacecraft means that the strain-equivalent noise curve begins to rise at frequencies above a few tens of mHz as the

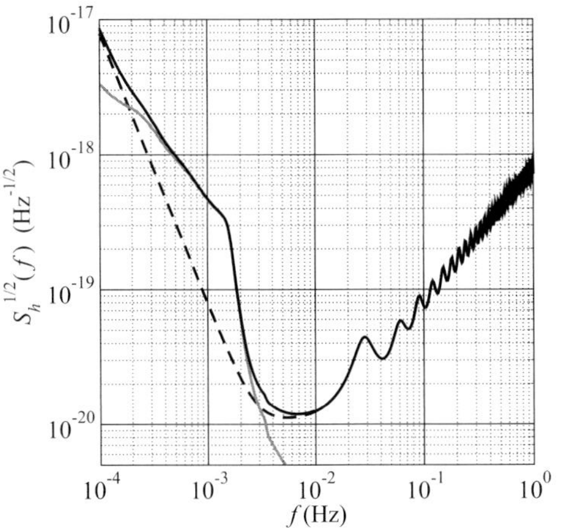

Figure 6.17 The LISA strain sensitivity curve $S_h^{1/2}(f)$, averaged over source sky positions and signal polarizations, for the Michelson X TDI variable. The dashed curve shows the instrumental sources of noise: spacecraft position and acceleration noise, and laser shot noise. The grey curve is the gravitational-wave background noise caused by unresolved Galactic binary systems. The black curve is the overall LISA sensitivity curve. (Source: Shane L. Larson *Online Sensitivity Curve Generator*,[9] based on Larson *et al.* (2000).)

9) http://www.srl.caltech.edu/~shane/sensitivity (last accessed 2011-01-03)

gravitational waves will no longer affect all spacecraft in concert and they begin to have self-cancelling effects. To determine the LISA response to a gravitational wave, one must substitute Eq. (6.206) into the formula for the particular TDI variable that is being considered and use a Fourier transform to relate the Fourier component of the TDI variable to the Fourier component of the gravitational wave. The resulting response function will depend on the sky position of the gravitational-wave source as well as on the TDI variable chosen. Figure 6.17 shows the sensitivity curve for LISA, averaged over source sky position and signal polarization for a particular TDI variable.

6.2.3
Decihertz Experiments

The frequency band between 0.1 and 1 Hz is inaccessible to either LISA or ground-based detectors: it is too high in frequency for LISA because of LISA's large inter-spacecraft distances, and it is too low in frequency for ground-based detectors because of terrestrial gravity gradient noise. However, this is a potentially important band for detection of a primordial background of gravitational waves because such a background is expected to fall off rapidly with increasing frequency, so they are more detectable in this band than at the higher-frequency band probed by ground-based detectors; yet the Galactic binaries may not be so numerous at these frequencies that they would create an unresolvable background that would mask the primordial background. There are two concepts for space-based detectors that would operate at frequencies $\sim$0.1 Hz: the Decihertz Interferometer Gravitational-wave Observatory (DECIGO) (Seto *et al.*, 2001; Kawamura *et al.*, 2006) and the Big Bang Observatory (BBO) (Crowder and Cornish, 2005), which would both have shorter arms than LISA.

6.3
Pulsar Timing Experiments

Operating at still lower frequencies, between 1 nHz and 1 mHz, are gravitational-wave experiments conducted through pulsar timing measurements (Lorimer, 2001).

Pulsars are rapidly rotating neutron stars whose magnetic axis passes near to the line-of-sight between the Earth and the pulsar. Each time the magnetic axis and the line-of-sight align we see a flash of electromagnetic radiation, which is produced by synchrotron emission from electrons in the magnetic field. Because of the large moment of inertia of neutron stars ($\sim 10^{38}\,\mathrm{kg\,m^2}$), the rotation of a neutron star is very stable; although each individual pulse light-curve can differ dramatically, the light curve averaged over many pulses (a process known as *folding*) will also be remarkably stable. Pulsars therefore make very good clocks, with an accuracy rivaling modern atomic clocks. Since gravitational waves affect clocks as they pass, we can use the times of arrival of pulses from pulsars as sensitive detectors of low-frequency gravitational waves (with frequencies on the order of one over the total duration of observations, which can be around ten years).

The uncertainty in the time of arrival (TOA) of a pulse depends on the strength of the radio emission from the pulsar, the sharpness of each pulse from that pulsar, the sensitivity of the radio receiver on Earth, and the integration time, which describes how many different pulses are folded together to get a single light-curve. Pulse TOAs can be measured to ~100 ns in some cases, where the TOA measurement involves integrating over several minutes. These TOAs are measured approximately every month.

The TOAs are converted from Earth-based atomic time to *solar system barycentre time*, as the solar system barycentre is nearly an inertial frame of reference. The solar system barycentre time, T, is related to the *topocentric time*, t, at the Earth by

$$T = t + \Delta_{\text{Roemer}} + \Delta_{\text{Shapiro}} + \Delta_{\text{Einstein}} , \tag{6.211}$$

where the correction factors are the classical *Roemer delay*, Δ_{Roemer}, which results from the light travel time from the solar system barycentre to the detector; the *Shapiro delay*, Δ_{Shapiro} which results from the curvature of spacetime near the Sun; and the *Einstein delay* Δ_{Einstein}, which results from the gravitational redshift of the signal caused by the Sun and other planets.

The largest correction is the Roemer delay,

$$\Delta_{\text{Roemer}} = -\frac{\boldsymbol{r} \cdot \hat{\boldsymbol{n}}}{c} + \frac{(\boldsymbol{r} \cdot \hat{\boldsymbol{n}})^2 - \|\boldsymbol{r}\|^2}{2cD} , \tag{6.212}$$

where $\boldsymbol{r}$ is the location of the detector relative to the solar system barycentre, $\hat{\boldsymbol{n}}$ is the direction unit vector from the pulsar to Earth, and D is the distance to the pulsar. The second term in the Roemer delay accounts for the curvature of the wavefront that must be included if the pulsar is relatively near. The Shapiro delay is given by

$$\Delta_{\text{Shapiro}} = -\frac{2GM_\odot}{c^3} \log\left(1 - \hat{\boldsymbol{n}} \cdot \boldsymbol{r}/\|\boldsymbol{r}\|\right) \tag{6.213}$$

and the Einstein delay is found by integrating the equation

$$\frac{d\Delta_{\text{Einstein}}}{dt} = \frac{GM_\odot}{c^2\|\boldsymbol{r}\|} + \frac{\|d\boldsymbol{r}/dt\|^2}{2c^2} . \tag{6.214}$$

The conversion from topocentric time (the time at the detector) and solar system barycentric time requires an accurate ephemeris of the position of the Earth, and it depends on the direction to the pulsar.

In solar system barycentre time, T, the phase of the neutron star $\varphi_{\text{obs}}(T)$ is computed from the TOAs, and can be fit to a model of the phase:

$$\varphi_{\text{model}}(t) = \varphi_0 + (T - T_0)\omega_0 + \frac{1}{2}(T - T_0)^2\dot{\omega}_0 , \tag{6.215}$$

where T is related to t via Eq. (6.211) and φ_0 is the pulsar phase at some fiducial time t_0.[10] This model represents a Taylor expansion of the neutron star's rotational angular velocity $\omega(T)$ (which changes with time due to spin-down of the neutron star) about some fixed value $\omega_0 = \omega(T_0)$. The difference between the observed pulse time and the predicted pulse time is the *timing residual,*

$$R(t) := \frac{\varphi_{\text{model}}(t) - \varphi_{\text{obs}}(t)}{2\pi\nu} , \tag{6.216}$$

where $\nu = \omega_0/(2\pi)$ is the frequency of the pulsar's pulses. The timing residuals are minimized by finding the best fit values of the parameters ω_0, $\dot{\omega}_0$, and the direction to and proper motion of the pulsar. The remaining residual is what we use to search for gravitational waves.

Young pulsars often exhibit considerable *timing noise* in the residuals, which likely comes from processes internal to the neutron star itself, for example temperature changes. However, old pulsars are often very stable, and the timing noise in the residuals is dominated by the measurement errors in the TOAs. Since these errors are independent – the error in measuring the TOA one month is independent of the error in the next month – the intrinsic noise in the pulsar timing residuals is *white.* (In fact, at low frequencies there is, for some pulsars, more noise than there is at high frequencies, so the timing noise is sometimes found to be somewhat *red,* but it is not known what the origin of the red noise is.)

A gravitational wave induces a Doppler shift on the signals from the pulsar, which is realized in the timing residual as

$$R(t) = \int_{t_0}^{t} \mathfrak{z}(t')dt' , \tag{6.217a}$$

where the gravitational-wave redshift

$$\mathfrak{z}(t) = \frac{\nu_{\text{pulsar}} - \nu_{\text{Earth}}(t)}{\nu_{\text{Earth}}} = \frac{1}{2}\frac{\hat{p}^i\hat{p}^j}{1 + \hat{\boldsymbol{p}}\cdot\hat{\boldsymbol{n}}}\left(h_{ij,\text{Earth}} - h_{ij,\text{pulsar}}\right) \tag{6.217b}$$

contains two terms: the first term is known as the *Earth term* and results from the metric perturbation at the time the pulse is received, and the second term is known as the *pulsar term* and results from the metric perturbation at the time the pulse was emitted. Here, $\hat{\boldsymbol{p}}$ is the unit vector directed from the Earth to the pulsar and $\hat{\boldsymbol{n}}$ is the unit vector in the direction of propagation of the gravitational wave. If we retain only the Earth term then we can relate the Fourier components of the

10) In fact, we do not wish to develop the model in terms of the solar system barycentre time, but instead in terms of the time coordinate at the pulsar. An additional correction due to the dispersion of the radio signal by the interstellar medium is therefore necessary to include in Eq. (6.211). The amount of dispersion can be deduced by viewing the pulse times at various radio frequencies; the difference in arrival time as a function of radio frequency allows one to determine the amount of interstellar dispersion and to correct for this factor.

timing residuals to the Fourier coefficients of the gravitational-wave signal by

$$\tilde{R}(f) = \frac{1}{2\pi i f}\tilde{\mathfrak{z}}(f) = \frac{1}{2\pi i f}\left[F_+(\hat{\boldsymbol{n}})\,\tilde{h}_+(f) + F_\times(\hat{\boldsymbol{n}})\,\tilde{h}_\times(f)\right], \tag{6.218a}$$

where antenna beam pattern functions $F_+(\hat{\boldsymbol{n}})$ and $F_\times(\hat{\boldsymbol{n}})$ are

$$F_+(\hat{\boldsymbol{n}}) := \frac{1}{2}\frac{\hat{p}^i\hat{p}^j}{1+\hat{\boldsymbol{p}}\cdot\hat{\boldsymbol{n}}}e^+_{ij}(\hat{\boldsymbol{n}}) \tag{6.218b}$$

and

$$F_\times(\hat{\boldsymbol{n}}) := \frac{1}{2}\frac{\hat{p}^i\hat{p}^j}{1+\hat{\boldsymbol{p}}\cdot\hat{\boldsymbol{n}}}e^\times_{ij}(\hat{\boldsymbol{n}})\,. \tag{6.218c}$$

These can be expressed in terms of a sensing response function for the timing residuals, $C(f)$, and an effective strain $\tilde{h}(f) := F_+\tilde{h}(f) + F_\times\tilde{h}(f)$, as

$$\tilde{R}(f) = C(f)\tilde{h}(f) \tag{6.219}$$

with

$$C(f) = \frac{1}{2\pi i f}\,. \tag{6.220}$$

As mentioned earlier, the noise in the timing residuals, $\sigma \sim 100$ ns, is, for stable pulsars, essentially a white noise source; it creates a shot noise

$$S_R(f) = 2\sigma^2\Delta t\,, \tag{6.221}$$

where Δt is the sampling interval. This means that the strain-equivalent pulsar timing noise is

$$S_h(f) = \frac{S_R(f)}{|C(f)|^2} = 8\pi^2\sigma^2 f^2\Delta t\,. \tag{6.222}$$

The gravitational-wave frequencies that a pulsar timing experiment can observe are $T^{-1} < f < \frac{1}{2}(\Delta t)^{-1}$ where T is the total observational time. Typically, pulsar timing residuals are measured approximately monthly, so $\Delta t \sim 3\times 10^6$ s and, if we assume ten years of pulsar timing, $T \sim 10$ years $\sim 3\times 10^8$ s, the frequency range of the experiment is $3\,\text{nHz} \lesssim f \lesssim 0.2\,\text{mHz}$, and

$$S_h^{1/2}(f) \approx 4\times 10^{-12}\,\text{Hz}^{-1/2}\left(\frac{f}{3\,\text{nHz}}\right)\left(\frac{\sigma}{100\,\text{ns}}\right)\left(\frac{\Delta t}{1\,\text{month}}\right)^{1/2} \tag{6.223}$$

in this frequency band. The lower bound of the frequency band can be reduced by longer observations, while the upper bound of the frequency band, as well as the strain sensitivity, can be increased by increasing the sampling rate.

In fact, the response function is more complicated than is described by $C(f)$ because the timing residuals have been fitted to a phase model. In particular, the fit to ω_0 and $\dot{\omega}_0$ (a linear and a quadratic piece) affects the low-frequency sensitivity, while the fit to unknown pulsar location and proper motion essentially removes sensitivity at the frequency $f = 1/(\text{sidereal year})$.

Currently, there are $N_{\text{pulsars}} \sim 20$ known pulsars that have timing accuracies of $\sim$100 nHz. Such a pulsar timing array will be a factor of $N_{\text{pulsars}}^{1/2}$ more sensitive in $S_h^{1/2}(f)$.

6.4 Resonant Mass Detectors

The original gravitational-wave detectors were cylindrical resonant mass detectors, also known as *bar detectors*. Gravitational waves will excite the acoustic modes of oscillation of the bar, and these oscillations are measured with a transducer. Typical bar detectors are constructed of high-quality materials and have masses on the order of several tonnes. They are suspended on stacks that provide seismic isolation and are cryogenically cooled to keep thermal noise low.

A simplified model of a resonant mass detector as a damped harmonic oscillator was presented in Example 3.6. We model the cylindrical bar as a harmonic oscillator with effective mass $\mu = M/2$ where M is the mass of the bar. The position variable, x, is the observable, and the equation of motion is

$$\frac{d^2x}{dt^2} + \frac{2\pi f_0}{Q}\frac{dx}{dt} + 4\pi^2 f_0^2 x = -\left(\frac{2}{\pi}\right)^2 \frac{1}{2}\frac{d^2h}{dt^2} L\,, \tag{6.224}$$

where f_0 is the natural frequency of the bar and Q is its quality factor. The factor $(2/\pi)^2$ is a geometric factor that arises because the bar is really a solid cylinder rather than two masses connected by a light spring. In the frequency domain we have

$$\tilde{x}(f) = G(f)\tilde{h}(f)\,, \tag{6.225}$$

where

$$G(f) = \frac{2L}{\pi^2}\frac{f^2}{(f_0^2 - f^2) + i f f_0/Q} \tag{6.226}$$

is the transfer function.

The two dominant noise sources at the resonant frequency of bar detectors are thermal noise of the bar and the sensor noise of the transducer and amplifier. The thermal noise can be computed via Eq. (6.130), and we obtain

$$S_{x,\text{thermal}}(f) = \frac{k_B T}{2\pi^3\left(\frac{1}{2}M\right)}\frac{1}{Q}\frac{f_0}{\left(f_0^2 - f^2\right)^2 + f_0^4/Q^2}\,. \tag{6.227}$$

This can be converted into a strain-equivalent noise power spectrum,

$$S_{h,\mathrm{thermal}}(f) = \frac{S_x(f)}{|G(f)|^2} = \frac{\pi}{8}\frac{1}{L^2}\frac{k_\mathrm{B}T}{\frac{1}{2}M}\frac{1}{Q}\frac{f_0}{f^4}\,. \tag{6.228}$$

For a $L \sim 3\,\mathrm{m}$ bar of mass $M \sim 2000\,\mathrm{kg}$ having resonance frequency $f_0 \sim 1\,\mathrm{kHz}$ and quality $Q \sim 10^6$, $S^{1/2}_{h,\mathrm{thermal}}(f_0) \sim 2\times10^{-21}\,\mathrm{Hz}^{-1/2}$ if it is cooled to liquid helium temperature, $T = 4.2\,\mathrm{K}$. The thermal noise determines the best sensitivity that can be achieved for a bar detector at a given temperature regardless of the readout scheme.

Rather than directly measuring the displacement of the end of the bar, the sensitivity of the readout scheme can be dramatically improved by using a *resonant transducer*. In this scheme an additional mass is mechanically coupled to the bar. If the mass of the transducer m is much smaller than the effective mass of the bar $m \ll M/2$, and if the coupling between the transducer and the bar has a natural frequency that is the same as the natural frequency of the acoustic oscillations in the bar, then these oscillations are mechanically amplified in the small mass. The readout now measures the position of the small mass rather than directly measuring the oscillation displacements in the bar. The coupled oscillators now have two normal mode frequencies,

$$f_\pm = f_0\left(1 \pm \frac{1}{2}\sqrt{\frac{m}{\frac{1}{2}M}}\right), \tag{6.229}$$

and the transfer function becomes

$$G(f) = \left(\frac{2}{\pi}\right)^2 \frac{L}{2}\frac{f^2 f_0^2}{\left(f_+^2 - f^2 + i f f_0/Q\right)\left(f_-^2 - f^2 + i f f_0/Q\right)}\,, \tag{6.230}$$

where we suppose that the quality factor of the resonant oscillations of the bar is approximately equal to the quality factor of the coupling with the transducer. There is also additional thermal noise for the transducer, so we now have

$$S_{h,\mathrm{thermal}}(f) = \frac{\pi}{8}\frac{1}{L^2}\frac{k_\mathrm{B}T}{\frac{1}{2}M}\frac{1}{Q}\frac{f_0}{f^4}\left\{1 + \frac{\frac{1}{2}M}{m}\frac{\left(f^2-f_0^2\right)^2}{f_0^4}\right\}. \tag{6.231}$$

Example 6.8 Coupled oscillators

Consider two coupled oscillators, the first being a mass $\frac{1}{2}M$ with displacement x, and spring constant $K = \frac{1}{2}M(2\pi f_0)^2$, and the second being a mass $m = \epsilon\frac{1}{2}M$ with position y, which is coupled to the first by a spring with spring constant $k =$

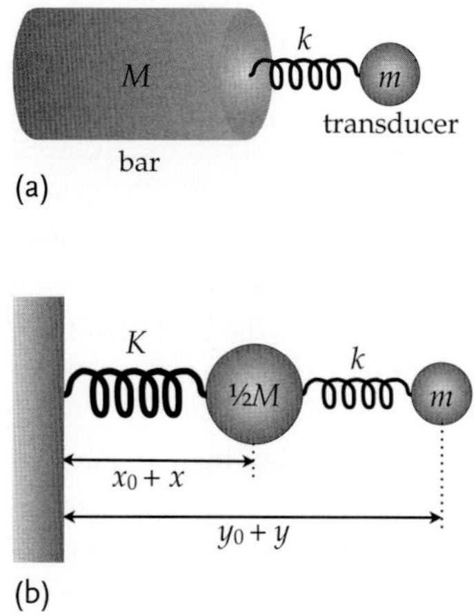

Figure 6.18 (a) A bar of mass M is coupled to a transducer of mass m with an effective coupling spring constant k. (b) The bar itself can be treated as an oscillator with effective mass $1/2M$ and spring constant K. Here, x_0 and y_0 are the equilibrium positions of the effective bar mass and the transducer, while x and y are the displacements of these masses with respect to the equilibrium.

$m(2\pi f_0)^2$. (See Figure 6.18.) Assuming that the mechanical losses in the springs are the same, so the quality factor for both oscillators individually are Q, then the equations of motion for the coupled system are

$$\frac{d^2x}{dt^2} + \frac{2\pi f_0}{Q}\frac{dx}{dt} + 4\pi^2 f_0^2 x + \frac{2\pi f_0}{Q}\frac{d(x-y)}{dt} + 4\pi^2 f_0^2 \frac{m}{\frac{1}{2}M}(x-y)$$
$$= -\left(\frac{2}{\pi}\right)^2 \frac{d^2h}{dt^2} L \,, \tag{6.232a}$$

$$\frac{d^2y}{dt^2} + \frac{2\pi f_0}{Q}\frac{d(y-x)}{dt} + 4\pi^2 f_0^2 (y-x) = \frac{F_{\text{ext}}}{m} \,, \tag{6.232b}$$

where F_{ext} is any external force (particularly thermal force fluctuations or back-action from the amplifier) on the transducer. From these equations with $F_{\text{ext}} = 0$ we obtain the transfer function,

$$G(f) := \frac{\tilde{y}(f)}{\tilde{h}(f)}$$
$$= \left(\frac{2}{\pi}\right)^2 \frac{L}{2} f^2 \frac{f_0^2 + i f f_0/Q}{\left(f_0^2 - f^2 + i f f_0/Q\right)^2 - \epsilon f^2 \left(f_0^2 + i f f_0/Q\right)} \,, \tag{6.233}$$

and with $h = 0$ we obtain the admittance,

$$Y(f) := 2\pi i f \frac{\tilde{y}(f)}{\tilde{F}_{\rm ext}}$$
$$= \frac{2\pi i f}{4\pi^2 m} \frac{(f_0^2 - f^2 + i f f_0/Q) + \epsilon (f_0^2 + i f f_0/Q)}{(f_0^2 - f^2 + i f f_0/Q)^2 - \epsilon f^2 (f_0^2 + i f f_0/Q)} . \tag{6.234}$$

From the transfer function and the admittance, we find the strain-equivalent thermal noise

$$S_{h,\rm thermal}(f) = \frac{S_{y,\rm thermal}(f)}{|G(f)|^2}$$
$$= \frac{k_{\rm B} T}{\pi^2 f^2} \frac{\pi^4}{4L^2} \frac{1}{2\pi m} \frac{f_0}{Q}$$
$$\times \frac{(f_0^2 - f^2)^2 + f^2 f_0^2/Q^2 + \epsilon (f_0^4 + f^2 f_0^2/Q^2)}{f^2 f_0^4 + f^4 f_0^2/Q^2}$$
$$\simeq \frac{\pi}{8} \frac{1}{L^2} \frac{k_{\rm B} T}{m} \frac{1}{Q} \frac{f_0}{f^4} \left\{ \frac{(f^2 - f_0^2)^2}{f_0^4} + \frac{m}{\frac{1}{2} M} \frac{2 f_0^2 - f^2}{f_0^2} \right\} \tag{6.235}$$

where in the last line we have discarded the small terms that were $O(Q^{-2})$; we see that this agrees with Eq. (6.228) for frequencies near the resonant frequency, $f \simeq f_0$.

The sensor noise depends (in part) on the electronics noise in the amplifier. If the amplifier noise is white noise with displacement-equivalent power $S_{x,\rm ampl}$ then the strain-equivalent sensor noise is

$$S_{h,\rm sensor}(f) = S_{x,\rm ampl} \frac{1}{|G(f)|^2} . \tag{6.236}$$

The sensor noise is strongly suppressed at the resonant frequency, but will dominate at frequencies substantially away from this frequency.

The sensitivity of the amplifier is often expressed either in terms of the number of phonons N that are required for measurement, or in terms of an effective noise temperature $T_{\rm eff}$ of the electronics. The ambient energy associated with the displacement power spectral density S_x is $E_{\rm noise} = m(2\pi f)^2 S_{x,\rm ampl} \Delta f$ where Δf is bandwidth of the readout (increasing the bandwidth increases the total amount of electronics noise from the amplifier). Each phonon has energy $2\pi\hbar f_0$ so the number of phonons required to exceed this noise energy level is $N = E_{\rm noise}/(2\pi\hbar f_0)$ or

$$N = \frac{2\pi f_0 m S_{x,\rm ampl} \Delta f}{\hbar} . \tag{6.237}$$

Typical readout systems require $N \gtrsim 100$ phonons when operating with a $\Delta f \sim 100$ Hz bandwidth. The noise energy can also be expressed in terms of effective

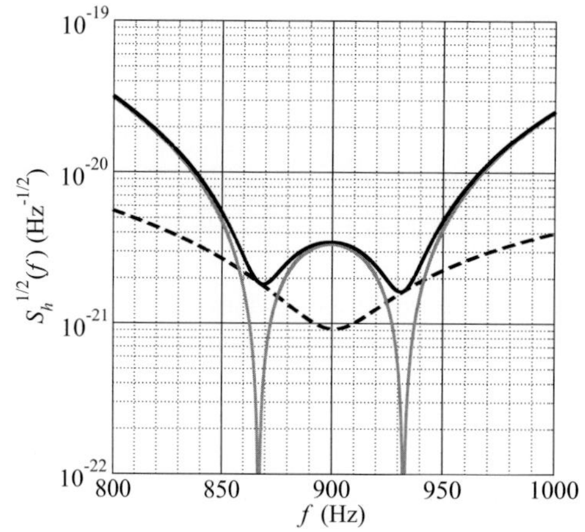

Figure 6.19 Amplitude spectral density of various noise components of resonant mass detector with typical parameter values for the AURIGA detector: bar length $L = 2.94$ m, bar mass $M = 2300$ kg, resonant frequency $f_0 = 900$ Hz and quality $Q = 4 \times 10^6$, bar temperature $T = 4.5$ K, transducer effective mass $m = 6.1$ kg, transducer quality $Q_t = 1.5 \times 10^6$, readout effective temperature $T_{\text{eff}} = 200\,\mu$K, readout bandwidth $\Delta f = 100$ Hz. The dashed line is the bar and transducer thermal noise, $S^{1/2}_{h,\text{thermal}}$, the grey line is the sensor electronics noise, $S^{1/2}_{h,\text{sensor}}$, and the black line is the total strain-equivalent detector noise $S^{1/2}_h$.

temperature, $T_{\text{eff}} := E_{\text{noise}}/k_B$

$$S_{x,\text{ampl}} = \frac{k_B T_{\text{eff}}}{m} \frac{1}{(2\pi f_0)^2} \frac{1}{\Delta f} . \tag{6.238}$$

For a readout system with effective temperature $T_{\text{eff}} \sim 100\,\mu$K, approximately $N \sim 2000$ phonons would be required to produce a measurable signal for $f_0 \sim 1$ kHz. A quantum limit on detection occurs when the effective temperature of the sensor is brought down to the $N = 1$ limit, $T_{\text{eff}} = 2\pi\hbar f_0/k_B$; improvements in sensitivity beyond this standard quantum limit require different readout schemes, but are possible in principle. Figure 6.19 shows a representative noise curve for a resonant mass detector with typical parameters.

Current resonant mass detectors have lengths of $L \sim 3$ m and resonant frequencies $f_0 \sim 1$ kHz, which correspond to gravitational waves with wavelengths $\lambda \sim 300$ m, so $\lambda \sim 100L$. We are therefore in the long-wavelength limit when it comes to determining the response of a resonant mass detector to a gravitational wave. Bar detectors are excited by the strain experienced relative to the principal axis,

$$h := \hat{p}^i \hat{p}^j h_{ij} , \tag{6.239}$$

where $\hat{\boldsymbol{p}}$ is the unit vector along the bar's axis. For a gravitational wave travelling in direction $\hat{\boldsymbol{n}}$, the strain is expressed in terms of the antenna beam pattern responses

$$h = F_+(\hat{\boldsymbol{n}}, \psi)h_+ + F_\times(\hat{\boldsymbol{n}}, \psi)h_\times \tag{6.240a}$$

with

$$F_+(\hat{\boldsymbol{n}}, \psi) := \hat{p}^i \hat{p}^j e^+_{ij}(\hat{\boldsymbol{n}}, \psi) \quad \text{and} \quad F_\times(\hat{\boldsymbol{n}}, \psi) := \hat{p}^i \hat{p}^j e^\times_{ij}(\hat{\boldsymbol{n}}, \psi) . \tag{6.240b}$$

These beam pattern functions can be expressed in terms of a gravitational-wave polarization angle ψ and the angle $\theta := \arccos(\hat{\boldsymbol{n}} \cdot \hat{\boldsymbol{p}})$ as

$$F_+(\theta, \psi) = -\sin^2\theta \cos 2\psi \quad \text{and} \quad F_\times(\theta, \psi) = +\sin^2\theta \sin 2\psi . \tag{6.240c}$$

Clearly the bar detectors are primarily sensitive to gravitational waves travelling orthogonal to the axis of the bar and polarized such that the stretching and squeezing is along the axis.

6.5 Problems

Problem 6.1

For a Fabry–Pérot cavity with a perfectly reflecting end test mass and an input test mass with amplitude reflectivity r_{ITM}, show that the average number of round trips that a photon will make before escaping the cavity is $\langle N_{\text{round-trips}}\rangle = r^2_{\text{ITM}}/(1-r^2_{\text{ITM}})$. If the cavity initially contains light power I_0 and there is no light incident on the cavity, show that the power contained in the cavity decreases with time as $I(t) = I_0 e^{-t/\tau_s}$ where $\tau_s = 2r^2_{\text{ITM}} L/[c(1-r^2_{\text{ITM}})]$.

Problem 6.2

Consider a Michelson interferometer with arms that are length ℓ_1 and ℓ_2. Suppose the laser has frequency noise so that the light frequency is $\omega = \omega_0 + \delta\omega \cos \omega_{\text{fm}} t$, and intensity noise, so the amplitude of the light is $E = E_0 + \delta E \cos \omega_{\text{am}} t$. Here $\delta\omega$ is the amplitude of the frequency noise and δE is the amplitude of the intensity noise while ω_{fm} and ω_{am} are the frequencies (in the audio band) of the noise fluctuations. The light produced by the laser therefore has the field

$$E = E_0 e^{i\omega_0 t}\left[1 + \frac{1}{2}\frac{\delta E}{E_0}\left(e^{i\omega_{\text{am}} t} + e^{-i\omega_{\text{am}} t}\right)\right] \times \left[1 + \frac{1}{2}\frac{\delta\omega}{\omega_{\text{fm}}}\left(e^{i\omega_{\text{fm}} t} - e^{-i\omega_{\text{fm}} t}\right)\right].$$

Determine the contributions of the frequency and intensity noise components to the symmetric and the antisymmetric output of the Michelson interferometer as a function of $\delta = \ell_1 - \ell_2$.

Problem 6.3

Rather than using a Fabry–Pérot cavity, a Michelson interferometer with arms having a very long *effective* arm length could be built by allowing the light to bounce back and forth along the arms for a fixed number N of journeys before coming back to the beam splitter (this is known as *folding* the arms). Suppose the Michelson interferometer has arms of length $L = 4\,\text{km}$ and light bounces back and forth along the arms $N = 125$ times so that the effective length of the Michelson interferometer's arms is $L_\text{eff} = 500\,\text{km}$. Such an apparatus behaves (in the long wavelength limit) exactly the same as a Michelson interferometer with arms of length L_eff. Compute the shot noise power spectrum in strain units, analogous to Eq. (6.99), for such a detector, and plot the sensitivity curve $S_h^{1/2}(f)$ of the detector as in Figure 6.9 assuming that the power on the beam splitter is $I_\text{BS} = 250\,\text{W}$.

Problem 6.4

Consider a Fabry–Pérot cavity of length L with an input mirror that has reflectivity r_ITM and transmissivity $t_\text{ITM} = (1 - r_\text{ITM}^2)^{1/2}$, and a perfectly reflecting end mirror. Suppose a radiation field

$$E_\text{I}(t) = E_\text{I,0} + \int_{-\infty}^{\infty} e^{2\pi i f t} \tilde{E}_\text{I}'(f) d f$$

is incident on the input mirror. Show that the light within in the cavity,

$$E_\text{FP}(t) = E_\text{FP,0} + \int_{-\infty}^{\infty} e^{2\pi i f t} \tilde{E}_\text{FP}'(f) d f ,$$

has

$$E_\text{FP,0} = \frac{t_\text{ITM}}{1 - r_\text{ITM}} E_\text{I,0} \quad \text{and} \quad E_\text{FP}' = \frac{t_\text{ITM}}{1 - r_\text{ITM}} \frac{1 - r_\text{ITM}}{1 - r_\text{ITM} e^{-4\pi i f L/c}} E_\text{I,0} .$$

Suppose that incident carrier light has intensity $I_0 = |E_\text{I,0}|^2$ and the quantum fluctuations of the input light at the sideband frequencies have a power spectrum $S_{E_\text{I}}(f) = \frac{1}{2}\hbar c k$ where

$$\frac{1}{2} S_{E_\text{I}}(f)\delta(f - f') = \langle \tilde{E}_\text{I}'^{*}(f')\tilde{E}_\text{I}'(f)\rangle .$$

The intensity of light within the cavity is $I_\text{FP}(t) = I_\text{FP,0} + I_\text{FP}'(t)$ where $I_\text{FP,0}$ is the constant intensity of the carrier light and $I_\text{FP}'(t)$ is the fluctuating light at the sideband frequencies. Show that the carrier light in the Fabry–Pérot cavity has intensity $I_\text{FP,0} = I_0 G_\text{arm}$ and the intensity power spectrum at the sideband frequencies is $S_{I_\text{FP}}(f) = 2\hbar c k I_0 G_\text{arm}^2 |\hat{C}(f)|^2$.

References

Allen, Z.A. *et al.* (2000) First search for gravitational wave bursts with a network of detectors. *Phys. Rev. Lett.*, **85**, 5046–5050. doi: 10.1103/PhysRevLett.85.5046.

Armstrong, J.W., Estabrook, F.B. and Tinto, M. (1999) Time-delay interferometry for space-based gravitational wave searches. *Astrophys. J*, **527**, 814–826. doi: 10.1086/308110.

Armstrong, J.W. (2006) Low-frequency gravitational wave searches using spacecraft doppler tracking. *Living Rev. Rel.*, **9**(1). http://www.livingreviews.org/lrr-2006-1 (last accessed 2011-01-03).

Buonanno, A. and Chen, Y. (2001) Quantum noise in second generation, signal recycled laser interferometric gravitational wave detectors. *Phys. Rev.*, **D64**, 042 006. doi: 10.1103/PhysRevD.64.042006.

Callen, H.B. and Greene, R.F. (1952) On a theorem of irreversible thermodynamics. *Phys. Rev.*, **86**(5), 702–710. doi: 10.1103/PhysRev.86.702.

Callen, H.B. and Welton, T.A. (1951) Irreversibility and generalized noise. *Phys. Rev.*, **83** (1), 34–40. doi: 10.1103/PhysRev.83.34.

Creighton, T. (2008) Tumbleweeds and airborne gravitational noise sources for LIGO. *Class. Quant. Grav.*, **25**, 125 011. doi: 10.1088/0264-9381/25/12/125011.

Crowder, J. and Cornish, N.J. (2005) Beyond LISA: Exploring future gravitational wave missions. *Phys. Rev.*, **D72**, 083 005. doi: 10.1103/PhysRevD.72.083005.

Gonzalez, G. (2000) Suspensions thermal noise in LIGO gravitational wave detector. *Class. Quant. Grav.*, **17**, 4409–4436. doi: 10.1088/0264-9381/17/21/305.

Hughes, S.A. and Thorne, K.S. (1998) Seismic gravity-gradient noise in interferometric gravitational-wave detectors. *Phys. Rev.*, **D58**, 122 002. doi: 10.1103/PhysRevD.58.122002.

Kawamura, S. *et al.* (2006) The Japanese space gravitational wave antenna DECIGO. *Class. Quant. Grav.*, **23**, S125–S132. doi: 10.1088/0264-9381/23/8/S17.

Larson, S.L., Hiscock, W.A. and Hellings, R.W. (2000) Sensitivity curves for spaceborne gravitational wave interferometers. *Phys. Rev.*, **D62**, 062 001. doi: 10.1103/PhysRevD.62.062001.

Lorimer, D.R. (2001) Binary and millisecond pulsars at the new millennium. *Living Rev. Rel.*, **4**(5). http://www.livingreviews.org/lrr-2001-5 (last accessed 2011-01-03).

Maggiore, M. (2007) *Gravitational Waves*, Vol. 1: Theory and Experiments, Oxford University Press.

Punturo, M. *et al.* (2010) The third generation of gravitational wave observatories and their science reach. *Class. Quant. Grav.*, **27**, 084 007. doi: 10.1088/0264-9381/27/8/084007.

Saulson, P.R. (1984) Terrestrial gravitational noise on a gravitational wave antenna. *Phys. Rev.*, **D30**, 732–736. doi: 10.1103/PhysRevD.30.732.

Saulson, P.R. (1994) *Fundamentals of Interferometric Gravitational Wave Detectors*, World Scientific, Singapore.

Seto, N., Kawamura, S. and Nakamura, T. (2001) Possibility of direct measurement of the acceleration of the universe using 0.1-Hz band laser interferometer gravitational wave antenna in space. *Phys. Rev. Lett.*, **87**, 221 103. doi: 10.1103/PhysRevLett.87.221103.

7
Gravitational-Wave Data Analysis

Even with the capabilities of the current generation (or even the next generation) of detectors, gravitational-wave signals are expected to be very weak, and so will require optimized statistical methods of signal extraction to identify the signals enveloped in detector noise. Optimal methods for detecting signals in detector noise are well developed for situations in which the detector noise is relatively well characterized. An excellent treatment of this problem is given in Wainstein and Zubakov (1971). We will give a review of the optimal detection statistics that are available for a variety of gravitational-wave source types, and illustrate how the general methods have been used to date. We will focus primarily on the applications to ground-based interferometric gravitational-wave data analysis, though in most cases the methods are quite general and the techniques can be used directly for the analysis of data from other sorts of detectors (pulsar timing data, space-borne interferometer detector data, and resonant mass detector data).

A more formal discussion of gravitational-wave data analysis techniques can be found in Jaranowski and Królak (2009).

7.1
Random Processes

A *random process* is a sequence of random variables; instrumental noise is an example of a random process representing some time series $x(t)$. Although we don't know in advance the exact realization of such a time series, it is often possible to determine the statistical properties of the time series. Suppose that p_x is the probability density function for the value of x at some arbitrary time t. Then the *expectation value* of x at that time is defined as an ensemble average

$$\langle x \rangle := \int x p_x(x) dx \,. \tag{7.1}$$

If the statistical properties of a random process do not change with time we call it a *stationary random process*. In this case the ensemble average is equivalent to a long

Gravitational-Wave Physics and Astronomy, First Edition. Jolien D. E. Creighton, Warren G. Anderson.

time average so the expectation value can also be written

$$\langle x \rangle = \lim_{T\to\infty} \frac{1}{T} \int_{-T/2}^{T/2} x(t)dt \,. \tag{7.2}$$

For most of our purposes we will be considering a special class of random processes known as *Gaussian random processes* that are also stationary. Such processes are (statistically) characterized by their expectation value and their *power spectrum.*

7.1.1
Power Spectrum

Consider some signal $x(t)$. Let us for simplicity assume that it has zero mean, that is $\langle x \rangle = 0$. We define the power in the signal by integrating $x^2(t)$ over some duration T and then dividing the result by T. If $x(t)$ is stationary then, by choosing T large enough, the time average of $x^2(t)$ is the same as the expectation value $\langle x^2 \rangle$:

$$\langle x^2 \rangle = \lim_{T\to\infty} \frac{1}{T} \int_{-T/2}^{T/2} x^2(t)dt \,. \tag{7.3}$$

It is useful to define a windowed signal $x_T(t)$

$$x_T(t) = \begin{cases} x(t) & -T/2 < t < T/2 \,, \\ 0 & \text{otherwise.} \end{cases} \tag{7.4}$$

Then

$$\begin{aligned} \langle x^2 \rangle &= \lim_{T\to\infty} \frac{1}{T} \int_{-\infty}^{\infty} x_T^2(t)dt \\ &= \lim_{T\to\infty} \frac{1}{T} \int_{-\infty}^{\infty} |\tilde{x}_T^2(f)|^2 df \\ &= \lim_{T\to\infty} \frac{2}{T} \int_{0}^{\infty} |\tilde{x}_T^2(f)|^2 df \\ &= \int_{0}^{\infty} S_x(f) df \,, \end{aligned} \tag{7.5}$$

where $S_x(f)$ is the *power spectral density* of the process $x(t)$. We used *Parseval's theorem,* $\int_{-\infty}^{\infty} |x_T^2(t)|^2 dt = \int_{-\infty}^{\infty} |\tilde{x}_T(f)|^2 df$, to obtain the second equality and, because of the reality of $x_T(t)$, negative frequency components are related to positive frequency components via $\tilde{x}_T(-f) = \tilde{x}_T^*(f)$, which is used to obtain the third equality. Here $\tilde{x}_T(f)$ is the Fourier transform of $x_T(t)$, which is defined by Eq. (0.1). The power spectral density of a stationary random process $x(t)$ is thus defined by

$$S_x(f) := \lim_{T\to\infty} \frac{2}{T} \left| \int_{-T/2}^{T/2} x(t) e^{-2\pi i f t} dt \right|^2 . \qquad (7.6)$$

For a stationary process, the power spectral density is twice the Fourier transform of the *autocorrelation function*

$$R_x(\tau) = \langle x(t)x(t+\tau)\rangle . \qquad (7.7)$$

(Note that the fact that the autocorrelation function depends only on the time shift τ and not on t results from the assumption that the random process is stationary.) To see this, recall the definition of the power spectral density, Eq. (7.6),

$$S_x(f) = \lim_{T\to\infty} \frac{2}{T} \int_{-T/2}^{T/2} x(t) e^{2\pi i f t} dt \int_{-T/2}^{T/2} x(t') e^{-2\pi i f t'} dt' , \qquad (7.8)$$

and perform the change of variables: $t = t' + \tau$. Then

$$S_x(f) = 2 \int_{-\infty}^{\infty} d\tau e^{-2\pi i f \tau} \left[\lim_{T\to\infty} \frac{1}{T} \int_{-T/2}^{T/2} x(t')x(t'+\tau) dt' \right] . \qquad (7.9)$$

The quantity in brackets is the autocorrelation function as expressed in terms of a time average rather than an ensemble average, so we have

$$S_x(f) = 2 \int_{-\infty}^{\infty} R_x(\tau) e^{-2\pi i f \tau} d\tau \qquad (7.10)$$

as desired.

One other useful expression for the power spectrum can be obtained by considering the expectation value of the frequency components $\tilde{x}(f)$:

$$\langle \tilde{x}^*(f')\tilde{x}(f)\rangle = \left\langle \int_{-\infty}^{\infty} x(t') e^{2\pi i f' t'} dt' \int_{-\infty}^{\infty} x(t) e^{-2\pi i f t} dt \right\rangle . \qquad (7.11)$$

Again, we perform the change of variable $t = t' + \tau$ to obtain

$$\begin{aligned} \langle \tilde{x}^*(f')\tilde{x}(f)\rangle &= \left\langle \int_{-\infty}^{\infty} x(t') e^{2\pi i f' t'} dt' \int_{-\infty}^{\infty} x(t'+\tau) e^{-2\pi i f(t'+\tau)} d\tau \right\rangle \\ &= \int_{-\infty}^{\infty} dt' e^{-2\pi i (f-f')t'} \int_{-\infty}^{\infty} d\tau e^{-2\pi i f \tau} \langle x(t')x(t'+\tau)\rangle . \end{aligned} \qquad (7.12)$$

The second integral is just $\frac{1}{2}S_x(f)$, which does not depend on t', and the first integral is then the Dirac delta function $\delta(f-f')$. Therefore,

$$\langle \tilde{x}^*(f')\tilde{x}(f)\rangle = \frac{1}{2}S_x(f)\delta(f-f')\,. \tag{7.13}$$

Example 7.1 Shot noise

A random process $x(t)$ is known as *shot noise* if it consists of a large number of pulses arriving at random times:

$$x(t) = \sum_{j=0}^{N-1} F(t-t_j)\,, \tag{7.14}$$

where the function $F(t)$ describes the shape of each pulse (we assume that each pulse has the same shape and we also assume that the pulses are short compared to the time between pulse arrivals) and $\{t_j\}$ are the random times of arrival of the N pulses that occur in some long time T. The mean rate of pulse arrivals is $R = N/T$.

To compute the power spectral density, consider Eq. (7.13): Let $\tilde{F}(f)$ be the Fourier transform of $F(t)$. Then

$$\tilde{x}(f) = \tilde{F}(f)\sum_{j=0}^{N-1} \mathrm{e}^{-2\pi i f t_j} \tag{7.15}$$

and, according to Eq. (7.13),

$$\begin{aligned}
\frac{1}{2}S_x(f)\delta(f-f') &= \langle \tilde{x}^*(f')\tilde{x}(f)\rangle \\
&= \tilde{F}^*(f')\tilde{F}(f)\left\langle \sum_{j=0}^{N-1}\sum_{k=0}^{N-1} \mathrm{e}^{2\pi i f' t_j}\mathrm{e}^{-2\pi i f t_k}\right\rangle \\
&= \tilde{F}^*(f')\tilde{F}(f)\left\{\sum_{j=0}^{N-1}\left\langle \mathrm{e}^{-2\pi i (f-f')t_j}\right\rangle + \sum_{j=0}^{N-1}\sum_{\substack{k=0\\k\neq j}}^{N-1}\left\langle \mathrm{e}^{2\pi i f' t_j}\right\rangle\left\langle \mathrm{e}^{-2\pi i f t_k}\right\rangle\right\} \\
&= \tilde{F}^*(f')\tilde{F}(f)\sum_{j=0}^{N-1}\left\{\left\langle \mathrm{e}^{-2\pi i (f-f')t_j}\right\rangle - \left\langle \mathrm{e}^{2\pi i f' t_j}\right\rangle\left\langle \mathrm{e}^{-2\pi i f t_j}\right\rangle\right\} \\
&\quad + \tilde{F}^*(f')\tilde{F}(f)\sum_{j=0}^{N-1}\left\langle \mathrm{e}^{2\pi i f' t_j}\right\rangle\sum_{k=0}^{N-1}\left\langle \mathrm{e}^{-2\pi i f t_k}\right\rangle \\
&= \tilde{F}^*(f')\tilde{F}(f)N\left\{T^{-1}\delta(f-f') - T^{-2}\delta(f')\delta(f)\right\} + \langle \tilde{x}^*(f')\rangle\langle \tilde{x}(f)\rangle\,,
\end{aligned} \tag{7.16}$$

where we have used

$$\langle \mathrm{e}^{-2\pi i f t_j}\rangle = \frac{1}{T}\int_{-T/2}^{T/2} e^{-2\pi i f t_j}\,dt_j \approx T^{-1}\delta(f) \tag{7.17}$$

for large T. If we assume $\langle \tilde{x}(f) \rangle = 0$ and ignore the zero-frequency components (i.e. take $\tilde{F}(0) = 0$) then we find

$$S_x(f) = 2R|\tilde{F}(f)|^2 , \tag{7.18}$$

where $R = N/T$ is the mean rate of pulse arrival.

If the pulses are very narrow then their spectrum is broad-band. We can approximate $F(t) \approx \sigma\delta(t)$ where σ is an amplitude (with the units of x times seconds). Then the power spectrum,

$$S_x(f) \approx 2R\sigma^2 , \tag{7.19}$$

is *white*, that is independent of frequency.

7.1.2 Gaussian Noise

Suppose a noise time-series of duration T, $x(t)$, $0 \le t < T$, is sampled at regular intervals Δt to produce a set of N samples $x_j = x(j\Delta t)$ where $j = 0, \ldots, N-1$ and $N = T/\Delta t$. If each of these samples x_j are independent Gaussian random variables then the joint probability distribution of obtaining this collection $\{x_j\}$ is

$$p_x(\{x_j\}) = \left(\frac{1}{\sqrt{2\pi}\sigma}\right)^N \exp\left\{-\frac{1}{2\sigma^2}\sum_{j=0}^{N-1} x_j^2\right\} , \tag{7.20}$$

where we have assumed that the samples have zero mean and variance of σ^2. The noise is known as *Gaussian noise*. We obtain a continuum limit by allowing $\Delta t \to 0$ while holding T fixed. Then

$$\lim_{\Delta t \to 0} \sum_{j=0}^{N-1} x_j^2 \Delta t = \int_0^T x^2(t)dt \approx \int_{-\infty}^{\infty} |\tilde{x}(f)|^2 df , \tag{7.21}$$

where the second (approximate) equality holds for large observation time T. Since the samples x_j are independent, we say the noise is *white noise*. Shot noise is an example of white noise (see Example 7.1). This means that the autocorrelation function $R_x(\tau) = \langle x(t)x(t+\tau) \rangle \propto \delta(\tau)$. The constant of proportionality is determined by noticing that $R_{jk} = \langle x_j x_k \rangle = \sigma^2\delta_{jk}$, so the power spectral density is

$$S_x(f) = 2\int_{-\infty}^{\infty} R_x(\tau)\,\mathrm{e}^{-2\pi i f\tau} d\tau = \lim_{\Delta t \to 0} 2\sigma^2\Delta t . \tag{7.22}$$

Notice that power spectral density is constant in frequency for white noise, so, for the moment we will write it as S_x rather than $S_x(f)$. We will soon generalize to the case of coloured noise in which the spectral density does depend on frequency.

In the limit $\Delta t \to 0$ the power $\sigma^2 \Delta t$ in any frequency band of width Δf is to be held fixed. (That is, the variance of the sampled time-series samples depends on the sampling interval.) Therefore,

$$\begin{aligned}
\lim_{\Delta t \to 0} \exp\left\{-\frac{1}{2\sigma^2}\sum_{j=0}^{N-1} x_j^2\right\} &= \lim_{\Delta t \to 0} \exp\left\{-\frac{1}{2\sigma^2 \Delta t}\sum_{j=0}^{N-1} x_j^2 \Delta t\right\} \\
&= \exp\left\{-\frac{1}{S_x}\int_0^T x(t)^2 dt\right\} \\
&\approx \exp\left\{-\int_{-\infty}^{\infty} \frac{|\tilde{x}(f)|^2}{S_x} df\right\} \\
&= \exp\left\{-\frac{1}{2} 4 \int_0^{\infty} \frac{|\tilde{x}(f)|^2}{S_x} df\right\} .
\end{aligned} \tag{7.23}$$

Hence the continuum limit probability density of the time-series $x(t)$ is

$$p_x[x(t)] \propto \exp\left\{-\frac{1}{2} 4 \int_0^{\infty} \frac{|\tilde{x}(f)|^2}{S_x} df\right\} . \tag{7.24}$$

In fact, this formula holds also for *coloured noise,* in which the power spectral density *does* depend on frequency! To see this, consider a linear process (a convolution) on the time-series $x(t)$ that produces a new time-series of non-white noise $y(t)$:

$$y(t) = \int_{-\infty}^{\infty} K(t-t')x(t')dt' , \tag{7.25}$$

where $K(t-t')$ is a kernel that determines the spectrum of y. In the frequency domain this operation is written (via the convolution theorem) as

$$\tilde{y}(f) = \tilde{K}(f)\tilde{x}(f) . \tag{7.26}$$

It is then straightforward to show that $S_y(f) = |\tilde{K}(f)|^2 S_x$ and

$$p_y[y(t)] \propto \exp\left\{-\frac{1}{2} 4 \int_0^{\infty} \frac{|\tilde{y}(f)|^2}{S_y(f)} df\right\} . \tag{7.27}$$

This motivates the definition of a noise-weighted inner product (a, b) of two time-series $a(t)$ and $b(t)$,

$$(a, b) := 4\,\mathrm{Re} \int_0^\infty \frac{\tilde{a}(f)\tilde{b}^*(f)}{S(f)}\, df \tag{7.28a}$$

$$= 2 \int_{-\infty}^{\infty} \frac{\tilde{a}(f)\tilde{b}^*(f)}{S(|f|)}\, df \tag{7.28b}$$

$$= \int_{-\infty}^{\infty} \frac{\tilde{a}(f)\tilde{b}^*(f) + \tilde{a}^*(f)\tilde{b}(f)}{S(|f|)}\, df\,, \tag{7.28c}$$

where $S(f)$ is the one-sided power spectral density of the noise. Note that the reality of $a(t)$ implies $\tilde{a}(-f) = \tilde{a}^*(f)$ and similarly with b. With this inner product, the probability density for a stationary Gaussian noise process $x(t)$, (whether coloured or not) can be written concisely as

$$p_x[x(t)] \propto \mathrm{e}^{-(x,x)/2}\,. \tag{7.29}$$

7.2
Optimal Detection Statistic

When the statistical properties of the noise process are known and the exact form of the signal is also known, it is possible to construct an *optimal detection statistic*. This optimal statistic is simply the quantity that expresses the value of the probability that the data contains the anticipated signal.

Suppose that the strain data, $s(t)$, recorded by a gravitational-wave detector consists of a noise random process, $n(t)$, and possibly a gravitational wave of known form, $h(t)$. We wish to distinguish between two hypotheses:

$$\begin{aligned} &\text{Null Hypothesis } \mathcal{H}_0: && s(t) = n(t) \\ &\text{Alternative Hypothesis } \mathcal{H}_1: && s(t) = n(t) + h(t)\,. \end{aligned} \tag{7.30}$$

This is done by computing the *odds ratio* $O(\mathcal{H}_1|s) = P(\mathcal{H}_1|s) : P(\mathcal{H}_0|s)$, that is, the ratio of the probability that the alternative hypothesis $\mathcal{H}_1$ is true, given the data $s(t)$, to the probability that the null hypothesis $\mathcal{H}_0$ is true given the data. To compute the odds ratio we use *Bayes's theorem*.

7.2.1
Bayes's Theorem

In order to describe Bayes's theorem, we must first recall some definitions from probability theory. We write $P(\mathcal{A})$ for the probability of $\mathcal{A}$ being true, $P(\mathcal{B})$ for the probability of $\mathcal{B}$ being true, and $P(\mathcal{A}, \mathcal{B})$ for the *joint probability* of both $\mathcal{A}$ being true and $\mathcal{B}$ being true. The *conditional probability* $P(\mathcal{A}|\mathcal{B})$, which is the probability that $\mathcal{A}$ is true *given that* $\mathcal{B}$ is true, is defined by

$$P(\mathcal{A}|\mathcal{B}) := \frac{P(\mathcal{A},\mathcal{B})}{P(\mathcal{B})}\,, \tag{7.31}$$

and similarly

$$P(\mathcal{B}|\mathcal{A}) := \frac{P(\mathcal{A},\mathcal{B})}{P(\mathcal{A})}\,, \tag{7.32}$$

is the probability that $\mathcal{B}$ is true given that $\mathcal{A}$ is true. Combining these two equations yields Bayes's theorem:

$$P(\mathcal{B}|\mathcal{A}) = \frac{P(\mathcal{B})\,P(\mathcal{A}|\mathcal{B})}{P(\mathcal{A})}\,. \tag{7.33}$$

In this formula the quantity $P(\mathcal{B})$ is called the *prior probability* or *marginal probability* of $\mathcal{B}$ being true; $P(\mathcal{A})$ is the marginal probability of $\mathcal{A}$, which is known as the *evidence*, and acts as a normalizing constant; $P(\mathcal{B}|\mathcal{A})$ is the *posterior probability* of $\mathcal{B}$ being true given that $\mathcal{A}$ is true; and $P(\mathcal{A}|\mathcal{B})$ is the probability that $\mathcal{A}$ is true given that $\mathcal{B}$ is true.

Bayes's theorem can be expressed in a form more convenient for our purposes by employing the *completeness* relationship: $P(\mathcal{A}) = P(\mathcal{A}|\mathcal{B})\,P(\mathcal{B}) + P(\mathcal{A}|\neg\mathcal{B})\,P(\neg\mathcal{B})$ where $P(\mathcal{A}|\neg\mathcal{B})$ is the probability of $\mathcal{A}$ given that $\mathcal{B}$ is *not* true and $P(\neg\mathcal{B}) = 1 - P(\mathcal{B})$ is the probability that $\mathcal{B}$ is not true. Using the completeness relation, we find

$$P(\mathcal{B}|\mathcal{A}) = \frac{P(\mathcal{B})\,P(\mathcal{A}|\mathcal{B})}{P(\mathcal{A}|\mathcal{B})\,P(\mathcal{B}) + P(\mathcal{A}|\neg\mathcal{B})\,P(\neg\mathcal{B})} = \frac{\Lambda(\mathcal{B}|\mathcal{A})}{\Lambda(\mathcal{B}|\mathcal{A}) + P(\neg\mathcal{B})/P(\mathcal{B})}\,, \tag{7.34}$$

where

$$\Lambda(\mathcal{B}|\mathcal{A}) := \frac{P(\mathcal{A}|\mathcal{B})}{P(\mathcal{A}|\neg\mathcal{B})} \tag{7.35}$$

is the *likelihood ratio*. Another form of this equation is

$$O(\mathcal{B}|\mathcal{A}) = O(\mathcal{B})\Lambda(\mathcal{B}|\mathcal{A}) \tag{7.36}$$

where $O(\mathcal{B}|\mathcal{A}) := P(\mathcal{B}|\mathcal{A}) : P(\neg\mathcal{B}|\mathcal{A})$ is the odds ratio of $\mathcal{B}$ being true given $\mathcal{A}$, and $O(\mathcal{B}) = P(\mathcal{B}) : P(\neg\mathcal{B})$ is the prior odds ratio of $\mathcal{B}$ being true.

7.2.2 Matched Filter

For the problem of detection, we wish to decide between the two hypotheses of Eq. (7.30): the null hypothesis $\mathcal{H}_0$ that there is no gravitational-wave signal $h(t)$ in the data $s(t)$, so that $s(t) = n(t)$ is noise alone, and the alternative hypothesis $\mathcal{H}_1$ that there is a gravitational-wave signal in the data $s(t) = n(t) + h(t)$. To do so we must compute the odds ratio for the alternative hypothesis given the observed data $s(t)$: $O(\mathcal{H}_1|s)$. Equation (7.36) shows this is proportional to the prior odds ratio, which does not depend in any way on the data, and the likelihood ratio

$$\Lambda(\mathcal{H}_1|s) = \frac{p(s|\mathcal{H}_1)}{p(s|\mathcal{H}_0)} . \tag{7.37}$$

(Here we have replaced the probabilities with probability densities.)

If the noise is Gaussian then we can compute the probability densities: under the null hypothesis $\mathcal{H}_0$, $n(t) = s(t)$ and so

$$p(s|\mathcal{H}_0) = p_n[s(t)] \propto \mathrm{e}^{-(s,s)/2} . \tag{7.38}$$

Under the alternative hypothesis $\mathcal{H}_1$, $n(t) = s(t) - h(t)$ and so

$$p(s|\mathcal{H}_1) = p_n[s(t) - h(t)] \propto \mathrm{e}^{-(s-h,s-h)/2} . \tag{7.39}$$

Therefore,

$$\Lambda(\mathcal{H}_1|s) = \frac{\mathrm{e}^{-(s-h,s-h)/2}}{\mathrm{e}^{-(s,s)/2}} = \mathrm{e}^{(s,h)}\mathrm{e}^{-(h,h)/2} . \tag{7.40}$$

We see that the likelihood ratio $\Lambda(\mathcal{H}_1|s)$ depends on the data $s(t)$ only through the inner product (s, h); furthermore the likelihood ratio (and hence the odds ratio $O(\mathcal{H}_1|s)$) is a monotonically increasing function of this inner product. Therefore this inner product,

$$(s, h) = 4\,\mathrm{Re} \int_0^\infty \frac{\tilde{s}(f)\tilde{h}^*(f)}{S_n(f)} df , \tag{7.41}$$

is the optimal detection statistic: any choice of threshold on the required odds ratio for accepting the alternative hypothesis can be translated to a threshold on the value of (s, h). We call (s, h) the *matched filter* since it is essentially a noise-weighted correlation of the anticipated signal with the data.

7.2.3
Unknown Matched Filter Parameters

We are normally in the situation in which there is a space of possible signals that is characterized by a set of parameters $\{\lambda_i\}$, $i = 1, \ldots, N$, which we represent as a vector $\boldsymbol{\lambda} = [\lambda_1, \ldots, \lambda_N]$ in the N-dimensional parameter space of signals. For example, for Newtonian chirps the set of parameters would be the signal amplitude, the coalescence time, the coalescence phase, and the chirp mass. The gravitational-wave signal is now written as $h(t; \boldsymbol{\lambda})$. To obtain the optimal detection statistic in this case we must integrate out or *marginalize* over the unknown parameters. The likelihood ratio for a particular signal in signal-parameter space is

$$\Lambda(\mathcal{H}_\lambda|s) = \frac{p(s|\mathcal{H}_\lambda)}{p(s|\mathcal{H}_0)}\,, \tag{7.42}$$

where $\mathcal{H}_\lambda$ is the hypothesis that a signal with parameters $\boldsymbol{\lambda}$ is present, and the *marginalized likelihood* is

$$\Lambda(\mathcal{H}_1|s) = \frac{\int p\,(s|\mathcal{H}_\lambda)\,p\,(\mathcal{H}_\lambda)\,d\boldsymbol{\lambda}}{p\,(s|\mathcal{H}_0)} = \int \Lambda\,(\mathcal{H}_\lambda|s)\,p\,(\mathcal{H}_\lambda)\,d\boldsymbol{\lambda}\,, \tag{7.43}$$

where $\mathcal{H}_1$ is the alternative hypothesis that some signal is present and $p(\mathcal{H}_\lambda)$ is a prior probability distribution describing which parameter values are intrinsically more likely. While this marginalization can sometimes be computed explicitly, a suitable approximation is usually possible: when a strong signal with parameters $\boldsymbol{\lambda}_{\text{true}}$ is present, the likelihood ratio $\Lambda(\mathcal{H}_\lambda|s)$ is normally a strongly peaked function in parameter space with a maximum value $\boldsymbol{\lambda}_{\max}$ very close to the true parameters. If the parameters take on discrete values, then it is possible to compute $p(\mathcal{H}_\lambda|s)$ for each choice of $\boldsymbol{\lambda}$ and choose the largest one as a test of the hypothesis $\mathcal{H}_1$. This is known as the *maximum likelihood* statistic. The discrete set of posterior probabilities $p(\mathcal{H}_\lambda|s)$ forms a *bank* of detection statistics. Even if the parameters are continuous, one can sample each parameter finely enough so that the maximum likelihood is well approximated.

When we consider Gaussian noise, the logarithm of the likelihood ratio for a particular choice of parameters is

$$\ln \Lambda\,(\mathcal{H}_\lambda|s) = (s, h(\boldsymbol{\lambda})) - \frac{1}{2}\,(h(\boldsymbol{\lambda}), h(\boldsymbol{\lambda}))\,. \tag{7.44}$$

Thus, the maximum value of the likelihood ratio is achieved when

$$\left.\left(s - h(\boldsymbol{\lambda}), \frac{\partial}{\partial \lambda_i} h(\boldsymbol{\lambda})\right)\right|_{\boldsymbol{\lambda}=\boldsymbol{\lambda}_{\max}} = 0. \tag{7.45}$$

By solving the system of equations given by Eq. (7.45) for $\boldsymbol{\lambda}_{\max}$, we see that the maximum likelihood statistic is given by Eq. (7.44) with $\boldsymbol{\lambda} = \boldsymbol{\lambda}_{\max}$.

Example 7.2 Unknown amplitude

Suppose that the gravitational-wave signal $h(t; A) = Ag(t)$ has a known signal form $g(t)$ with an unknown amplitude A. The logarithm of the likelihood ratio is

$$\ln \Lambda(\mathcal{H}_A|s) = (s, h(A)) - \frac{1}{2}(h(A), h(A)) = A(s, g) - \frac{1}{2}A^2(g, g)\,. \tag{7.46}$$

This is maximized for an amplitude $A_{\max}$ where

$$A_{\max} = \frac{(s, g)}{(g, g)}\,. \tag{7.47}$$

Therefore, the maximum log-likelihood is

$$\ln \Lambda\left(\mathcal{H}_{A_{\max}}|s\right) = \frac{1}{2}\frac{(s,g)^2}{(g,g)} . \tag{7.48}$$

This is the maximum likelihood detection statistic.

7.2.4 Statistical Properties of the Matched Filter

In Example 7.2 we saw that in the case that a gravitational-wave signal $h(t;A) = Ag(t)$ has known form $g(t)$ but unknown amplitude A, the maximum log-likelihood ratio is $\ln \Lambda(\mathcal{H}_{A_{\max}}|s) = \frac{1}{2}(s,g)^2/(g,g)$ where $A_{\max} = (s,g)/(g,g)$ is the most likely amplitude. The function $g(t)$, which is proportional to the anticipated signal, is known as a *template*; we also refer to the inner product $x = (s,g)$ as the matched filter since (s,g) and (s,h) differ only by a constant factor. We now investigate the statistical properties of the quantity x:

If no signal is present in the strain data so that $s(t) = n(t)$ is purely noise, which we assume has zero mean, $\langle x \rangle = 0$, then

$$\begin{aligned}
\langle x^2 \rangle &= \left\langle \left(2\int_{-\infty}^{\infty} \frac{\tilde{n}(f)\tilde{g}^*(f)}{S_n(|f|)} df \right) \left(2\int_{-\infty}^{\infty} \frac{\tilde{n}^*(f')\tilde{g}(f')}{S_n(|f'|)} df' \right) \right\rangle \\
&= 4\int_{-\infty}^{\infty} df \int_{-\infty}^{\infty} df' \frac{\langle \tilde{n}^*(f')\tilde{n}(f)\rangle \tilde{g}^*(f)\tilde{g}(f')}{S_n(|f|)S_n(|f'|)} \\
&= 4\int_{-\infty}^{\infty} df \int_{-\infty}^{\infty} df' \frac{\frac{1}{2}S_n(|f|)\delta(f-f')\tilde{g}^*(f)\tilde{g}(f')}{S_n(|f|)S_n(|f'|)} \\
&= 2\int_{-\infty}^{\infty} \frac{|\tilde{g}(f)|^2}{S_n(|f|)} df \\
&= (g,g) ,
\end{aligned} \tag{7.49}$$

where Eq. (7.13) was used to obtain the third line. Therefore, the variance $\sigma^2 = \mathrm{Var}(x) = \langle x^2 \rangle$ of the matched filter is $\sigma^2 = (g,g)$. That is, when there is no signal present, the matched filter is a zero-mean Gaussian random variable with variance $\sigma^2 = (g,g)$. (It is a Gaussian random variable since we have assumed that $n(t)$ is a Gaussian random process and x is formed from a linear operation on $n(t)$.)

Now consider the case when the strain data $s(t) = n(t) + h(t;A) = n(t) + Ag(t)$ contains, in addition to the noise $n(t)$, a signal $h(t;A) = Ag(t)$ of amplitude A relative to the template $g(t)$. Now we see that

$$\langle x \rangle = \langle (s,g) \rangle = \langle (n,g) \rangle + (h,g) = A(g,g) = A\sigma^2 . \tag{7.50}$$

Similarly,

$$\begin{aligned}\langle x^2\rangle &= \left\langle (s,g)^2\right\rangle = \left\langle \left[(n,g)+(h,g)\right]^2\right\rangle \\ &= \left\langle (n,g)^2\right\rangle + 2(h,g)\langle (n,g)\rangle + (h,g)^2 \\ &= \sigma^2 + A^2\sigma^4 \end{aligned} \tag{7.51}$$

so

$$\mathrm{Var}(x) = \langle x^2\rangle - \langle x\rangle^2 = \left(\sigma^2 + A^2\sigma^4\right) - (A\sigma^2)^2 = \sigma^2\,. \tag{7.52}$$

This shows that, in the presence of a signal with amplitude A relative to the template, the matched filter statistic $x = (s,g)$ is a Gaussian random variable with mean $A\sigma^2$ and variance σ^2. Recall that the maximum likelihood value for the amplitude was $A_{\max} = (s,g)/(g,g)$ so we see that $\langle A_{\max}\rangle = A$.

It is customary to define the *signal-to-noise ratio*, $\rho := x/\sigma$, to be a normalized matched filter so that, when Gaussian noise alone is present, the signal-to-noise ratio ρ is a normally distributed random variable (a Gaussian random variable with zero mean, $\langle\rho\rangle = 0$ and unit variance, $\mathrm{Var}(\rho) = 1$), and when a signal $h(t;A) = Ag(t)$ is present then the mean becomes $\varrho = \langle\rho\rangle = A\sigma = (h,h)^{1/2}$. Sometimes ϱ is called the signal-to-noise ratio of a signal, characterizing its strength; other times ρ is called the signal-to-noise ratio, describing the result of filtering the data. This often causes confusion.

Note on usage In this book, we will attempt to disambiguate the term "signal-to-noise-ratio" as follows: we refer to the quantity, $\varrho := (h,h)^{1/2}$, which characterizes the detectability of a signal in a detector with a given noise power spectrum (which is buried in the definition of the inner product) as the *characteristic signal-to-noise ratio*, while we refer to the normalized matched filter output $\rho(t)$ as the signal-to-noise ratio. Notice that both of these are *amplitude* signal-to-noise ratios as they increase linearly with the amplitude of the signal. Sometimes signal-to-noise ratios are defined so that they increase linearly with the *power* of the signal (quadratically with the amplitude). We do not use such power signal-to-noise ratios in this book.

Example 7.3 Sensitivity of a matched filter gravitational-wave search

In a matched filter search for a signal of known form $g(t)$, we describe the sensitivity of our search as the value of the amplitude $A_{\min}$ (or related parameter) that would be required in order to achieve some particular characteristic signal-to-noise ratio $\varrho_{\min}$: $A_{\min} = \varrho_{\min}/\sigma$. Therefore, the quantity $\sigma = (g,g)^{1/2}$ sets a scale for the sensitivity of the search.

In many cases, the signal of interest is band-limited with a fairly narrow frequency band about some central frequency f_0. If this is the case then we can approximate the detector noise power spectrum as $S_h(f_0)$ over the frequency band of the signal, and the characteristic signal-to-noise ratio of the signal is

$$
\begin{aligned}
\varrho^2 = (h, h) &= 4 \int_0^\infty \frac{|\tilde{h}(f)|^2}{S_h(f)} df \\
&\simeq \frac{1}{2\pi^2 f_0^2 S_h(f_0)} \int_{-\infty}^{\infty} \left|2\pi i f \tilde{h}(f)\right|^2 df = \frac{1}{2\pi^2 f_0^2 S_h(f_0)} \int \dot{h}^2(t) dt \\
&\leq \frac{8}{\pi} \frac{G}{c^3} \frac{1}{f_0^2 S(f_0)} \Phi_{\mathrm{GW}} ,
\end{aligned}
\tag{7.53}
$$

where

$$
\Phi_{\mathrm{GW}} = \int \left| \frac{d E_{\mathrm{GW}}}{dt dA} \right| dt = \frac{c^3}{16\pi G} \int \left[\dot{h}_+^2(t) + \dot{h}_\times^2(t) \right] dt \tag{7.54}
$$

is the gravitational-wave fluence (gravitational-wave flux integrated over the duration of the signal). The inequality arises because the strain on the detector h is less than or equal to only one of the polarization components of the incident radiation.

For example, the Initial LIGO detectors have $S_h(f_0) \sim 6 \times 10^{-46}\,\mathrm{Hz}^{-1}$ at $f_0 \sim 150\,\mathrm{Hz}$, and the minimum gravitational-wave fluence required to achieve a characteristic signal-to-noise ratio $\varrho = 8$ would be $\Phi_{\mathrm{GW,min}} \sim 10^{-4}\,\mathrm{J\,m^{-2}}$, and this fluence would be sufficient only if the gravitational waves were linearly polarized and the source of gravitational waves was optimally located (directly above the detector, so that the detector lies in the transverse plane) and optimally oriented (so that the detector is sensitive to the single polarization of the gravitational wave).

7.2.5
Matched Filter with Unknown Arrival Time

Suppose that the signal has a known form but an unknown amplitude *and* arrival time. That is, suppose that the true signal is

$$
h(t) = A g(t - t_0) , \tag{7.55}
$$

where A is the unknown amplitude, t_0 is the unknown arrival time, but $g(t)$ is the known form of the signal, which is our waveform template. Note that the Fourier transform of this signal is

$$
\tilde{h}(f) = A \tilde{g}(f) e^{-2\pi i f t_0} . \tag{7.56}
$$

The matched filter is therefore

$$
(s, h) = 2A \int_{-\infty}^{\infty} \frac{\tilde{s}(f) \tilde{g}^*(f)}{S_n(|f|)} e^{2\pi i f t_0} df = A x(t_0) , \tag{7.57}
$$

where

$$x(t) := 2 \int_{-\infty}^{\infty} \frac{\tilde{s}(f)\tilde{g}^*(f)}{S_n(|f|)} e^{2\pi i f t} d f \tag{7.58}$$

is a time-series representing the application of the matched filter at various possible arrival times (parameterized by t). The signal-to-noise ratio can also be expressed as a function of time: $\rho(t) = x(t)/\sigma$. The maximum likelihood detection statistic is simply the largest value of ρ in the time period of interest. The maximum likelihood estimate of the arrival time is the time, t_{peak}, of peak signal-to-noise ratio. The maximum likelihood estimate of the amplitude is $A_{\max} = \rho(t_{\text{peak}})/\sigma$.

7.2.6 Template Banks of Matched Filters

If it is not possible to compute an explicit form for the maximum likelihood statistic, it is necessary to grid up parameter space sufficiently finely that the maximum likelihood ratio can be approximately identified as the best match from among the collection of waveforms. That is, a set of templates $\{u(t;\boldsymbol{\lambda})\}$ must be constructed such that the templates are so densely packed that any true signal will lie close enough to one of the templates. Each template $u(t;\boldsymbol{\lambda})$ is normalized, $(u(\boldsymbol{\lambda}), u(\boldsymbol{\lambda})) = 1$, so that $\rho(\boldsymbol{\lambda}) = (s, u(\boldsymbol{\lambda}))$ is the signal-to-noise ratio for that template. The maximum likelihood detection statistic is $\max_{\boldsymbol{\lambda}} \rho(\boldsymbol{\lambda})$.

We wish to assess the meaning of "sufficiently finely" and "close enough" in the previous paragraph. What we mean is that the fractional loss in the expected value of the signal-to-noise ratio in the filters that results from the parameters of a true signal not agreeing exactly with the parameters of the nearest template in the discretized bank is below some level of tolerance, say a few percent. Suppose that $\boldsymbol{\lambda}$ represents the parameters of some signal, so that $h(t) = \varrho u(t;\boldsymbol{\lambda})$ is the signal, while the nearest template in the bank, $u(t;\boldsymbol{\lambda} + \Delta\boldsymbol{\lambda})$, has parameters that differ by $\Delta\boldsymbol{\lambda}$. The expected value of the signal-to-noise ratio in this template will be $\varrho' = (h, u(\boldsymbol{\lambda} + \Delta\boldsymbol{\lambda})) = \varrho(u(\boldsymbol{\lambda}), u(\boldsymbol{\lambda} + \Delta\boldsymbol{\lambda}))$ where ϱ is the expected signal-to-noise ratio of the signal in a perfectly matched template – that is the characteristic signal-to-noise ratio. Therefore, the fractional loss in the expected signal-to-noise ratio is $(\varrho - \varrho')/\varrho = 1 - (u(\boldsymbol{\lambda}), u(\boldsymbol{\lambda} + \Delta\boldsymbol{\lambda})) = 1 - \mathcal{A}$ where

$$\mathcal{A}(\boldsymbol{\lambda}; \boldsymbol{\lambda} + \Delta\boldsymbol{\lambda}) := (u(\boldsymbol{\lambda}), u(\boldsymbol{\lambda} + \Delta\boldsymbol{\lambda})) \tag{7.59}$$

is known as the *ambiguity function*.

If the parameter mismatch $\Delta\boldsymbol{\lambda}$ is small then we can expand $u(\boldsymbol{\lambda} + \Delta\boldsymbol{\lambda})$ in the power series

$$u(\boldsymbol{\lambda} + \Delta\boldsymbol{\lambda}) = u(\boldsymbol{\lambda}) + \Delta\lambda_i \frac{\partial u}{\partial \lambda_i}(\boldsymbol{\lambda}) + \frac{1}{2}\Delta\lambda_i \Delta\lambda_j \frac{\partial^2 u}{\partial\lambda_i \partial\lambda_j}(\boldsymbol{\lambda}) + O\left((\Delta\boldsymbol{\lambda})^3\right). \tag{7.60}$$

The ambiguity function is

$$\begin{aligned}\mathcal{A} &= \left(u(\boldsymbol{\lambda}), u(\boldsymbol{\lambda}) + \Delta\lambda_i \frac{\partial u}{\partial \lambda_i}(\boldsymbol{\lambda}) + \frac{1}{2}\Delta\lambda_i \Delta\lambda_j \frac{\partial^2 u}{\partial\lambda_i \partial\lambda_j}(\boldsymbol{\lambda}) \right) + O\left((\Delta\boldsymbol{\lambda})^3\right) \\ &= 1 + \frac{1}{2}\Delta\lambda_i\Delta\lambda_j \left(u(\boldsymbol{\lambda}), \frac{\partial^2 u}{\partial\lambda_i\,\partial\lambda_j}(\boldsymbol{\lambda}) \right) + O\left((\Delta\boldsymbol{\lambda})^3\right), \end{aligned} \tag{7.61}$$

where the linear term in the series expansion does not contribute because the ambiguity function is maximized when $\Delta\boldsymbol{\lambda} = 0$. If the parameter mismatch $d\boldsymbol{\lambda}$ is infinitesimal then we write

$$\mathcal{A} = 1 - ds^2 , \tag{7.62}$$

where ds^2 is a squared metrical distance where distance is measured in terms of the fractional loss of expected signal-to-noise ratio,

$$ds^2 = g^{ij} d\lambda_i d\lambda_j , \tag{7.63}$$

with the metric on parameter space

$$g^{ij}(\boldsymbol{\lambda}) = -\frac{1}{2}\left(u(\boldsymbol{\lambda}), \frac{\partial^2 u}{\partial\lambda_i\partial\lambda_j}(\boldsymbol{\lambda}) \right). \tag{7.64}$$

As we have seen, for example, in Section 7.2.5, some parameters such as the arrival time of a signal can be searched over without resorting to a bank of templates. In the case of the unknown arrival time, a Fourier transform can be used to apply a matched filter template to the data for all possible arrival times – there is no need to construct a bank of templates all with different arrival times. Parameters for which we can algebraically maximize the matched filter are known as *extrinsic parameters*. On the other hand, the parameters that we need to cover with our template bank are known as *intrinsic parameters*. Suppose that there is a single extrinsic parameter, the arrival time t_0, so that we can write $\boldsymbol{\lambda} = [t_0, \boldsymbol{\mu}]$ where $\lambda_0 = t_0$ and $\lambda_i = \mu_i$ ($i > 0$) are the remaining intrinsic parameters. Then we define an *overlap* between two different templates as

$$\begin{aligned}\mathcal{O}(\boldsymbol{\mu}; \boldsymbol{\mu} + \Delta\boldsymbol{\mu}) &:= \max_{t_0} \mathcal{A}\left(t_0, \boldsymbol{\mu}; t_0, \boldsymbol{\mu} + \Delta\boldsymbol{\mu}\right) \\ &= \max_{t_0} \left(u(t_0, \boldsymbol{\mu}), u(t_0, \boldsymbol{\mu} + \Delta\boldsymbol{\mu})\right). \end{aligned} \tag{7.65}$$

A metric can be constructed from the overlap, rather than the ambiguity function,

$$\mathcal{O} = 1 - d\varsigma^2 = 1 - \gamma^{ij} d\mu_i d\mu_j , \tag{7.66}$$

where

$$\begin{aligned}\gamma^{ij}(\boldsymbol{\mu}) &= -\frac{1}{2}\max_{t_0}\left(u(t_0, \boldsymbol{\mu}), \frac{\partial^2 u}{\partial\mu_i\,\partial\mu_j}(t_0, \boldsymbol{\mu}) \right) \\ &\quad - g^{ij} - g^{i0}g^{0j}/g^{00} \qquad (i, j > 0) . \end{aligned} \tag{7.67}$$

In constructing a bank of templates, one must place the templates close enough together that there is not too much loss in signal-to-noise ratio due to the mismatch between a signal and the nearest template. For example, if we only tolerate a 3% loss in signal-to-noise ratio, then there must be some template in the bank within a squared metrical distance of $(\Delta\varsigma_{\max})^2 = 3\%$ of every point in parameter space. The quantity $1 - (\Delta\varsigma_{\max})^2$ is known as the *minimal match* of the template bank, or, in other words, $(\Delta\varsigma_{\max})^2$ is the maximum mismatch of the bank. The number of templates needed to cover the entire parameter space $\Omega(\boldsymbol{\mu})$ of dimension $\dim\Omega$ will be

$$N \sim \frac{1}{(\Delta\varsigma_{\max})^{\dim\Omega}} \int\limits_{\Omega(\boldsymbol{\mu})} \sqrt{\det\gamma^{ij}}\, d\boldsymbol{\mu}\,, \tag{7.68}$$

where the first factor is the inverse of the volume covered by one template and the integral is the volume of the entire parameter space to be covered.

Example 7.4 Unknown phase

Suppose that the gravitational-wave signal is a linear combination of two known waveforms, a cosine-phase waveform $p(t)$ and a sine-phase waveform $q(t)$, $h(t;\theta) = p(t)\cos\theta + q(t)\sin\theta$, where θ is the unknown waveform phase and where the waveforms $p(t)$ and $q(t)$ are orthogonal, $(p,q) = 0$, and have the same amplitude, $(p,p) = (q,q) = (h,h)$. Let $x = (s,p)$ and $y = (s,q)$ be the two filter outputs for the template signals $p(t)$ and $q(t)$. Then

$$\begin{aligned}\ln\Lambda(\mathcal{H}_\theta|s) &= (s,h(\theta)) - \frac{1}{2}(h(\theta),h(\theta))\\ &= x\cos\theta + y\sin\theta - \frac{1}{2}(h,h)\\ &= z\cos(\Theta-\theta) - \frac{1}{2}(h,h)\,,\end{aligned} \tag{7.69}$$

where $z^2 = x^2 + y^2$ and $\Theta = \arctan(y/x)$ so that $x = z\cos\Theta$ and $y = z\sin\Theta$. It is clear that this is maximized for $\theta = \theta_{\max} = \Theta$,

$$\ln\Lambda(\mathcal{H}_{\theta_{\max}}|s) = z - \frac{1}{2}(h,h)\,, \tag{7.70}$$

so the maximum likelihood detection statistic is therefore the quadrature sum of matched filter with the cosine- and sine-phase templates: $z^2 = (s,p)^2 + (s,q)^2$.

We can also compute the marginalized likelihood ratio:

$$\begin{aligned}\Lambda(\mathcal{H}_1|s) &= \frac{1}{2\pi}\int_0^{2\pi} \exp\left[(s,h(\theta)) - \frac{1}{2}(h(\theta),h(\theta))\right] d\theta\\ &= \frac{1}{2\pi}e^{-(h,h)/2}\int_0^{2\pi} e^{z\cos(\Theta-\theta)}\,d\theta\\ &= e^{-(h,h)/2} I_0(z)\,,\end{aligned} \tag{7.71}$$

where $I_0(z)$ is the modified Bessel function of the first kind of order zero, which is a monotonically increasing function of its argument z. Now we see that z is not only the maximum likelihood statistic but also that the marginalized likelihood ratio is a monotonically increasing function of z. For large z, $I_0(z) \sim e^z/(2\pi z)^{1/2}$, so $\ln \Lambda(\mathcal{H}_1) \sim z + \frac{1}{2}\ln(2\pi z) - \frac{1}{2}(h,h)$ and the difference between the logarithm of the maximum likelihood $\ln \Lambda(\mathcal{H}_{\theta_{\max}})$ and the logarithm of the marginalized likelihood is the relatively small $\frac{1}{2}\ln(2\pi z)$ term.

Clearly we do not need a template bank to find the maximum likelihood detection statistic in the case of an unknown phase; however, it is illustrative to consider how such a bank could be created. Let $u(t;\theta) = h(t;\theta)/(h,h)^{1/2}$ be a normalized template (so that $(u(\theta), u(\theta)) = 1$). Then the ambiguity function is

$$\begin{aligned}
\mathcal{A} &= (u(\theta), u(\theta + \Delta\theta)) \\
&= (p\cos\theta + q\sin\theta, p\cos(\theta+\Delta\theta) + q\sin(\theta+\Delta\theta))/(h,h) \\
&= \cos\theta\cos(\theta+\Delta\theta) + \sin\theta\sin(\theta+\Delta\theta) \\
&= \cos\Delta\theta = 1 - \frac{1}{2}\Delta\theta^2 + O\left(\Delta\theta^4\right).
\end{aligned} \tag{7.72}$$

Since $ds^2 = g^{\theta\theta}d\theta^2 = 1 - \mathcal{A}$, we see that $g_{\theta\theta} = 1/2$. Templates must be placed on a unit circle at even steps in θ. The number of templates required depends on the largest allowed loss in signal-to-noise ratio due to the discretization, $\Delta s_{\max}$. The number of templates required is

$$N \sim \frac{1}{\Delta s_{\max}} \int_0^{2\pi} \sqrt{g^{\theta\theta}}\, d\theta = \frac{2\pi}{\sqrt{2}} \frac{1}{\Delta s_{\max}}. \tag{7.73}$$

If a $(\Delta s_{\max})^2 = 1\%$ loss in signal-to-noise ratio is tolerated, this corresponds to $N \sim 44$ templates. It is clearly better to simply compute the two filters $x = (s,p)$ and $y = (s,q)$ and construct the maximum likelihood detection statistic $z = (x^2 + y^2)^{1/2}$ directly rather than compute all 44 filter outputs $(s, u(\theta_i))$ for $i \in [1,44]$ and select the maximum one.

Until now we have supposed that the signal is accurately modelled by *some* template within the parameter space, though perhaps not one of the templates that is in the bank (which covers the parameter space only to the level of the minimal match). However, if the waveform model is not perfect then a true signal may not lie in the parameter space that is covered by the templates. The *fitting factor*,

$$\mathcal{F} := \max_{\boldsymbol{\lambda}} \frac{(h, u(\boldsymbol{\lambda}))}{\sqrt{(h,h)}}, \tag{7.74}$$

is a measure of how well the parameterized waveforms $u(t;\boldsymbol{\lambda})$ will match a true signal $h(t)$.

7.3
Parameter Estimation

For the optimal detection statistic, we compute the probability of a signal being present in the data, given the data, $P(\mathcal{H}_1|s)$, where $\mathcal{H}_1$ is the alternative hypothesis (the hypothesis that a signal is present). When the signal is parameterized by a set of parameters $\boldsymbol{\lambda}$, we can construct the probability density $p(\mathcal{H}_{\boldsymbol{\lambda}}|s)$ as a function of the parameters $\boldsymbol{\lambda}$. To estimate the true value of the parameter $\boldsymbol{\lambda}$, we look for the peak, or mode, of this probability density, which will occur at the value $\boldsymbol{\lambda}_{\text{max}}$. That is we solve the equations

$$\frac{\partial}{\partial \lambda_i} p(\mathcal{H}_{\boldsymbol{\lambda}}|s)\bigg|_{\boldsymbol{\lambda}=\boldsymbol{\lambda}_{\text{max}}} = 0 \tag{7.75}$$

for $\boldsymbol{\lambda}_{\text{max}}$. Note that the parameters $\boldsymbol{\lambda}$ appear in the probability density $p(\mathcal{H}_{\boldsymbol{\lambda}}|s)$ both via the prior probability density over parameter space, $p(\mathcal{H}_{\boldsymbol{\lambda}})$, and in the likelihood ratio $\Lambda(\mathcal{H}_{\boldsymbol{\lambda}})$. For simplicity, let us assume that the prior probability density is a relatively constant function over the parameter ranges of interest (as it would be if we have little prior knowledge of the likely values) and focus instead on finding the maximum of the likelihood ratio, that is by solving (cf. Eq. (7.45))

$$\left(s - h(\boldsymbol{\lambda}_{\text{max}}), \frac{\partial h}{\partial \lambda_i}(\boldsymbol{\lambda}_{\text{max}})\right) = 0 \tag{7.76}$$

for $\boldsymbol{\lambda}_{\text{max}}$.

7.3.1
Measurement Accuracy

We can now estimate what our measurement accuracy will be. The data $s(t)$ contains both noise, $n(t)$, and the gravitational wave $h(t;\boldsymbol{\lambda})$ with the *true* parameter values $\boldsymbol{\lambda}$: $s(t) = n(t) + h(t;\boldsymbol{\lambda})$. Therefore, the parameters $\boldsymbol{\lambda}_{\text{max}}$ that solve Eq. (7.76) will satisfy the equations

$$\left(h(\boldsymbol{\lambda}) - h(\boldsymbol{\lambda}_{\text{max}}), \frac{\partial h}{\partial \lambda_i}(\boldsymbol{\lambda}_{\text{max}})\right) = -\left(n, \frac{\partial h}{\partial \lambda_i}(\boldsymbol{\lambda}_{\text{max}})\right). \tag{7.77}$$

The right hand side of these equations comprises a set of zero mean multivariate Gaussian random variables (assuming Gaussian noise). Let us denote them by

$$\nu^i := \left(n, \frac{\partial h}{\partial \lambda_i}(\boldsymbol{\lambda}_{\text{max}})\right). \tag{7.78}$$

We will need the distribution function of $\boldsymbol{\nu}$ in order to calculate the measurement accuracy.

Since the distribution $\boldsymbol{\nu}$ is a multivariate Gaussian distribution with zero mean, $\langle \nu^i \rangle = 0$, the distribution of the random variables is entirely described by the *Fisher information matrix*

$$\Gamma^{ij} := \langle \nu^i \nu^j \rangle = \left\langle \left(n, \frac{\partial h}{\partial \lambda_i}(\boldsymbol{\lambda}_{\max}) \right) \left(\frac{\partial h}{\partial \lambda_j}(\boldsymbol{\lambda}_{\max}), n \right) \right\rangle$$
$$= \left(\frac{\partial h}{\partial \lambda_i}(\boldsymbol{\lambda}_{\max}), \frac{\partial h}{\partial \lambda_j}(\boldsymbol{\lambda}_{\max}) \right), \tag{7.79}$$

where, to obtain the second line, we used the relationship $\langle (n, g)(g, n) \rangle = (g, g)$, which was shown in Eq. (7.49). Therefore the probability density function for the random variables $\boldsymbol{\nu}$ is

$$p(\boldsymbol{\nu}) = \frac{1}{\sqrt{\det(2\pi \Gamma^{kl})}} \exp\left(-\frac{1}{2} (\Gamma^{-1})_{ij} \nu^i \nu^j \right), \tag{7.80}$$

where $(\Gamma^{-1})_{ij}$ is the inverse of the Fisher information matrix.

Suppose that the signal is strong enough that the maximum likelihood estimate of the parameters, $\boldsymbol{\lambda}_{\max}$, is reasonably close to the actual value of the parameters, $\boldsymbol{\lambda}_{\max}$. That is, the measurement error,

$$\Delta\boldsymbol{\lambda} = \boldsymbol{\lambda}_{\max} - \boldsymbol{\lambda} \tag{7.81}$$

is small. Then,

$$h(\boldsymbol{\lambda}) = h(\boldsymbol{\lambda}_{\max}) - \Delta\lambda_i \frac{\partial h}{\partial \lambda_i}(\boldsymbol{\lambda}_{\max}) + O\left((\Delta\boldsymbol{\lambda})^2\right). \tag{7.82}$$

We substitute this expansion into Eq. (7.77) to obtain

$$\left(\frac{\partial h}{\partial \lambda_j}(\boldsymbol{\lambda}_{\max}), \frac{\partial h}{\partial \lambda_i}(\boldsymbol{\lambda}_{\max}) \right) \Delta\lambda_j + O\left((\Delta\boldsymbol{\lambda})^2\right) = \nu^i \tag{7.83}$$

or

$$\Gamma^{ij} \Delta\lambda_j \approx \nu^i, \tag{7.84}$$

where we have dropped terms of quadratic order or higher in the measurement errors. Therefore, the measurement errors arise from the noise according to the linear relationship $\Delta\boldsymbol{\lambda} \approx \boldsymbol{\Gamma}^{-1}\boldsymbol{\nu}$; the distribution function for these measurement errors is

$$p(\Delta\boldsymbol{\lambda}) \approx \sqrt{\det\left(\frac{\Gamma^{kl}}{2\pi} \right)} \exp\left(-\frac{1}{2} \Gamma^{ij} \Delta\lambda_i \Delta\lambda_j \right) \tag{7.85}$$

in the strong signal limit. This probability function shows us that the inverse of the Fisher information matrix contains the information about variances of the measurement error in the parameters as well as how the measurement errors for different parameters are correlated. In particular, if we wished to compute the root-mean-squared error in parameter λ_i, it would be

$$(\Delta\lambda_i)_{\text{rms}} = \left\langle (\Delta\lambda_i)^2 \right\rangle^{1/2} = (\text{Var}\Delta\lambda_i)^{1/2} = \sqrt{(\Gamma^{-1})_{ii}} \quad \text{(no summation)}. \tag{7.86}$$

Example 7.5 Measurement accuracy of signal amplitude and phase

Suppose there is just one unknown parameter: the amplitude A of the signal, so $h(t; A) = Ag(t)$ where $g(t)$ is a known time-series. We have seen in Example 7.2 that the maximum likelihood estimate of the amplitude is $A_{\max} = (s, g)/(g, g)$. To compute the measurement uncertainty, we must compute the Fisher information "matrix"

$$\Gamma = \left(\frac{\partial h}{\partial A}, \frac{\partial h}{\partial A}\right) = (g, g) = \sigma^2 . \tag{7.87}$$

The root-mean-squared uncertainty in the amplitude is therefore $(\Delta A)_{\text{rms}} = \sqrt{\Gamma^{-1}} = \sigma^{-1}$. It is more common, though, to quote the *fractional* uncertainty in the amplitude. This is

$$(\Delta \ln A)_{\text{rms}} = \frac{1}{A_{\max}} \sqrt{\Gamma^{-1}} = \frac{1}{A_{\max}\sigma} = \frac{1}{\rho} \tag{7.88}$$

so the fractional uncertainty in the amplitude is equal to the inverse of the measured signal-to-noise ratio: for a signal-to-noise ratio $\rho = 10$ detection, the amplitude is determined to a precision of 10%; for a signal-to-noise ratio $\rho = 100$ detection, the amplitude is determined to a precision of 1%.

Suppose instead that the only unknown parameter is a phase, so the signal has the form $h(t; \theta) = p(t) \cos \theta + q(t) \sin \theta$ with $(p, p) = (q, q) = (h, h)$ and $(p, q) = 0$. We have

$$\begin{aligned} \Gamma &= \left(\frac{\partial h}{\partial \theta}, \frac{\partial h}{\partial \theta}\right) = (-p \sin \theta_{\max} + q \cos \theta_{\max}, -p \sin \theta_{\max} + q \cos \theta_{\max}) \\ &= (h, h) , \end{aligned} \tag{7.89}$$

where the maximum likelihood estimate of the phase is $\theta_{\max} = \arctan((s, q)/(s, p))$. Therefore the root-mean-squared uncertainty in the phase measurement is inversely proportional to the signal's characteristic signal-to-noise ratio, $(\Delta\theta)_{\text{rms}} = (h, h)^{-1/2} = 1/\varrho$. For a signal with a characteristic signal-to-noise ratio of $\varrho = 10$, the phase is obtained to a precision of 0.1 radians or about 6°.

Now consider the situation in which the amplitude and the phase are both unknown parameters so that $h(t; A, \theta) = A[p(t) \cos \theta + q(t) \sin \theta]$ and $\boldsymbol{\lambda} = [\lambda_1, \lambda_2] = [\ln A, \theta]$. (Notice that we take $\ln A$ as the parameter rather than A so that we will directly obtain the fractional uncertainty in the amplitude rather than the absolute uncertainty in the amplitude.) We find

$$\begin{aligned} \Gamma^{11} &= A^2_{\max} (p \cos \theta_{\max} + q \sin \theta_{\max}, p \cos \theta_{\max} + q \sin \theta_{\max}) \\ &= \rho^2 \end{aligned} \tag{7.90}$$

$$\begin{aligned} \Gamma^{22} &= A^2_{\max} (-p \sin \theta_{\max} + q \cos \theta_{\max}, -p \sin \theta_{\max} + q \cos \theta_{\max}) \\ &= \rho^2 \end{aligned} \tag{7.91}$$

$$\begin{aligned}\Gamma^{12} &= \Gamma^{21} \\ &= A_{\max}^2 (p \cos\theta_{\max} + q \sin\theta_{\max}, -p \sin\theta_{\max} + q \cos\theta_{\max}) \\ &= 0 .\end{aligned} \tag{7.92}$$

Therefore the inverse of the Fisher information matrix is

$$\boldsymbol{\Gamma}^{-1} = \begin{bmatrix} 1/\rho^2 & 0 \\ 0 & 1/\rho^2 \end{bmatrix} . \tag{7.93}$$

Notice that the fractional uncertainty in the amplitude and the uncertainty in the phase are both $1/\rho$ and the measurement errors in these parameters are uncorrelated.

7.3.2
Systematic Errors in Parameter Estimation

Suppose that there is some systematic uncertainty in the waveform which is not parameterized and so is not part of the maximization procedure. Then, for any set of parameters $\boldsymbol{\lambda}$, the difference between the true waveform $h(t;\boldsymbol{\lambda})$ and the approximate waveform family that we are using for parameter estimation, now written $h'(t;\boldsymbol{\lambda})$, is $\delta h(t;\boldsymbol{\lambda}) = h'(t;\boldsymbol{\lambda}) - h(t;\boldsymbol{\lambda})$. Now Eq. (7.77) becomes

$$\left(h(\boldsymbol{\lambda}) - h'(\boldsymbol{\lambda}_{\max}), \frac{\partial h'}{\partial \lambda_i}(\boldsymbol{\lambda}_{\max}) \right) = - \left(n, \frac{\partial h'}{\partial \lambda_i}(\boldsymbol{\lambda}_{\max}) \right) . \tag{7.94}$$

Repeating the previous analysis we find that Eq. (7.84) becomes

$$\left(\delta h(\boldsymbol{\lambda}), \frac{\partial h'}{\partial \lambda_j}(\boldsymbol{\lambda}_{\max}) \right) + \Gamma^{ij} \Delta\lambda_j \approx \nu^i . \tag{7.95}$$

The additional term results in a systematic error component to the overall parameter uncertainty:

$$\begin{aligned}(\Delta\lambda_i)_{\text{syst}} &= (\Gamma^{-1})_{ij} \left(\delta h(\boldsymbol{\lambda}), \frac{\partial h'}{\partial \lambda_j}(\boldsymbol{\lambda}_{\max}) \right) \\ &\approx (\Gamma^{-1})_{ij} \left(\delta h(\boldsymbol{\lambda}_{\max}), \frac{\partial h'}{\partial \lambda_j}(\boldsymbol{\lambda}_{\max}) \right) .\end{aligned} \tag{7.96}$$

This parameter estimation error is independent of the strength of the signal, so it will dominate over the noise-induced rms error for sufficiently strong signals.

Example 7.6 Systematic error in estimate of signal amplitude

Suppose a true signal has the form $h(t; A, \theta) = A[p(t) \cos\theta + q(t) \sin\theta]$ where θ is a small angle parameter, which is unmodelled, and A is the (modelled) signal amplitude. The family of waveforms that are used for parameter estimation is therefore $h'(t; A) = Ap(t)$. We have $\Gamma = (p, p)$ and therefore

$$(\Delta A)_{\text{syst}} = \Gamma^{-1}(Ap(1 - \cos\theta) - Aq\sin\theta, p) \approx \frac{1}{2}\theta^2 A \tag{7.97}$$

so the fractional uncertainty in the amplitude is

$$(\Delta \ln A)_{\text{syst}} \approx \frac{1}{2}\theta^2 . \tag{7.98}$$

This systematic uncertainty is independent of the signal amplitude. It will dominate over the noise-induced error, $(\Delta \ln A)_{\text{rms}} = 1/\rho$, when $\rho > 2/\theta^2$. For example, if there is a systematic phase error in the template used of $\theta = 0.1$ radians then this systematic error will dominate the fractional measurement error in the amplitude for signal-to-noise ratios $\rho > 200$.

While Eq. (7.96) describes the systematic error that is produced for a *particular* parameter given an error in the model waveform, there is also a useful bound on the size of any systematic error associated with an error in the model waveform. To obtain this, imagine the scenario in which a single parameter λ that we are hoping to measure happens to be the one that interpolates the difference between the true waveform and the model waveform:

$$h(t;\lambda) = h(t;\lambda = 0) + \lambda\delta h , \tag{7.99}$$

where $\delta h = h'(t;\lambda = 0) - h(t;\lambda = 0)$. Then $\Gamma = (\delta h, \delta h)$ and $(\Delta\lambda)_{\text{syst}} = 1$. The requirement that $(\Delta\lambda)_{\text{rms}} > (\Delta\lambda)_{\text{syst}}$ where $(\Delta\lambda)_{\text{rms}} = 1/\Gamma$ (i.e. that the systematic error is less than the random error due to the noise) therefore corresponds to

$$(\delta h, \delta h) < 1 . \tag{7.100}$$

If the waveform difference satisfies this inequality then the two waveforms are said to be *indistinguishable*, so Eq. (7.100) is called the *indistinguishability criterion.* It is amplitude dependent: two waveforms that are indistinguishable at low signal-to-noise ratio will eventually become distinguishable at high signal-to-noise ratio. An amplitude-independent measure of the systematic error in a parameter estimate due to waveform uncertainty is

$$\epsilon_{\text{syst}} := \frac{(\delta h, \delta h)}{(h, h)} \tag{7.101}$$

and the systematic errors may become significant for signals where

$$\epsilon_{\text{syst}} \gtrsim \frac{1}{\varrho^2} . \tag{7.102}$$

That is, if we observe a signal with $\varrho \sim 10$, we require our model waveforms to be sufficiently accurate, with $\epsilon_{\text{syst}} < 1\%$, or else our parameter estimations may be affected by the systematic error in the model.

Another bound on the magnitude of the systematic error in terms of the random error is also provided by the indistinguishability criterion. We can apply Schwartz's inequality on Eq. (7.96) to obtain an upper bound on the size of the systematic error,

$$\left|(\Delta\lambda_i)_{\text{syst}}\right| \leq (\Delta\lambda_i)_{\text{rms}}(\delta h, \delta h)^{1/2} . \tag{7.103}$$

Note that this bound is still amplitude-independent, as expected.

7.3.3 Confidence Intervals

Beyond finding the parameter values $\boldsymbol{\lambda}_{\max}$ that maximize the posterior probability density function, $p(\mathcal{H}_{\boldsymbol{\lambda}}|s)$ – the mode of the distribution – we may wish to report a region of parameter space that we expect contains the true values of the parameters with some probability. This is called a *confidence region*, and in the case of a single parameter of interest, we call this a *confidence interval*. For illustration we will consider the problem of setting a confidence interval for a single parameter, λ.

There are various types of confidence intervals, which differ in their construction and interpretation. A *Bayesian confidence interval*, which is often called a *credible interval* to avoid confusion with the frequentist confidence intervals described below, is obtained directly from the posterior probability density function, $p(\mathcal{H}_\lambda|s)$, as follows: let us fix some *confidence level*, α. Then we wish to solve the equation

$$\alpha = \int_{\lambda_1}^{\lambda_2} p(\mathcal{H}_\lambda|s)d\lambda \tag{7.104}$$

for the values λ_1 and λ_2, which implies that the parameter λ is in the range $\lambda_1 < \lambda < \lambda_2$ with a probability α. The interval (λ_1, λ_2) is the confidence interval for λ at the particular confidence level α (for example, if $\alpha = 0.95$ then the confidence level is 95%). However, there are two parameters, λ_1 and λ_2, that must be obtained from the single equation (7.104) and to obtain a single solution we require an additional supplementary equation on λ_1 and λ_2. If we seek an upper limit on λ, we simply set $\lambda_1 = -\infty$ (or the smallest allowed value of λ) and solve Eq. (7.104) for the value of λ_2 (the upper limit). Similarly, a lower limit on λ is obtained by setting $\lambda_2 = \infty$ (or the largest allowed value of λ) and solving for λ_1. Another possible supplementary condition would be to require that the two limits of the interval have the same value of the probability density function: $p(\mathcal{H}_{\lambda_1}|s) = p(\mathcal{H}_{\lambda_2}|s)$. However, if the probability density function is multi-modal, this may result in a disconnected set of confidence intervals, and an alternative is to choose an interval that has the same probability of having $\lambda < \lambda_1$ as $\lambda > \lambda_2$:

$$\frac{1-\alpha}{2} = \int_{-\infty}^{\lambda_1} p\,(\mathcal{H}_\lambda|s)\,d\lambda = \int_{\lambda_2}^{\infty} p\,(\mathcal{H}_\lambda|s)\,d\lambda . \tag{7.105}$$

Note that such an interval will always include the median value of $p(\mathcal{H}_\lambda|s)$, though it may not include the mode.

A *frequentist confidence interval* begins with the probability distribution $p(s|\mathcal{H}_\lambda)$ rather than the posterior probability distribution $p(\mathcal{H}_\lambda|s)$. According to Bayes's theorem, Eq. (7.33), the two are simply related by the prior probability factor, $p(\mathcal{H}_\lambda)$, and a normalization constant $1/p(s)$ which does not depend on λ. Nevertheless, the frequentist and Bayesian confidence intervals have very different meanings, as we shall see.

The frequentist confidence interval at confidence level α is constructed so that for any true value of λ (regardless of what it is) then the intervals obtained by repeated experiments will contain λ in a fraction α of those experiments. The construction is achieved by the *Neyman method*: define some statistic x that is derived from $s(t)$, for example it could be the signal-to-noise ratio $x = \rho$. Because x is derived from s, the probability distribution $p(s|\mathcal{H}_\lambda)$ can be re-expressed as a probability distribution $p(x;\lambda)$. Note that the distribution of x depends on the unknown parameter λ. For each value of λ, we produce an interval, (x_1, x_2), such that

$$\alpha = \int_{x_1}^{x_2} p(x;\lambda)dx \,. \tag{7.106}$$

These intervals, which depend on λ, define a belt in the λ–x plane, and for any *observed* value of x in an experiment, the confidence interval on λ constitutes those values of λ that exist on the belt at that fixed value of x. Note that Eq. (7.106) guarantees the coverage required for a frequentist confidence interval: for any value of λ, the observed value of x will lie in the acceptance belt with probability α by construction.

As with the Bayesian confidence intervals, construction of the frequentist confidence belts requires a supplementary condition since Eq. (7.106) is a single equation that depends on the two values x_1 and x_2. To construct a belt that is appropriate for an upper limit on λ, we take $x_2 = \infty$ and solve Eq. (7.106) for x_1, while to construct a belt that is appropriate for a lower limit on λ we take $x_1 = -\infty$ and solve Eq. (7.106) for x_2. A central confidence interval is one for which x_1 and x_2 satisfy

$$\frac{1-\alpha}{2} = \int_{-\infty}^{x_1} p(x;\lambda)dx = \int_{x_2}^{\infty} p(x;\lambda)dx \,. \tag{7.107}$$

Frequentist confidence intervals may be empty in some situations, as seen in Example 7.7.

Example 7.7 Frequentist upper limits

Consider a simple situation in which the observed statistic x is distributed with a normal Gaussian distribution with unit variance and mean λ, the parameter we are trying to bound, which is always non-negative. Suppose we wish to set an upper limit on λ given the results of an experiment: a measurement that samples the distribution and yields a value x_{obs}.

Using the Neyman procedure, we construct a confidence belt appropriate for setting an upper limit on λ. For each value of λ we find the value x_1 that solves

$$\alpha = \int_{x_1}^{\infty} p(x;\lambda)dx = \frac{1}{\sqrt{2\pi}} \int_{x_1}^{\infty} \mathrm{e}^{-(x-\lambda)^2/2} dx = \frac{1}{2}\mathrm{erfc}\left(\frac{x_1 - \lambda}{\sqrt{2}}\right) \tag{7.108}$$

so $x_1 = \lambda + \sqrt{2}\mathrm{erfc}^{-1}(2\alpha)$. To obtain, say, a 90%-confidence upper limit, $\alpha = 0.9$, we construct a confidence belt $x_1(\lambda) \simeq \lambda - 1.28$. Then, for any observed value x_{obs} we find that the confidence interval in λ is $0 \leq \lambda < 1.28 + x_{\mathrm{obs}}$.

Notice that when $x_{\mathrm{obs}} < -1.28$ is observed, the confidence interval in λ is *empty*. This situation will arise 10% of the time when the true value of λ is zero.

Feldman and Cousins (1998) describe a unified approach to setting frequentist confidence intervals that addresses issues such as the possibility of empty intervals, and also allows for transitions from one-sided (upper limit) intervals to two-sided intervals depending on the value of the observation.

7.4 Detection Statistics for Poorly Modelled Signals

Often we do not have enough information about anticipated signals to be able to construct the matched filters required for optimal detection statistics. We may, nevertheless, have some *limited* information about the signal, perhaps an idea of what the signal's duration and frequency band is, and this information can be used to construct a detection statistic.

7.4.1 Excess-Power Method

The *excess-power* detection statistic is useful when a signal has a few known parameters, but is otherwise arbitrary. Suppose that there exists some set of N basis functions $\hat{e}_i(t)$ which spans the space of possible signals. We further assume that these basis functions are orthonormal in the inner product so that $(\hat{e}_i, \hat{e}_j) = \delta_{ij}$. Then, the true gravitational wave would be some linear combination of these basis functions,

$$h(t) = \sum_{i=1}^{N} c_i \hat{e}_i(t) \tag{7.109}$$

for some set of constants c_i. Of course, we do not know what the constants c_i are – otherwise we would have the matched filter – so instead we insert this model of the waveform into the likelihood ratio and treat the constants as unknown parameters as we did in Section 7.2.3 to be obtained by maximizing the likelihood ratio. The logarithm of the likelihood ratio, Eq. (7.44), is

$$\ln \Lambda \left(\mathcal{H}_{\{c_i\}}|s\right) = \sum_{i=1}^{N} \left\{c_i(s, \hat{e}_i) - \frac{1}{2}c_i^2\right\} \tag{7.110}$$

and the coefficients that maximize this are

$$c_{i,\max} = (s, \hat{e}_i) . \tag{7.111}$$

Once these coefficients have been determined, they are substituted back into the expression for the logarithm of the likelihood, which yields

$$\mathcal{E} = 2\ln \Lambda(\mathcal{H}_{\{c_{i,\max}\}}|s) = \sum_{i=1}^{N}(s, \hat{e}_i)^2 , \tag{7.112}$$

where $\mathcal{E}$ is known as the excess-power statistic. Furthermore, the coefficients can be used to reconstruct the maximum likelihood estimate of the signal,

$$h_{\text{est}}(t) = \sum_{i=1}^{N}(s, \hat{e}_i)\hat{e}_i(t) . \tag{7.113}$$

Notice that the values $(s, \hat{e}_i)$ are the projections of the data stream onto the basis vectors of the space of possible signals. If we write

$$s(t) = s_{\|}(t) + s_{\perp}(t) \tag{7.114}$$

where $s_{\|}(t)$ is the portion of the data stream that lives in the space of signals, so that $s_{\|}(t)$ can be expanded in terms of the basis functions, and $s_{\perp}(t)$ is the remainder, the part of the stream that is orthogonal to the space of signals, so that $(s_{\perp}, \hat{e}_i) = 0$ for all basis functions $\hat{e}_i(t)$, then the excess-power statistic is simply

$$\mathcal{E} = (s_{\|}, s_{\|}) . \tag{7.115}$$

If the noise is Gaussian, then in the absence of any gravitational-wave signal the excess-power statistic $\mathcal{E}$ is the sum of the squares of N independent random variables, $(s, \hat{e}_i)$, (cf. Eq. (7.112)), and hence it has a chi-squared distribution with N degrees of freedom. The chi-squared distribution with N degrees of freedom has a mean of N and a variance of $2N$.

If, however, a signal $h(t)$ is present, then the power contained in this signal is

$$\varrho^2 = (h, h) = \sum_{i=1}^{N}(h, \hat{e}_i)^2 \tag{7.116}$$

provided that the signal lies entirely within the space spanned by the basis functions $\hat{e}_i$. For Gaussian noise, the distribution of the excess-power statistic $\mathcal{E}$ becomes a *non-central* chi-squared distribution with N degrees of freedom and a non-central parameter ϱ^2. This distribution has a mean of $N + \varrho^2$ and a variance of

$2N + 4\varrho^2$. The presence of the signal therefore shifts the mean of the distribution of $\mathcal{E}$ by an amount ϱ^2, so a detectable signal would have a characteristic signal-to-noise ratio ϱ that is a few times $\sim (2N)^{1/4}$. We see, therefore, that the effectiveness of the excess-power statistic is limited by the degree of uncertainty about the signal: as the number of basis functions N required to span the space of possible signals grows, the stronger a signal would have to be in order to be detectable.

Example 7.8 Time-frequency excess-power statistic

An important application of the excess-power statistic is the case when a signal's duration and frequency band are known (or approximately known), but little else about the signal is understood. In such a situation, the natural basis functions are the finite-time Fourier modes that span the desired time-frequency region.

A straightforward way to construct $s_{\parallel}$ is to (i) window the original time series so that only the desired time range, say between t_0 and $t_0 + T$, remains,

$$s_T(t) = \begin{cases} s(t) & t_0 < t < t_0 + T \\ 0 & \text{otherwise,} \end{cases} \tag{7.117}$$

and (ii) retain only those frequency components in the desired frequency range, say between f_0 and $f_0 + F$,

$$\tilde{s}_{\parallel}(f) = \begin{cases} \tilde{s}_T(f) & f_0 < f < f_0 + F \\ 0 & \text{otherwise.} \end{cases} \tag{7.118}$$

When this procedure is carried out, the excess-power statistic, $\mathcal{E} = (s_{\parallel}, s_{\parallel})$, has a chi-squared distribution (or a non-central chi-squared distribution if there is a signal present) with $N = 2TF$ degrees of freedom for Gaussian noise.

7.5
Detection in Non-Gaussian Noise

Until now, our focus has been on the derivation of detection statistics under the assumptions that the detector noise is stationary and Gaussian. Unfortunately, these assumptions seldom apply, and detector noise often has a non-Gaussian component of transient noise artefacts – some of which can be quite significant. A search designed to be optimal under conditions of stationary and Gaussian noise can have poor performance for real detector data, perhaps even mis-identifying detector glitches as gravitational-wave events. Therefore, we must supplement our "optimal" detection statistics with robust measures that can reject instrumental artefacts while retaining true signals. In addition, because we cannot be sure that the statistical behaviour of our optimal detection strategy will be exactly what is predicted under the assumption of Gaussian noise, we must devise methods for assessing such quantities of interest as the false alarm probability and the detection efficiency for our detection statistic.

To distinguish between true gravitational waves and the vagaries of the detector noise, we need a method to corroborate a gravitational-wave event candidate. If we are looking for a signal of a known form, we can examine the morphology of the candidate to verify that the signal appears as expected. For example, for gravitational waves from binary inspiral, we know how the signal is expected to evolve in time, and we can construct waveform consistency tests that would accept signals of the anticipated form while rejecting those of an unknown form. As another example, consider continuous signals of nearly constant frequency that might be produced by pulsars: if the position on the sky of the pulsar is known, then we can make a detailed prediction of the phase and amplitude modulation that would be produced by the Earth's motion relative to the source, and if what we see does not have these precise modulations then it is not consistent with the anticipated signal.

On the flip side, we may attempt to identify environmental origins of a potential event, and, if it can be shown that the event likely arose from environmental causes, then we *veto* the event. The environment of ground-based gravitational-wave detectors is continually monitored for such things as seismic, atmospheric, or magnetic activity which may produce a signal that can mimic a gravitational-wave signal. Interferometric detectors also monitor the light in various parts of the interferometer, and these readouts, which are often quite insensitive to true gravitational-wave signals, may be used to identify other causes for an event such as a fluctuation in light power from the laser, or mis-alignment of one of the interferometer's optical cavities.

Signals that are always on, such as continuous-wave signals and stochastic background signals, can be corroborated with additional observation. If a stochastic background signal is seen in one year of observation, then it should be seen in the next year of observation too.

For transient, short-lived bursts of gravitational waves, another powerful method is available when there are multiple detectors: the signal must be seen in all detectors at roughly the same time. This requirement of *coincidence* can be very powerful at rejecting noise transients, which are not normally expected to occur simultaneously in multiple detectors (especially amongst detectors that are not co-located, and so are not subject to the same environmental disturbances). A related method is to create a combination of the various detector data streams in which true gravitational waves would cancel but noise transients would likely persist. If a candidate event is also seen in such a null-combination, it is not likely to be a gravitational wave. These multiple-detector methods are described more in the next section.

Finally, if the non-Gaussian nature of the detector noise is understood, it is possible to repeat the construction of the optimal detection statistics using this non-Gaussian noise model to obtain a robust detection statistic (Kassam, 1987).

In order to make a confident detection of gravitational waves we must be able to estimate the false alarm probability for any candidate event we encounter. To do this, we must know the statistical properties of our detection statistic, for example the signal-to-noise ratio of the matched filter output in a search for a signal of known form, in the presence of noise alone. That is, we want to find the cumulative probability, $P(\rho' > \rho|\mathcal{H}_0)$, that is the chance of getting a value of the signal-to-noise ratio ρ' (or whatever the detection statistic is) greater than some fixed value ρ (perhaps the observed value in the search, or a threshold value for detection), in the case that there is no gravitational-wave signal. This is known as the *false alarm probability*, and it is an important characteristic of the noise *background*.

The difficulty in estimating the false alarm probability is in knowing that there is no signal present. Unlike other experiments where one can simply remove a source and so determine the properties of the detector noise when there is noise alone, we cannot "shield" a gravitational-wave detector from gravitational waves. Nevertheless, there are several strategies that we take to obtain false alarm probabilities.

One straightforward approach, when we are targeting a particular signal (or anticipated signal), is simply to search the data for signals with parameters that are in a vicinity of the target signal's parameters but not having the exact parameter values of the target signal. For example, in searches for gravitational waves associated with gamma-ray bursts, we might expect that the gravitational waves will have a close temporal association with the observed gamma-ray burst. Therefore, by examining the behaviour of the detection statistic at times *other than* near the observed burst, we can be confident that there is no gravitational-wave signal, so we are measuring the false alarm probability. A similar situation arises in searches for continuous waves produced by known pulsars: the frequency of the anticipated signal is known to high precision, and so a search of nearby frequencies, which do *not* contain the anticipated signal, can tell us about the properties of the noise. In both these cases it is a matter of judgement to determine how close one needs to be – either in time or in frequency – to the true signal in order to get an accurate reading of the properties of the noise at the time or the frequency of the signal.

When data from several detectors are brought together for analysis then there is another approach that can be used in searches for short-lived gravitational-wave bursts: by introducing an artificial time shift in the data streams obtained from the various detectors – shifts that are longer than the expected duration and offsets of the anticipated signals – true gravitational waves, which must arrive at all the detectors at nearly the same time, can be excluded. The *time-slide method* for determining the background properties of the noise can be quite effective, but requires multiple detectors.

In addition to the false alarm probability, we need to evaluate the *detection probability* of our search, as this is critical to the interpretation of the search result. The detection probability, $P(\rho' > \rho|\mathcal{H}_1)$, which is simply related to the *false dismissal* probability, $P(\rho' < \rho|\mathcal{H}_1) = 1 - P(\rho' > \rho|\mathcal{H}_1)$, is the same as the false alarm probability but evaluated under the alternative hypothesis $\mathcal{H}_1$ that a signal is present rather than the null hypothesis $\mathcal{H}_0$ that a signal is absent. The detection probability can be determined by repeating a gravitational-wave search on data that has been altered by the addition of simulated signals. Simulated signals can be added to the detector output either by actuation of the detector, which are known as *hardware injections*, or by adding the signals into the data stream recorded by the detector post facto, which are called *software injections*.

7.6 Networks of Gravitational-Wave Detectors

Until now, we have considered the detection and measurement problem for a single detector alone. Here we consider the case in which multiple detectors are operating as a network. The approach is a generalization of Section 7.2. We first consider the simple network consisting of two co-located detectors with the same beam pattern response (so that they are sensitive to the same gravitational-wave polarization), such as the two interferometers at the LIGO Hanford Observatory, and then we consider an arbitrary network of detectors.

7.6.1 Co-located and Co-aligned Detectors

The LIGO Hanford Observatory houses two interferometers; an impinging gravitational wave has exactly the same physical strain effect on each of the interferometers. This means that the data $s_1(t)$ and $s_2(t)$ that are produced by each detector will contain a common signal $h(t)$ in addition to the (largely independent) random noise $n_1(t)$ and $n_2(t)$: $s_1(t) = n_1(t) + h(t)$ and $s_2(t) = n_2(t) + h(t)$. If we assume that the noise processes are independent Gaussian noise (in fact, there are common noise sources for the two Hanford detectors, so this assumption is not entirely appropriate) then the joint noise probability density is

$$p_{n_1,n_2}\left[n_1(t), n_2(t)\right] = p_{n_1}\left[n_1(t)\right] p_{n_2}\left[n_2(t)\right] \propto \mathrm{e}^{-(n_1,n_1)_1/2}\mathrm{e}^{-(n_2,n_2)_2/2} , \tag{7.119}$$

where the inner products $(a, b)_1$ and $(a, b)_2$ are different because the noise power spectral densities $S_1(f)$ and $S_2(f)$ are different:

$$(a, b)_1 := 4\,\mathrm{Re}\int_0^\infty \frac{\tilde{a}(f)\tilde{b}^*(f)}{S_1(f)}\,df \quad \text{and} \quad (a, b)_2 := 4\,\mathrm{Re}\int_0^\infty \frac{\tilde{a}(f)\tilde{b}^*(f)}{S_2(f)}\,df . \tag{7.120}$$

Again we construct the likelihood ratio $\Lambda(\mathcal{H}_1|s_1, s_2)$ as

$$\Lambda\left(\mathcal{H}_1|s_1, s_2\right) = \frac{p\left(s_1, s_2|\mathcal{H}_1\right)}{p\left(s_1, s_2|\mathcal{H}_0\right)} , \tag{7.121}$$

where

$$p\left(s_1, s_2|\mathcal{H}_0\right) = p_{n_1,n_2}\left[s_1(t), s_2(t)\right] \propto \mathrm{e}^{-(s_1,s_1)_1/2}\mathrm{e}^{-(s_2,s_2)_2/2} \tag{7.122}$$

and

$$p(s_1, s_2|\mathcal{H}_1) = p_{n_1,n_2}\left[s_1(t) - h(t), s_2(t) - h(t)\right] \\ \propto \mathrm{e}^{-(s_1-h,s_1-h)_1/2}\mathrm{e}^{-(s_2-h,s_2-h)_2/2}\,. \tag{7.123}$$

The log of the likelihood ratio is therefore

$$\ln \Lambda(\mathcal{H}_1|s_1, s_2) = (s_1, h)_1 + (s_2, h)_2 - \frac{1}{2}\left[(h, h)_1 + (h, h)_2\right]. \tag{7.124}$$

The first two terms alone contain the detector data; the network matched filter is therefore

$$(s_1, h)_1 + (s_2, h)_2 = 4\,\mathrm{Re}\int_0^\infty \left[\frac{\tilde{s}_1(f)}{S_1(f)} + \frac{\tilde{s}_2(f)}{S_2(f)}\right]\tilde{h}^*(f)df\,. \tag{7.125}$$

This equation can be written in a more convenient form if one identifies a combination of the detector data streams:

$$\tilde{s}_{1+2}(f) := \frac{\tilde{s}_1(f)/S_1(f) + \tilde{s}_2(f)/S_2(f)}{1/S_1(f) + 1/S_2(f)}\,. \tag{7.126}$$

The combination $\tilde{s}_{1+2}(f)$, or its time domain representation $s_{1+2}(t)$, is known as the *coherent combination* of the data streams. Notice that when both signal and noise are present, $s_1(t) = n_1(t) + h(t)$ and $s_2(t) = n_2(t) + h(t)$, and the coherent combination is

$$\tilde{s}_{1+2}(f) = \tilde{h}(f) + \frac{\tilde{n}_1(f)/S_1(f) + \tilde{n}_2(f)/S_2(f)}{1/S_1(f) + 1/S_2(f)} \tag{7.127}$$

so the coherent stream is an estimator of the strain $h(t)$. If, on the other hand, there is no signal present so that $s_1(t) = n_1(t)$ and $s_2(t) = n_2(t)$, it follows from Eq. (7.13) that the power spectrum of the noise in the coherent data stream $s_{1+2}(t)$ is the harmonic sum of the power spectra of the noise in the two detectors:

$$S_{1+2}^{-1}(f) := S_1^{-1}(f) + S_2^{-1}(f)\,. \tag{7.128}$$

Then, Eq. (7.125) can be written as

$$(s_1, h)_1 + (s_2, h)_2 = (s_{1+2}, h)_{1+2} := 4\,\mathrm{Re}\int_0^\infty \frac{\tilde{s}_{1+2}(f)\tilde{h}^*(f)}{S_{1+2}(f)}df\,. \tag{7.129}$$

We see that the two-detector matched filter is exactly the same as the single-detector matched filter but with the coherently combined data stream $s_{1+2}(t)$. Similarly, the log of the likelihood ratio is

$$\ln \Lambda(\mathcal{H}_1|s_1, s_2) = (s_{1+2}, h)_{1+2} - \frac{1}{2}(h, h)_{1+2}\,. \tag{7.130}$$

It follows that we can now treat the network of two co-located and co-aligned detectors as a single detector with the coherent stream.

In addition to the coherent combination of the two detectors' data streams, a useful combination, known as the *null combination* of the data streams, is

$$s_{1-2}(t) := s_1(t) - s_2(t) . \tag{7.131}$$

Note that this stream has the feature that any signal present in the individual detector data streams will cancel when forming the null combination. Therefore this null stream will never contain any gravitational-wave signal content. The null stream is useful as a diagnostic for discriminating between true gravitational waves (which do not affect the null stream) and instrumental glitches or other transients which most likely will show up in the null stream.

7.6.2
General Detector Networks

Now we consider the case of some number N of detectors that are not necessarily at the same locations or even aligned. The response of these detectors to a gravitational wave will depend on the direction to the source, and different detectors may observe different polarizations of the incident wave. Furthermore, if we are not in the long-wavelength limit, the response of an individual detector to a signal arriving from a particular direction may be frequency dependent (see Section 6.1.10). We write the effective strain (in the frequency domain) on the ith detector in the network as

$$\tilde{h}_i(f) = \mathrm{e}^{-2\pi i f \tau_i(\hat{\boldsymbol{n}})} \left[G_{+,i}(\hat{\boldsymbol{n}}, \psi, f)\, \tilde{h}_+(f) + G_{\times,i}(\hat{\boldsymbol{n}}, \psi, f)\, \tilde{h}_\times(f) \right] , \tag{7.132}$$

where $G_{+,i}$ and $G_{\times,i}$ are the frequency-dependent beam pattern responses for the ith detector to the plus- and cross-polarization states of the gravitational wave, respectively, and $\tau_i = \hat{\boldsymbol{n}} \cdot \boldsymbol{r}_i / c$ is the time delay for the arrival of the gravitational wave at the ith detector compared to some fiducial point (say, the centre of the Earth), and $\boldsymbol{r}_i$ is the position of the detector relative to the fiducial point. In the long-wavelength limit, we can replace $G_{+,i}(\hat{\boldsymbol{n}}, \psi, f)$ and $G_{+,i}(\hat{\boldsymbol{n}}, \psi, f)$ with $F_{+,i}(\hat{\boldsymbol{n}}, \psi)$ and $F_{\times,i}(\hat{\boldsymbol{n}}, \psi)$.

The strain-calibrated readout from the ith detector contains that detector's noise plus any signal-induced strain that may be present:

$$\begin{aligned} \tilde{s}_i(f) &= \tilde{n}_i(f) + \tilde{h}_i(f) \\ &= \tilde{n}_i(f) + \mathrm{e}^{-2\pi i f \tau_i(\hat{\boldsymbol{n}})} \left[G_{+,i}(\hat{\boldsymbol{n}}, \psi, f) \tilde{h}_+(f) + G_{\times,i}(\hat{\boldsymbol{n}}, \psi, f) \tilde{h}_\times(f) \right] . \end{aligned} \tag{7.133}$$

As before, the logarithm of the likelihood ratio can be constructed (under the assumption that each of the detectors' noise is independent and Gaussian), and we find

$$\ln \Lambda\left(\mathcal{H}_1 | \{s_i\}\right) = \sum_{i=1}^{N} \left\{ (s_i, h_i)_i - \frac{1}{2}(h_i, h_i) \right\}. \tag{7.134}$$

We will focus for now on just terms $(s_i, h_i)_i$ as these are the only terms that carry the detector data. It is straightforward to show that

$$\sum_{i=1}^{N} (s_i, h_i)_i = (s_+, h_+)_+ + (s_\times, h_\times)_\times , \tag{7.135}$$

where $s_+(t)$ and $s_\times(t)$ are two coherent combinations of the data streams, which are, in the frequency domain,

$$\tilde{s}_+(f) := \frac{\sum_{i=1}^{N} e^{2\pi i f \tau_i(\hat{\boldsymbol{n}})} G_{+,i}^{*}(\hat{\boldsymbol{n}}, \psi, f)\, \tilde{s}_i(f)/S_i(f)}{\sum_{i=1}^{N} |G_{+,i}(\hat{\boldsymbol{n}}, \psi, f)|^2/S_i(f)} \tag{7.136a}$$

and

$$\tilde{s}_\times(f) := \frac{\sum_{i=1}^{N} e^{2\pi i f \tau_i(\hat{\boldsymbol{n}})} G_{\times,i}^{*}(\hat{\boldsymbol{n}}, \psi, f)\, \tilde{s}_i(f)/S_i(f)}{\sum_{i=1}^{N} |G_{\times,i}(\hat{\boldsymbol{n}}, \psi, f)|^2/S_i(f)} \tag{7.136b}$$

and these coherently combined streams have noise power spectra $S_+(f)$ and $S_\times(f)$

$$\frac{1}{S_+(f)} := \sum_{i=1}^{N} \frac{|G_{+,i}(\hat{\boldsymbol{n}}, \psi, f)|^2}{S_i(f)} \quad \text{and} \quad \frac{1}{S_\times(f)} := \sum_{i=1}^{N} \frac{|G_{\times,i}(\hat{\boldsymbol{n}}, \psi, f)|^2}{S_i(f)}, \tag{7.137}$$

while the inner products $(a, b)_+$ and $(a, b)_\times$ are simply

$$(a, b)_+ := 4\,\mathrm{Re} \int_0^\infty \frac{\tilde{a}(f)\tilde{b}^*(f)}{S_+(f)} df \quad \text{and} \quad (a, b)_\times := 4\,\mathrm{Re} \int_0^\infty \frac{\tilde{a}(f)\tilde{b}^*(f)}{S_\times(f)} df . \tag{7.138}$$

Unlike the case of aligned detectors considered previously, there are two coherent combinations that must be formed and these are subjected to matched filters with the two different waveform polarizations according to Eq. (7.135).

Null streams can also be found for general networks of detectors, though unless two detectors have the same response to a gravitational wave then the null stream will depend on the direction to the source. To compute the null combination of the detector outputs, we formulate the general detector network, in the long-wavelength limit, as

$$\begin{bmatrix} s_1(t+\tau_1) \\ s_2(t+\tau_2) \\ \vdots \\ s_N(t+\tau_N) \end{bmatrix} = \begin{bmatrix} n_1(t+\tau_1) \\ n_2(t+\tau_2) \\ \vdots \\ n_N(t+\tau_N) \end{bmatrix} + \begin{bmatrix} F_{1,+} & F_{1,\times} \\ F_{2,+} & F_{2,\times} \\ \vdots & \vdots \\ F_{N,+} & F_{N,\times} \end{bmatrix} \cdot \begin{bmatrix} h_+(t) \\ h_\times(t) \end{bmatrix} \tag{7.139}$$

or, more concisely, $\boldsymbol{s} = \boldsymbol{n} + \mathbf{F}\boldsymbol{h}$. We seek a set of coefficients $\boldsymbol{c} = [c_1, c_2, \ldots, c_N]$ that specifies the linear combination of the detector streams $s_0(t) = \boldsymbol{c} \cdot \boldsymbol{s}$ for which there is no signal present. This is done by finding a vector $\boldsymbol{c}$ for which $\boldsymbol{c}\mathbf{F} = 0$, or, equivalently, $\mathbf{F}^{\mathrm{T}}\boldsymbol{c}^{\mathrm{T}} = 0$:

$$\begin{bmatrix} F_{1,+} & F_{2,+} & \cdots & F_{N,+} \\ F_{1,\times} & F_{2,\times} & \cdots & F_{N,\times} \end{bmatrix} \cdot \begin{bmatrix} c_1 \\ c_2 \\ \vdots \\ c_N \end{bmatrix} = \begin{bmatrix} 0 \\ 0 \end{bmatrix} . \tag{7.140}$$

The vector of coefficients $\boldsymbol{c}^{\mathrm{T}}$ belongs to the *nullspace* of the matrix $\mathbf{F}^{\mathrm{T}}$.

Note that the matrix of antenna beam pattern functions $\mathbf{F}$ depends on the sky position of the source and also on the polarization angle ψ. However, a change in the value of ψ simply corresponds to a rotation of the vector $\boldsymbol{h}$, which does not change the nullspace of $\mathbf{F}$. Therefore, when computing the coefficients for the null stream, any convenient value of the polarization angle (say, $\psi = 0$) may be used. Also, the null stream will *always* depend on $\hat{\boldsymbol{n}}$ through $s_i(t_i + \tau_i) = s_i(t_i + \hat{\boldsymbol{n}} \cdot \boldsymbol{r}_i/c)$ unless the detectors are co-located.

Example 7.9 Nullspace of two co-aligned, co-located detectors

For two aligned detectors, $F_{+,1} = F_{+,2} = F_+$ and $F_{\times,1} = F_{\times,2} = F_\times$, so we seek the coefficients $\boldsymbol{c} = [c_1, c_2]$ that satisfies

$$\begin{bmatrix} F_+ & F_+ \\ F_\times & F_\times \end{bmatrix} \cdot \begin{bmatrix} c_1 \\ c_2 \end{bmatrix} = 0 . \tag{7.141}$$

Clearly this is solved if $c_1 = -c_2$. The overall scale is arbitrary, so let $c_1 = 1$, and therefore $\boldsymbol{c} = [+1, -1]$. We thus recover the co-aligned null stream:

$$s_0(t) = [+1, -1] \cdot \begin{bmatrix} s_1(t) \\ s_2(t) \end{bmatrix} = s_1(t) - s_2(t) . \tag{7.142}$$

Notice that this null stream is an *all-sky* null stream: the actual values of F_+ and $F_\times$, which depend on the location of the source on the sky, don't matter.

However, if the two detectors are not co-aligned, then the matrix $\mathbf{F}^{\mathrm{T}}$ will not have a nullspace, and there will be no null stream. If the detectors are non-aligned then three are required in order to produce a null stream.

Example 7.10 Nullspace of three non-aligned detectors

For the LIGO-Virgo network, there are three interferometer sites, the LIGO Hanford Observatory (H), the LIGO Livingston Observatory (L), and the Virgo Observatory (V). To compute a null stream that is insensitive to gravitational waves incident from a particular sky location, the coefficients c_{H}, c_{L} and c_{V} need to be determined, where

$$\begin{bmatrix} F_{+,\mathrm{H}} & F_{+,\mathrm{L}} & F_{+,\mathrm{V}} \\ F_{\times,\mathrm{H}} & F_{\times,\mathrm{L}} & F_{\times,\mathrm{V}} \end{bmatrix} \cdot \begin{bmatrix} c_\mathrm{H} \\ c_\mathrm{L} \\ c_\mathrm{V} \end{bmatrix} = 0 \,. \tag{7.143}$$

The solution to this system of equations is

$$\begin{aligned} c_\mathrm{H} &= -c_\mathrm{V} \frac{F_{+,\mathrm{V}} F_{\times,\mathrm{L}} - F_{\times,\mathrm{V}} F_{+,\mathrm{L}}}{F_{+,\mathrm{H}} F_{\times,\mathrm{L}} - F_{\times,\mathrm{H}} F_{+,\mathrm{L}}} \\ c_\mathrm{L} &= +c_\mathrm{V} \frac{F_{+,\mathrm{V}} F_{\times,\mathrm{H}} - F_{\times,\mathrm{V}} F_{+,\mathrm{H}}}{F_{+,\mathrm{H}} F_{\times,\mathrm{L}} - F_{\times,\mathrm{H}} F_{+,\mathrm{L}}} \,, \end{aligned} \tag{7.144}$$

where c_V is arbitrary. A natural choice would be $c_\mathrm{V} = F_{+,\mathrm{H}} F_{\times,\mathrm{L}} - F_{\times,\mathrm{H}} F_{+,\mathrm{L}}$, in which case the coefficients are

$$\begin{aligned} c_\mathrm{H} &= F_{+,\mathrm{L}} F_{\times,\mathrm{V}} - F_{\times,\mathrm{L}} F_{+,\mathrm{V}} \\ c_\mathrm{L} &= F_{+,\mathrm{V}} F_{\times,\mathrm{H}} - F_{\times,\mathrm{V}} F_{+,\mathrm{H}} \\ c_\mathrm{V} &= F_{+,\mathrm{H}} F_{\times,\mathrm{L}} - F_{\times,\mathrm{H}} F_{+,\mathrm{L}} \,. \end{aligned} \tag{7.145}$$

The null stream is $s_0(t) = c_\mathrm{H} s_\mathrm{H}(t + \tau_\mathrm{H}) + c_\mathrm{L} s_\mathrm{L}(t + \tau_\mathrm{L}) + c_\mathrm{V} s_\mathrm{V}(t + \tau_\mathrm{V})$.

Note that if one of the coefficients c_H, c_L, or c_V vanishes (which will occur for certain sky positions) then $\mathbf{F}^\mathrm{T}$ does not have a nullspace and there is no null stream.

For networks of non-aligned detectors, three detectors are required for there to be a null stream; if there are more than three detectors, one can typically form more than one null stream.

7.6.3 Time-Frequency Excess-Power Method for a Network of Detectors

In Example 7.8 we constructed an excess-power statistic for the detection of signals whose form is unknown except for their duration and frequency band. Here we generalize this to the case of a network of detectors. Following Sutton *et al.* (2010) (see also Klimenko *et al.*, 2005), an elegant treatment of the general detector network can be formulated in terms of the following matrices:

$$\tilde{\boldsymbol{h}}(f) := \begin{bmatrix} \tilde{h}_+(f) \\ \tilde{h}_\times(f) \end{bmatrix} \tag{7.146}$$

is a vector with of the two polarization components of a signal,

$$\tilde{\boldsymbol{w}}(f) := \begin{bmatrix} e^{2\pi i f \tau_1} \tilde{s}_1(f)/S_1^{1/2}(f) \\ e^{2\pi i f \tau_2} \tilde{s}_2(f)/S_2^{1/2}(f) \\ \vdots \\ e^{2\pi i f \tau_N} \tilde{s}_N(f)/S_N^{1/2}(f) \end{bmatrix} \tag{7.147}$$

is a vector of the detector data whitened by the factors of $S_i^{-1/2}(f)$ and delayed according to propagation times, and

$$\mathbf{G}(f) := \begin{bmatrix} G_{+,1}(f)/S_1^{1/2}(f) & G_{\times,1}(f)/S_1^{1/2}(f) \\ G_{+,2}(f)/S_2^{1/2}(f) & G_{\times,2}(f)/S_2^{1/2}(f) \\ \vdots & \vdots \\ G_{+,N}(f)/S_N^{1/2}(f) & G_{\times,N}(f)/S_N^{1/2}(f) \end{bmatrix} \tag{7.148}$$

is a matrix of weighting factors. Note that $\mathbf{G}$ and $\boldsymbol{\tau}$ depend on $\hat{\boldsymbol{n}}$, but we have suppressed explicit indication of this. In terms of these quantities, the log of the likelihood ratio is

$$\ln \Lambda\,(\mathcal{H}_1|\boldsymbol{w}) = 4\int_{f_0}^{f_1} \left[\frac{1}{2}\tilde{\boldsymbol{w}}^\dagger(f)\mathbf{G}(f)\tilde{\boldsymbol{h}}(f) + \frac{1}{2}\tilde{\boldsymbol{h}}^\dagger(f)\mathbf{G}^\dagger(f)\tilde{\boldsymbol{w}}(f) - \frac{1}{2}\tilde{\boldsymbol{h}}^\dagger(f)\mathbf{G}^\dagger(f)\mathbf{G}(f)\tilde{\boldsymbol{h}}(f)\right] df\,, \tag{7.149}$$

where we assume that the signal has support only between frequencies f_0 and $f_1 = f_0 + F$.

By employing a variation of the frequency components of the signal, we obtain the maximum likelihood estimate of these components:

$$\tilde{\boldsymbol{h}}_{\text{est}}(f) = \left[\mathbf{G}^\dagger(f)\mathbf{G}(f)\right]^{-1}\mathbf{G}^\dagger(f)\tilde{\boldsymbol{w}}(f)\,. \tag{7.150}$$

Then, by substituting this back into the equation for the logarithm of the likelihood ratio, we obtain the excess-power statistic,

$$\mathcal{E} = 2\ln \Lambda\,(\mathcal{H}_1|\boldsymbol{w})|_{\tilde{\boldsymbol{h}}(f)=\tilde{\boldsymbol{h}}_{\text{est}}(f)} = 4\int_{f_0}^{f_1} \tilde{\boldsymbol{w}}^\dagger(f)\mathbf{P}(f)\tilde{\boldsymbol{w}}(f)df\,, \tag{7.151}$$

where

$$\mathbf{P}(f) := \mathbf{G}(f)[\mathbf{G}^\dagger(f)\mathbf{G}(f)]^{-1}\mathbf{G}^\dagger(f) \tag{7.152}$$

projects the data vector $\boldsymbol{w}(f)$ onto the space that is occupied by gravitational-wave signals, that is it is a matrix that satisfies $\mathbf{P}(f)\mathbf{P}(f) = \mathbf{P}(f)$ (it is a projection operator) and $\mathbf{P}(f)\mathbf{G}(f)\tilde{\boldsymbol{h}}(f) = \mathbf{G}(f)\tilde{\boldsymbol{h}}(f)$. The quantity $\mathcal{E}$ is known as the *coherent excess-power statistic.*

In addition to the coherent excess-power statistic, it is possible to compute the energy in the null stream. Notice that since $\mathbf{P}(f)$ projects a vector into the space occupied by potential signals, the matrix $\mathbf{1}-\mathbf{P}(f)$ therefore projects into the nullspace where there is no gravitational-wave signal content. Thus,

$$\mathcal{E}_{\text{null}} = 4\int_{f_0}^{f_1} \tilde{\boldsymbol{w}}^\dagger(f)\left[\mathbf{1} - \mathbf{P}(f)\right]\tilde{\boldsymbol{w}}(f)df \tag{7.153}$$

is the null energy. Note that $\mathcal{E}_{\text{null}} = \mathcal{E}_{\text{tot}} - \mathcal{E}$ where

$$\mathcal{E}_{\text{tot}} = 4 \int_{f_0}^{f_1} \tilde{\boldsymbol{w}}^{\dagger}(f) \tilde{\boldsymbol{w}}(f) d f \tag{7.154}$$

is the total signal energy. The null energy is useful for distinguishing between true gravitational-wave signals (which produce energy only in the $\mathcal{E}$ statistic) and spurious detector noise transients (which will generically affect $\mathcal{E}$, $\mathcal{E}_{\text{tot}}$ and $\mathcal{E}_{\text{null}}$).

7.6.4
Sky Position Localization for Gravitational-Wave Bursts

Unlike telescopes, gravitational-wave detectors are sensitive to gravitational waves from nearly all directions, although the sensitivity varies somewhat due to the antenna beam pattern function. For gravitational-wave bursts one cannot deduce the location of the source of the burst in the sky using a single detector (unless the exact amplitude of the gravitational waves is known). However, with more than one detector at different locations, the sky position of the source *can* be determined: with two detectors, the location of a source can be restricted to an annulus in the sky simply by triangulation using the difference in arrival times at the two detectors; with three detectors, which define a plane, triangulation based on arrival time can locate the position of the source to two possible sky regions that are mirror images of each other about the plane defined by the three detectors. Four detectors at distinct locations can, in principle, locate the position of a source by triangulation alone. However, the precision of the sky localization depends on the precision with which the arrival times can be recorded.

Suppose a gravitational wave is travelling a direction given by the unit vector $\hat{\boldsymbol{n}}$. Then the time of arrival of the gravitational wave at detector i will be

$$t_i = t_0 + \hat{\boldsymbol{n}} \cdot \boldsymbol{r}_i / c , \tag{7.155}$$

where t_0 is the arrival time of the gravitational wave at the centre of the Earth and $\boldsymbol{r}_i$ is the position vector of detector i relative to the centre of the Earth. The *measured* times t_i will also include timing errors σ_i arising from the way the noise affects our estimation of the arrival times in each detector. Therefore, there will be a residual, χ^2, which is given by

$$\chi^2 := \sum_{i=1}^{N} \frac{(t_0 - t_i + \hat{\boldsymbol{n}} \cdot \boldsymbol{r}_i / c)^2}{\sigma_i^2} . \tag{7.156}$$

Generally, the source location is not known, nor is the geocentric arrival time t_0. An estimate of the values of t_0 is found by minimizing χ^2 over t_0, which yields

$$t_0 = \frac{\sum_{i=1}^{N} (t_i - \hat{\boldsymbol{n}} \cdot \boldsymbol{r}_i / c) / \sigma_i^2}{\sum_{i=1}^{N} 1/\sigma_i^2} . \tag{7.157}$$

Inserting this estimate into Eq. (7.156) gives the expression for the timing residual, minimized over the (unknown) geocentric arrival time, which can be expressed in the form

$$\min_{t_0} \chi^2 = \sum_{i=1}^{N} \frac{\left[-(t_i - \bar{t}) + \hat{\boldsymbol{n}} \cdot (\boldsymbol{r}_i - \bar{\boldsymbol{r}})/c\right]^2}{\sigma_i^2} , \tag{7.158}$$

where

$$\bar{t} := \frac{\sum_{i=1}^N t_i/\sigma_i^2}{\sum_{i=1}^N 1/\sigma_i^2} \quad \text{and} \quad \bar{\boldsymbol{r}} := \frac{\sum_{i=1}^N \boldsymbol{r}_i/\sigma_i^2}{\sum_{i=1}^N 1/\sigma_i^2} . \tag{7.159}$$

The residual can also be written concisely as

$$\min_{t_0} \chi^2 = \|\mathbf{M}\hat{\boldsymbol{n}} - \boldsymbol{\tau}\|^2 , \tag{7.160}$$

where the ith row of the matrix $\mathbf{M}$ is the 3-vector $\sigma_i^{-1}(\boldsymbol{r}_i - \bar{\boldsymbol{r}})/c$, and $\boldsymbol{\tau}$ is an N-dimensional column vector with components $\tau_i = \sigma_i^{-1}(t_i - \bar{t})$. To estimate $\hat{\boldsymbol{n}}$, this residual is minimized with respect to $\hat{\boldsymbol{n}}$ subject to the constraint $\|\hat{\boldsymbol{n}}\| = 1$. This can be done using the method of Lagrange multipliers, and the result is the system of equations

$$\left(\mathbf{M}^{\mathrm{T}}\mathbf{M} + \lambda \mathbf{1}\right)\hat{\boldsymbol{n}} = \mathbf{M}^{\mathrm{T}}\boldsymbol{\tau} \tag{7.161a}$$

$$\|\hat{\boldsymbol{n}}\| = 1 , \tag{7.161b}$$

where λ is a Lagrange multiplier. These equations must be solved together. A singular value decomposition can be used to express $\mathbf{M} = \mathbf{USV}^{\mathrm{T}}$ where $\mathbf{U}$ (a $N \times N$ matrix) and $\mathbf{V}$ (a 3×3 matrix) are orthogonal ($\mathbf{U}^{\mathrm{T}}\mathbf{U} = \mathbf{1}$ and $\mathbf{V}^{\mathrm{T}}\mathbf{V} = \mathbf{1}$) and $\mathbf{S}$ is an $N \times 3$ matrix with non-zero components $S_{11} = s_1$, $S_{22} = s_2$, and $S_{33} = s_3$ that are known as the singular values. For $N = 3$ detectors, $\lambda = 0$, so Eq. (7.161a) can be solved directly, but $s_3 = 0$ (which is the manifestation of degeneracy of the location of a source across the plane spanned by three detectors) and so $\mathbf{M}^{\mathrm{T}}\mathbf{M}$ is singular: the component of $\hat{\boldsymbol{n}}$ normal to the plane of the detectors is undetermined. The normalization of $\hat{\boldsymbol{n}}$ then gives the *magnitude* of this third component, but not its sign. For $N > 3$ detectors, the value of the Lagrange multiplier λ can be found by substitution of Eq. (7.161a) into Eq. (7.161b). In terms of the rotated vectors $\boldsymbol{\tau}' = \mathbf{U}^{\mathrm{T}}\boldsymbol{\tau}$ and $\hat{\boldsymbol{n}}' = \mathbf{V}^{\mathrm{T}}\hat{\boldsymbol{n}}$, we have $\hat{n}'_i = s_i\tau'_i/(s_i^2 + \lambda)$ ($i = \{1, 2, 3\}$) with λ determined by finding the largest root of the equation $0 = f(\lambda) = 1 - \sum_{i=1}^3 [s_i\tau'_i/(s_i^2 + \lambda)]^2$. Finally, $\hat{\boldsymbol{n}} = \mathbf{V}\hat{\boldsymbol{n}}'$ is computed.

Once the estimate of $\hat{\boldsymbol{n}}$ is found, it can be used in Eq. (7.157) to obtain the estimate of the geocentric arrival time t_0, and then these two together can be used in Eq. (7.156) to obtain the value of the residual, χ^2, that is a minimum over possible propagation directions and arrival time.

If the value of χ^2 is much bigger than N then there is no location of the source on the sky and geocentric arrival time that is consistent with the observed times of arrival and their measurement uncertainties. This can be a very powerful tool for discriminating between real gravitational-wave signals and random glitches in the various detectors. We see this already for the case of two detectors. Suppose that the rate of false events (due to noise) in detectors 1 and 2 are R_1 and R_2, respectively, and suppose that the maximum range of arrival time differences that two events can have is $\tau \simeq 2\Delta t + \sigma_1 + \sigma_2$ where $\Delta t = \|\boldsymbol{r}_1 - \boldsymbol{r}_2\|/c$ is the time-of-flight separation between the detectors, then the rate of coincident events, that is those with $|t_1 - t_2| < \tau$, occurring in both detectors is

$$R_{12} = \tau R_1 R_2 \,. \tag{7.162}$$

For example, for the LIGO Hanford Observatory and the LIGO Livingston Observatory, $\Delta t = 10\,\text{ms}$, and if $\sigma_{\text{LHO}} = \sigma_{\text{LLO}} = 1\,\text{ms}$ then $\tau \simeq 22\,\text{ms}$. If the false rate in each detector was one event per day then the coincident false rate would be around one event every ten thousand years.

7.7 Data Analysis Methods for Continuous-Wave Sources

Systems that constantly emit radiation at a nearly fixed frequency, for example rotating, non-axisymmetric isolated neutron stars, are called continuous-wave sources. The data analysis challenge to detect these sources seems straightforward: the optimal detection strategy is simply to take the Fourier transform of the data and look for an excess of power at the frequency of the source. The more data that is accumulated, the greater the sensitivity of such a search as the signal power grows linearly with the observation time while the energy in the noise at a given frequency bin only grows with the square root of the observation time.

However, this approach does not work in practice because the signal will not be truly fixed in frequency (or monochromatic). For one thing, most sources will not have a fixed intrinsic frequency of emission: rotating neutron stars will lose energy – most likely via electromagnetic emission, but certainly in gravitational waves – and this will cause the rotation to spin-down. The effect is slight, but in an observation time T_{obs} the frequency bins of a Fourier transform of the data will have a spacing of $1/T_{\text{obs}}$, which, for a four-month-long observation time, would be $\sim 10^{-7}$ Hz. If the change in the intrinsic frequency is greater than this amount, then the signal will drift out of a particular frequency bin during the observation time and will not be fully recovered.

A bigger problem arises from the motion of the Earth. The Earth rotates and it orbits about the Sun, each of which introduces a Doppler modulation of the gravitational wave. For example, the Doppler shift in frequency due to the Earth's rotation is on the order of $\Delta f/f \sim 2\pi R_\oplus/(c \times 1\,\text{day}) \sim 10^{-6}$, and so the gravitational-wave frequency, as observed on Earth, will drift sinusoidally over the course of a day.

To correct for the motion of the Earth (and other effects), it is necessary to represent the continuous-wave signal in terms of the solar system barycentric time, T, which we recall is related to the topocentric time, t, at the detector by Eq. (6.211). The phase of the continuous-wave signal is

$$\Phi(t) = 2\pi \left[f(T - T_0) + \frac{1}{2}\dot{f}(T - T_0)^2 + \ldots \right], \tag{7.163}$$

where T is given by Eq. (6.211), T_0 is a reference epoch, and f and $\dot{f}$ are the intrinsic frequency and the change in the frequency of gravitational waves, also evaluated at $T = T_0$. In principle we could have kept more terms in the Taylor expansion of the phase, but we will omit them here, supposing $\ddot{f}$ is very small. The gravitational signal experienced at a detector is

$$\begin{aligned} h(t) = {} & G_+(t, \hat{\boldsymbol{n}}, \psi, f)\, h_{0,+} \cos\left[\Phi(t) + \Phi_0\right] \\ & + G_\times(t, \hat{\boldsymbol{n}}, \psi, f)\, h_{0,\times} \sin\left[\Phi(t) + \Phi_0\right], \end{aligned} \tag{7.164}$$

where $h_{0,+}$ and $h_{0,\times}$ are the amplitudes of the two gravitational-wave polarizations and Φ_0 is the (normally unknown) phase at the epoch $T = T_0$. Notice that the amplitude of the strain in the detector is modulated by the time-dependent antenna beam pattern factors (these are time-dependent because the detector orientation changes as the Earth rotates). For example, in Section 3.4 we showed that for a rotating triaxial ellipsoid rotating with angular velocity ω,

$$h_{0,+} = -\frac{1}{2}(1 + \cos^2\iota) h_0 \quad \text{and} \quad h_{0,\times} = \cos\iota\, h_0\,, \tag{7.165}$$

where ι is the inclination of the rotational axis of the source to the direction to the Earth and

$$h_0 = \frac{4G}{c^4}\frac{I_3\omega^2}{r}\varepsilon \tag{7.166}$$

is the amplitude of the signal in a detector that is optimally oriented and optimally located relative to the source (e.g. a detector that is located on the spin axis of the rotating ellipsoid, so that $\iota = 0$, and oriented, say, so that $G_+ = 1$). Here, r is the distance to the source, I_3 is the principal moment of inertia about the rotation axis, and $\varepsilon = (I_1 - I_2)/I_3$ is the ellipticity. The gravitational-wave frequency is related to the rotational frequency by $\omega = \pi f$.

For a binary system in a circular orbit with constant orbital frequency, Eqs. (7.164) and (7.165) still hold but the overall amplitude factor is

$$h_0 = \frac{4G}{c^4}\frac{\mu a^2\omega^2}{r}, \tag{7.167}$$

where μ is the reduced mass of the binary system, a is the semi-major axis of the orbit, and ω is the orbital angular frequency, which is again related to the gravitational-wave frequency by $\omega = \pi f$. The gradual change in frequency (e.g. due to orbital decay from gravitational radiation, which would *increase* the gravitational-wave frequency) will need to be included in the phase model if the binary system evolves perceptibly during the observation time. In addition, a search for a rotating ellipsoidal body (e.g. a neutron star) within a binary system can be conducted in the same way as a search for such an isolated body, but a more complex phase model would be required to account for the orbital motion of the body in the binary.

7.7.1 Search for Gravitational Waves from a Known, Isolated Pulsar

In the case that a particular source of continuous waves is identified, for example a known, isolated pulsar (though in low-frequency gravitational-wave searches it could be a white dwarf binary in the Galaxy or a supermassive black hole binary at cosmological distances), the location of the source is known and the conversion between detector time and solar system barycentric time can be performed. For a pulsar, the rotational frequency and its spin-down rate can also be measured from electromagnetic observations. This then gives a complete phase model $\Phi(t)$ for the system (up to the unknown constant Φ_0) and a matched filter search is possible.

7.7.1.1 Heterodyne Method

An efficient technique for computing the matched filter, described in detail by Dupuis and Woan (2005), is to *heterodyne* the signal: the gravitational-wave detector data $s(t)$ is multiplied by a complex factor $\exp[-i\Phi(t)]$ containing the known phase evolution, and the resulting time series is low-pass filtered. The effect on a signal is the following: by multiplying by the complex phase factor, the signal becomes

$$h(t)e^{-i\Phi(t)} = A(t)e^{i\Phi_0} + A^*(t)e^{-i\Phi_0-2i\Phi(t)} , \qquad (7.168)$$

where

$$A(t) = \frac{1}{2}G_+(t, \hat{\boldsymbol{n}}, \psi, f)\, h_{0,+} - i\frac{1}{2}G_\times(t, \hat{\boldsymbol{n}}, \psi, f)\, h_{0,\times} . \qquad (7.169)$$

The first term of the heterodyned signal varies slowly as $A(t)$ changes only due to the slow (over the period of a day) change in the antenna beam pattern functions. The second term is rapidly varying at twice the gravitational-wave frequency; this term is discarded when the heterodyned data (which includes the heterodyned signal) is low-pass filtered so that only frequencies below some $f_{\max}$ are retained. This eliminates the rapidly varying term and all that remains is the slowly varying piece. Typically, for searches for known pulsars, $f_{\max} = (1\,\text{min})^{-1}$ is chosen as a convenient value. (The resampling interval $1/f_{\max}$ should be short enough that the data is relatively stationary during the interval and the antenna beam pattern function is relatively constant.)

After heterodyning and low-pass filtering, the data is integrated (averaged) for a duration $\Delta t = 1/f_{\max}$, again typically $\Delta t = 1\,\text{min}$ for pulsar searches with ground-based detectors. This results in a set of $N = T_{\text{obs}}/\Delta t$ data samples, $\{B_j\}$ ($j = 0, \ldots, N-1$), which are sampled at intervals Δt. In the presence of both a gravitational-wave signal and noise, these samples are

$$B_j - A_j e^{i\Phi_0} + n_j , \qquad (7.170)$$

where $A_j = A(t_j)$ are the sample values of the heterodyned signal, $t_j = t_0 + j\Delta t$ is the time corresponding to the jth sample (t_0 is the time at the start of the observation) and n_j is the noise contribution to B_j, which we will assume is well described by a Gaussian distribution (in both the real and the imaginary components) with zero mean and variance $\sigma_j^2/2$. The probability distribution for the set of samples $\{B_j\}$ under the hypothesis $\mathcal{H}_1$ that a signal of the form given by $\{A_j\}$ and with phase Φ_0 is

$$p(\{B_j\}|\mathcal{H}_1) = \prod_{j=0}^{N-1} \frac{1}{\pi\sigma_j^2} \exp\left(-\frac{1}{2}\frac{|B_j - A_j e^{i\Phi_0}|^2}{\sigma_j^2}\right). \tag{7.171}$$

In practice, it is difficult to estimate the noise variances, $\sigma_j^2/2$, though it is expected that they are roughly constant over moderately long timescales, τ. Typically, $\tau \sim 1\,\mathrm{h}$ for interferometer data. Let $M = \tau/\Delta t$ be the number of consecutive samples of $\{B_j\}$ that can be treated as having the same noise variance. We then divide the data into N/M intervals of duration τ, which have M points so that segment k is the set of samples $\{B_j : j \in [kM, (k+1)M)\}$, and $k = 0, \ldots, N/M - 1$. For this segment, we assume that the noise variance is the unknown, but fixed, value $\sigma_k^2/2$, and

$$\begin{aligned} &p\left(\{B_j : j \in [kM, (k+1)M)\}\,|\mathcal{H}_1, \sigma_k\right) \\ &= \frac{1}{\left(\pi\sigma_k^2\right)^M} \exp\left(-\frac{1}{2\sigma_k^2}\sum_{j=kM}^{(k+1)M} |B_j - A_j e^{i\Phi_0}|^2\right) \end{aligned} \tag{7.172}$$

and

$$p(\{B_j\}|\mathcal{H}_1, \{\sigma_k\}) = \prod_{k=0}^{N/M-1} p\left(\{B_j : j \in [kM, (k+1)M)\}\,|\mathcal{H}_1, \sigma_k\right). \tag{7.173}$$

We now marginalize over the unknown noise variances $\sigma_k^2/2$ by integrating the probability distributions for each segment, Eq. (7.172), over σ_k with a suitable prior distribution, $p(\sigma_k)$. An appropriate prior distribution is the *Jeffreys prior*, which is $p(\sigma_k) \propto 1/\sigma_k$ for the variance of a Gaussian distribution. We find

$$\begin{aligned} &p\left(\{B_j : j \in [kM, (k+1)M)\}\,|\mathcal{H}_1\right) \\ &= \int_0^\infty p\left(\{B_j : j \in [kM, (k+1)M)\}\,|\mathcal{H}_1, \sigma_k\right) p(\sigma_k) d\sigma_k \\ &\propto \int_0^\infty \frac{1}{(\pi\sigma_k^2)^M} \exp\left(-\frac{1}{2\sigma_k^2}\sum_{j=kM}^{(k+1)M} |B_j - A_j e^{i\Phi_0}|^2\right) \frac{d\sigma_k}{\sigma_k} \\ &= \frac{(M-1)!}{2}\left(\frac{2}{\pi}\right)^M \left(\sum_{j=kM}^{(k+1)M} |B_j - A_j e^{i\Phi_0}|^2\right)^{-M} \end{aligned} \tag{7.174}$$

and so we have

$$p\left(\{B_j\}|\mathcal{H}_1\right) \propto \prod_{k=0}^{N/M-1} \left(\sum_{j=kM}^{(k+1)M} |B_j - A_j e^{i\Phi_0}|^2\right)^{-M} . \tag{7.175}$$

This form of the probability distribution is now independent of the unknown noise variance in each segment.

Using Bayes's theorem, the probability distribution for the observed data given a signal model with a particular set of parameters, Eq. (7.175), can be expressed as a probability distribution for those parameters given the gravitational-wave data. The parameters in the signal model that are usually unknown are the phase Φ_0, the polarization angle ψ of the waves and the amplitudes $h_{+,0}$ and $h_{\times,0}$ of the two polarizations, or alternatively the amplitude h_0 and the inclination angle ι. Normally one wishes to obtain a probability distribution for h_0 as this is the parameter of primary interest. If we adopt uniform prior distributions on Φ_0, ψ and $\cos\iota$ then the posterior distribution for h_0 is obtained by integrating Eq. (7.175) over these unknown parameters:

$$p(h_0|\{B_j\}) \propto \int_{-1}^{1} d\cos\iota \int_{0}^{2\pi} d\psi \int_{0}^{2\pi} d\Phi_0 \prod_{k=0}^{N/M-1} \left(\sum_{j=kM}^{(k+1)M} |B_j - A_j e^{i\Phi_0}|^2\right)^{-M} , \tag{7.176}$$

where $A_j = A_j(h_0, \iota, \psi)$ is a function of the parameters h_0, ι and ψ and the normalization is set by

$$\int_{0}^{\infty} p(h_0|\{B_j\}) dh_0 = 1 . \tag{7.177}$$

Often we are interested in establishing an upper bound on the strength of the gravitational-wave emission h_0. For example, a 90% confidence upper bound on the characteristic amplitude parameter, $h_{0,90\%}$, is the value for which

$$\int_{0}^{h_{0,90\%}} p(h_0|\{B_j\})\, dh_0 = 90\% . \tag{7.178}$$

7.7.1.2 **Maximum Likelihood Method**

The heterodyne method discussed above applies a matched filter for every point in the $\{h_0, \iota, \psi\}$ parameter space and then marginalizes the "nuisance" parameters (typically ι, ψ and Φ_0). An alternative approach is one based on the maximum likelihood method (Jaranowski *et al.*, 1998).

First we rewrite Eq. (7.164) in the following form:

$$\begin{aligned} h(t) = {} & G_+(t, \hat{\boldsymbol{n}}, \psi, f) h_{0,+} \cos \Phi_0 \cos \Phi(t) \\ & - G_+(t, \hat{\boldsymbol{n}}, \psi, f) h_{0,+} \sin \Phi_0 \sin \Phi(t) \\ & + G_\times(t, \hat{\boldsymbol{n}}, \psi, f) h_{0,\times} \sin \Phi_0 \cos \Phi(t) \\ & + G_\times(t, \hat{\boldsymbol{n}}, \psi, f) h_{0,\times} \cos \Phi_0 \sin \Phi(t) . \end{aligned} \tag{7.179}$$

Now, the time-dependent amplitudes G_+ and $G_\times$ still depend on ψ. However, because the polarization angle can be defined about an arbitrary reference, we can write

$$\begin{aligned} G_+ (t, \hat{\boldsymbol{n}}, \psi, f) &= G_{+,0} (t, \hat{\boldsymbol{n}}, f) \cos 2\psi + G_{\times,0} (t, \hat{\boldsymbol{n}}, f) \sin 2\psi \\ G_\times (t, \hat{\boldsymbol{n}}, \psi, f) &= G_{\times,0} (t, \hat{\boldsymbol{n}}, f) \cos 2\psi - G_{+,0} (t, \hat{\boldsymbol{n}}, f) \sin 2\psi , \end{aligned} \tag{7.180}$$

where

$$\begin{aligned} G_{+,0} (t, \hat{\boldsymbol{n}}, f) &= G_+ (t, \hat{\boldsymbol{n}}, \psi = 0, f) \\ G_{\times,0} (t, \hat{\boldsymbol{n}}, f) &= G_\times (t, \hat{\boldsymbol{n}}, \psi = 0, f) . \end{aligned} \tag{7.181}$$

Then the waveform can be expressed as the sum of four different time series, which depend on the sky position of the source and the frequency and phase evolution of the gravitational waves, with four different amplitudes, which depend on the unknown parameters $h_{0,+}$, $h_{0,\times}$, ψ and Φ_0:

$$\begin{aligned} h(t) = {} & A_1(h_{0,+}, h_{0,\times}, \psi, \Phi_0) g_1(t, \hat{\boldsymbol{n}}, f) \\ & + A_2(h_{0,+}, h_{0,\times}, \psi, \Phi_0) g_2(t, \hat{\boldsymbol{n}}, f) \\ & + A_3(h_{0,+}, h_{0,\times}, \psi, \Phi_0) g_3(t, \hat{\boldsymbol{n}}, f) \\ & + A_4(h_{0,+}, h_{0,\times}, \psi, \Phi_0) g_4(t, \hat{\boldsymbol{n}}, f) , \end{aligned} \tag{7.182a}$$

where

$$\begin{aligned} g_1(t, \hat{\boldsymbol{n}}, f) &:= G_{+,0}(t, \hat{\boldsymbol{n}}, f) \cos \Phi(t) \\ g_2(t, \hat{\boldsymbol{n}}, f) &:= G_{\times,0}(t, \hat{\boldsymbol{n}}, f) \cos \Phi(t) \\ g_3(t, \hat{\boldsymbol{n}}, f) &:= G_{+,0}(t, \hat{\boldsymbol{n}}, f) \sin \Phi(t) \\ g_4(t, \hat{\boldsymbol{n}}, f) &:= G_{\times,0}(t, \hat{\boldsymbol{n}}, f) \sin \Phi(t) \end{aligned} \tag{7.182b}$$

are the four different template waveforms (which are known for a particular source) and

$$\begin{aligned} A_1(h_{0,+}, h_{0,\times}, \psi, \Phi_0) &:= h_{0,+} \cos 2\psi \cos \Phi_0 - h_{0,\times} \sin 2\psi \sin \Phi_0 \\ A_2(h_{0,+}, h_{0,\times}, \psi, \Phi_0) &:= h_{0,+} \sin 2\psi \cos \Phi_0 + h_{0,\times} \cos 2\psi \sin \Phi_0 \\ A_3(h_{0,+}, h_{0,\times}, \psi, \Phi_0) &:= -h_{0,+} \cos 2\psi \sin \Phi_0 - h_{0,\times} \sin 2\psi \cos \Phi_0 \\ A_4(h_{0,+}, h_{0,\times}, \psi, \Phi_0) &:= -h_{0,+} \sin 2\psi \sin \Phi_0 + h_{0,\times} \cos 2\psi \cos \Phi_0 \end{aligned} \tag{7.182c}$$

are the four different amplitudes (which are unknown). The goal is to find the values of these four amplitudes that maximize the likelihood statistic.

Before we proceed to evaluate the likelihood ratio, however, it is important to evaluate the degree of covariance of the four different template waveforms, $g_1(t)$, $g_2(t)$, $g_3(t)$ and $g_4(t)$. Because the observation period will contain many gravitational-wave cycles (though perhaps not so many diurnal or yearly cycles) the cosinusoidal waveforms $g_1(t)$ and $g_2(t)$ will be orthogonal (with respect to the inner product of Eq. (7.28)) to the sinusoidal waveforms, $g_3(t)$ and $g_4(t)$:

$$(g_1, g_3) = (g_1, g_4) = (g_2, g_3) = (g_2, g_4) = 0 . \tag{7.183}$$

On the other hand, the slowly varying amplitude factors vary on the diurnal and yearly timescales, so $g_1(t)$ and $g_2(t)$ will have a non-zero covariance, as will $g_3(t)$ and $g_4(t)$. We define the following constants:

$$\begin{aligned}
\mathcal{A} &:= (g_1, g_1) = (g_3, g_3) \\
\mathcal{B} &:= (g_2, g_2) = (g_4, g_4) \\
\mathcal{C} &:= (g_1, g_2) = (g_3, g_4) .
\end{aligned} \tag{7.184}$$

We can now write the log of the likelihood ratio (recall Eq. (7.40)) given observed data $s(t)$,

$$\begin{aligned}
\ln \Lambda\left(\mathcal{H}_1|s\right) &= (s, h) - \frac{1}{2}(h, h) \\
&= A_1(s, g_1) + A_2(s, g_2) + A_3(s, g_3) + A_4(s, g_4) \\
&\quad - \frac{1}{2}\left[A_1^2(g_1, g_1) + A_2^2(g_2, g_2) + A_3^2(g_3, g_3) + A_4^2(g_4, g_4)\right. \\
&\quad \left. +2A_1A_2(g_1, g_2) + 2A_3A_4(g_3, g_4)\right] \\
&= A_1(s, g_1) + A_2(s, g_2) + A_3(s, g_3) + A_4(s, g_4) \\
&\quad - \frac{1}{2}\mathcal{A}\left(A_1^2 + A_3^2\right) - \frac{1}{2}\mathcal{B}\left(A_2^2 + A_4^2\right) - \mathcal{C}\left(A_1A_2 + A_3A_4\right) .
\end{aligned} \tag{7.185}$$

We maximize this function over possible values of A_1, A_2, A_3 and A_4 (since these are the unknown parameters), and find that we must solve the system of equations

$$\begin{aligned}
0 &= \left.\frac{\partial \ln \Lambda(\mathcal{H}_1|s)}{\partial A_1}\right|_{\max} = (s, g_1) - \mathcal{A}A_{1,\max} - \mathcal{C}A_{2,\max} \\
0 &= \left.\frac{\partial \ln \Lambda(\mathcal{H}_1|s)}{\partial A_2}\right|_{\max} = (s, g_2) - \mathcal{B}A_{2,\max} - \mathcal{C}A_{1,\max} \\
0 &= \left.\frac{\partial \ln \Lambda(\mathcal{H}_1|s)}{\partial A_3}\right|_{\max} = (s, g_3) - \mathcal{A}A_{3,\max} - \mathcal{C}A_{4,\max} \\
0 &= \left.\frac{\partial \ln \Lambda(\mathcal{H}_1|s)}{\partial A_1}\right|_{\max} = (s, g_4) - \mathcal{B}A_{4,\max} - \mathcal{C}A_{3,\max} ,
\end{aligned} \tag{7.186}$$

which yields

$$
\begin{aligned}
A_{1,\max} &= \frac{\mathcal{B}(s,g_1) - \mathcal{C}(s,g_2)}{\mathcal{A}\mathcal{B} - \mathcal{C}^2} \\
A_{2,\max} &= \frac{\mathcal{A}(s,g_2) - \mathcal{C}(s,g_1)}{\mathcal{A}\mathcal{B} - \mathcal{C}^2} \\
A_{3,\max} &= \frac{\mathcal{B}(s,g_3) - \mathcal{C}(s,g_4)}{\mathcal{A}\mathcal{B} - \mathcal{C}^2} \\
A_{4,\max} &= \frac{\mathcal{A}(s,g_4) - \mathcal{C}(s,g_3)}{\mathcal{A}\mathcal{B} - \mathcal{C}^2} \, .
\end{aligned}
\tag{7.187}
$$

We now replace A_1, A_2, A_3 and A_4 in Eq. (7.185) with these maximum likelihood values, and we obtain the maximum likelihood statistic,

$$
\begin{aligned}
2\mathcal{F} &:= \max_{A_1,A_2,A_3,A_4} 2\ln \Lambda(\mathcal{H}_1|s) \\
&= \frac{\mathcal{B}(s,g_1)^2 + \mathcal{A}(s,g_2)^2 - 2\mathcal{C}(s,g_1)(s,g_2)}{\mathcal{A}\mathcal{B} - \mathcal{C}^2} \\
&+ \frac{\mathcal{B}(s,g_3)^2 + \mathcal{A}(s,g_4)^2 - 2\mathcal{C}(s,g_3)(s,g_4)}{\mathcal{A}\mathcal{B} - \mathcal{C}^2} \, .
\end{aligned}
\tag{7.188}
$$

This is known as the $\mathcal{F}$-statistic.[1)]

Notice that the $\mathcal{F}$-statistic depends on the four inner products (s,g_1), (s,g_2), (s,g_3) and (s,g_4) of the data $s(t)$ with the four template waveforms $g_1(t)$, $g_2(t)$, $g_3(t)$ and $g_4(t)$. For a source with a known sky position and known phase model $\Phi(t)$ (up to an unknown initial phase Φ_0), these waveforms are known. The *unknown* parameters, $h_{0,+}$, $h_{0,\times}$, ψ and Φ_0 (or, equivalently, h_0, ι, ψ and Φ_0) do not appear in the $\mathcal{F}$-statistic, though they can be recovered from the maximum likelihood amplitudes given in Eq. (7.187). If the data $s(t)$ consists of stationary Gaussian noise alone, $2\mathcal{F}$ is a random variable that has a χ^2 distribution with four degrees of freedom. If a signal $h(t)$ is present, it is a non-central χ^2 distribution with four degrees of freedom and a non-centrality parameter $\lambda = (h,h)$.

From these distributions it is possible to construct a threshold on $2\mathcal{F}$ required for a detection to be of a pre-set confidence level. The confidence level, α, here is interpreted in a frequentist sense: if there is *no* signal then the value of $2\mathcal{F}$ will exceed the threshold with a probability of $100\% - \alpha$, assuming Gaussian noise. Similarly, an upper limit on the amplitude of continuous gravitational waves, h_0, can be obtained from the probability distribution $2\mathcal{F}$ given its non-central chi-squared distribution. Again these are normally constructed as frequentist confidence intervals (see Section 7.3.3), and to relate the value of the non-central parameter $\lambda = (h,h)$ to the gravitational-wave amplitude, the "worst-case scenario" with the emitter inclined at $\iota = \pi/2$ and having the most pessimistic polarization angle ψ is normally assumed.

1) Not to be confused with Student's *F*-statistic.

Example 7.11 Sensitivity of the known-pulsar search

For either the matched filter heterodyne method or the maximum-likelihood method, the detectability of the pulsar signal depends on the characteristic signal-to-noise ratio $\varrho^2 = (h, h)$. The sensitivity of the search can be stated in terms of the amplitude h_0 that would produce a particular value of the characteristic signal-to-noise ratio ϱ. We take the most pessimistic value for the inclination angle, $\iota = \pi/2$, so that $h_{0,+} = -h_0/2$ and $h_{0,\times} = 0$; then we have

$$\begin{aligned}\varrho^2 = (h, h) &= \frac{2}{S_h(f)} \int_0^{T_{obs}} h^2(t) dt \\ &\geq \frac{h_0^2}{S_h(f)} \int_0^{T_{obs}} G_+^2(t, \hat{\boldsymbol{n}}, \psi, f) \cos^2 \left[\Phi(t) + \Phi_0 \right] dt \\ &> \frac{T_{obs}}{2} \frac{h_0^2}{S_h(f)} \langle G^2 \rangle_{min}(\hat{\boldsymbol{n}}, f) , \end{aligned} \tag{7.189}$$

where

$$\begin{aligned}\langle G^2 \rangle_{min}(\hat{\boldsymbol{n}}, f) &= \min_{\psi} \frac{1}{T_{sid}} \int_0^{T_{sid}} G_+^2 (t, \hat{\boldsymbol{n}}, \psi, f) \, dt \\ &= \min \left\{ \frac{1}{T_{sid}} \int_0^{T_{sid}} G_+^2(t, \hat{\boldsymbol{n}}, 0, f) dt, \frac{1}{T_{sid}} \int_0^{T_{sid}} G_+^2 (t, \hat{\boldsymbol{n}}, \pi/4, f) \, dt \right\} , \end{aligned} \tag{7.190}$$

where we note that G_+ is periodic over a sidereal day, T_{sid}, and we have assumed that the observation time T_{obs} contains many sidereal days. (Alternatively we could write $\varrho^2 > (h_0^2/S_h(f)) \min\{\mathcal{A}, \mathcal{B}\}$.) In the long-wavelength limit, $G_+ = F_+$ has no frequency dependence and $\langle G^2 \rangle_{min}$ depends only on the declination δ of the source; this function is plotted in Figure 7.1 for GEO, LIGO Hanford, LIGO Livingston, and Virgo.

The Crab pulsar has $\delta = 22°$ and is expected to emit gravitational waves at $f = 59.56$ Hz. The 4 km LIGO Hanford Observatory had $S_h(f) \sim 2 \times 10^{-44}$ Hz^{-1} at the gravitational-wave frequency and has $\langle G^2 \rangle_{min} = 0.125$ at the declination of the Crab pulsar. For a $T_{obs} = 1$ year search, the characteristic signal-to-noise ratio of gravitational waves from the Crab pulsar if it were radiating at the spin-down limit of $h_0 = 1.4 \times 10^{-24}$ (that is if the observed spin-down were entirely due to gravitational waves) would be

$$\varrho > h_0 \sqrt{\frac{\langle G^2 \rangle_{min} T_{obs}}{2 S_h(f)}} \simeq 14 , \tag{7.191}$$

which would be easily detectable.

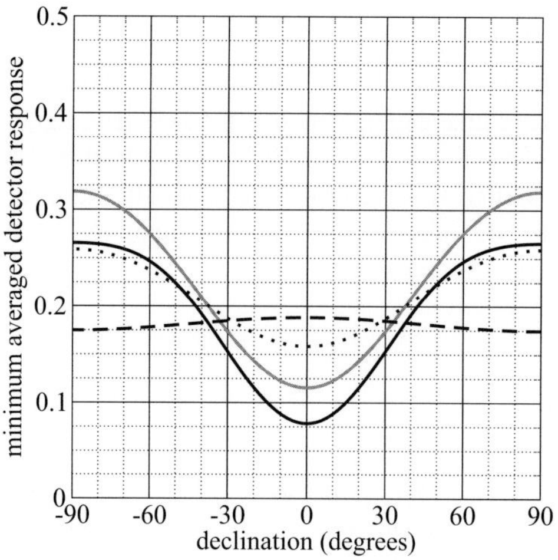

Figure 7.1 The squared detector response averaged over a sidereal day and minimized over polarization angles as a function of the declination of the gravitational-wave source. The curves are for GEO (grey), LIGO Hanford Observatory (black solid), LIGO Livingston Observatory (black dashed), and Virgo (black dotted).

7.7.2 All-Sky Searches for Gravitational Waves from Unknown Pulsars

To discover unknown pulsars in an all-sky search for continuous-wave signals is a far more challenging problem. The reason is the following: suppose that a certain amount of data, T, is Fourier-transformed. This yields a frequency series with bins that cover a frequency range $\Delta f = 1/T$. Meanwhile, the Doppler shift of a signal is $\Delta f = -f\boldsymbol{v}\cdot\hat{\boldsymbol{n}}/c$, and if we consider values of $\hat{\boldsymbol{n}}$ that are nearly orthogonal to the motion $\boldsymbol{v}$ of the detector, then $\boldsymbol{v}\cdot\hat{\boldsymbol{n}} \sim v\Delta\theta$ where $\Delta\theta$ is the change in the angle between $\boldsymbol{v}$ and $\hat{\boldsymbol{n}}$ over the course of the observation time T. (The value of v to be used is the Earth's rotational speed, $v \sim 460\,\mathrm{m\,s^{-1}}$, for $T \sim 1$ day, or the Earth's orbital speed, $v \sim 30\,\mathrm{km\,s^{-1}}$ for $T \sim 1$ year.) This means that different propagation directions will result in resolvably different Doppler shifts if $1/T \sim fv\Delta\theta/c$. Each patch on the sky with area $\sim (\Delta\theta)^2$ steradians must be individually corrected for the Earth's motion. Fortunately, the Doppler-demodulation of the data for a particular sky position is not a frequency-dependent process, so that demodulated data can be used to compute the matched filter for all frequencies for that sky position. This means that there must be

$$N_{\mathrm{sky}} \sim \frac{4\pi}{(\Delta\theta)^2} \sim 4\pi T^2 f_{\max}^2 (v/c)^2 \tag{7.192}$$

different sky positions searched over in a search for signals up to a frequency $f_{\max}$. For one day of observation, and $f_{\max} = 1\,\mathrm{kHz}$, this is $N_{\mathrm{sky}} \sim 10^5$ sky positions; for one year of observation it becomes $N_{\mathrm{sky}} \sim 10^{14}$ sky positions! The search itself will require a Fourier transform of the data for each sky position; for a fast Fourier transform this has a computational cost of $O(N_{\rm points} \ln N_{\rm points})$, where $N_{\rm points} \sim T f_{\rm max}$, so the computation cost scales as $\sim T^3 f_{\rm max}^3 (v/c)^2$. If, in addition, one must account for one pulsar spin-down parameter during the observation time, the number of templates will pick up an additional factor of T^2, which makes the computational cost scale as T^5. The upshot is that the computational cost of an all-sky search for unknown pulsars grows steeply with the observation time T, and, ultimately, the search is limited not by the length of observation but rather by the computation resources that are available.

On the other hand, the strain-sensitivity of the search grows only as $T^{1/2}$ for any given template: If the search is performed with the $\mathcal{F}$-statistic, $2\mathcal{F}$ has a non-central chi-squared distribution with four degrees of freedom and a non-central parameter λ that is quadratic in the gravitational-wave amplitude h_0, $\lambda \sim h_0^2 T$, and, for large values of λ, the distribution of $(2\mathcal{F})^{1/2}$ becomes approximately Gaussian with unit variance and mean $\lambda^{1/2}$, so the sensitivity grows as $\lambda^{1/2} \sim h_0 T^{1/2}$. Therefore, the sensitivity grows extremely slowly with the computation cost of the search.

Given fixed computational resources, there is a longest observational time, $T_{\rm max}$, that can be analyzed on roughly the same timescale (thereby keeping up with the data). This essentially sets the longest amount of data that can be coherently analyzed in an all-sky search for unknown pulsars. However, a number of coherently analyzed data stretches of duration $T_{\rm max}$, or *stacks*, can be combined incoherently (i.e. by adding the power in each frequency bin); to do so, however, the frequency bins might have to be *slid* with respect to each other to account for a signal drifting from one bin to another. The resulting *stack-slide method* can increase the sensitivity of the overall all-sky search by a factor of ~ 2 compared to the purely coherent search, see Brady and Creighton (2000) and Pletsch and Allen (2009).

7.8 Data Analysis Methods for Gravitational-Wave Bursts

Unlike the case of continuous gravitational-wave signals described above, bursts of gravitational waves are short-lived signals whose durations are much shorter than the observational timescale and can be identified by a distinct arrival time. The morphology of the burst signal can be known (or well modelled), as is the case for the late stages of coalescence of compact binary sources (for example, the inspiral and eventual collision of a binary neutron star pair), or they may be poorly modelled or largely unknown, as is the case for gravitational waves from core-collapse supernovae or from unanticipated sources. In either case, the central assumption that goes into a burst search is that these signals are sporadic and occupy only a fraction of the dataset under consideration.

Besides the distinction between modelled and unmodelled, burst searches may be omni-directional or directed, and they may be all-time or triggered. For example, a search for inspiral gravitational waves associated with a short gamma-ray burst (which are thought to be produced by the disruption of a neutron star by another neutron star or black hole) would be an externally triggered and directed burst search for a modelled signal because we know the sky position and time of the event; such a search is an example of an *externally triggered burst search*. Not all gravitational-wave events will have associated electromagnetic radiation, so omni-directional, all-time (i.e. not restricted to a single short interval of data) burst searches are also performed.

For directed searches one can combine the data from the various elements in a network of detectors together to form coherent combinations that can be used in a matched filter search if the gravitational waveform is well-modelled, or a time-frequency excess-power search if the waveform is not well-modelled (see Section 7.6). The null stream then provides an effective means of discriminating between gravitational-wave signals, which affect all the detectors in the network, and single-detector noise glitches. Unless the network consists of detectors that are co-located and co-aligned, however, the coherent and null combinations of the detector data depend on the sky position of the source. For an omni-directional search, then, there are two possibilities: (i) the coherent and null streams are computed for many positions covering the sky (a template bank covering the parameter space of sky positions) and the search is performed for each sky location with the requirement that any burst found in the coherent stream should not be seen in the null stream, or (ii) single-detector searches are performed on the data from each detector and events that are not found (nearly) simultaneously in all detectors are rejected.

For externally triggered burst searches we seek a single event associated with our external trigger, while for all-time burst searches we wish to measure the rate of burst events and to use this detection rate to understand details about the population of the possible sources of the gravitational-wave bursts. In any type of burst search, should a detection be made, we also wish to estimate the signal parameters, for example its amplitude, its arrival time, and its polarization, among others.

7.8.1
Searches for Coalescing Compact Binary Sources

In Section 3.5 we described the gravitational waves radiated from a binary system, and we obtained expressions for the phase and amplitude evolution of such systems and the time until coalescence from some starting orbital frequency. At high frequencies, binary systems formed of two massive compact stars – either black holes or neutron stars with masses on the order of a solar mass – will sweep up in frequency and amplitude as the orbit decays, and ultimately the companions collide and form a single object. For ground-based detectors with sensitivity to gravitational waves with frequency greater than ~ 10 Hz, the total duration of the late stages of inspiral will be tens of seconds or less. The signal is a burst signal: its duration is much shorter than the observation time.

The waveform for these inspiral signals is known from post-Newtonian calculations up to the point where the orbital motion becomes highly relativistic as was described in Section 4.1.3. The very late stages of inspiral, plunge and merger of the compact objects, however, must be computed numerically by the methods detailed in Section 4.3. The post-merger object is normally a black hole (though binary neutron star collisions occasionally produce a semi-stable massive neutron star which eventually collapses into a black hole), and the radiation produced by the settling of the initially distorted black hole can be described as the black hole ringdown radiation that was introduced in Section 4.2.2. The upshot is that the waveform for coalescing binaries is generally known, if only approximately, from post-Newtonian, numerical, or perturbation calculation.

7.8.1.1 Matched Filter for Binary Inspiral Signals

Here we will consider just the portion of the inspiral that is well-modelled by post-Newtonian calculations. This is the part of the waveform that is most important for the detection of binary neutron stars and low-mass binary black holes by ground-based gravitational-wave detectors. Because the waveform is well-modelled, we know that the optimal search (at least in Gaussian noise) will be one that uses matched filtering. For non-spinning bodies in non-eccentric orbits, the waveform parameter space is composed of the two masses, m_1 and m_2, the time and orbital phase of coalescence, t_c and φ_c, the inclination of the orbit to the plane of the sky ι, and the polarization angle ψ, the direction to the source $-\hat{\boldsymbol{n}}$, and its distance r.

The dominant radiation from the binary inspiral is emitted in the $\ell = 2$, $m = \pm 2$ quadrupolar mode. If we focus attention on this mode alone, we write the gravitational wave as

$$h_+(t) - i h_\times(t) = {}_{-2}Y_{2,+2}(\iota, 0) h_{2,+2}(t) + {}_{-2}Y_{2,-2}(\iota, 0) h_{2,-2}(t) \,, \tag{7.193}$$

which gives, *to lowest post-Newtonian order in the amplitude evolution*, the waveforms

$$h_+(t) = -\frac{G\mathcal{M}}{c^2 r}\frac{1+\cos^2\iota}{2}\left(\frac{t_c - t}{5G\mathcal{M}/c^3}\right)^{-1/4}\cos 2\varphi \tag{7.194a}$$

$$h_\times(t) = -\frac{G\mathcal{M}}{c^2 r}\cos\iota\left(\frac{t_c - t}{5G\mathcal{M}/c^3}\right)^{-1/4}\sin 2\varphi \,, \tag{7.194b}$$

where φ is a function of time that depends on the parameters m_1, m_2, t_c, and φ_c. Recall $\mathcal{M} = m_1^{3/5} m_2^{3/5}(m_1 + m_2)^{-1/5}$ is the chirp mass. This expression of the waveform keeps only the leading order (Newtonian) amplitude evolution term, though the phase evolution, φ, can be kept to higher post-Newtonian orders – for signal processing, it is more important to correctly model the phase evolution than the amplitude evolution. These waveforms that keep only the leading order quadrupole amplitude terms are known as *restricted post-Newtonian* waveforms.

A given detector experiences a strain given by

$$h(t+\tau) = F_+(\hat{\boldsymbol{n}}, \psi)\, h_+(t) + F_\times(\hat{\boldsymbol{n}}, \psi)\, h_\times(t) \,, \tag{7.195}$$

where τ is the delay between the time of arrival of the waves at the centre of the Earth and the time of arrival of the same waves at the detector and where we have restricted ourselves to the long-wavelength limit of the detector response (which is reasonable for binary inspiral signals in ground-based detectors). This strain can be written as

$$h(t) = -\frac{G\mathcal{M}}{c^2 D_{\text{eff}}}\left(\frac{t_0 - t}{5G\mathcal{M}/c^3}\right)^{-1/4} \cos 2\left[\varphi(t) + \Delta\varphi\right], \tag{7.196}$$

where $t_0 = t_c + \tau$ is the time of coalescence at the location of the detector, $\varphi_0 = \varphi_c + \Delta\varphi$ is the phase of coalescence observed in the detector,

$$2\Delta\varphi := -\arctan\left(\frac{F_\times(\hat{\boldsymbol{n}}, \psi)}{F_+(\hat{\boldsymbol{n}}, \psi)}\frac{2\cos\iota}{1+\cos^2\iota}\right) \tag{7.197}$$

and

$$D_{\text{eff}} := r\left[F_+^2(\hat{\boldsymbol{n}}, \psi)\left(\frac{1+\cos^2\iota}{2}\right)^2 + F_\times^2(\hat{\boldsymbol{n}}, \psi)\cos^2\iota\right]^{-1/2} \tag{7.198}$$

is the *effective distance* to the source. The effective distance D_{eff} is related to the physical distance r to the source by geometrical factors. An *optimally oriented* and *optimally located* source is one for which the inclination angle is zero, $\iota = 0$, and is situated in the sky so that $F_+^2 + F_\times^2 = 1$; for such a source, the effective distance is equal to the physical distance, but otherwise the effective distance is larger than the physical distance.

The utility of Eq. (7.196) is that the waveform parameters that describe the strain experienced by a detector have been reduced to t_0, φ_0, D_{eff}, m_1 and m_2 (the parameter φ_c can be absorbed into φ_0). For any mass pair (m_1, m_2) we are confronted with the problem of searching for a signal whose waveform is known up to an unknown arrival time (t_0), amplitude ($\propto D_{\text{eff}}^{-1}$) and phase ($\varphi_0$). We have seen in Section 7.2.5 how one can efficiently use a Fourier transform to apply a matched filter where the amplitude and arrival time is unknown, and we have seen in Example 7.4 how a search over the unknown phase can be performed algebraically. For restricted post-Newtonian waveform templates, the optimal detection statistic is constructed from the matched filters

$$x(t) = 4\,\text{Re}\int_0^\infty \frac{\tilde{s}(f)\tilde{p}^*(f)}{S_h(f)} e^{2\pi i f t} df \tag{7.199a}$$

$$y(t) = 4\,\text{Re}\int_0^\infty \frac{\tilde{s}(f)\tilde{q}^*(f)}{S_h(f)} e^{2\pi i f t} df \tag{7.199b}$$

with the two phases of the matched filter template being

$$p(t) = -\frac{G\mathcal{M}}{c^2}\left(-\frac{c^3 t}{5G\mathcal{M}}\right)^{-1/4} \cos 2\varphi(t) \tag{7.200a}$$

$$q(t) = -\frac{G\mathcal{M}}{c^2}\left(-\frac{c^3 t}{5G\mathcal{M}}\right)^{-1/4} \sin 2\varphi(t)\,. \tag{7.200b}$$

Here we set $t_c = 0$ and $\varphi_c = 0$ in our expression for φ. The variance of these two matched filters are both $\sigma^2 = (p, p) = (q, q)$ to a good approximation, and also $(p, q) = 0$. A search for binary inspirals looks for the time at which the signal-to-noise ratio

$$\rho(t) = \sqrt{\frac{x^2(t) + y^2(t)}{\sigma^2}} \tag{7.201}$$

is maximum – call the time $t_{\max}$ and the maximum value of the signal-to-noise ratio $\rho_{\max} = \rho(t_{\max})$.

A small simplification can be found in constructing the template waveforms directly in the frequency domain using the stationary phase approximation, see Problem 3.7. The stationary phase template is

$$\tilde{g}(f) = -\left(\frac{5\pi}{24}\right)^{1/2} \frac{G^2 \mathcal{M}^2}{c^5} \left(\frac{\pi G \mathcal{M} f}{c^3}\right)^{-7/6} e^{-i\Psi(f)}\,, \tag{7.202}$$

where $\Psi(f)$ is the stationary phase function. To second post-Newtonian order it is

$$\begin{aligned}\Psi(f) = -\pi/4 + \frac{3}{128\eta}\left[x^{-5/2} + \left(\frac{3715}{756} + \frac{55}{9}\eta\right)x^{-3/2} - 16\pi x^{-1}\right.\\ \left. + \left(\frac{15\,293\,365}{508\,032} + \frac{27\,145}{504}\eta + \frac{3085}{72}\eta^2\right)x^{-1/2}\right],\end{aligned} \tag{7.203}$$

where $\eta := m_1 m_2/(m_1+m_2)^2$ is the symmetric mass ratio and $x := (\pi G M f/c^3)^{2/3}$ is the post-Newtonian parameter. Note that if we take $\tilde{p} = \tilde{g}$ then $\tilde{q} = -i\tilde{g}$. Using the stationary phase template, a complex matched filter is constructed as

$$z(t) := x(t) + i y(t) = 4 \int_0^\infty \frac{\tilde{s}(f)\tilde{g}^*(f)}{S_h(f)} e^{2\pi i f t}\, df \tag{7.204}$$

and the signal-to-noise ratio is the normalized version of this

$$\rho(t) := \frac{|z(t)|}{\sigma}\,, \tag{7.205}$$

where

$$\sigma^2 := (g, g) = \frac{5}{6\pi}\left(\frac{G\mathcal{M}}{c^2}\right)^2 \left(\frac{\pi G \mathcal{M}}{c^3}\right)^{-1/3} \int_0^\infty \frac{f^{-7/3}}{S_h(f)}\, df\,. \tag{7.206}$$

A true gravitational wave will produce a waveform $\tilde{h}(f) \simeq D_{\text{eff}}^{-1} \exp(-2\pi i f t_0 + 2i\varphi_0)\tilde{g}(f)$ (to the extent that the restricted post-Newtonian stationary phase approximate waveform is accurate). Such a signal will produce a matched filter output $z(t_0) = (h, g) = e^{2i\varphi_0}(\sigma^2/D_{\text{eff}})$, so our estimates of φ_0 and D_{eff} are $2\varphi_0 = \arg z(t_{\text{peak}})$ and $D_{\text{eff}} = \sigma^2/|z(t_{\text{peak}})| = \sigma/\rho(t_{\text{peak}})$, assuming that the signal-to-noise ratio indeed peaks at $t_{\text{peak}} = t_0$.

Example 7.12 Horizon distance and range

The sensitivity of ground-based interferometers is often characterized by the distance to which they can detect binary neutron star signals. This is somewhat vague. To make such a statement precise, suppose we stipulate that a representative binary neutron star inspiral is produced by two neutron stars with masses $m_1 = m_2 = 1.4\,M_\odot$. By convention, we say that the system is detectable if it produces a characteristic signal-to-noise ratio of $\varrho = 8$.

The *horizon distance* for a detector is the farthest distance that such a binary neutron star could be and still produce the required signal-to-noise ratio in the detector. That is, the horizon distance is the effective distance of a representative binary neutron star inspiral that produces $\varrho = 8$:

$$D_{\text{hor}} := \left.\frac{\sigma}{\varrho}\right|_{\varrho=8,\mathcal{M}=1.22\,M_\odot} = \left[\frac{1}{\varrho}\frac{G\mathcal{M}}{c^2}\sqrt{\frac{5}{6\pi}\left(\frac{\pi G\mathcal{M}}{c^3}\right)^{-1/3}\int_0^\infty \frac{f^{-7/3}}{S_h(f)}}\right]_{\varrho=8,\mathcal{M}=1.22\,M_\odot} . \quad (7.207)$$

Often one is interested in the distance to which "typical" binary neutron star signals can be detected, rather than the farthest we could see such a system if it happened to be optimally located and oriented. Suppose binary neutron star systems are homogeneously distributed in the universe. Within the horizon distance, we would see only some fraction X of these – that is, assuming a homogeneous spatial distribution of sources that have random orientations, X is the fraction of these sources with $D_{\text{eff}} \le D_{\text{hor}}$. It turns out that $X \simeq 1/(2.2627)^3$. The *sense-monitor range*, D_{range}, (named after the LIGO data monitor SENSEMONITOR) is the radius of a sphere that contains the same number of sources as those that have $D_{\text{eff}} \le D_{\text{hor}}$, $D_{\text{range}} = X^{1/3} D_{\text{hor}} \simeq D_{\text{hor}}/2.2627$. If $\mathcal{R}$ is the rate of binary neutron star coalescences per unit volume in the universe then the rate of events detected with $\rho > 8$ will be $\frac{4}{3}\pi D_{\text{range}}^3 \mathcal{R}$.

For the idealized Initial LIGO interferometer model (see Section A.2), the horizon distance and sense-monitor range are $D_{\text{hor}} = 43$ Mpc and $D_{\text{range}} = 19$ Mpc. For the typical LIGO S5 sensitivity shown in Figure A.2 these values are $D_{\text{hor}} = 35$ Mpc and $D_{\text{range}} = 15$ Mpc.

7.8.1.2 **Bank of Inspiral Templates**

Until now we have described a search for a single template, that is for a specified mass pair (m_1, m_2). To cover the mass-parameter space, a bank of templates is required (recall Section 7.2.6), spaced sufficiently close together that there will be only a small loss in signal-to-noise ratio due to filtering with a slightly incorrect template.

As an illustration of the construction of an inspiral template bank, we consider a Newtonian chirp in the stationary phase approximation,

$$\tilde{h}(f) = -\frac{1}{D_{\text{eff}}}\left(\frac{5\pi}{24}\right)^{1/2}\frac{G^2\mathcal{M}^2}{c^5}\left(\frac{\pi G\mathcal{M}f}{c^3}\right)^{-7/6} \mathrm{e}^{-2\pi i f t_0}\mathrm{e}^{2i\varphi_0}\mathrm{e}^{-i\Psi(f;\mathcal{M})} \tag{7.208a}$$

with

$$\Psi(f;\mathcal{M}) = -\frac{\pi}{4} + \frac{3}{128}\left(\frac{\pi G\mathcal{M}f}{c^3}\right)^{-5/3} . \tag{7.208b}$$

Notice that the mass parameters m_1 and m_2 appear only in the combination $\mathcal{M}$, so at Newtonian order the bank need only be one-dimensional to explore the entire space of mass pairs. Post-Newtonian corrections of second order and higher are required for searches for binary inspirals from stellar-mass compact objects for ground-based interferometric detectors, so the illustration we give here certainly does not provide a sufficient template bank for actual searches. See Owen and Sathyaprakash (1999) for details on the construction of a two-dimensional template bank that is sufficient for searching for binary inspiral signals – for systems in which the components are not highly spinning.

It is customary to choose the Newtonian coalescence time τ_0 rather than the chirp mass $\mathcal{M}$ as the parameter to be used in template bank construction. This is because, at first post-Newtonian order, the two-dimensional parameter space, $\{\tau_0, \tau_1\}$, where τ_1 is the first post-Newtonian contribution to the coalescence time, turns out to be flat in the metric that describes signal-to-noise ratio loss on the parameter space, which is helpful for positioning the templates on the parameter space. The Newtonian coalescence time from frequency f_0 is

$$\tau_0 = \frac{5}{256}\frac{G\mathcal{M}}{c^3}\left(\frac{\pi G\mathcal{M}f_0}{c^3}\right)^{-8/3} \tag{7.209}$$

and, in terms of this parameter,

$$\Psi(f;\mathcal{M}) = -\frac{\pi}{4} + \frac{3}{5}(2\pi f_0)\left(\frac{f}{f_0}\right)^{-5/3} \tau_0 . \tag{7.210}$$

We will choose an explicit value for f_0 below.

Within the stationary phase approximation, to compute the metric on parameter space we only need to consider the derivatives of the *normalized* template with respect to the phase function Ψ – derivatives with respect to the amplitude have no effect. The extrinsic parameters are t_0 and φ_0, while the intrinsic parameter is $\mathcal{M}$. In fact, we can ignore the extrinsic parameter φ_0 as it is removed altogether simply by redefining the inner product of Eq. (7.28) to take the complex modulus of the integral rather than the real part. The ambiguity function is then given by

$$\mathcal{A}(t_0, \tau_0; t_0 + \Delta t_0, \tau_0 + \Delta\tau_0) = \left| \frac{1}{I_{-7/3}} \int_0^\infty \frac{f^{-7/3}}{S_h(f)} \mathrm{e}^{i\omega^0 \Delta t_0}\mathrm{e}^{i\omega^1 \Delta\tau_0}\, d f \right| , \tag{7.211}$$

where I_α is the integral

$$I_\alpha := \int_0^\infty \frac{f^\alpha}{S_h(f)} df \tag{7.212}$$

and the derivatives of the phase of the template with respect to the parameters t_0 and τ_0 are, respectively,

$$\omega^0 = -(2\pi f_0)\left(\frac{f}{f_0}\right) \quad \text{and} \quad \omega^1 = \frac{3}{5}(2\pi f_0)\left(\frac{f}{f_0}\right)^{-5/3} . \tag{7.213}$$

Therefore, to quadratic order in variations of the parameters $\lambda_0 = t_0$ and $\lambda_1 = \tau_0$, we have

$$\begin{aligned}
\mathcal{A}^2 &\simeq \Bigg| 1 + i\Delta\lambda_i \frac{1}{I_{-7/3}} \int_0^\infty \frac{f^{-7/3}}{S_h(f)} \omega^i df \\
&\quad - \frac{1}{2}\Delta\lambda_i \Delta\lambda_j \frac{1}{I_{-7/3}} \int_0^\infty \frac{f^{-7/3}}{S_h(f)} \omega^i \omega^j df \Bigg|^2 \\
&\simeq 1 - \Delta\lambda_i \Delta\lambda_j \frac{1}{I_{-7/3}} \int_0^\infty \frac{f^{-7/3}}{S_h(f)} \omega^i \omega^j df \\
&\quad + \Delta\lambda_i \Delta\lambda_j \left(\frac{1}{I_{-7/3}} \int_0^\infty \frac{f^{-7/3}}{S_h(f)} \omega^i df\right)\left(\frac{1}{I_{-7/3}} \int_0^\infty \frac{f^{-7/3}}{S_h(f)} \omega^j df\right) .
\end{aligned} \tag{7.214}$$

Since we also have $\mathcal{A}^2 \simeq 1 - 2g^{ij}\Delta\lambda_i\Delta\lambda_j$ we obtain a formula for the metric

$$\begin{aligned}
g^{ij} &= \frac{1}{2}\frac{1}{I_{-7/3}} \int_0^\infty \frac{f^{-7/3}}{S_h(f)} \omega^i \omega^j df \\
&\quad - \frac{1}{2}\left(\frac{1}{I_{-7/3}} \int_0^\infty \frac{f^{-7/3}}{S_h(f)} \omega^i df\right)\left(\frac{1}{I_{-7/3}} \int_0^\infty \frac{f^{-7/3}}{S_h(f)} \omega^j df\right) .
\end{aligned} \tag{7.215}$$

The components of the metric are

$$g^{00} = (2\pi f_0)^2 \frac{1}{2} f_0^{-2} \left[\frac{I_{-1/3}}{I_{-7/3}} - \left(\frac{I_{-4/3}}{I_{-7/3}}\right)^2\right] \tag{7.216a}$$

$$g^{01} = -(2\pi f_0)^2 \frac{3}{10} f_0^{-2/3} \left[\frac{I_{-3}}{I_{-7/3}} - \frac{I_{-4/3}}{I_{-7/3}} \frac{I_{-4}}{I_{-7/3}}\right] \tag{7.216b}$$

$$g^{11} = (2\pi f_0)^2 \frac{9}{50} f_0^{10/3} \left[\frac{I_{-17/3}}{I_{-7/3}} - \left(\frac{I_{-4}}{I_{-7/3}}\right)^2\right] . \tag{7.216c}$$

We now compute the metric on the one-dimensional parameter space that results after the explicit maximization over the arrival time t_0. The projected metric component is

$$\gamma^{11} = g^{11} - (g^{01})^2/g^{00} . \tag{7.217}$$

In order to compute this metric component, we need the various moments of the detector noise power spectral density given by the integrals $I_{-1/3}$, $I_{-4/3}$, $I_{-7/3}$, I_{-3}, I_{-4} and $I_{-17/3}$. To proceed we need to know the detector's noise power spectral density $S_h(f)$. However, to gain some insight, we note that $S_h(f)$ is generically some U-shaped function, so we suppose that $S_h^{-1}(f)$ is a relatively constant function over some frequency band Δf about a maximum f_0. We therefore write $I_\alpha = \hat{I}_\alpha S_h^{-1}(f_0) f_0^\alpha \Delta f$ where $\hat{I}_\alpha$ is some dimensionless constant of order unity. Furthermore, all of the moments appear in a ratio with the moment $I_{-7/3}$, so it is useful to write $I_\alpha / I_\beta = f_0^{\alpha-\beta} \hat{I}_{\alpha:\beta}$ with $\hat{I}_{\alpha:\beta} := \hat{I}_\alpha / \hat{I}_\beta$,

$$\hat{I}_{\alpha:\beta} = \frac{\int_0^\infty \frac{x^\alpha}{S_h(x f_0)} dx}{\int_0^\infty \frac{x^\beta}{S_h(x f_0)} dx} . \tag{7.218}$$

Then we find

$$\gamma^{11} = (2\pi f_0)^2 \frac{9}{50} \left[\hat{I}_{-17/3:-7/3} - \left(\hat{I}_{-4:-7/3}\right)^2 - \frac{\left(\hat{I}_{-3:-7/3} - \hat{I}_{-4/3:-7/3} \hat{I}_{-4:-7/3}\right)^2}{\hat{I}_{-1/3:-7/3} - \left(\hat{I}_{-4/3:-7/3}\right)^2} \right] . \tag{7.219}$$

The quantity in the square brackets is some constant on the order of unity; therefore for the purposes of this illustration, we will just take

$$\gamma^{11} \sim (2\pi f_0)^2 . \tag{7.220}$$

This determines the spacing of the templates in the chirp time parameter τ_0.

The number of templates required is determined by the minimum match $(\Delta\varsigma_{\max})^2$

$$N \sim \frac{1}{\sqrt{(\Delta\varsigma_{\max})^2}} \int_{\tau_{0,\min}}^{\tau_{0,\max}} \sqrt{\gamma^{11}} d\tau_0 , \tag{7.221}$$

where the limits of the integration, $\tau_{0,\min}$ and $\tau_{0,\max}$, are set by the range of chirp masses $\mathcal{M}$ that we wish to search over. Suppose that we are interested in systems with component masses greater than $1\,M_\odot$, so $\mathcal{M} > 0.87\,M_\odot$. Then, if we take $f_0 = 100$ Hz, $\tau_{0,\max} = 3.8$ s and a maximum mismatch of $(\Delta\varsigma_{\max})^2 = 3\%$, we have

$N \sim 2\pi f_0 \tau_{0,\max}/\sqrt{(\Delta\varsigma_{\max})^2}$ (we take $\tau_{0,\min} = 0$), or

$$N \sim \frac{5}{128}\frac{1}{\sqrt{(\Delta\varsigma_{\max})^2}}\left(\frac{\pi G \mathcal{M}_{\min} f_0}{c^3}\right)^{-5/3}$$
$$\sim 10\,000\left(\frac{\Delta\varsigma^2_{\max}}{3\%}\right)^{-1/2}\left(\frac{\mathcal{M}_{\min}}{0.87\,M_\odot}\right)^{-5/3}\left(\frac{f_0}{100\,\mathrm{Hz}}\right)^{-5/3}. \tag{7.222}$$

Even for our simple one-dimensional parameter space of Newtonian chirps, we require tens of thousands of templates to cover a reasonable mass range!

When post-Newtonian terms are added to the phase, the two masses m_1 and m_2 decouple: the post-Newtonian terms cannot be expressed in terms of the chirp mass alone. The result is that the parameter space becomes two-dimensional. As we remarked earlier, at first post-Newtonian order the parameter space is flat in the two chirp times τ_0 (the Newtonian chirp time) and τ_1 (the post-Newtonian correction to the chirp time). We can *estimate* that the factor by which the number of templates is increased by introduction of a second dimension would be $\sim 2\pi f_0 \tau_{1,\max}/\sqrt{(\Delta\varsigma_{\max})^2}$. The post-Newtonian correction to the chirp time is

$$\tau_1 = \frac{5}{192\eta}\left(\frac{743}{336}+\frac{11}{4}\eta\right)\left(\frac{\pi G M f_0}{c^3}\right)^{-2}\frac{G M}{c^3} \tag{7.223}$$

so for minimum component masses of one solar mass, and with $f_0 = 100\,\mathrm{Hz}$, $\tau_{1,\max} = 0.3\,\mathrm{s}$ and there could be $\sim$1000 times as many templates. It turns out that this is a vast overestimate as the region of the τ_0–τ_1 plane that actually needs to be covered is a small fraction of the area $\tau_{0,\max} \times \tau_{1,\max}$, see Owen (1996).

7.8.1.3 **Parameter Estimation**

We now apply the methods described in Section 7.3 to obtain an estimate of our ability to extract waveform parameters from a signal. For simplicity we continue to restrict ourselves to Newtonian chirps in the stationary phase approximation given in Eq. (7.208). The four waveform parameters are $\{\lambda_0, \lambda_1, \lambda_2, \lambda_3\} = \{t_0, \varphi_0, \ln\mathcal{M}, \ln D_{\mathrm{eff}}\}$; notice that only the chirp mass appears in the waveform – there is no way to separate the individual component masses m_1 and m_2. At post-Newtonian order, however, the symmetric mass ratio η becomes a distinct parameter in the phase function and the individual masses can be resolved.

To compute the Fisher matrix, Γ^{ij}, we compute the following derivatives (Finn and Chernoff, 1993):

$$\frac{\partial\tilde{h}}{\partial\lambda_0} = \frac{\partial\tilde{h}}{\partial t_0} = -2\pi i f\tilde{h} \tag{7.224}$$

$$\frac{\partial\tilde{h}}{\partial\lambda_1} = \frac{\partial\tilde{h}}{\partial\varphi_0} = 2i\tilde{h} \tag{7.225}$$

$$\frac{\partial\tilde{h}}{\partial\lambda_2} = \frac{\partial\tilde{h}}{\partial\ln\mathcal{M}} \simeq -i\tilde{h}\frac{\partial\Psi}{\partial\ln\mathcal{M}} = i\frac{5}{128}\left(\frac{\pi G\mathcal{M} f}{c^3}\right)^{-5/3}\tilde{h} \tag{7.226}$$

$$\frac{\partial\tilde{h}}{\partial\lambda_3} = \frac{\partial\tilde{h}}{\partial\ln D_{\mathrm{eff}}} = -\tilde{h}\,. \tag{7.227}$$

(For the derivative with respect to the chirp mass, we neglect the term that would arise from the derivative of the amplitude factor and keep only the term that arises from the derivative of the phase.)

The Fisher matrix is formed from the inner products $\Gamma^{ij} = (\partial h/\partial\lambda_i, \partial h/\partial\lambda_j)$. The components related to parameter $\lambda_3 = \ln D_{\text{eff}}$ are

$$\Gamma^{03} = \Gamma^{30} = \Gamma^{13} = \Gamma^{31} = \Gamma^{23} = \Gamma^{32} = 0 \quad \text{and} \quad \Gamma^{33} = \varrho^2 , \tag{7.228}$$

where $\varrho^2 = (h, h)$ is the characteristic signal-to-noise ratio. The Fisher matrix is therefore block-diagonal with the effective distance parameter decoupling from the other parameters. We have

$$\frac{\Delta D_{\text{eff}}}{D_{\text{eff}}} = \sqrt{(\Gamma^{-1})_{33}} = \frac{1}{\varrho} . \tag{7.229}$$

The fractional error in the effective distance estimate is the inverse of the characteristic signal-to-noise ratio.

The remaining components of the Fisher matrix are:

$$\Gamma^{00} = 4\pi^2 \frac{I_{-1/3}}{I_{-7/3}} \varrho^2 \tag{7.230a}$$

$$\Gamma^{01} = -4\pi \frac{I_{-4/3}}{I_{-7/3}} \varrho^2 \tag{7.230b}$$

$$\Gamma^{02} = -\frac{5\pi}{64} \left(\frac{\pi G \mathcal{M}}{c^3}\right)^{-5/3} \frac{I_{-3}}{I_{-7/3}} \varrho^2 \tag{7.230c}$$

$$\Gamma^{11} = 4\varrho^2 \tag{7.230d}$$

$$\Gamma^{12} = \frac{5}{64} \left(\frac{\pi G \mathcal{M}}{c^3}\right)^{-5/3} \frac{I_{-4}}{I_{-7/3}} \varrho^2 \tag{7.230e}$$

$$\Gamma^{22} = \frac{25}{16\,384} \left(\frac{\pi G \mathcal{M}}{c^3}\right)^{-10/3} \frac{I_{-17/3}}{I_{-7/3}} \varrho^2 , \tag{7.230f}$$

where I_α is the integral given in Eq. (7.212). As in the previous section, we let f_0 be the most sensitive frequency of the detector and define the noise moment ratios $\hat{I}_{\alpha:\beta} = f_0^{\alpha-\beta} I_\alpha / I_\beta$. Then we have

$$\Gamma^{ij} = \varrho^2 \begin{bmatrix} 4\pi^2 f_0^2 A_{00} & -4\pi f_0 A_{01} & -\frac{5\pi}{64} f_0 \left(\frac{\pi G \mathcal{M} f_0}{c^3}\right)^{-5/3} A_{02} \\ \cdot & 4 & \frac{5}{64} \left(\frac{\pi G \mathcal{M} f_0}{c^3}\right)^{-5/3} A_{12} \\ \cdot & \cdot & \frac{25}{16\,384} \left(\frac{\pi G \mathcal{M} f_0}{c^3}\right)^{-10/3} A_{22} \end{bmatrix} , \tag{7.231}$$

where **A** is a symmetric matrix of dimensionless constants

$$A_{ij} := \begin{bmatrix} \hat{I}_{-1/3:-7/3} & \hat{I}_{-4/3:-7/3} & \hat{I}_{-3:-7/3} \\ \cdot & \hat{I}_{-7/3:-7/3} & \hat{I}_{-4:-7/3} \\ \cdot & \cdot & \hat{I}_{-17/3:-7/3} \end{bmatrix} . \tag{7.232}$$

The inverse of the Fisher matrix is

$$(\Gamma^{-1})_{ij} = \frac{1}{\varrho^2} \begin{bmatrix} \frac{1}{4\pi^2 f_0^2} B^{00} & -\frac{1}{4\pi f_0} B^{01} & -\frac{64}{5\pi f_0}\left(\frac{\pi G \mathcal{M} f_0}{c^3}\right)^{5/3} B^{02} \\ \cdot & \frac{1}{4} B^{11} & \frac{64}{5}\left(\frac{\pi G \mathcal{M} f_0}{c^3}\right)^{5/3} B^{12} \\ \cdot & \cdot & \frac{16\,384}{25}\left(\frac{\pi G \mathcal{M} f_0}{c^3}\right)^{10/3} B^{22} \end{bmatrix}, \quad (7.233)$$

where $\mathbf{B} = \mathbf{A}^{-1}$. We then arrive at the following expressions for the measurement accuracy for the remaining parameters:

$$\Delta t_0 = \sqrt{(\Gamma^{-1})_{00}} = \frac{1}{\varrho}\frac{B^{00}}{2\pi f_0}, \quad (7.234a)$$

$$\Delta \varphi_0 = \sqrt{(\Gamma^{-1})_{11}} = \frac{1}{\varrho}\frac{B^{11}}{2}, \quad (7.234b)$$

$$\frac{\Delta \mathcal{M}}{\mathcal{M}} = \sqrt{(\Gamma^{-1})_{22}} = \frac{1}{\varrho}\frac{128}{5}\left(\frac{\pi G \mathcal{M} f_0}{c^3}\right)^{5/3} B^{22}. \quad (7.234c)$$

If we take $f_0 \sim 100\,\text{Hz}$ and set $B^{00} \approx B^{11} \approx B^{22} \approx 1$, we obtain the following rough values

$$\Delta t_0 \sim 0.1\,\text{ms}\left(\frac{10}{\varrho}\right), \quad (7.235a)$$

$$\Delta \varphi_0 \sim 1^\circ \left(\frac{10}{\varrho}\right), \quad (7.235b)$$

$$\frac{\Delta \mathcal{M}}{\mathcal{M}} \sim 10^{-4}\left(\frac{10}{\varrho}\right)\left(\frac{\mathcal{M}}{1.22\,M_\odot}\right)^{5/3}. \quad (7.235c)$$

Be aware that the actual values depend on the values of B^{00}, B^{11} and B^{22}, which in turn depend on the shape of the detector noise curve! (See Problem 7.4.)

So far we have only estimated the measurability of the parameters t_0, φ_0, $\mathcal{M}$ and D_{eff} that appear in the Newtonian chirp waveforms. The masses m_1 and m_2 can be individually resolved by post-Newtonian corrections to the phase evolution. However, the precision with which individual masses are resolved is much worse than for the chirp mass. The remaining parameters – the sky location, the polarization angle, the inclination, and the true distance to the source, which were combined together as the effective distance – cannot be resolved using a single detector. However, we have seen in Section 7.6.4 that a network of detectors can determine the sky position of the source through the times of arrival of the gravitational wave at the various elements of the detector network; the accuracy to which this can be achieved is set by measurability of the arrival times, Δt_0, as a fraction of the light-travel times between the detectors in the network. Four different detector sites are required to unambiguously identify a patch in the sky that contains the source, but with three different sites the source can typically be located in one of two patches in the sky. Similarly, from the relative amplitudes of the signal received in the various detectors, one can deduce the polarization details (which depend on the inclination of the source to the plane of the sky and on the polarization angle).

7.8.1.4 Waveform Consistency Test

Detector noise is often riddled with sharp impulsive glitches that trigger the matched filter and cause spurious events. We have seen how multiple detectors can provide consistency checks that are powerful in discriminating between true gravitational waves and instrumental artefacts; however, it is also possible to construct a waveform consistency test for a single detector's output that tests how similar the event is to the anticipated signal waveform.

For inspiral signals, an effective discriminant tests whether the observed signal-to-noise ratio accumulates with time and frequency in the expected way. A time-frequency chi-squared waveform consistency test can be constructed as follows (Allen, 2005): the matched filter $\tilde{g}(f)$ is divided into p sub-templates, $\{\tilde{g}_i(f)\}$, with support in different frequency bands, in such a way that each of the sub-templates contributes equally to the signal-to-noise ratio in the case that a true signal is present. Furthermore, these bands do not overlap so that each sub-template is orthogonal. That is, $(g_i, g_j) = \sigma^2 \delta_{ij}/p$ where $(g, g) = \sigma^2$. The bands are $0 \leq f < f_1$, $f_1 \leq f < f_2, \ldots, f_{p-1} \leq f < \infty$, and these must satisfy

$$4 \int_0^{f_1} \frac{|\tilde{g}(f)|^2}{S_h(f)} df = 4 \int_{f_1}^{f_2} \frac{|\tilde{g}(f)|^2}{S_h(f)} df = \cdots = 4 \int_{f_{p-1}}^{\infty} \frac{|\tilde{g}(f)|^2}{S_h(f)} df = \frac{\sigma^2}{p} . \tag{7.236}$$

The sub-templates, $\{\tilde{g}_1(f), \tilde{g}_2(f), \ldots, \tilde{g}_p(f)\}$ are then given by

$$\tilde{g}_i(f) := \begin{cases} \tilde{g}(f) & f_{i-1} \leq f < f_i \\ 0 & \text{otherwise} , \end{cases} \tag{7.237}$$

where $f_0 = 0$ and $f_p = \infty$. The data is then matched filtered against each of these sub-templates, which gives a set of sub-band matched filter outputs $\{z_i(t)\}$ with

$$z_i(t) := 4 \int_0^{\infty} \frac{\tilde{s}(f) \tilde{g}_i^*(f)}{S_h(f)} e^{2\pi i f t} df = 4 \int_{f_{i-1}}^{f_i} \frac{\tilde{s}(f) \tilde{g}^*(f)}{S_h(f)} e^{2\pi i f t} df . \tag{7.238}$$

Note that

$$z(t) = \sum_{i=1}^{p} z_i(t) . \tag{7.239}$$

For an event occurring at time t_0, we compute a chi-squared value

$$\chi^2(t_0) = \sum_{i=1}^{p} \frac{|z_i(t_0) - z(t_0)/p|^2}{\sigma^2/p} . \tag{7.240}$$

The chi-squared value will be small for signals that match the template well, but can be quite large for noise glitches that might have a large signal-to-noise ratio but do not match the template well.

If the data consists of Gaussian noise plus a gravitational wave that is exactly matched by the template, then $\chi^2(t_0)$ will have a chi-squared distribution with $\nu = 2p - 2$ degrees of freedom. This distribution has a mean of ν, so the reduced chi-squared, $\chi^2(t_0)/\nu$, will be near unity. If there is a mismatch between the signal (or glitch) and the template then the distribution of $\chi^2(t_0)$ becomes a non-central chi-squared distribution. For example, suppose that there is an extremely loud glitch ($\rho \gg 1$), but it is band-limited so it only contributes in the first frequency band $0 \leq f < f_1$, say. Then we have $z_1(t_0) = z(t_0)$ with $|z_1(t_0)|^2 = \rho^2\sigma^2 \gg \sigma^2$ and $|z_i(t_0)|^2 \sim \sigma^2/p$ for $i \neq 1$. If we ignore the values $z_i(t_0)$ for $i \neq 1$, then we find $\chi^2(t_0)/\nu = \rho^2/2$ which is much larger than would be expected for a true signal.

The templates in the filter bank do not exactly correspond to the waveforms of true gravitational waves because (i) the finite sampling of the parameter space of component masses means that a true signal may lie "between" two templates (recall that the template bank is designed to achieve a specified value of minimal match) and (ii) the model waveforms may not agree exactly with the true waveforms (recall that the fitting factor characterizes how well the templates represent the true waveform). Let δ represent the *mismatch* between the template, g, and the true waveform, h, defined by

$$\delta = 1 - \frac{(h, g)}{\sqrt{(h, h)(g, g)}} . \tag{7.241}$$

Then it can be shown that $\chi^2(t_0)$ has a non-central chi-squared distribution with $\nu = 2p - 2$ degrees of freedom and a non-central parameter $\lambda \lesssim 2\rho^2\delta$. Very large signal-to-noise ratio gravitational-wave signals can produce a large value of $\chi^2(t_0)$ owing to the mismatch between the signal and the template; however, if δ is kept reasonably small, then the chi-squared veto remains highly effective at discriminating between true gravitational-wave signals and common noise glitches.

7.8.1.5 Bounds on Binary Coalescence Rate

The observed rate of binary coalescence will provide critical information about the astrophysical population of compact binary systems – information that will reveal details of the formation of compact binaries, which depends on stellar evolution scenarios. To date, since no detections have been made, rate constraints have been upper limits, but when detections become routine the event rate bounds will be intervals with a lower and an upper bound. The important quantity to constrain is the rate density, which could be the rate per unit volume (expressed in units of events per year per cubic-megaparsec) or the rate per Milky-Way-like galaxy (events per year per Milky Way Equivalent Galaxy), or the rate per blue-light luminosity (events per year per luminosity in B-band).[2)]

2) The B-band luminosity is often used as an indicator of star formation and, by extension, a measure of the number density of compact binary coalescences. However, it may not be a very good measure: the coalescences themselves may lag the star formation by a billion years, and globular clusters, which have little blue-light luminosity, may harbour many close compact binaries due to stellar interactions.

Whatever the units that we wish to express our rate density in, we need to establish the *efficiency* of the search for a population of signals. Suppose we count the number of events observed above some signal-to-noise ratio threshold, ρ_{thresh}, that is set high enough that only true compact binary coalescences could produce such an event. Our hypothetical population of sources produces events at a (to be determined) rate R, but only some fraction, given by the search efficiency ϵ, of these will be detected by the search. The probability of there being n *detected* events in observation time T from our population of sources follows the *Poisson distribution*,

$$p(n;\lambda) = \frac{1}{n!}\lambda^n e^{-\lambda} , \tag{7.242}$$

where $\lambda = \epsilon R T$. Suppose we seek a frequentist upper limit on the rate of events at a 90% confidence level, $R_{90\%}$. Then given N observed events in a search, we must solve

$$10\% = \sum_{n=0}^{N} p(n;\lambda_{90\%}) \tag{7.243}$$

for $\lambda_{90\%} = \epsilon R_{90\%} T$ (see Section 7.3.3 and Problem 7.3). If $N = 0$, this reduces to solving

$$0.1 = e^{-\epsilon R_{90\%} T} \quad (N = 0) , \tag{7.244}$$

which gives

$$R_{90\%} \simeq \frac{2.303}{\epsilon T} \quad (N = 0) . \tag{7.245}$$

The factor ϵ must be determined for the search procedure. Typically this is done through simulations in which artificial signals drawn from the hypothetical population distribution are added to detector noise and processed through the search procedure; then ϵ is the fraction of the events that are counted as being detected.

An alternative approach to the counting method described above for setting upper limits that does not require a threshold for detection is known as the *loudest event* method. A search will normally produce many event candidates, especially at low signal-to-noise ratio, as a result of the detector noise. If we are agnostic as to whether any event is a result of a true gravitational wave or a result of the detector noise alone, we can simply ask what is the probability that a population of sources will produce no events in our search with a signal-to-noise ratio greater than the largest signal-to-noise ratio that we obtained, ρ_{loudest}. This probability is $p(0;\epsilon_{\text{loudest}} R T)$ where $\epsilon_{\text{loudest}}$ is the fraction of signals produced by the population of sources that would be detected with a signal-to-noise greater than the loudest signal-to-noise ratio observed in the search, ρ_{loudest}. As before, a frequentist upper limit on rate, R_α, at confidence level α can be obtained by solving $1 - \alpha = p(0;\epsilon_{\text{loudest}} R_\alpha T)$ for R_α:

$$R_\alpha = \frac{-\ln(1-\alpha)}{\epsilon_{\text{loudest}} T} . \tag{7.246}$$

For a $\alpha = 90\%$ confidence upper limit,

$$R_{90\%} \simeq \frac{2.303}{\epsilon_{\text{loudest}} T} . \qquad (7.247)$$

7.8.2
Searches for Poorly Modelled Burst Sources

When a gravitational-wave burst signal is not known in detail then it is called a poorly modelled burst. For example, simulations of gravitational waves from stellar core collapse – one progenitor of supernovae – are difficult to model numerically (owing, in part, to the complexity of the physical processes involved). Poorly modelled signals such as these are not entirely *unmodelled*; however, typically we have some information about the nature of the signal such as its characteristic frequency and the frequency band of the radiation, and also the timescale over which radiation is produced. However, we do not know the waveform with sufficient accuracy for matched filtering to be a viable detection method.

Searches for such poorly modelled signals typically make use of some variant of the time-frequency excess-power method described in Example 7.8 for the case of a single detector and in Section 7.6.3 for the case of multiple detectors. In Section 7.6.3 we constructed not only the coherent excess-power detection statistic, $\mathcal{E}$, given by Eq. (7.151), but also the null energy, $\mathcal{E}_{\text{null}}$, given by Eq. (7.153), which provides an effective method of discriminating between true gravitational waves (which should not contribute to the null energy) and single-detector noise glitches (which almost certainly will contribute to the null energy). In addition to these we also constructed the maximum likelihood estimate of the gravitational waveform, given by Eq. (7.150).

The coherent excess-power statistic is our principal detection statistic. If the detector noise is stationary and Gaussian, the coherent excess-power statistic is chi-squared distributed with $2NTF$ degrees of freedom, where N is the number of detectors and TF is the product of the duration and the frequency bandwidth of the signal, when there is no signal, and a non-central chi-squared distribution with non-central parameter $\lambda = \sum_{i=1}^{N} \varrho_i^2$ where ϱ_i is the characteristic signal-to-noise ratio produced by the signal in detector i. However, since detector data often contains noise artefacts, which can be difficult to distinguish from gravitational-wave events when the anticipated gravitational waveform is not well known, the statistical properties of the coherent excess power must be measured. One way to do this is to reprocess the data with artificial time shifts in the detector data so that true gravitational waves will not occur simultaneously in the different detectors. Noise artefacts are not expected to occur simultaneously in two detectors (except by chance) so such time-shift analyses can be used to measure a background (i.e. with no gravitational-wave content) distribution for the coherent excess-power statistic. Given this distribution, a threshold may be found for $\mathcal{E}$ that will yield the desired false alarm probability for the actual search (in which there is no relative time-shifting of the data streams). Typical burst searches then count the number of foreground (i.e. those occurring in the non-time-slid data) events and from this number a rate limit is obtained.

The search efficiency can also be assessed by reprocessing data into which simulated signals have been injected. However, because the gravitational waveform is not well known, it is not clear what waveforms should be added to the data in order to determine the detection efficiency. The time-frequency excess-power statistic depends only weakly on the exact morphology of signal, though, provided it is within the designed frequency and time intervals, so one often adopts representative waveforms such as *sine-Gaussian* waveforms,

$$h_+(t) + i h_\times(t) = \left(\frac{2}{\pi}\right)^{1/4} \tau^{-1/2} h_{\mathrm{rss}} \mathrm{e}^{-(t/\tau)^2} \mathrm{e}^{2\pi i f_c t} , \tag{7.248}$$

where τ characterizes the duration of the signal, f_c is its central frequency, and the amplitude is characterized by the root-sum-square amplitude h_{rss} which is defined (for any waveform, not just for sine-Gaussians) by

$$h_{\mathrm{rss}}^2 := \int \left[h_+^2(t) + h_\times^2(t)\right] dt . \tag{7.249}$$

The efficiency of a burst search as a function of the h_{rss} amplitude of a signal, $\epsilon(h_{\mathrm{rss}})$, is simply the fraction of events with a given value of h_{rss} that would be detected by the search. The efficiency characterizes the sensitivity of the search, and upper limits from burst searches are usually expressed in "rate-versus-strain" upper limit plots: for example, if zero events are detected, the rate-versus-strain plot for a 90% (frequentist) confidence upper limit is a plot of the function

$$R_{90\%}(h_{\mathrm{rss}}) = \frac{2.303}{\epsilon(h_{\mathrm{rss}}) T} , \tag{7.250}$$

where T is the observation time.

In the case that a burst is detected, Eq. (7.150) gives the maximum likelihood estimate of the gravitational waveform, from which one may deduce such things as the fluence of the burst in gravitational radiation, the polarization of the gravitational radiation, and even the morphology of the gravitational wave.

7.9
Data Analysis Methods for Stochastic Sources

A stochastic background of gravitational waves will introduce an excess of random noise within a detector. If the gravitational-wave background noise is weak compared to the instrumental noise, it is essentially impossible to distinguish the gravitational-wave noise from the instrumental noise. However, if there are several detectors operating, the gravitational-wave background radiation can introduce correlated (but still random) noise amongst the detectors; by cross-correlating the data from two detectors the gravitational-wave background can be measured, provided that there is no significant amount of correlated non-gravitational-wave noise between the two detectors. (For example, the two LIGO Hanford detectors, H1 and H2, share a common vacuum environment and thus have common environmental noise; it would be difficult to distinguish between a gravitational-wave background noise and a common environmental noise.)

We assume that the stochastic background of gravitational waves creates metric perturbations that are stationary and Gaussian random processes with no preferred direction, that is they are isotropic and unpolarized. The background has some frequency spectrum, and because it is often regarded as cosmological in origin, it is normally expressed in terms of the dimensionless energy density spectrum, $\Omega_{\text{GW}}(f)$, which is defined so that the energy density in gravitational waves, $c^2 d\rho_{\text{GW}}(f)$, in the frequency range f to $f + df$ is

$$c^2 d\rho_{\text{GW}}(f) = c^2 \rho_{\text{crit}} \Omega_{\text{GW}}(f) \frac{df}{f} , \tag{7.251}$$

where

$$c^2 \rho_{\text{crit}} := \frac{3c^2 H_0^2}{8\pi G} \tag{7.252}$$

is known as the critical energy density. The stochastic background is unpolarized and isotropic, so it can be decomposed into gravitational plane waves coming from every direction and polarization with the statistical properties

$$\begin{aligned} \left\langle \tilde{h}_+^*(f', \hat{\boldsymbol{n}}') \tilde{h}_+(f, \hat{\boldsymbol{n}}) \right\rangle &= \left\langle \tilde{h}_\times^*(f', \hat{\boldsymbol{n}}') \tilde{h}_\times(f, \hat{\boldsymbol{n}}) \right\rangle \\ &= \frac{3H_0^2}{32\pi^3} \frac{\Omega_{\text{GW}}(f)}{f^3} \delta^2(\hat{\boldsymbol{n}}, \hat{\boldsymbol{n}}') \delta(f - f') \\ \left\langle \tilde{h}_+^*(f', \hat{\boldsymbol{n}}') \tilde{h}_\times(f, \hat{\boldsymbol{n}}) \right\rangle &= 0 , \end{aligned} \tag{7.253}$$

where $\delta^2 (\hat{\boldsymbol{n}}, \hat{\boldsymbol{n}}') = \delta(\cos\theta - \cos\theta')\delta(\phi - \phi')$. (Since the polarization angle ψ merely defines $+$ and $\times$ polarizations, we simply set it to some fixed value, say $\psi = 0$, and do not explicitly include it in the equations.) The strain noise in a gravitational-wave detector is

$$\frac{1}{2} S_{h,\text{GW}}(f) \delta(f - f') = \int d\hat{\boldsymbol{n}} \int d\hat{\boldsymbol{n}}' \left\langle \tilde{h}^*(f', \hat{\boldsymbol{n}}') \tilde{h}(f, \hat{\boldsymbol{n}}) \right\rangle \tag{7.254}$$

with

$$\begin{aligned} \left\langle \tilde{h}^* (f', \hat{\boldsymbol{n}}') \tilde{h} (f, \hat{\boldsymbol{n}}) \right\rangle = {} & G_+ (f, \hat{\boldsymbol{n}}) G_+^* (f', \hat{\boldsymbol{n}}') \left\langle \tilde{h}_+^* (f', \hat{\boldsymbol{n}}') \tilde{h}_+ (f, \hat{\boldsymbol{n}}) \right\rangle \\ & + G_\times (f, \hat{\boldsymbol{n}}) G_\times^* (f', \hat{\boldsymbol{n}}') \left\langle \tilde{h}_\times^* (f', \hat{\boldsymbol{n}}') \tilde{h}_\times (f, \hat{\boldsymbol{n}}) \right\rangle . \end{aligned} \tag{7.255}$$

In particular, in an interferometric gravitational-wave detector with orthogonal arms and in the long-wavelength limit, the resulting noise power spectral density is

$$S_{h,\mathrm{GW}}(f) = \frac{3H_0^2}{10\pi^2}\frac{\Omega_{\mathrm{GW}}(f)}{f^3} \,. \tag{7.256}$$

For example, inflationary cosmology predicts a stochastic gravitational-wave background with a flat (constant in frequency) spectrum $\Omega_{\mathrm{GW}}(f) = \Omega_{\mathrm{GW},0}$ for gravitational-wave frequencies that are detected by current ground-based detectors. For the value of the Hubble constant $H_0 = 70\,\mathrm{km\,s^{-1}\,Mpc^{-1}}$,

$$S_{h,\mathrm{GW}}^{-1/2}(f) = 4\times 10^{-22}\,\mathrm{Hz}^{-1/2}\,\Omega_{\mathrm{GW},0}^{1/2}\left(\frac{f}{100\,\mathrm{Hz}}\right)^{3/2} \,. \tag{7.257}$$

This is comparable to the noise level in initial generation interferometers for $\Omega_{\mathrm{GW},0} \sim 0.01$, but the bound on primordial gravitational waves placed by Big Bang nucleosynthesis constraints limits $\Omega_{\mathrm{GW},0} \lesssim 10^{-5}$, and the background is likely to be substantially smaller than this, so the extra noise in a single gravitational-wave detector is unnoticeably small.

For a pair of detectors, the stochastic background gravitational-wave signal is correlated. If the two detectors are co-located and co-aligned then they both experience the same gravitational-wave strain, so

$$\left\langle \tilde{h}_1^*(f')\tilde{h}_2(f)\right\rangle = \left\langle \tilde{h}_1^*(f')\tilde{h}_1(f)\right\rangle = \left\langle \tilde{h}_2^*(f')\tilde{h}_2(f)\right\rangle \tag{7.258}$$

and therefore the *cross-spectral density* of the gravitational-wave background noise is

$$S_{12}(f) = S_{h,\mathrm{GW},1}(f) = S_{h,\mathrm{GW},2}(f) \quad \text{(co-located and co-aligned).} \tag{7.259}$$

However, if the two detectors are not located at the same place, or if they do not have the same orientation (i.e. they have a different response to gravitational waves) then the cross-spectral density is

$$\frac{1}{2}S_{12}(f)\delta(f-f') = \int d\hat{\boldsymbol{n}} \int d\hat{\boldsymbol{n}}' \left\langle \tilde{h}_1^*\left(f', \hat{\boldsymbol{n}}'\right)\tilde{h}_2(f, \hat{\boldsymbol{n}})\right\rangle \tag{7.260}$$

which can be expressed in terms of Ω_{GW} as

$$S_{12}(f) = \beta^{-1}\gamma_{12}(f)\frac{3H_0^2}{4\pi^2}\frac{\Omega_{\mathrm{GW}}(f)}{f^3} \,, \tag{7.261}$$

where

$$\begin{aligned}\gamma_{12}(f) := \beta\frac{1}{4\pi}\int d\hat{\boldsymbol{n}} e^{2\pi i f \hat{\boldsymbol{n}}\cdot\boldsymbol{r}_{12}/c} \\ \cdot\left[G_{+,1}^*(f,\hat{\boldsymbol{n}})G_{+,2}(f,\hat{\boldsymbol{n}}) + G_{\times,1}^*(f,\hat{\boldsymbol{n}})G_{\times,2}(f,\hat{\boldsymbol{n}})\right]\end{aligned} \tag{7.262}$$

is known as the *overlap reduction function* between detectors 1 and 2, which are separated by the vector $\boldsymbol{r}_{12} = \boldsymbol{r}_1 - \boldsymbol{r}_2$ where $\boldsymbol{r}_1$ and $\boldsymbol{r}_2$ are the locations of the two detectors. Here β is a normalization constant that is typically chosen so that $\gamma_{12}(f) = 1$ when two detectors are coaligned and colocated. (For interferometric detectors with orthogonal arms in the long-wavelength limit, $\beta = 5/2$.)

Example 7.13 Overlap reduction function in the long-wavelength limit

The response of a detector to a gravitational wave travelling in the direction $\hat{\boldsymbol{n}}$ in the long-wavelength limit can be expressed as

$$h(t) = D^{ij} h_{ij}^{\mathrm{TT}}(t, \hat{\boldsymbol{n}}) \,, \tag{7.263}$$

where D^{ij} is a detector response tensor which depends only on the geometry of the detector. Note that $F_+(\hat{\boldsymbol{n}}) = D^{ij} e_{ij}^+$ and $F_\times(\hat{\boldsymbol{n}}) = D^{ij} e_{ij}^\times$. For example, for an interferometer with its two arms along the $\hat{\boldsymbol{p}}$ and $\hat{\boldsymbol{q}}$ unit vectors, $D^{ij} = \frac{1}{2}(\hat{p}^i \hat{p}^j - \hat{q}^i \hat{q}^j)$, while for a bar detector with its axis along the unit vector $\hat{\boldsymbol{p}}$, $D^{ij} = \hat{p}^i \hat{p}^j$. Consider two gravitational-wave detectors with detector response tensors D_1^{ij} and D_2^{ij} that are separated by the vector $\boldsymbol{r}_{12} = \boldsymbol{r}_1 - \boldsymbol{r}_2$ where $\boldsymbol{r}_1$ and $\boldsymbol{r}_2$ are the position vectors of the two detectors. Then the overlap reduction function is given by (Allen and Romano, 1999)

$$\begin{aligned}\beta^{-1}\gamma_{12}(f) = &\left[2 j_0(\alpha) - 4\frac{j_1(\alpha)}{\alpha} + 2\frac{j_2(\alpha)}{\alpha^2}\right] D_1^{ij} D_{2,ij} \\ &+ \left[-4 j_0(\alpha) + 16\frac{j_1(\alpha)}{\alpha} - 20\frac{j_2(\alpha)}{\alpha^2}\right] D_1^{ij} D_{2,ik} \hat{s}_j \hat{s}^k \\ &+ \left[j_0(\alpha) - 10\frac{j_1(\alpha)}{\alpha} + 35\frac{j_2(\alpha)}{\alpha^2}\right] D_1^{ij} D_2^{kl} \hat{s}_i \hat{s}_j \hat{s}_k \hat{s}_l \,,\end{aligned} \tag{7.264}$$

where $\alpha = 2\pi f r_{12}/c$, $\boldsymbol{r}_{12} = r_{12}\hat{\boldsymbol{s}}_{12}$, and $j_0(\alpha) = \sin(\alpha)/\alpha$, $j_1(\alpha) = \sin(\alpha)/\alpha^2 - \cos(\alpha)/\alpha$, and $j_2(\alpha) = 3\sin(\alpha)/\alpha^3 - 3\cos(\alpha)/\alpha^2 - \sin(\alpha)/\alpha$ are spherical Bessel functions. Recall that β is a normalization constant that is chosen so that $\gamma_{12}(f)$ is unity for detectors that are coaligned and colocated; for interferometric detectors with orthogonal arms, $\beta = 5/2$.

For the LIGO interferometers at Hanford (H) and Livingston (L), the overlap reduction function can be evaluated as

$$\gamma_{\mathrm{HL}}(f) \simeq -0.1080 j_0(\alpha) - 3.036\frac{j_1(\alpha)}{\alpha} + 3.443\frac{j_2(\alpha)}{\alpha^2} \,, \tag{7.265}$$

where $\alpha = 2\pi f r_{12}/c \approx 2\pi(f/100\,\mathrm{Hz})$ since the separation between the two sites is $r_{12} \simeq 3002\,\mathrm{km}$. For low-frequency gravitational waves, $f \ll c/r_{12}$, we note that $j_n(\alpha)/\alpha^n \to 1/(2n+1)!!$ and so this function approaches the constant value $\gamma_{\mathrm{HL}}(0) \simeq -0.1080 - 3.036/3 + 3.443/15 \simeq -0.8907$. That is, the Hanford and Livingston detectors are reasonably well (anti)-aligned. At higher frequencies, however, the overlap reduction function becomes a decaying function that oscillates about zero, see Figure 7.2. This means that high-frequency stochastic gravitational-wave background radiation becomes difficult to detect because it is not strongly correlated between the two detectors.

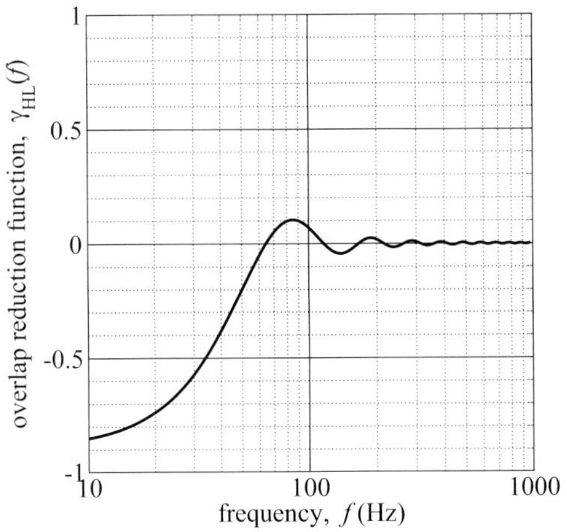

Figure 7.2 The overlap reduction function $\gamma_{\mathrm{HL}}(f)$ for the LIGO Hanford Observatory (H) and the LIGO Livingston Observatory (L).

Example 7.14 Hellings–Downs curve

For gravitational-wave detection via pulsar timing experiments, the detector consists of a single baseline given by the vector $\boldsymbol{p}$ from the Earth to the pulsar. For such a detector, the detector response tensor (*not* in the long-wavelength limit, as we are now interested in waves that have periods shorter than the light travel time between the source and the Earth) is

$$D^{ij} = \frac{1}{2}\frac{\hat{p}^i \hat{p}^j}{1+\hat{\boldsymbol{p}}\cdot\hat{\boldsymbol{n}}} \,. \tag{7.266}$$

Now suppose that there are two such detectors with baseline vectors $\boldsymbol{p}_1$ and $\boldsymbol{p}_2$ and having detector response tensors D_1^{ij} and D_2^{ij}. The detectors are colocated in the sense that the signals from both baselines are received at the Earth, so the overlap reduction function has no frequency dependence. The overlap reduction function,

$$\gamma_{12} = 1 + 3\frac{1-\hat{\boldsymbol{p}}_1\cdot\hat{\boldsymbol{p}}_2}{2}\left[\ln\left(\frac{1-\hat{\boldsymbol{p}}_1\cdot\hat{\boldsymbol{p}}_2}{2}\right) - \frac{1}{6}\right] \tag{7.267}$$

is also known as the *Hellings–Downs curve*. The normalization factor $\beta = 3$ has been chosen to make $\gamma_{12} = 1$ when $\hat{\boldsymbol{p}}_1 \cdot \hat{\boldsymbol{p}}_2 = 1$. See Figure 7.3.

Normally we assume that the stochastic gravitational-wave spectrum is some power-law in frequency,

$$\Omega_{\mathrm{GW}}(f) = \Omega_\alpha (f/f_{\mathrm{ref}})^\alpha \,, \tag{7.268}$$

where f_{ref} is some reference frequency and Ω_α is the value of the dimensionless energy spectrum at that frequency. For a flat spectrum, $\Omega_{\mathrm{GW}}(f) = \Omega_0$, which

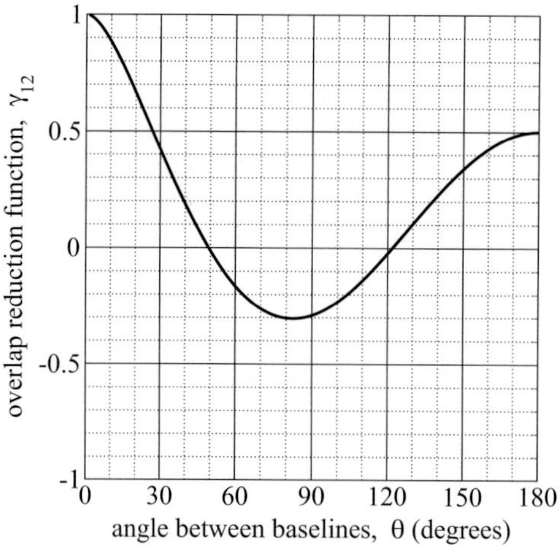

Figure 7.3 The overlap reduction function γ_{12} for two pulsar timing gravitational-wave detectors as a function of the angle between the two pulsars, $\theta = \arccos(\hat{\boldsymbol{p}}_1 \cdot \hat{\boldsymbol{p}}_2)$, where $\hat{\boldsymbol{p}}_1$ and $\hat{\boldsymbol{p}}_2$ are unit vectors in the direction of pulsar 1 and 2, respectively. This is known as the Hellings–Downs curve.

is a small value, and in general Ω_α is a small number (compared to unity) for reference frequencies in the band of any gravitational-wave detector: that is, Ω_α is a small parameter that encodes the power in the stochastic gravitational-wave background. It is the value of Ω_α that we hope to determine or constrain with a search for stochastic background gravitational radiation, so we introduce a scaled cross-spectral density $\hat{S}_{12}$,

$$\hat{S}_{12} := \beta^{-1}\gamma_{12}(f)\frac{3H_0^2}{4\pi^2}\frac{f^{\alpha-3}}{f_{\mathrm{ref}}^{\alpha}} \tag{7.269}$$

and write

$$S_{12} = \Omega_\alpha \hat{S}_{12} \,. \tag{7.270}$$

To obtain an optimal detection statistic, consider the effect of the stochastic gravitational-wave background on a network of N detectors. The detector data forms a vector of time series $\boldsymbol{s}(t) = [s_1(t), s_2(t), \ldots, s_N(t)]$. We assume that there is no correlated instrumental noise between the detectors and we suppose the instrumental noise and the stochastic background noise are stationary and Gaussian. Then, at any given frequency,

$$p\left[\tilde{\mathbf{s}}(f)\right] \propto \frac{1}{\sqrt{\det \boldsymbol{\Sigma}}}\exp\left\{-\tilde{\mathbf{s}}^{\dagger}(f)\boldsymbol{\Sigma}^{-1}(f)\tilde{\mathbf{s}}(f)\right\} \tag{7.271}$$

is the probability density function for the frequency components $\tilde{s}(f) = [\tilde{s}_1(f), \tilde{s}_2(f), \ldots, \tilde{s}_N(f)]$ and

$$\boldsymbol{\Sigma}(f) := \begin{bmatrix} S_1(f) & S_{12}(f) & \cdots & S_{1N}(f) \\ S_{21}(f) & S_2(f) & \cdots & S_{2N}(f) \\ \vdots & \vdots & \ddots & \vdots \\ S_{N1}(f) & S_{N2}(f) & \cdots & S_N(f) \end{bmatrix} \tag{7.272}$$

encodes the auto- and cross-correlations between the detector data. For each detector i the power spectral density $S_i(f)$ is a combination of the instrumental power spectral density $S_{h,\text{inst}}(f)$ and the stochastic gravitational-wave background noise $S_{h,\text{GW}}(f)$, $S_i(f) = S_{h,\text{inst}}(f) + S_{h,\text{GW}}(f)$. Given a stochastic background with power Ω_α, the probability distribution function for the network data s, $p(s|\Omega_\alpha)$, is given by

$$\ln p(s|\Omega_\alpha) = -\frac{1}{2}4\int_0^\infty \left\{\tilde{s}^\dagger(f)\boldsymbol{\Sigma}^{-1}(f)\tilde{s}(f) + \frac{1}{2}T\ln\det\boldsymbol{\Sigma}(f)\right\} df \tag{7.273}$$

up to an irrelevant constant. Here T is the observation time, which appears when converting the sum over all frequency bins into an integral. Notice that we need to invert the matrix $\boldsymbol{\Sigma}$ to compute this probability density. This can be done by expansion in the small parameter Ω_α: we write

$$\boldsymbol{\Sigma} = \begin{bmatrix} S_1(f) & 0 & \cdots & 0 \\ 0 & S_2(f) & \cdots & 0 \\ \vdots & \vdots & \ddots & \vdots \\ 0 & 0 & \cdots & S_N(f) \end{bmatrix} + \Omega_\alpha \begin{bmatrix} 0 & \hat{S}_{12}(f) & \cdots & \hat{S}_{1N}(f) \\ \hat{S}_{21}(f) & 0 & \cdots & \hat{S}_{2N}(f) \\ \vdots & \vdots & \ddots & \vdots \\ \hat{S}_{N1}(f) & \hat{S}_{N2}(f) & \cdots & 0 \end{bmatrix} \tag{7.274}$$

from which we obtain

$$\boldsymbol{\Sigma}^{-1} = \boldsymbol{\Sigma}_0^{-1} + \Omega_\alpha\boldsymbol{\Sigma}_1^{-1} + \Omega_\alpha^2\boldsymbol{\Sigma}_2^{-1} + O\left(\Omega_\alpha^3\right) \tag{7.275}$$

with

$$\boldsymbol{\Sigma}_0^{-1} = \begin{bmatrix} S_1^{-1}(f) & 0 & \cdots & 0 \\ 0 & S_2^{-1}(f) & \cdots & 0 \\ \vdots & \vdots & \ddots & \vdots \\ 0 & 0 & \cdots & S_N^{-1}(f) \end{bmatrix}, \tag{7.276}$$

$$\boldsymbol{\Sigma}_1^{-1} = -\begin{bmatrix} 0 & \frac{\hat{S}_{12}(f)}{S_1(f)S_2(f)} & \cdots & \frac{\hat{S}_{1N}(f)}{S_1(f)S_N(f)} \\ \frac{\hat{S}_{21}(f)}{S_2(f)S_1(f)} & 0 & \cdots & \frac{\hat{S}_{2N}(f)}{S_2(f)S_N(f)} \\ \vdots & \vdots & \ddots & \vdots \\ \frac{\hat{S}_{N1}(f)}{S_N(f)S_1(f)} & \frac{\hat{S}_{N2}(f)}{S_N(f)S_2(f)} & \cdots & 0 \end{bmatrix}, \tag{7.277}$$

and

$$\left(\Sigma_2^{-1}\right)_{ij} = \sum_{\substack{k=1 \\ k\neq i,j}}^{N} \frac{\hat{S}_{ik}(f)\hat{S}_{kj}(f)}{S_i(f)S_k(f)S_j(f)}\,. \tag{7.278}$$

We also have

$$\ln\det\boldsymbol{\Sigma} = \sum_{i=1}^{N}\ln S_i(f) - \Omega_\alpha^2 \sum_{i=1}^{N}\sum_{\substack{j=1 \\ j<i}}^{N} \frac{\hat{S}_{ij}(f)\hat{S}_{ji}(f)}{S_i(f)S_j(f)} + O\left(\Omega_\alpha^3\right). \tag{7.279}$$

With these results we can express the logarithm of the likelihood ratio as

$$\ln\Lambda = \ln p\left(s|\Omega_\alpha\right) - \ln p(s|0) = \Omega_\alpha\mathcal{S} - \frac{1}{2}\Omega_\alpha^2\mathcal{N}^2 + O\left(\Omega_\alpha^3\right), \tag{7.280}$$

where

$$\mathcal{S} := \frac{1}{2}4\int_0^\infty \tilde{s}^\dagger(f)\boldsymbol{\Sigma}_1^{-1}(f)\tilde{s}(f)df = \sum_{i=1}^{N}\sum_{\substack{j=1 \\ j<i}}^{N} 4\,\mathrm{Re}\int_0^\infty \frac{\tilde{s}_i^*(f)\hat{S}_{ij}(f)\tilde{s}_j(f)}{S_i(f)S_j(f)}df \tag{7.281}$$

is the factor in the $O(\Omega_\alpha)$ term of the logarithm of the likelihood ratio and

$$\mathcal{N}^2 := \sum_{i=1}^{N}\sum_{\substack{j=1 \\ j<i}}^{N} 4\int_0^\infty \left\{ -T\frac{\hat{S}_{ij}(f)\hat{S}_{ji}(f)}{S_i(f)S_j(f)} + 2\,\mathrm{Re}\sum_{\substack{k=1 \\ k\neq i,j}}^{N} \frac{\tilde{s}_i^*(f)\hat{S}_{ik}(f)\hat{S}_{kj}(f)\tilde{s}_j(f)}{S_i(f)S_k(f)S_j(f)} \right\} df \tag{7.282}$$

is the factor in the $O(\Omega_\alpha^2)$ term.

Unlike the situation with deterministic signals, for which the likelihood depends on the data only via the matched filter statistic and it increases monotonically with signal amplitude, the likelihood ratio for the stochastic gravitational-wave background search depends on the data and the presumptive amplitude of the stochastic background in a non-trivial way. That is, the detector data appears in both the factor $\mathcal{S}$ and in the factor $\mathcal{N}^2$, and these two factors combine in a manner that depends on characteristic strength of the background, Ω_α. This means that there is some ambiguity as to what the "optimal" detection statistic is in the case of a stochastic signal: the most sensitive search will depend on how strong the stochastic background signal is.

If we assume that the stochastic background is *weak*, $\Omega_\alpha \ll 1$, then we are led to a *locally optimal detection statistic*,

$$\lim_{\Omega_\alpha \to 0} \frac{d \ln \Lambda}{d \Omega_\alpha} = \mathcal{S} \,. \tag{7.283}$$

The locally optimal statistic is the best performing statistic (in the sense of having the greatest detection probability for fixed false alarm probability) in the limit of infinitely weak signals (Kassam, 1987). On the other hand, if we wish to determine the strength of a stochastic background, we construct the maximum likelihood estimator, $\Omega_{\alpha,\text{est}}$, for which

$$\left. \frac{d \ln \Lambda}{d \Omega_\alpha} \right|_{\Omega_\alpha = \Omega_{\alpha,\text{est}}} = 0 \,, \tag{7.284}$$

and we obtain

$$\Omega_{\alpha,\text{est}} \simeq \frac{\mathcal{S}}{\mathcal{N}^2} \,, \tag{7.285}$$

where we have ignored the $O(\Omega_\alpha^3)$ terms. By substituting this maximum likelihood estimate into Eq. (7.280) we obtain the maximum likelihood detection statistic

$$\max_{\Omega_\alpha} \ln \Lambda \simeq \frac{1}{2} \frac{\mathcal{S}^2}{\mathcal{N}^2} \,, \tag{7.286}$$

where we continue to ignore the $O(\Omega_\alpha^3)$ terms. This statistic would be more appropriate for situations where we have no prior expectation that the stochastic background is weak.

To simplify the discussion we now specialize to the case of $N = 2$ detectors. The locally optimal detection statistic $\mathcal{S}$ is simply a cross-correlation of the data from the two detectors with appropriate frequency weighting:

$$\begin{aligned} \mathcal{S} &= 4 \,\text{Re} \int_0^\infty \frac{\tilde{s}_1^*(f) \hat{S}_{12}(f) \tilde{s}_2(f)}{S_1(f) S_2(f)} d f \\ &= \beta^{-1} \frac{3 H_0}{4\pi^2} 2 \int_{-\infty}^{\infty} \frac{\gamma_{12}(|f|) |f/f_{\text{ref}}|^\alpha}{|f|^3 S_1(|f|) S_2(|f|)} \tilde{s}_1^*(f) \tilde{s}_2(f) d f \,. \end{aligned} \tag{7.287}$$

By convention we write this cross-correlation as

$$Y := \alpha \int_{-\infty}^{\infty} \frac{\gamma_{12}(|f|) |f/f_{\text{ref}}|^\alpha}{|f|^3 S_1(|f|) S_2(|f|)} \tilde{s}_1^*(f) \tilde{s}_2(f) d f \tag{7.288}$$

so that

$$\mathcal{S} = \frac{2}{\alpha\beta}\frac{3H_0^2}{4\pi^2}Y\,, \tag{7.289}$$

where the normalization constant α, which is defined by

$$2\alpha^{-1} = \beta^{-1}\frac{3H_0^2}{4\pi^2}\int\limits_{-\infty}^{\infty}\frac{\gamma_{12}^2(f)|f/f_{\mathrm{ref}}|^{2\alpha}}{|f|^6 S_1(|f|)S_2(|f|)}df\,, \tag{7.290}$$

is chosen so that if a stochastic gravitational-wave background with characteristic power Ω_α is present then the expectation value of Y is

$$\langle Y\rangle = T\Omega_\alpha\,, \tag{7.291}$$

where T is the observation time. Note that $\langle\tilde{s}_1^*(f)\tilde{s}_2(f)\rangle = \langle\tilde{h}_1^*(f)\tilde{h}_2(f)\rangle = \frac{1}{2}S_{12}(f)\delta(f-f)$ is formally divergent because of the vanishing argument to the delta-function, but if data for a finite observation period of duration T is used then $\delta(f-f) = T$, and consequently $\langle\tilde{s}_1^*(f)\tilde{s}_2(f)\rangle = \frac{1}{2}TS_{12}(f)$. The variance of Y is

$$\sigma^2 = \operatorname{Var}Y = \frac{1}{4}\alpha^2 T\int\limits_{-\infty}^{\infty}\frac{\gamma_{12}^2(f)|f/f_{\mathrm{ref}}|^{2\alpha}}{|f|^6 S_1(|f|)S_2(|f|)}df \tag{7.292}$$

and we can compute a characteristic signal-to-noise ratio for the cross-correlation statistic

$$\begin{aligned}\varrho = \frac{\langle Y\rangle}{\sqrt{\operatorname{Var}Y}} &= T^{1/2}\beta^{-1}\frac{3H_0^2}{4\pi^2}\sqrt{2\int\limits_0^{\infty}\frac{\gamma_{12}^2(f)}{S_1(f)S_2(f)}\frac{\Omega_{\mathrm{GW}}^2(f)}{f^6}df}\\ &= T^{1/2}\sqrt{2\int\limits_0^{\infty}\frac{S_{12}^2(f)}{S_1(f)S_2(f)}df}\,.\end{aligned} \tag{7.293}$$

A detectable signal would require $\varrho \sim$ a few; notice that ϱ grows with observation time as $T^{1/2}$ so longer observations allow for the detection of weaker signals.

The characteristic signal-to-noise ratio also determines the expected value of the likelihood ratio in the presence of a true gravitational-wave signal. Up to quadratic order in Ω_α, the logarithm of the likelihood ratio depends on the locally optimal statistic $\mathcal{S}$ and also on the quantity $\mathcal{N}^2$, so we compute the expectation value of both of these. The expectation value of the locally optimal statistic is

$$\Omega_\alpha\langle\mathcal{S}\rangle = \varrho^2\,. \tag{7.294}$$

For two detectors,

$$\begin{aligned}\mathcal{N}^2 = 8\,\mathrm{Re}&\int\limits_0^{\infty}\frac{\hat{S}_{12}^2(f)}{S_1(f)S_2(f)}\left[\frac{|\tilde{s}_1(f)|^2}{S_1(f)} + \frac{|\tilde{s}_2(f)|^2}{S_2(f)}\right]df\\ &- 4T\int\limits_0^{\infty}\frac{\hat{S}_{12}^2(f)}{S_1(f)S_2(f)}df\end{aligned} \tag{7.295}$$

so we find

$$\Omega_\alpha^2 \langle \mathcal{N}^2 \rangle = \varrho^2 \,. \tag{7.296}$$

Therefore we have

$$\langle \ln \Lambda \rangle \simeq \frac{1}{2} \varrho^2 \,, \tag{7.297}$$

where the $O(\Omega_\alpha^3)$ terms continue to be dropped.

Example 7.15 Sensitivity of a stochastic background search

The sensitivity of a stochastic background search can be obtained by computing the characteristic signal-to-noise ratio, ϱ, given in Eq. (7.293). For example, a cosmological stochastic background of gravitational waves with a flat-spectrum would have $\Omega_{\mathrm{GW}} = \Omega_0$. For a $T \sim 1$ year search using the LIGO Hanford 4 km detector (H1) and the LIGO Livingston 4 km detector (L1), the value of Ω_0 required to produce a characteristic signal-to-noise ratio of $\varrho = 3$ would be

$$\begin{aligned}\Omega_0 &= \varrho T^{-1/2} \left(\frac{3 H_0^2}{10\pi^2} 2 \int_0^\infty \frac{\gamma_{\mathrm{HL}}^2(f)}{f^6 S_{\mathrm{H1}}(f) S_{\mathrm{L1}}(f)} df \right)^{-1/2} \\ &\sim 2 \times 10^{-6} \,,\end{aligned} \tag{7.298}$$

where we have taken the noise power spectral densities $S_{\mathrm{H1}}(f)$ and $S_{\mathrm{L1}}(f)$ to be those of the idealized Initial LIGO detector model described in Section A.2. During LIGO's fifth science run, S5, the sensitivity was considerably poorer than the idealized model at low frequencies; using a typical S5 noise spectrum (see Figure A.2) instead gives a value $\Omega_0 \sim 6 \times 10^{-6}$.

A stochastic background search is normally conducted by computing the cross-correlation statistic Y for intervals of duration Δt over which the detector noise is relatively stationary, for example $\Delta t \sim 15$ min is a typical timescale for searches with ground-based interferometric detectors. This yields a number $N_{\mathrm{intervals}} = T/\Delta t$ intervals, Y_i for $i \in [1, N_{\mathrm{intervals}}]$, each having a variance σ_i^2 given by Eq. (7.292) with T replaced by Δt. These are then combined optimally (given a varying background noise level) as

$$Y = \sigma^2 \sum_{i=1}^{N_{\mathrm{intervals}}} Y_i / \sigma_i^2 \,, \tag{7.299a}$$

where

$$\sigma^{-2} = \sum_{i=1}^{N_{\mathrm{intervals}}} \sigma_i^{-2} \tag{7.299b}$$

is the variance in Y. This construction of Y naturally favours those intervals in which the detectors are most sensitive, and de-emphasizes those intervals in which the detectors are not performing well.

The cross-correlation search is a measure of *all* cross-correlated noise sources, not just the stochastic gravitational-wave background. The overlap reduction function is designed so that the stochastic gravitational-wave background will produce a positive expectation value for Y, but other types of correlated noise can be either correlated or anti-correlated between the detectors, and the expectation value for Y can be either positive or negative. This makes it difficult to bound the stochastic gravitational-wave background if it is possible that there are correlated non-gravitational-wave noise sources. However, if we assume that there is no common non-gravitational-wave noise between two detectors then we can place a bound on Ω_α from the measured value of Y. For example, if we also assume that Y is a Gaussian random variable with mean $T\Omega_\alpha$ and variance σ^2, then a 90% confidence frequentist upper limit would be

$$\Omega_{\alpha,90\%} = \frac{Y}{T} + 1.28\frac{\sigma}{T} . \tag{7.300}$$

(Notice that this upper limit may become unphysically negative if Y happens to have a large negative value.)

7.9.1 Stochastic Gravitational-Wave Point Sources

A targeted search for stochastic gravitational waves is useful for detecting continuous random processes coming from a particular sky location (e.g. the Galactic core), or for searching for continuous-wave sources at known sky positions but where the phase evolution is not known (e.g. a search for gravitational waves from a neutron star in an X-ray binary system) so that the methods of matched filtering cannot be used. In such a situation, the procedure described previously continues to hold but now the overlap reduction function is simply

$$\gamma_{12}(f, \hat{\boldsymbol{n}}, t) = \mathrm{e}^{2\pi i f \hat{\boldsymbol{n}}\cdot \boldsymbol{r}_{12}/c} \left[G^*_{+,1}(f, \hat{\boldsymbol{n}})\, G_{+,2}(f, \hat{\boldsymbol{n}}) + G^*_{\times,1}(f, \hat{\boldsymbol{n}})\, G_{\times,2}(f, \hat{\boldsymbol{n}})\right] . \tag{7.301}$$

Notice that the overlap reduction function is now time-dependent (with a period of one sidereal day for ground-based detectors) so the cross-correlation detection statistic Y must be computed over short enough periods of time that the overlap reduction function does not change significantly. Fortunately, we have already seen how cross-correlation statistics computed for short intervals, Y_i, can be combined to produce a single statistic Y over the entire observation time, cf. Eq. (7.299) (the earlier goal was to handle non-stationarities in the detector noise properties). Therefore, to target a particular time and direction, one must re-compute the overlap reduction function for each interval analyzed, but the analysis is essentially unchanged. Note that since $\hat{\boldsymbol{n}}$ is known in the directed search, the factor $\mathrm{e}^{2\pi i f \hat{\boldsymbol{n}}\cdot \boldsymbol{r}_{12}/c}$

gives a frequency-dependent phase-shift to the overlap reduction function, but the overlap reduction function does *not* damp-out at high frequency: directed searches are not hampered at high frequency by the distance between the detectors.

7.10 Problems

Problem 7.1

Consider a noise model in which the noise n present in an observed outcome, s, of an experiment is not Gaussian but rather has a distribution $p(n) = (2\pi)^{-1}\mathrm{sech}(n)$ where sech is the hyperbolic secant function. The null hypothesis $\mathcal{H}_0$ is that the observed outcome is noise, $s = n$, while the alternative hypothesis is that the observed outcome is noise plus a signal of amplitude h, $s = n + h$. Show that the optimal statistic (the likelihood ratio Λ) depends on the expected value of the signal amplitude, that is, the optimal statistic for detecting a signal of amplitude $h = 1$ is different from the optimal statistic for detecting a signal of amplitude $h = 4$. Suppose that a population of signals has amplitudes with a distribution $p(h) \propto \mathrm{e}^{-|h|}$ (h may be either positive or negative). Find the optimal detection statistic (the marginalized likelihood ratio) for such a signal population. Plot the marginalized likelihood Λ as a function of observed outcome values s.

Problem 7.2

Consider the network of four detectors, GEO-600 (G), LIGO Hanford Observatory (H), LIGO Livingston Observatory (L) and Virgo (V). For some source (located directly above the prime meridian and the equator) we have

$$\mathbf{F}^{\mathrm{T}} = \begin{bmatrix} F_{+,\mathrm{G}} & F_{+,\mathrm{H}} & F_{+,\mathrm{L}} & F_{+,\mathrm{V}} \\ F_{\times,\mathrm{G}} & F_{\times,\mathrm{H}} & F_{\times,\mathrm{L}} & F_{\times,\mathrm{V}} \end{bmatrix} = \begin{bmatrix} +0.35 & +0.25 & +0.19 & -0.65 \\ -0.50 & -0.46 & +0.36 & -0.38 \end{bmatrix}. \tag{7.302}$$

Find the coefficients c_1 and c_2 for *two* null streams for this network of detectors.

Answer: your solutions for c_1 and c_2 should be two linear combinations of $e_1 = [0, 0.351, 0.854, 0.384]$ and $e_2 = [0.671, -0.692, 0.213, 0.157]$.

Problem 7.3

Consider a Poisson process with an unknown rate R. The probability of obtaining n events in an observation time T is given by

$$p(n;\lambda) = \frac{1}{n!}\lambda^n e^{-\lambda} ,$$

where $\lambda = RT$. Use the Neyman method described in Section 7.3.3 to show that the upper limit on the rate, R_α, at confidence level α, given N observed events is found by solving

$$1 - \alpha = \sum_0^N p(n;\lambda_\alpha)$$

for $\lambda_\alpha = R_\alpha T$. Suppose in $T =$ one year of observation $N = 0$ events are found. What is the (frequentist) $\alpha = 90\%$ confidence upper limit on the rate, $R_{90\%}$? What is the limit if $N = 1$ events are found?

Problem 7.4

Equation (7.234) gives the accuracies with which an inspiral search can measure various parameters (specifically, the coalescence time t_0, the coalescence phase φ_0 and the chirp mass $\mathcal{M}$) in terms of the matrix $\mathbf{B}$ whose inverse $\mathbf{A}$ is given in Eq. (7.232). Use the idealized Initial LIGO model presented in Appendix A.2 to compute the components of the matrix B^{ij} and thus obtain values for Δt_0, $\Delta\varphi_0$ and $(\Delta\mathcal{M})/\mathcal{M}$ for a $\varrho = 10$ signal.

Problem 7.5

The energy density of gravitational waves in a stochastic gravitational-wave background is given by $c^2\rho_{\text{GW}} = T_{00}^{\text{GW}}/c^2$. Expand the metric perturbation into plane waves

$$h_{ij}(t) = \int d\hat{\boldsymbol{\Omega}} \int d f\, e^{2\pi i f t} \left\{h_+\left(f, \hat{\boldsymbol{\Omega}}\right) e_{ij}^+ + h_\times\left(f, \hat{\boldsymbol{\Omega}}\right) e_{ij}^\times\right\}$$

(here $\hat{\boldsymbol{\Omega}} = -\hat{\boldsymbol{n}}$) and use Eq. (3.79) with Eqs. (7.251) and (7.252) to validate the relations given in Eq. (7.253). See Allen and Romano (1999).

Problem 7.6

Derive the formula for the Hellings–Downs curve, Eq. (7.267).

References

Allen, B. (2005) A χ^2 time-frequency discriminator for gravitational-wave detection. *Phys. Rev.*, **D71**, 062 001. doi: 10.1103/PhysRevD.71.062001.

Allen, B. and Romano, J.D. (1999) Detecting a stochastic background of gravitational radiation: signal processing strategies and sensitivities. *Phys. Rev.*, **D59**, 102 001. doi: 10.1103/PhysRevD.59.102001.

Brady, P.R. and Creighton, T. (2000) Searching for periodic sources with LIGO. II: hierarchical searches. *Phys. Rev.*, **D61**, 082 001. doi: 10.1103/PhysRevD.61.082001.

Dupuis, R.J. and Woan, G. (2005) Bayesian estimation of pulsar parameters from gravitational wave data. *Phys. Rev.*, **D72**, 102 002. doi: 10.1103/PhysRevD.72.102002.

Feldman, G.J. and Cousins, R.D. (1998) A unified approach to the classical statistical analysis of small signals. *Phys. Rev.*, **D57**, 3873–3889. doi: 10.1103/PhysRevD.57.3873.

Finn, L.S. and Chernoff, D.F. (1993) Observing binary inspiral in gravitational radiation: one interferometer. *Phys. Rev.*, **D47**, 2198–2219. doi: 10.1103/PhysRevD.47.2198.

Jaranowski, P. and Królak, A. (2009) *Analysis of Gravitational-Wave Data*, Cambridge University Press.

Jaranowski, P., Krolak, A. and Schutz, B.F. (1998) Data analysis of gravitational-wave signals from spinning neutron stars. I: the signal and its detection. *Phys. Rev.*, **D58**, 063 001. doi: 10.1103/PhysRevD.58.063001.

Kassam, S.A. (1987) *Signal Detection in Non-Gaussian Noise*, Springer.

Klimenko, S., Mohanty, S., Rakhmanov, M., and Mitselmakher, G., (2005) Constraint likelihood analysis for a network of gravitational wave detectors. *Phys. Rev.*, **D72**, 122002. doi:10.1103/PhysRevD.72.122002.

Owen, B.J. (1996) Search templates for gravitational waves from inspiraling binaries: choice of template spacing. *Phys. Rev.*, **D53**, 6749–6761. doi: 10.1103/PhysRevD.53.6749.

Owen, B.J. and Sathyaprakash, B.S. (1999) Matched filtering of gravitational waves from inspiraling compact binaries: computational cost and template placement. *Phys. Rev.*, **D60**, 022 002. doi: 10.1103/PhysRevD.60.022002.

Pletsch, H.J. and Allen, B. (2009) Exploiting global correlations to detect continuous gravitational waves. *Phys. Rev. Lett.*, **103**, 181 102. doi: 10.1103/PhysRevLett.103.181102.

Sutton, P.J., Jones, G., Chatterji, S., Kalmus, P.M., Leonor, I., Poprocki, S., Rollins, J., Searle, A., Stein, L., Tinto, M. and Was, M. (2010) X-Pipeline: an analysis package for autonomous gravitational-wave burst searches. *New J. Phys.*, **12**, 053 034. doi: 10.1088/1367-2630/12/5/053034.

Wainstein, L.A. and Zubakov, V.D. (1971) *Extraction of Signals from Noise*, Dover.

8
Epilogue: Gravitational-Wave Astronomy and Astrophysics

At the time of writing – the end of 2010, the end of Initial LIGO/Virgo operation – there has been no direct detection of gravitational waves. Gravitational waves have nevertheless been an important part of astrophysical modelling of relativistic systems for decades. The well-studied Hulse–Taylor binary pulsar system gives a very precise measurement of the orbital decay due to energy loss to gravitational waves, but even before that gravitational radiation was found to be important in understanding the balance of angular momentum in certain cataclysmic variable binary systems (where a white dwarf primary star pulls matter off of the secondary star). The true promise of gravitational-wave astronomy, however, lies in the direct detection of gravitational waves; as an astronomical tool, gravitational-wave observations offer the power to probe the dynamical processes of gravitating systems, even when those dynamics are obscured from electromagnetic observations, such as in the core of a star exploding in a supernovae or in the collisions of black holes or in processes occurring early in the history of the universe when it was still opaque to electromagnetic radiation.

Because gravitational-wave astronomy is an evolving subject – and there is every potential for stunning revelations in the years to come – it is impossible to detail all avenues in which we expect gravitational-wave astronomy to impact our outlook on the Universe. We will speculate here on some ways we expect gravitational-wave observations might provide crucial physical and astrophysical understanding, but some of these ways will be more speculative than others. Since our field is dynamic and this textbook is being written at an early stage, a better up-to-date presentation of our current understanding and expectations would be in the form of a living review; we refer the reader to Sathyaprakash and Schutz (2009) for such a review.

8.1
Fundamental Physics

General Relativity is one of the foundational theories of modern physics and, as such, experimental tests of the predictions of General Relativity have a particular importance. In fact, General Relativity is an extremely well tested theory, with confirmation of critical predictions such as the perihelion advance of Mercury and

Gravitational-Wave Physics and Astronomy, First Edition. Jolien D. E. Creighton, Warren G. Anderson.

the bending of light by the Sun giving compelling initial support. Most alternative theories of gravity have been strongly constrained by observations of the orbital dynamics of bodies in the solar system and also by the orbits of binary pulsar systems, most notably the Hulse–Taylor binary pulsar (see Example 3.15). These observations give solid indications that General Relativity is correct – at least in the regimes that have been probed. For the most part, the observations have been limited to the equations of motion at the first post-Newtonian order, and of the quadrupole formula for gravitational radiation. The exact nature of the radiation, and the dynamics of highly relativistic systems is yet untested.

General Relativity makes specific predictions about the nature of gravitational radiation. The wave equation for the metric perturbation indicates that gravitational radiation should propagate at the speed of light, c. If the graviton has a small but non-zero mass, however, one expects that gravitational waves would travel somewhat less than the speed of light. Observations of gravitational waves in conjunction with electromagnetic observations of the radiating system will allow us to test whether gravitational waves do indeed travel at the same speed as light. The Einstein field equations restrict the number of possible gravitational-wave polarization states: we have seen in Example 3.4 that there are six polarizations available in metrical theories of gravity, but only the two transverse spin-2 $+$ and $\times$ modes are allowed in General Relativity. For example, in Brans–Dicke theory (a scalar-tensor theory of gravity), in addition to the spin-2 transverse modes of General Relativity, one expects there to be a spin-0 transverse "breathing" mode and also (for a massive graviton) a spin-0 longitudinal mode. With a number of detectors observing a gravitational wave, the exact polarization state can be determined, and we may be able to observe non-spin-2 polarizations if they exist.

Testing the highly relativistic and strong field predictions of General Relativity will require a laboratory in which such conditions exist – for example, in the collisions of highly compact objects such as neutron stars and black holes. The late stages of inspiral during binary orbital decay for systems of neutron stars of black holes will reveal the dynamics of highly relativistic systems, and will provide a sensitive test of higher post-Newtonian effects on orbital motion and gravitational-wave production. The detailed dynamics of colliding black holes can be modelled by numerical relativity and these simulations can be confronted with gravitational-wave observations to see if the strong-field predictions of General Relativity agree with nature.

General Relativity also makes a specific prediction about the nature of black holes. The generic vacuum black hole solution is the Kerr black hole, which is characterized by only two parameters: its mass and its spin. Motions of small bodies in the vicinity of a black hole will allow the determination of the black hole's gravitational multipole moments, which again are uniquely determined by the hole's mass and spin, and it may be possible to detect deviations from the Kerr solution in the case that General Relativity is not correct. Additionally, General Relativity predicts a spectrum of natural modes of black hole oscillations – the quasi-normal modes described in Section 4.2.2 – which is determined by the black hole's mass and spin; a different spectrum of oscillations may be observed in other theories of gravity.

Beyond tests of General Relativity, gravitational-wave observations may uncover other aspects of fundamental physics. For example, discovery of a gravitational-wave background from the early universe may reveal physical processes during phase transitions at various energy scales (e.g. quantum chromodynamics energy scales, electroweak energy scales, or supersymmetry energy scales), or during the re-heating of the universe after inflation (see Section 5.3.1). The gravitational waveform produced by the merger of binary neutron stars will encode not only the masses of the neutron stars but also their tidal deformability, and measurements of the tidal interaction will constrain the possible equations of state of neutron star interiors – which describes the nature of cold matter at the most extreme densities.

8.2 Astrophysics

The population of binary neutron stars and stellar-mass ($M < 100\ M_\odot$) black holes in the Galaxy is highly uncertain: only a handful of binary neutron star systems have been observed and no neutron-star+black-hole or black-hole+black-hole binaries are known. Stellar evolution models can be used to predict the frequency of such systems, but it is not known whether such models correctly account for all the variety of physical phenomena expected to describe the evolution of binary star systems. Furthermore, the population that is formed within a galactic field is expected to be quite different from the population that is formed in stellar clusters, where multiple dynamical interactions between stars are important. Our limited understanding of the population of compact objects that are near enough to merge under the effects of gravitational radiation means that our estimates of the rate of events we expect in gravitational-wave detectors are highly uncertain. On the other hand, when we *do* begin to observe these systems, the rate at which we see binary coalescences will be a critical observation that will inform stellar evolutionary models.

Supermassive black holes ($M > 10^4\ M_\odot$) are found to inhabit the cores of many galaxies, including our own. Just as galaxies have been observed to merge, we expect the supermassive black holes in their cores to merge too, and space-based detectors such as LISA as well as pulsar timing arrays should detect these mergers. In addition, smaller bodies (neutron stars or stellar mass black holes) falling into supermassive black holes will produce *extreme mass ratio inspiral* (EMRI) signals, which should be detectable by space-based detectors. A possible population of intermediate-mass black holes ($100\ M_\odot < M < 10^4\ M_\odot$) between the stellar-mass black holes and the supermassive black holes may also exist. By studying the merger rates of intermediate mass black holes and supermassive black holes, we hope to learn about the processes that are important during early galaxy formation, for example did galaxies form hierarchically, from collisions of smaller objects, each containing a small black hole? or did large galaxies (and their supermassive black holes) form directly?

Neutron star and black hole coalescences occurring when galaxies are first forming may produce a detectable astrophysical background of gravitational waves, and observations of the spectrum of such a background may also yield information about the process of galaxy formation.

Black hole formation is expected to power the central engine that is the progenitor of most gamma-ray bursts, but there is likely more than one process by which these bursts are produced. The "long" gamma-ray bursts are likely formed in the gravitational collapse of the cores of massive stars that result in a black hole, while the "short" gamma-ray bursts may be produced by the disruption of neutron stars during binary coalescence of neutron-star + neutron-star binaries or neutron-star + black-hole binaries. Gravitational-wave observations of a binary inspiral in association with a short gamma-ray burst would confirm this progenitor model. (Indeed, the *failure* to observe a binary inspiral signal associated with the short gamma-ray burst GRB 070201, which likely originated in the nearby Andromeda galaxy, lends support to the hypothesis that at least some short gamma-ray bursts are produced by soft-gamma-ray repeaters rather than coalescing binaries.)

Associating gravitational-wave binary inspiral signals with optical transients can provide a means for precision measurement of cosmological parameters. The distance-redshift relation, which is central to the determination of Hubble's constant and the measurement of the current acceleration of the Universe, requires the combined measurement of the distance and the redshift of distant objects such as supernovae. Distance measurements are often difficult to make. The establishment of a "standard candle", such as the luminosity of a particular type of supernova, can be used to relate the observed brightness of an object with its distance, but these standards are difficult to construct. The gravitational waves from a binary inspiral, however, have a directly measurable distance standard: the amplitude of the wave is determined by the chirp mass of the system (which is easily determined by the frequency evolution) and the distance to the system. If the galactic host of an observed binary inspiral can be identified, and the redshift of the host galaxy is measured, then this will yield a direct measurement of a point on the distance-redshift relation.

Another astrophysical mystery that gravitational-wave observations may help resolve is the mechanism by which supernovae form. Supernovae are produced in massive stars when their cores collapse to form neutron stars. Infalling matter impacts on the nascent neutron star and rebounds to form a shock. However, this initial shock stalls before it can explode the surrounding star; various models for rejuvenating the shock have been proposed. Supercomputer simulations of supernovae have shed some light on the viability of various proposals, however the complexity of the physical processes that are present make it difficult to obtain a definitive answer to the question of which mechanism (or perhaps there is more than one mechanism) is responsible for the explosion. Electromagnetic observations cannot resolve this issue as the dynamical processes that occur during core collapse are entirely obscured by the surrounding star. However, the different proposed mechanisms are expected to produce very different gravitational-wave signatures, so gravitational-wave observations of a supernova in our Galaxy or a nearby galaxy may complement numerical simulations in determining the origin of supernovae.

Newly formed neutron stars, forged in supernovae, will be hot, rapidly rotating, and pulsating with their various modes of oscillation. For rotating neutron stars, certain modes of oscillation are unstable to gravitational radiation: that is, the gravitational radiation generated by the oscillations amplifies the oscillations rather than dissipates them. This instability, which occurs when the pattern-speed of the oscillation appears prograde with respect to the rotation for inertial observers at a large distance from the star (who receive the gravitational radiation from the oscillations) but retrograde with respect to the direction of rotation relative to the neutron star matter (so that the negative angular momentum radiated amplifies the negative angular momentum of the mode of oscillation), is known as the Chandrasekhar–Friedman–Schutz instability. Viscosity of the neutron star matter is known to dissipate the oscillation modes, but nascent, hot neutron stars will have less viscosity, so gravitational waves produced by oscillations of these young neutron stars may possibly be observed. Of particular interest are the *r*-mode oscillations, which are driven by current-quadrupole radiation (rather than mass-quadrupole radiation which normally dominates gravitational radiation).

Such *r*-mode oscillations may also play an important role in low-mass X-ray binary (LMXB) systems, in which old neutron stars are spun-up through mass transfer from a companion (which produces the observed X-rays); gravitational waves (and most likely from *r*-modes) may be responsible for balancing the angular momentum gained from the mass transfer to keep the rotational period within the relatively narrow range observed. If this is the case, then gravitational waves from the LMXB system Scorpius X-1 would be marginally detectable with advanced ground-based interferometric detectors. Neutron star oscillation modes may also be excited during neutron star glitches (when the shape of a neutron star suddenly changes) and during magnetar (neutron stars with strong magnetic fields) flares. The observation of one of these modes of oscillation would help constrain the neutron star equation of state.

Every survey of gravitational-wave sources must conclude with the possibility that some new, unexpected source may be found. Gravitational waves are generated by very different mechanisms than electromagnetic waves, so it is quite possible that there is a component of the universe that is entirely unexplored thus far. By the very virtue of being "unexpected", we should not count on such surprises. But if we do find some totally new phenomenon, it sure would be fun.

References

Sathyaprakash, B.S. and Schutz, B.F. (2009) Physics, astrophysics and cosmology with gravitational waves. *Living Rev. Rel.*, **12**(2). http://www.livingreviews.org/lrr-2009-2 (last accessed 2011-01-03).

Appendix A
Gravitational-Wave Detector Data

A.1
Gravitational-Wave Detector Site Data

Table A.1 gives the site location information, $\boldsymbol{r}$, of the current gravitational-wave detectors. These vectors are expressed in Earth-fixed coordinates: the origin is the centre of the Earth, the x-axis pierces the Earth at the point where the Prime Meridian intersects the equator, the y-axis pierces the Earth on the equator at 90° East longitude, and the z-axis pierces the Earth at the North pole.

The orientations of the detectors determine their response tensors and thus their antenna beam pattern functions. Resonant mass bar detectors operate in the long-wavelength limit, and their response tensor is given by

$$D^{ij} = \hat{p}^i \hat{p}^j \,, \tag{A.1}$$

where $\hat{\boldsymbol{p}}$ is the unit vector along the axis of the bar. Interferometric detectors have a response tensor

$$D^{ij}(\hat{\boldsymbol{n}}, f) = \frac{1}{2}\hat{p}^i \hat{p}^j D\left(\hat{\boldsymbol{p}} \cdot \hat{\boldsymbol{n}}, f\|\boldsymbol{p}\|/c\right) - \frac{1}{2}\hat{q}^i \hat{q}^j D\left(\hat{\boldsymbol{q}} \cdot \hat{\boldsymbol{n}}, f\|\boldsymbol{q}\|/c\right), \tag{A.2}$$

where $\hat{\boldsymbol{p}}$ is the unit vector along one arm of the interferometer, $\hat{\boldsymbol{q}}$ is the unit vector along the other arm of the interferometer, $\|\boldsymbol{p}\|$ and $\|\boldsymbol{q}\|$ are the lengths of these arms, and

$$\begin{aligned} D(\mu, x) := \frac{1}{2}\mathrm{e}^{2\pi i x}\{&\mathrm{e}^{i\pi x(1-\mu)}\operatorname{sinc}[\pi x(1+\mu)] \\ &+ \mathrm{e}^{-i\pi x(1+\mu)}\operatorname{sinc}[\pi x(1-\mu)]\}. \end{aligned} \tag{A.3}$$

In the long-wavelength limit, $f L/c \ll 1$, $D \to 1/2$ and the response tensor for an interferometric detector is

$$D^{ij}(\hat{\boldsymbol{n}}, f) = \frac{1}{2}\left(\hat{p}^i \hat{p}^j - \hat{q}^i \hat{q}^j\right). \tag{A.4}$$

Table A.2 gives the orientation, $\hat{\boldsymbol{p}}$, of the axis of resonant mass bar detectors. These unit vectors are expressed in Earth-fixed coordinates.

Table A.1 Gravitational-wave detector site information.

Observatory	r (m)		
	Resonant mass detectors		
ALLEGRO	− 113 259	−5 504 083	+3 209 895
AURIGA	+4 392 467	+ 929 509	+4 515 029
EXPLORER	+4 376 454	+ 475 435	+4 599 853
NAUTILUS	+4 644 110	+1 044 253	+4 231 047
NIOBE	−2 359 489	+4 877 216	−3 354 160
	Interferometric detectors		
GEO	+3 856 310	+ 666 599	+5 019 641
LHO	−2 161 415	−3 834 695	+4 600 350
LLO	− 74 276	−5 496 284	+3 224 257
TAMA	−3 946 409	+3 366 259	+3 699 151
Virgo	+4 546 374	+ 842 990	+4 378 577

Table A.2 Resonant mass detector orientation data.

Detector	$\hat{p}$		
ALLEGRO (A1)	−0.6347	+0.4009	+0.6606
AURIGA (O1)	−0.6445	+0.5737	+0.5055
EXPLORER (E1)	−0.6279	+0.5648	+0.5354
NAUTILUS (N1)	−0.6204	+0.5725	+0.5360
NIOBE (B1)	−0.2303	+0.4761	+0.8486

Table A.3 gives the orientations, $\hat{\boldsymbol{p}}$ and $\hat{\boldsymbol{q}}$, of the two arms of interferometric gravitational-wave detectors. These unit vectors are expressed in Earth-fixed coordinates.

A gravitational wave is described in terms of the Earth-fixed coordinates as follows: suppose that the waves are coming from a sky position (θ, ϕ) where θ is the polar angle relative to the z-axis and ϕ is the azimuthal angle from the x-axis along the x–y plane; that is, the unit vector that points toward the source is $\hat{\boldsymbol{\Omega}} = \{\sin\theta\cos\phi, \sin\theta\sin\phi, \cos\theta\}$ in the Earth-fixed coordinates. Note that the direction of propagation of the waves is $\hat{\boldsymbol{n}} = -\hat{\boldsymbol{\Omega}}$. The angles θ and ϕ are related to the coordinates of the source in the equatorial system, right-ascension α and declination δ, by $\theta = \pi/2 - \delta$ and $\phi = \alpha - \mathrm{GMST}$ where GMST is the Greenwich mean sidereal time of arrival of the signal. The metric perturbation h_{ij} in the transverse-traceless gauge is defined in a coordinate system with basis vectors $\boldsymbol{e}_1$, $\boldsymbol{e}_2$ and $\boldsymbol{e}_3 = \hat{\boldsymbol{n}}$, where $\boldsymbol{e}_1$ and $\boldsymbol{e}_2$ lie on the transverse plane; in terms of the

Table A.3 Interferometric detector orientation data.

Detector	$\hat{p}$			$\hat{q}$		
GEO 600 m (G1)	−0.4453	+0.8665	+0.2255	−0.6261	−0.5522	+0.5506
LHO 4 km (H1) [a]	−0.2239	+0.7998	+0.5569	−0.9140	+0.0261	−0.4049
LHO 2 km (H2) [b]	−0.2239	+0.7998	+0.5569	−0.9140	+0.0261	−0.4049
LLO 4 km (L1) [c]	−0.9546	−0.1416	−0.2622	+0.2977	−0.4879	−0.8205
TAMA 300 m (T1)	+0.6490	+0.7608	+0.0000	−0.4437	+0.3785	−0.8123
Virgo 3 km (V1)	−0.7005	+0.2085	+0.6826	−0.0538	−0.9691	+0.2408

a 1. H1 has arm lengths $\|\boldsymbol{p}\| = 3995.084\,\text{m}$ and $\|\boldsymbol{q}\| = 3995.044\,\text{m}$
b 2. H2 has arm lengths $\|\boldsymbol{p}\| = \|\boldsymbol{q}\| = 2009\,\text{m}$
c 3. L1 has arm lengths $\|\boldsymbol{p}\| = \|\boldsymbol{q}\| = 3995.15\,\text{m}$

Earth-fixed coordinates, their components are

$$\begin{aligned} \boldsymbol{e}_1 = \{ &+ \cos\psi \sin\phi - \sin\psi \cos\phi \cos\theta\,, \\ &- \cos\psi \cos\phi - \sin\psi \sin\phi \cos\theta\,, \\ &+ \sin\psi \sin\theta \} \end{aligned} \tag{A.5a}$$

$$\begin{aligned} \boldsymbol{e}_2 = \{ &- \sin\psi \sin\phi - \cos\psi \cos\phi \cos\theta\,, \\ &+ \sin\psi \cos\phi - \cos\psi \sin\phi \cos\theta\,, \\ &+ \cos\psi \sin\theta \}\,. \end{aligned} \tag{A.5b}$$

Here, ψ is an angle that specifies the $\boldsymbol{e}_1$ and $\boldsymbol{e}_2$ basis vectors on the transverse plane, and therefore it specifies the plus- and cross-polarization states and so it is known as the polarization angle. Specifically it is the angle counter-clockwise about $\boldsymbol{e}_3$ from the line of nodes to the axis $\boldsymbol{e}_1$. The gravitational-wave polarization tensors are $\mathbf{e}_+ := \boldsymbol{e}_1 \otimes \boldsymbol{e}_1 - \boldsymbol{e}_2 \otimes \boldsymbol{e}_2$ and $\mathbf{e}_\times := \boldsymbol{e}_1 \otimes \boldsymbol{e}_2 + \boldsymbol{e}_2 \otimes \boldsymbol{e}_1$, or, explicitly,

$$e^+_{ij} := (e_1)_i (e_1)_j - (e_2)_i (e_2)_j \tag{A.6a}$$

$$e^\times_{ij} := (e_1)_i (e_2)_j + (e_2)_i (e_1)_j\,. \tag{A.6b}$$

The metric perturbation h_{ij} in the TT-gauge can therefore be expressed in terms of the two polarization components, h_+ and $h_\times$, as

$$h_{ij} = h_+ e^+_{ij} + h_\times e^\times_{ij} \tag{A.7a}$$

with

$$h_+ := \frac{1}{2} e^{ij}_+ h_{ij} \tag{A.7b}$$

$$h_\times := \frac{1}{2} e^{ij}_\times h_{ij}\,. \tag{A.7c}$$

The strain produced on a detector with response tensor D^{ij} by the metric perturbation h_{ij} in the TT-gauge is

$$h := D^{ij} h_{ij} \,. \tag{A.8}$$

This can be expressed in terms of polarization components h_+ and $h_\times$ as

$$h = G_+ h_+ + G_\times h_\times \tag{A.9a}$$

where

$$G_+ := D^{ij} e^+_{ij} \tag{A.9b}$$

$$G_\times := D^{ij} e^\times_{ij} \tag{A.9c}$$

are the detector antenna beam pattern response functions. In the long-wavelength limit, D^{ij} becomes a constant tensor (in the Earth-fixed frame) with no frequency or direction dependence, and the antenna beam pattern functions depend only on the direction to the source in the Earth-fixed coordinates. In the long-wavelength limit, we write $F_+ := G_+|_{f=0}$ and $F_\times := G_\times|_{f=0}$.

Example A.1 Antenna response beam patterns for interferometer detectors

For interferometric detectors with orthogonal arms, the long-wavelength limit of the beam pattern functions can be given in terms of the horizontal coordinate system for the detector: in a coordinate system with the x-axis being along one arm, the y-axis along the other arm, and the z-axis being the upward normal to the x–y plane, we take θ and ϕ to be the normal spherical angles. Then we have $\hat{p} = [1, 0, 0]$, $\hat{q} = [0, 1, 0]$ and we find

$$F_+(\theta, \phi, \psi) = -\frac{1}{2}(1 + \cos^2\theta)\cos 2\phi \cos 2\psi - \cos\theta \sin 2\phi \sin 2\psi \tag{A.10}$$

$$F_\times(\theta, \phi, \psi) = +\frac{1}{2}(1 + \cos^2\theta)\cos 2\phi \sin 2\psi - \cos\theta \sin 2\phi \cos 2\psi \,. \tag{A.11}$$

These beam pattern functions are shown in Figure A.1.

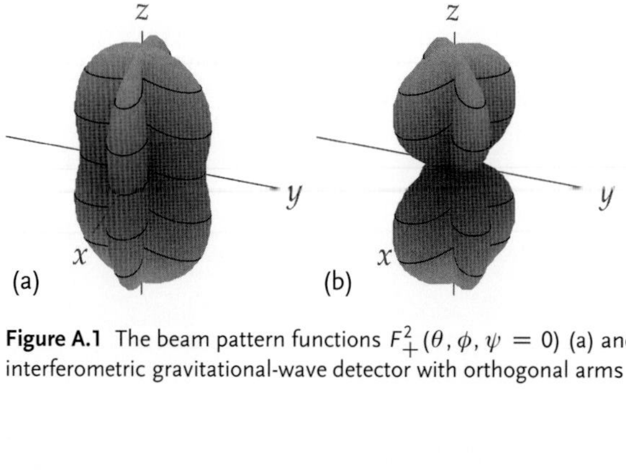

Figure A.1 The beam pattern functions $F_+^2(\theta, \phi, \psi = 0)$ (a) and $F_\times^2(\theta, \phi, \psi = 0)$ (b) for an interferometric gravitational-wave detector with orthogonal arms along the x- and y-axes.

A.2 Idealized Initial LIGO Model

In various examples throughout this book, we have used the LIGO detector for illustration. In Table A.4 we provide parameter values that describe an "idealized" initial LIGO detector model. Note that these are approximate representative values, and some of the parameters varied between the LIGO detectors and over the course of data taking runs.

To obtain a sensitivity curve for our idealized initial LIGO detector model, we focus on the main broadband noise sources. These are:

- Shot noise. The shot noise is given by

$$S_{h,\mathrm{shot}}(f) = \frac{3}{\eta_q}\frac{1}{L^2}\frac{\lambda}{2\pi}\frac{\hbar c}{2 I_0 G_{\mathrm{prc}} G_{\mathrm{arm}}^2}\frac{1}{|\hat{C}_{\mathrm{FP}}(f)|^2}, \tag{A.12}$$

where the normalized sensing transfer function for the Fabry–Pérot cavity is

$$\hat{C}_{\mathrm{FP}} = \frac{1 - r_{\mathrm{ITM}}}{1 - r_{\mathrm{ITM}} e^{-4\pi i f L/c}}. \tag{A.13}$$

Note that our shot noise estimate includes the quantum efficiency of the photodiodes $\eta_q = 0.9$ and the effects of non-stationary shot noise for a heterodyne RF readout scheme, along with the folding of frequencies (the factor of 3).

Table A.4 Approximate values for various parameters of the 4 km idealized initial LIGO model, as used in the text.

Laser wavelength λ	1 μm
Laser power at input I_0	5 W
ITM and ETM mirror mass M	11 kg
ITM and ETM internal vibrational frequency f_{int}	10 kHz
ITM and ETM internal vibrational quality $Q_{\mathrm{int}} = 1/\phi_{\mathrm{int}}$	10^6
ITM and ETM suspension pendulum frequency f_{pend}	0.76 Hz
ITM and ETM suspension pendulum quality $Q_{\mathrm{pend}} = 1/\phi_{\mathrm{pend}}$	10^6
ITM power transmissivity t_{ITM}^2	2.8%
Recycling cavity gain G_{prc}	50
Light power incident on beam splitter $I_0 G_{\mathrm{prc}}$	250 W
Arm length L	4000 m
Arm cavity finesse $\mathcal{F}$	220
Arm cavity gain G_{arm}	140
Arm cavity pole frequency f_{pole}	85 Hz
Arm cavity zero frequency f_{zero}	12 kHz
Arm cavity free spectral range f_{FSR}	37 kHz
Arm cavity light storage time τ_{store}	0.94 ms

- Suspension thermal noise:

$$S_{h,\text{pend}}(f) = \frac{1}{L^2}\frac{2k_B T}{\pi^3 M}\frac{f_{\text{pend}}^2}{Q_{\text{pend}}}\frac{1}{f^5}\,. \tag{A.14}$$

- Internal mirror vibrational noise:

$$S_{h,\text{int}}(f) = \frac{1}{L^2}\frac{2k_B T}{\pi^3 M}\frac{1}{f_{\text{int}}^2 Q_{\text{int}}}\frac{1}{f}\,. \tag{A.15}$$

- Seismic noise:

$$S_{h,\text{seis}}(f) = \frac{1}{L^2} S_{X,\text{ground}} |A(f)|^2 |A_{\text{stack}}(f)|^2\,, \tag{A.16}$$

where

$$S_{X,\text{ground}} = 10^{-18}\,\text{m}^2\,\text{Hz}^{-1}\begin{cases}1 & 1\,\text{Hz} < f \le 10\,\text{Hz}\\ \left(\frac{10\,\text{Hz}}{f}\right)^4 & f > 10\,\text{Hz}\end{cases} \tag{A.17}$$

is the approximate power spectrum of seismic ground motion,

$$|A(f)| = \frac{1}{1-(f/f_{\text{pend}})^2} \quad (f > f_{\text{pend}}) \tag{A.18}$$

is the magnitude of the transfer function for the pendulum suspension, and

$$|A_{\text{stack}}(f)| = \left(\frac{10\,\text{Hz}}{f}\right)^8 \tag{A.19}$$

is the transfer function for the isolation stack, which contains four alternating mass-spring layers.

The strain amplitude sensitivity curve for our idealized initial LIGO detector is formed by combining these individual noise components:

$$S_h^{1/2}(f) = \sqrt{S_{h,\text{shot}}(f) + S_{h,\text{pend}}(f) + S_{h,\text{int}}(f) + S_{h,\text{seis}}(f)}\,. \tag{A.20}$$

A significant noise source that we have ignored is actuator noise from the electronic currents in the length control and alignment servo system. The actuator noise during S5 was roughly equal to the suspension thermal noise (Abbott *et al.*, 2009). Additional noise sources that were significant in the 50–100 Hz frequency band possibly arose from up-conversion of low-frequency actuator noise into broadband noise and broadband noise caused by electric charge buildup on the surfaces of the mirrors.

We have also focused entirely on broadband noise and have neglected narrow-band sources of noise that are seen in the LIGO spectrum. These occur at power line harmonics – multiples of 60 Hz – as well as at the violin vibrational frequencies of the suspension wires.

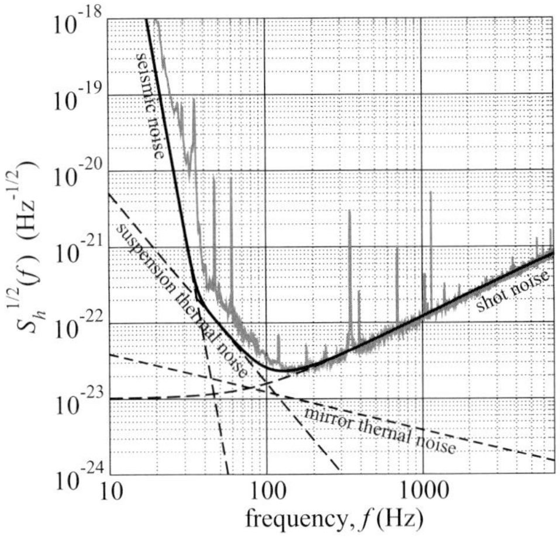

Figure A.2 The sensitivity curve $S_h^{1/2}(f)$ for the idealized Initial LIGO detector model (solid black) and a representative spectrum from the LHO 4 km detector H1 taken during LIGO's fifth science run (S5) (grey) (4 km Hanford interferometer on 18 March 2007[1]). Also shown (dashed lines) are the components of the idealized LIGO detector model including the shot noise component $S_{h,\text{shot}}^{1/2}$, the suspension thermal noise component $S_{h,\text{pend}}^{1/2}$, the internal mirror vibrational noise component $S_{h,\text{int}}^{1/2}$, and the seismic noise component $S_{h,\text{seis}}^{1/2}$.

In Figure A.2 we show the sensitivity curve $S_h^{1/2}(f)$ compared to a representative measured noise curve for the LHO 4 km (H1) interferometer taken during the fifth science run (S5).

References

Abbott, B. *et al.* (2009) LIGO: the laser interferometer gravitational-wave observatory. *Rept. Prog. Phys.*, **72**, 076 901. doi: 10.1088/0034-4885/72/7/076901.

1) http://www.ligo.caltech.edu/~jzweizig/distribution/LSC_Data (last accessed 2011-01-03).

Appendix B
Post-Newtonian Binary Inspiral Waveform

Here we give expressions for the gravitational waveforms from binary systems with component masses m_1 and m_2 in the point-particle post-Newtonian approximation. In the following expressions we have the total mass $M := m_1 + m_2$, the reduced mass $\mu := m_1 m_2 / M$ and the symmetric mass ratio $\eta := \mu / M$. The expressions below are given in terms of the post-Newtonian parameter $x := (G M \omega / c^3)^{2/3}$ where ω is the orbital angular momentum.

The gravitational waveform is written in terms of the modes $h_{\ell m}$ as

$$h_+ - i h_\times - \sum_{\ell=2}^{\infty} \sum_{m=-\ell}^{\ell} {}_{-2}Y_{\ell m}(\iota, \phi)\, h_{\ell m} \,, \tag{B.1}$$

where ${}_{-2}Y_{\ell m}$ are the spin-2 weighted spherical harmonics, which satisfy

$${}_{-2}Y_{\ell,m}(\theta, \phi) = (-1)^{\ell}\, {}_{-2}Y_{\ell,-m}(\pi - \theta, \phi) \tag{B.2}$$

and the complex modes satisfy

$$h_{\ell,m} = (-1)^{\ell}\, h^*_{\ell,-m} \,. \tag{B.3}$$

The dominant mode is the $\ell = 2$, $m = \pm 2$ mode, which is given to 3 post-Newtonian order by

$$\begin{aligned} h_{22} = -8\sqrt{\frac{5}{\pi}}\frac{G\mu}{c^2 r}\mathrm{e}^{-2i\varphi}x \Bigg\{ 1 - \left(\frac{107}{42} - \frac{55}{42}\eta\right)x \\ + \left[2\pi + 6i\ln\left(\frac{x}{x_0}\right)\right]x^{3/2} \\ - \left(\frac{2173}{1512} + \frac{1096}{216}\eta - \frac{2047}{1512}\eta^2\right)x^2 \\ - \left[\left(\frac{107}{21} - \frac{34}{21}\eta\right)\pi + 24i\eta + i\left(\frac{107}{7} - \frac{34}{7}\eta\right)\ln\left(\frac{x}{x_0}\right)\right]x^{5/2} \\ + \Bigg[\frac{27\,027\,409}{646\,800} - \frac{856}{105}\gamma_{\mathrm{E}} + \frac{2}{3}\pi^2 - \frac{1712}{105}\ln 2 - \frac{428}{105}\ln x \\ - 18\left[\ln\left(\frac{x}{x_0}\right)\right]^2 - \left(\frac{278\,185}{33\,264} - \frac{41}{96}\pi^2\right)\eta - \frac{20\,261}{2772}\eta^2 + \frac{114\,635}{99\,792}\eta^3 \\ + i\frac{428}{105}\pi + 12i\pi\ln\left(\frac{x}{x_0}\right)\Bigg]x^3\Bigg\}; \end{aligned} \tag{B.4}$$

expressions for the other modes are found in Kidder (2008). Here, x_0 is a freely-specifiable constant that arises in the gravitational-wave tails in relating the origin of time for the orbital dynamics to the origin of time for the radiation. The $\ell = 2$, $m = \pm 2$ spin-2 weighted spherical harmonics are

$$_{-2}Y_{2,\pm 2}(\theta, \phi) = \frac{1}{8}\sqrt{\frac{5}{\pi}}(1 \pm \cos\theta)\mathrm{e}^{\pm 2i\phi}\,. \tag{B.5}$$

Other $s = -2$ spin-weighted spherical harmonics are given in Table B.1.

To compute the phase evolution, we require the energy and flux functions,

$$\begin{aligned} \mathcal{E}(x) = -\frac{1}{2}\eta x \Bigg\{ 1 - \left(\frac{3}{4} + \frac{1}{12}\eta\right)x \\ - \left(\frac{27}{8} - \frac{19}{8}\eta + \frac{1}{24}\eta^2\right)x^2 \\ - \left[\frac{675}{64} - \left(\frac{34\,445}{576} - \frac{205}{96}\pi^2\right)\eta + \frac{155}{96}\eta^2 + \frac{35}{5184}\eta^3\right]x^3 \Bigg\} \end{aligned} \tag{B.6a}$$

Table B.1 The $\ell = 2, \ell = 3$ and $\ell = 4$ spin-weighted spherical harmonics for spin-weight $s = -2$.

$\ell = 2$	${}_{-2}Y_{2,+2} = \frac{1}{8}\sqrt{\frac{5}{\pi}}(1+\cos\theta)^2 e^{+2i\phi}$
	${}_{-2}Y_{2,+1} = \frac{1}{4}\sqrt{\frac{5}{\pi}}(1+\cos\theta)\sin\theta\, e^{+i\phi}$
	${}_{-2}Y_{2,0} = \frac{1}{8}\sqrt{\frac{30}{\pi}}\sin^2\theta$
	${}_{-2}Y_{2,-1} = \frac{1}{4}\sqrt{\frac{5}{\pi}}(1-\cos\theta)\sin\theta\, e^{-i\phi}$
	${}_{-2}Y_{2,-2} = \frac{1}{8}\sqrt{\frac{5}{\pi}}(1-\cos\theta)^2 e^{-2i\phi}$
$\ell = 3$	${}_{-2}Y_{3,+3} = -\frac{1}{16}\sqrt{\frac{42}{\pi}}(1+\cos\theta)^2\sin\theta\, e^{+3i\phi}$
	${}_{-2}Y_{3,+2} = \frac{1}{8}\sqrt{\frac{7}{\pi}}(1+\cos\theta)^2(3\cos\theta-2)e^{+2i\phi}$
	${}_{-2}Y_{3,+1} = \frac{1}{16}\sqrt{\frac{70}{\pi}}(1+\cos\theta)(3\cos\theta-1)\sin\theta\, e^{+i\phi}$
	${}_{-2}Y_{3,0} = \frac{1}{8}\sqrt{\frac{210}{\pi}}\cos\theta\sin^2\theta$
	${}_{-2}Y_{3,-1} = \frac{1}{16}\sqrt{\frac{70}{\pi}}(1-\cos\theta)(3\cos\theta+1)\sin\theta\, e^{-i\phi}$
	${}_{-2}Y_{3,-2} = \frac{1}{8}\sqrt{\frac{7}{\pi}}(1-\cos\theta)^2(3\cos\theta+2)e^{-2i\phi}$
	${}_{-2}Y_{3,-3} = \frac{1}{16}\sqrt{\frac{42}{\pi}}(1-\cos\theta)^2\sin\theta\, e^{-3i\phi}$
$\ell = 4$	${}_{-2}Y_{4,+4} = \frac{3}{16}\sqrt{\frac{7}{\pi}}(1+\cos\theta)^2\sin^2\theta\, e^{+4i\phi}$
	${}_{-2}Y_{4,+3} = -\frac{3}{16}\sqrt{\frac{14}{\pi}}(1+\cos\theta)^2(2\cos\theta-1)\sin\theta\, e^{+3i\phi}$
	${}_{-2}Y_{4,+2} = \frac{3}{8}\sqrt{\frac{1}{\pi}}(1+\cos\theta)^2(7\cos^2\theta-7\cos\theta+1)e^{+2i\phi}$
	${}_{-2}Y_{4,+1} = \frac{3}{16}\sqrt{\frac{2}{\pi}}(1+\cos\theta)(14\cos^2\theta-7\cos\theta-1)\sin\theta\, e^{+i\phi}$
	${}_{-2}Y_{4,0} = \frac{3}{16}\sqrt{\frac{10}{\pi}}(7\cos^2-1)\sin^2\theta$
	${}_{-2}Y_{4,-1} = \frac{3}{16}\sqrt{\frac{2}{\pi}}(1-\cos\theta)(14\cos^2\theta+7\cos\theta-1)\sin\theta\, e^{-i\phi}$
	${}_{-2}Y_{4,-2} = \frac{3}{8}\sqrt{\frac{1}{\pi}}(1-\cos\theta)^2(7\cos^2\theta+7\cos\theta+1)e^{-2i\phi}$
	${}_{-2}Y_{4,-3} = \frac{3}{16}\sqrt{\frac{14}{\pi}}(1-\cos\theta)^2(2\cos\theta+1)\sin\theta\, e^{-3i\phi}$
	${}_{-2}Y_{4,-4} = \frac{3}{16}\sqrt{\frac{7}{\pi}}(1-\cos\theta)^2\sin^2\theta\, e^{-4i\phi}$

and

$$\begin{aligned}
\mathcal{F}(x) = \frac{32}{5}\eta^2 x^5 \Bigg\{ & 1 - \left(\frac{1247}{336} + \frac{35}{12}\eta\right) x + 4\pi x^{3/2} \\
& - \left(\frac{44\,711}{9072} - \frac{9271}{504}\eta - \frac{65}{18}\eta^2\right) x^2 - \left(\frac{8191}{672} + \frac{583}{24}\eta\right)\pi x^{5/2} \\
& + \Bigg[\frac{6\,643\,739\,519}{69\,854\,400} + \frac{16}{3}\pi^2 - \frac{1712}{105}\gamma_E - \frac{856}{105}\ln(16x) \\
& + \left(\frac{41}{48}\pi^2 - \frac{134\,543}{7776}\right)\eta - \frac{94\,403}{3024}\eta^2 - \frac{775}{324}\eta^3\Bigg] x^3 \\
& - \left(\frac{16\,285}{504} - \frac{214\,745}{1728}\eta - \frac{193\,385}{3024}\eta^2\right)\pi x^{7/2}\Bigg\},
\end{aligned} \qquad \text{(B.6b)}$$

where $\gamma_{\rm E} \approx 0.577\,216$ is the Euler constant. These functions are known to 3 post-Newtonian order and to 3.5 post-Newtonian order respectively, see Blanchet (2002).[1] The energy and flux equations are used to obtain the orbital phase as a function of time (or frequency) by various schemes. The most common schemes are presented below (see Buonanno *et al.*, 2009).

B.1 TaylorT1 Orbital Evolution

In the TaylorT1 method, we obtain the orbital phase $\varphi(t)$ by numerically integrating the system of ordinary differential equations

$$\frac{dx}{dt} = -\frac{c^3}{GM}\frac{\mathcal{F}}{d\mathcal{E}/dx} \tag{B.7a}$$

$$\frac{d\varphi}{dt} = \frac{c^3}{GM}x^{3/2}\,. \tag{B.7b}$$

B.2 TaylorT2 Orbital Evolution

The TaylorT2 method obtains a parametric solution for the orbital phase and time as a function of the post-Newtonian parameter x. These are obtained from the equations

$$t(x) = t_{\rm c} + \frac{GM}{c^3}\int_x^{x_{\rm c}} \frac{1}{\mathcal{F}}\frac{d\mathcal{E}}{dx}dx \tag{B.8a}$$

$$\varphi(x) = \varphi_{\rm c} + \int_x^{x_{\rm c}} x^{3/2}\frac{1}{\mathcal{F}}\frac{d\mathcal{E}}{dx}dx \tag{B.8b}$$

1) The regularization constants λ and θ that appear in Blanchet (2002) have the values $\lambda = -1987/3080$ and $\theta = -11\,831/9240$.

by expanding the ratio $\mathcal{F}/(d\mathcal{E}/dx)$ in a power series in x and performing the integral. The result is

$$
\begin{aligned}
t(x) = t_\mathrm{c} - \frac{5}{256\eta}\frac{GM}{c^3}x^{-4}\Bigg\{ & 1 + \left(\frac{743}{252} + \frac{11}{3}\eta\right)x - \frac{32}{5}\pi x^{3/2} \\
& + \left(\frac{3\,058\,673}{508\,032} + \frac{5429}{504}\eta + \frac{617}{72}\eta^2\right)x^2 - \left(\frac{7729}{252} - \frac{13}{3}\eta\right)\pi x^{5/3} \\
& + \left[-\frac{10\,052\,469\,856\,691}{23\,471\,078\,400} + \frac{128}{3}\pi^2 + \frac{6848}{105}\gamma_\mathrm{E} + \frac{3424}{105}\ln(16x)\right. \\
& \left. + \left(\frac{3\,147\,553\,127}{3\,048\,192} - \frac{451}{12}\pi^2\right)\eta - \frac{15\,211}{1728}\eta^2 + \frac{25\,565}{1296}\eta^3\right]x^3 \\
& + \left(-\frac{15\,419\,335}{127\,008} - \frac{75\,703}{756}\eta + \frac{14\,809}{378}\eta^2\right)\pi x^{7/2}\Bigg\}
\end{aligned} \tag{B.9a}
$$

and

$$
\begin{aligned}
\varphi(x) = \varphi_\mathrm{c} - \frac{1}{32\eta}x^{-5/2}\Bigg\{ & 1 + \left(\frac{3715}{1008} + \frac{55}{12}\eta\right)x - 10\pi x^{3/2} \\
& \left(\frac{15\,293\,365}{1\,016\,064} + \frac{27\,145}{1008}\eta + \frac{3085}{144}\eta^2\right)x^2 \\
& + \left(\frac{38\,645}{1344} - \frac{65}{16}\eta\right)\ln\left(\frac{x}{x_0}\right)\pi x^{5/2} \\
& + \left[\frac{12\,348\,611\,926\,451}{18\,776\,862\,720} - \frac{160}{3}\pi^2 - \frac{1712}{21}\gamma_\mathrm{E} - \frac{856}{21}\ln(16x)\right. \\
& \left. + \left(-\frac{15\,737\,765\,635}{12\,192\,768} + \frac{2255}{48}\pi^2\right)\eta + \frac{76\,055}{6912}\eta^2 - \frac{127\,825}{5184}\eta^3\right]x^3 \\
& + \left(\frac{77\,096\,675}{2\,032\,128} + \frac{378\,515}{12\,096}\eta - \frac{74\,045}{6048}\eta^2\right)\pi x^{7/2}\Bigg\}.
\end{aligned} \tag{B.9b}
$$

B.3 TaylorT3 Orbital Evolution

The TaylorT3 approximation expresses the orbital phase as a function of time, and is obtained from the parametric form of TaylorT2 by reversion of the power series for $t(x)$, which yields $x(t)$, and then insertion of this into the expression for $\varphi(x)$. The resulting expressions of the post-Newtonian parameter and the orbital phase, written in terms of the surrogate time variable

$$\Theta := \frac{\eta}{5}\frac{c^3(t_\mathrm{c} - t)}{GM} \tag{B.10}$$

are

$$x = \frac{1}{4}\Theta^{-1/4}\left\{1 + \left(\frac{743}{4032} + \frac{11}{48}\eta\right)\Theta^{-1/4} - \frac{1}{5}\pi\Theta^{-3/8} + \left(\frac{19\,583}{254\,016} + \frac{24\,401}{193\,536}\eta + \frac{31}{288}\eta^2\right)\Theta^{-1/2} + \left(-\frac{11\,891}{53\,760} + \frac{109}{1920}\eta\right)\pi\Theta^{-5/8} + \left[-\frac{10\,052\,469\,856\,691}{6\,008\,596\,070\,400} - \frac{1}{6}\pi^2 + \frac{107}{420}\gamma_E - \frac{107}{3360}\ln\left(\frac{\Theta}{256}\right) + \left(\frac{3\,147\,553\,127}{780\,337\,155} - \frac{451}{3072}\pi^2\right)\eta - \frac{15\,211}{442\,368}\eta^2 + \frac{25\,565}{331\,776}\eta^3\right]\Theta^{-3/4} + \left(-\frac{113\,868\,647}{433\,520\,640} - \frac{31\,821}{143\,360}\eta + \frac{294\,941}{3\,870\,720}\eta^2\right)\pi\Theta^{-7/8}\right\} \tag{B.11a}$$

and

$$\varphi = \varphi_c - \frac{1}{\eta}\Theta^{5/8}\left\{1 + \left(\frac{3715}{8064} + \frac{55}{96}\eta\right)\Theta^{-1/4} - \frac{3}{4}\pi\Theta^{-3/8} + \left(\frac{9\,275\,495}{14\,450\,688} + \frac{284\,875}{258\,048}\eta + \frac{1855}{2048}\eta^2\right)\Theta^{-1/2} + \left(-\frac{38\,645}{172\,032} + \frac{65}{2048}\eta\right)\ln\left(\frac{\Theta}{\Theta_0}\right)\pi\Theta^{-5/8} + \left[\frac{831\,032\,450\,749\,357}{57\,682\,522\,275\,840} - \frac{53}{40}\pi^2 - \frac{107}{56}\gamma_E + \frac{107}{448}\ln\left(\frac{\Theta}{256}\right) + \left(-\frac{126\,510\,089\,885}{4\,161\,798\,144} + \frac{2255}{2048}\pi^2\right)\eta + \frac{154\,565}{1\,835\,008}\eta^2 - \frac{1\,179\,625}{1\,769\,472}\eta^3\right]\Theta^{-3/4} + \left(\frac{188\,516\,689}{173\,408\,256} + \frac{488\,825}{516\,096}\eta - \frac{141\,769}{516\,096}\eta^2\right)\pi\Theta^{-7/8}\right\}. \tag{B.11b}$$

B.4
TaylorT4 Orbital Evolution

The TaylorT4 evolution scheme is the same as TaylorT1 except that the ratio $\mathcal{F}/(d\mathcal{E}/dx)$ is expanded in a power series, which results in the coupled set of ordinary differential equations

$$\begin{aligned}\frac{dx}{dt} = \frac{64}{5}\eta\frac{c^3}{GM}x^5 \Bigg\{ 1 &- \left(\frac{743}{336} + \frac{11}{4}\eta\right)x + 4\pi x^{3/2} \\ &+ \left(\frac{34\,103}{18\,144} + \frac{13\,661}{2016}\eta + \frac{59}{18}\eta^2\right)x^2 \\ &- \left(\frac{4159}{672} + \frac{189}{8}\eta\right)\pi x^{5/2} \\ &+ \Bigg[\frac{16\,447\,322\,263}{139\,708\,800} + \frac{16}{3}\pi^2 - \frac{1712}{105}\gamma_{\mathrm{E}} - \frac{856}{105}\ln(16x) \\ &+ \left(-\frac{56\,198\,689}{217\,728} + \frac{451}{48}\pi^2\right)\eta + \frac{541}{896}\eta^2 - \frac{5605}{2592}\eta^3\Bigg]x^3 \\ &- \left(\frac{4415}{4032} - \frac{358\,675}{6048}\eta - \frac{91\,495}{1512}\eta^2\right)\pi x^{7/2}\Bigg\} \end{aligned} \tag{B.12a}$$

and

$$\frac{d\varphi}{dt} = \frac{c^3}{GM}x^{3/2} . \tag{B.12b}$$

These equations are evolved by integration.

B.5 TaylorF2 Stationary Phase

The stationary phase approximation can be used to represent the inspiral waveform in the frequency domain. The TaylorF2 method is similar to the TaylorT2 method described above, but in the stationary phase approximation. The waveform is

$$\tilde{h}_{+}(f) \approx -\frac{1+\cos^2\iota}{2}\left(\frac{5\pi}{24}\right)^{1/2}\eta^{1/2}\frac{G^2M^2}{c^5 r}x^{-7/4}\mathrm{e}^{-2\pi i f t_{\mathrm{c}}}\mathrm{e}^{2\pi i\varphi_{\mathrm{c}}}\mathrm{e}^{-i\Psi(f)} \tag{B.13a}$$

$$\tilde{h}_{\times}(f) \approx i\cos\iota\left(\frac{5\pi}{24}\right)^{1/2}\eta^{1/2}\frac{G^2M^2}{c^5 r}x^{-7/4}\mathrm{e}^{-2\pi i f t_{\mathrm{c}}}\mathrm{e}^{2\pi i\varphi_{\mathrm{c}}}\mathrm{e}^{-i\Psi(f)} , \tag{B.13b}$$

where $\Psi(f)$ is the stationary phase function

$$\begin{aligned}
\Psi = & -\frac{\pi}{4} + \frac{3}{128}\frac{1}{\eta}x^{-5/2}\Bigg\{1 + \left(\frac{3715}{756} + \frac{55}{9}\eta\right)x - 16\pi x^{3/2} \\
& + \left(\frac{15\,293\,365}{508\,032} + \frac{27\,145}{504}\eta + \frac{3085}{72}\eta^2\right)x^2 \\
& + \left(\frac{38\,645}{756} - \frac{65}{9}\eta\right)\left[1 + \frac{3}{2}\ln\left(\frac{x}{x_0}\right)\right]\pi x^{5/2} \\
& \Bigg[\frac{11\,583\,231\,236\,531}{4\,694\,215\,680} - \frac{640}{3}\pi^2 - \frac{6848}{21}\gamma_{\mathrm{E}} - \frac{3424}{21}\ln(16x) \\
& + \left(-\frac{15\,737\,765\,635}{3\,048\,192} + \frac{2255}{12}\pi^2\right)\eta + \frac{76\,055}{1728}\eta^2 - \frac{127\,825}{1296}\eta^3\Bigg]x^3 \\
& + \left(\frac{77\,096\,675}{254\,016} + \frac{378\,515}{1512}\eta - \frac{74\,045}{756}\eta^2\right)\pi x^{7/2}\Bigg\}.
\end{aligned} \tag{B.14}$$

Here, $x = (\pi G M f/c^3)^{2/3}$.

References

Blanchet, L. (2002) Gravitational radiation from post-Newtonian sources and inspiralling compact binaries. *Living Rev. Rel.*, 5(3). http://www.livingreviews.org/lrr-2002-3 (last accessed 2011-01-03).

Buonanno, A., Iyer, B., Ochsner, E., Pan, Y. and Sathyaprakash, B.S. (2009) Comparison of post-Newtonian templates for compact binary inspiral signals in gravitational-wave detectors. *Phys. Rev.*, **D80**, 084 043. doi: 10.1103/PhysRevD.80.084043.

Kidder, L.E. (2008) Using full information when computing modes of Post-Newtonian waveforms from inspiralling compact binaries in circular orbit. *Phys. Rev.*, **D77**, 044 016. doi: 10.1103/PhysRevD.77.044016.

Index

Gravitational-Wave Physics and Astronomy, First Edition. Jolien D. E. Creighton, Warren G. Anderson.